THE IMMUNE SYSTEM SECOND EDITION

Preface

In the twentieth century immunology was often taught as a form of epic saga, a progression of elegant experiments, breakthrough discoveries, and intrepid investigators. The merit of the approach was its simplicity, as in tracing history using the dates of monarchs, ministers, conflicts, and conquests. It also added spice and personality in an era when knowledge was hard won, incomplete and only tenuously related to the human condition. That situation has clearly changed, the direct consequence of technical innovations implemented during the latter part of the twentieth century. Today new information literally pours from the research pipeline, onto the printed page and into the World Wide Web. Elegant experiments, breakthrough discoveries, and intrepid investigators are now in such abundance that they in themselves can no longer provide a simplifying principle for explaining immunology. However, as a result of all this industry immunologists now have a more coherent and intellectually more satisfying picture of the immune system and how it works, not always successfully, to protect the body from infection.

This book is aimed at students who are coming to immunology for the first time. Throughout the book the emphasis is upon the human immune system and how its successes, failures, and compromises affect the lives of each and every one of us. The first eight chapters of the book describe the cells and molecules of the immune system and how they work together in providing defenses against invading microorganisms. This part of the book is built around the biology of the B and T lymphocytes, the cells that adapt to infection and provide long-lasting protective immunity, but also emphasizes the importance of other cells—macrophages, dendritic cells, granulocytes, mast cells, and natural killer cells—that provide both the front-line defenses of innate immunity and the inflammation necessary to stimulate B and T cells. The first chapter introduces the different types of cells that make up the immune system and the functions they perform. Antigen recognition, repertoire development, and effector functions of B and T cells are dealt with in turn in the next six chapters. Completing this part of the book is a chapter describing how innate and adaptive immunity work together to battle common types of infection.

The next three chapters focus upon diseases that arise from inadequacies of the human immune system. The first describes situations where infections fail to be controlled, often because the microorganism actively evades, exploits, or subverts the immune response. The following chapter examines conditions in which the immune system overreacts to innocuous substances in the environment and causes chronic inflammatory diseases such as allergies and asthma. The third of these chapters examines the autoimmune diseases, such as Graves' disease, insulin-dependent diabetes, and rheumatoid arthritis, in which the immune system attacks healthy cells and causes tissue damage and loss of function. The last chapter of the book illustrates how the immune system can be manipulated to improve human health, by consideration of the well-established practices of vaccination and transplantation as well as the emerging field of cancer immunology.

In this second edition of *The Immune System* the changes made to the text and figures of the first edition had two aims. First were changes aimed at improving the clarity and coherence of the material presented in the first edition and stem in large part from the suggestions and criticisms of instructors using the book. An almost universal request was for questions that would help students to learn and revise the material. Such questions are now included at the end of each chapter with a full set of answers included at the back of the book. In making revisions one guiding principle was to replace discussion of animal models with examples drawn from human immunology and clinical practice; a second was to include more of the genetics of human immunity. On a more specific issue the text pertaining to the complement system in chapters 7 and 8 has been rewritten and many figures have been redrawn using a modified iconography. The second category of revision embraces advances in knowledge and understanding of the immune system. Amongst the many new sections are ones describing pathogen receptors of innate immune cells, the role of dendritic cells in triggering adaptive immunity, mechanisms that contribute to immunological self-tolerances, and autoimmunity including regulatory T cells and rational approaches to cancer vaccination. As well as adding such new material, the revision has involved the elimination of text or figures that were judged unnecessary or digressive.

The end-of-chapter questions new to this edition, as well as the answers to those questions, were composed by Sheryl L. Fuller-Espie of Cabrini College, Radnor, Pennsylvania. I am grateful for Sherry's contribution to the book. She wrote the first draft of questions, made significant collaborations during the revision process, and was meticulous in editing the final drafts.

I thank and acknowledge the authors of *Immunobiology* and of *Case Studies in Immunology* for giving me license with the text and figures of their books. I am most grateful to the many instructors and their students whose encouragement and criticism prompted this revision of *The Immune System*. In particular, Frances Brodsky made innumerable and ineffable contributions at each stage of the revision. Eleanor Lawrence not only edited all aspects of the manuscript and figures but contributed much more. Her unwavering sympathy for the prospective reader ensured that much jargon, complexity and other nonsense never made it to the printed page. Nigel Orme created many new illustrations for this edition, as well as the image for the front cover photographed by Richard Denyer. Emma Hunt was creative with the layout and assiduous in proofing and production. The orchestration of all these efforts has been most ably and good humoredly done by Mike Morales. I am indebted to Denise Schanck for her consistent support and encouragement of this project. Finally, I would like to thank my sister Joyce and her husband Rowland for providing a welcoming place to stay in London when I am have been working there on the book.

Reviewer Acknowledgments

The author and publisher would like to thank the following reviewers for their thoughtful comments and guidance: Kent L. Buchanan, Tulane University Health Sciences Center; David K. Burnham, Oklahoma State University; Eunice C. Carlson, Michigan Technological University; Michael J. Chorney, The Pennsylvania State University College of Medicine; Howard B. Fleit, State University of New York at Stony Brook; Michael L. Misfeld, University of Missouri; and Virginia L. Thomas, University of Texas Health Science Center at San Antonio.

Contents

Instructor and Student Resources

Case Studies in Immunology, Fourth Edition
by Fred Rosen and Raif Geha

Case Studies in Immunology reviews major topics of immunology as the background to a selection of real clinical cases that serve to reinforce and extend the basic science. This new edition vividly illustrates the importance of an understanding of immunology in diagnosis and therapy. This book can be used as a clinical companion alongside *The Immune System,* Second Edition. A correlation guide matching relevant sections of the texts is packaged with this book.

The Art of The Immune System (IS2) (CD-ROM)

The Art of IS2 CD-ROM contains the figures from the book for presentation purposes. The figures are available in both PowerPoint® and JPEG formats.

Garland Science Classwire™

Available at: www.classwire.com/garlandscience, the Classwire course management system allows instructors to build websites for their courses easily. It also serves as an online archive for instructors' resources. After registering for Classwire, you will be able to download all of the figures from *The Immune System,* which are available in JPEG and PowerPoint formats. Additionally, instructors may download resources from other Garland Science textbooks. Garland Science Classwire is offered free of charge to all instructors who adopt *The Immune System.* Please contact science@garland.com for additional information on accessing the Classwire system. (Classwire™ is a trademark of Chalkfree, Inc.)

Detailed Contents

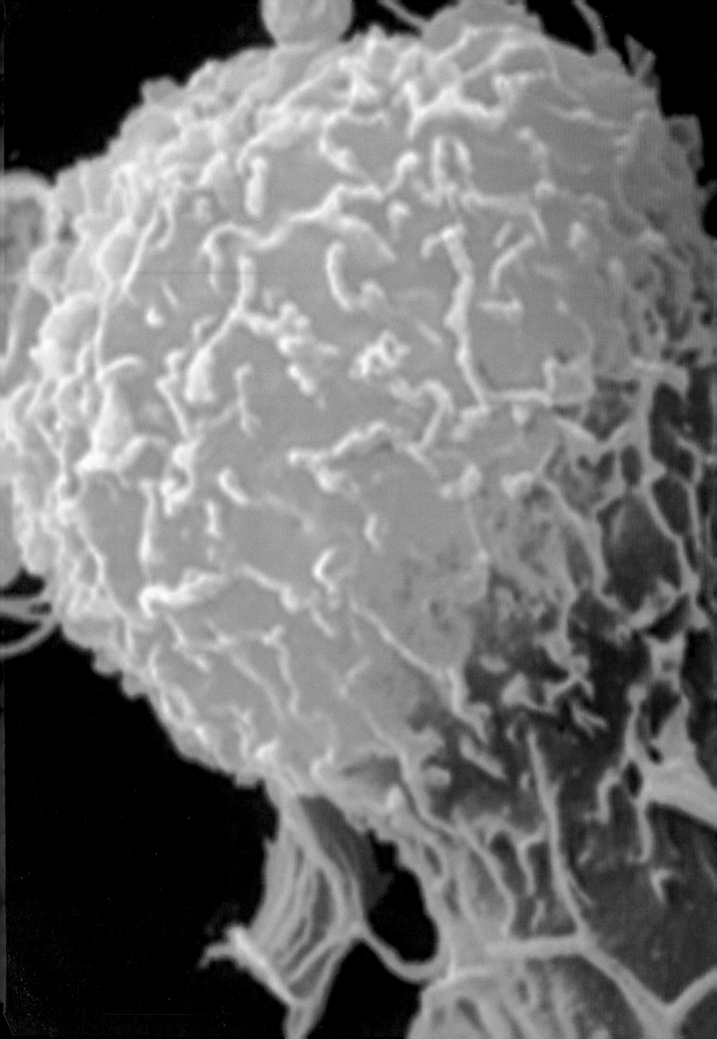

Chapter 1

Elements of the Immune System and their Roles in Defense

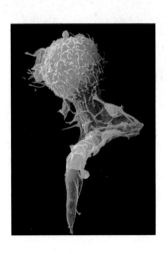

Immunology is the study of the physiological mechanisms that humans and other animals use to defend their bodies from invasion by other organisms. The origins of the subject lie in the practice of medicine and in historical observations that people who survived the ravages of epidemic disease were untouched when faced with that same disease again—they had become **immune** to infection. Infectious diseases are caused by microorganisms, which have the advantage of reproducing and evolving much more rapidly than do their human hosts. During the course of an infection, the microorganism can pit enormous populations of its species against an individual *Homo sapiens*. In response, the human body invests heavily in cells dedicated to defense, which collectively form the **immune system**.

The immune system is crucial to human survival. In the absence of a working immune system, even minor infections can take hold and prove fatal. Without intensive treatment, children born without a functional immune system die in early childhood from the effects of common infections. However, in spite of their immune systems, all humans suffer from infectious diseases, especially when young. This is because the immune system takes time to build up a strong response to an invading microorganism, time during which the invader can multiply and cause disease. To provide **protective immunity** for the future, the immune system must first do battle with the microorganism. This places people at highest risk during their first infection with a microorganism and, in the absence of modern medicine, leads to substantial child mortality, as witnessed in the developing world. When entire populations face a completely new infection, the outcome can be catastrophic, as experienced by indigenous Americans who were decimated by European diseases to which they were suddenly exposed after 1492. Today, infection with human immunodeficiency virus (HIV) and the acquired immune deficiency syndrome (AIDS) it causes are having a similarly tragic impact on the populations of several African countries.

In medicine the greatest triumph of immunology has been **vaccination**, or **immunization**, a procedure whereby severe disease is prevented by prior exposure to the infectious agent in a form that cannot cause disease. Vaccination provides the opportunity for the immune system to gain the experience needed to make a protective response with little risk to health or life. Vaccination was first used against smallpox, a viral scourge that once ravaged populations and disfigured the survivors. In Asia, small amounts of smallpox virus had been used to induce protective immunity for hundreds of years before 1721, when Lady Mary Wortley Montagu introduced the method into western Europe. Subsequently, in 1796, Edward Jenner, a doctor in rural England, showed how inoculation with cowpox virus offered protection against the related smallpox virus with less risk than the earlier methods. Jenner called his procedure vaccination, after *vaccinia*, the name given to the mild disease produced by cowpox, and he is generally credited with its invention. Since his time, vaccination dramatically reduced the incidence of smallpox worldwide, with the last cases being seen by physicians in the 1970s (Figure 1.1).

Effective vaccines have been made from only a fraction of the agents that cause disease and some are of limited availability because of their cost. Most of the widely used vaccines were first developed many years ago by processes of trial and error, before very much was known about the workings of the immune system. That approach is no longer so successful for making new vaccines, perhaps because all the easily won vaccines have been made. But deeper understanding of the mechanisms of immunity is spawning new ideas for vaccines against infectious diseases and even against other types of disease such as cancer. Much is now known about the molecular and cellular components of the immune system and what they can do in the laboratory. Current research aims at understanding their contributions to fighting infections in the world at large. The new knowledge is also being used to find better ways of manipulating the immune system to prevent the unwanted immune responses that cause allergies, autoimmune diseases, and rejection of organ transplants.

In the first part of this chapter we consider the microorganisms that infect human beings and the defenses they must overcome to start an infection. These include physical barriers, chemical barriers, and the fixed defenses of innate immunity, which are ready and waiting to halt infections before they can barely start. The individual cells and tissues of the immune system will also be described, and how they integrate their functions with the rest of the human body. The second part of the chapter focuses on the more flexible and forceful defenses of adaptive immunity. These mechanisms are only brought into play if and when an infection is established. The adaptive response is always targeted to the specific problem at hand and is made and refined during the course of the infection. When successful, it clears the infection and provides long-lasting immunity that prevents its recurrence.

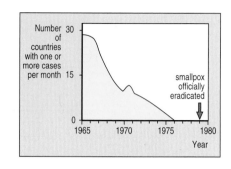

Figure 1.1 **The eradication of smallpox by vaccination.** Smallpox vaccination was started in 1796. In 1979, after 3 years in which no case of smallpox was recorded, the World Health Organization announced that the virus had been eradicated. Since then the proportion of the human population that has been vaccinated against smallpox, or has acquired immunity from an infection, has steadily decreased. The result is that the human population has become increasingly vulnerable should the virus emerge again, either naturally or as a deliberate act of human malevolence.

Defenses facing invading pathogens

The purpose of the vertebrate immune system is to recognize invading foreign organisms, to prevent their spread, and ultimately to clear them from the body. It consists of billions of cells of various types, which interact with the infectious agent and with each other to fight the infection. All cells of the immune system originate in the **bone marrow**, but at some time in their life they leave that site to circulate in the blood, to enter other tissues, and to form part of specialized **lymphoid** tissues. This part of the chapter introduces the cells and tissues of the immune system and some of the ways in which they carry out their functions. But first we shall consider the organisms and infections that the immune system evolved to protect us against.

1-1 Pathogens are infectious organisms that cause disease

Numerous species of microorganism colonize the human body in large numbers and rarely produce symptoms of disease, for example the benign strains of the bacterium *Escherichia coli* that normally inhabit the gastrointestinal tract. However, for some microorganisms, such as the influenza virus or the

The immune system protects against four classes of pathogen		
Type of pathogen	**Examples**	**Diseases**
Bacteria	*Salmonella enteritidis* *Mycobacterium tuberculosis*	Food poisoning Tuberculosis
Viruses	Variola Influenza HIV	Smallpox Flu AIDS
Fungi	*Epidermophyton floccosum* *Candida albicans*	Ringworm Thrush, systemic candidiasis

Parasites
 	
    protozoa
 	
    worms
	Trypanosoma brucei
	Leishmania donovani
	Plasmodium falciparum
	Ascaris lumbricoides
	Schistosoma mansoni
		Sleeping sickness
		Leishmaniasis
		Malaria
		Ascariasis
		Schistosomiasis

Figure 1.2 The four kinds of pathogen that cause human disease. Examples of the types of pathogen are listed, along with the diseases they cause.

typhoid bacillus, infection habitually causes a disease. Any organism with the potential to cause disease is known as a **pathogen**. This definition includes not only microorganisms that generally cause disease if they enter the body, but also ones that can colonize the human body to no ill effect for much of the time, but cause illness if the body's defenses are weakened or if the microbe gets into the 'wrong' place. The latter kinds of pathogen are known as **opportunistic pathogens**.

Pathogens can be divided into four kinds: **bacteria**, **viruses**, and **fungi**, which are each a group of related microorganisms, and internal **parasites**, a less precise term used to embrace a heterogeneous collection of unicellular protozoa and multicellular invertebrates, mainly worms (Figure 1.2). In this book we consider the functions of the human immune system principally in the context of controlling infections. For some pathogens this necessitates their complete elimination, but for others it is sufficient to limit the size and location of the pathogen population within the human host. Figure 1.3 gives examples of pathogens from the four classes.

Over evolutionary time, the relationship between a pathogen and its human hosts can change, affecting the severity of the disease produced. Most pathogenic organisms have evolved special adaptations that enable them to invade their hosts, replicate in them, and be transmitted. However, the rapid death of its host is rarely in a microbe's interest, because this destroys both its home and its source of food. Consequently, those organisms with the potential to cause severe and rapidly fatal disease often tend to evolve towards an accommodation with their hosts. In complementary fashion, host populations have evolved a degree of in-built genetic resistance to common disease-causing organisms, as well as acquiring lifetime immunity to endemic diseases as a result of infection in childhood. Because of the interplay between host and pathogen, the nature and severity of infectious diseases in the human population are always changing.

Influenza is a good example of a common viral disease that, although severe in its symptoms, is usually overcome successfully by the immune system. The fever, aches, and lassitude that accompany infection can be overwhelming, and it is difficult to imagine overcoming foes or predators at the peak of a bout of influenza. However, despite the severity of the symptoms, most strains of influenza pose no great danger to healthy people in populations where

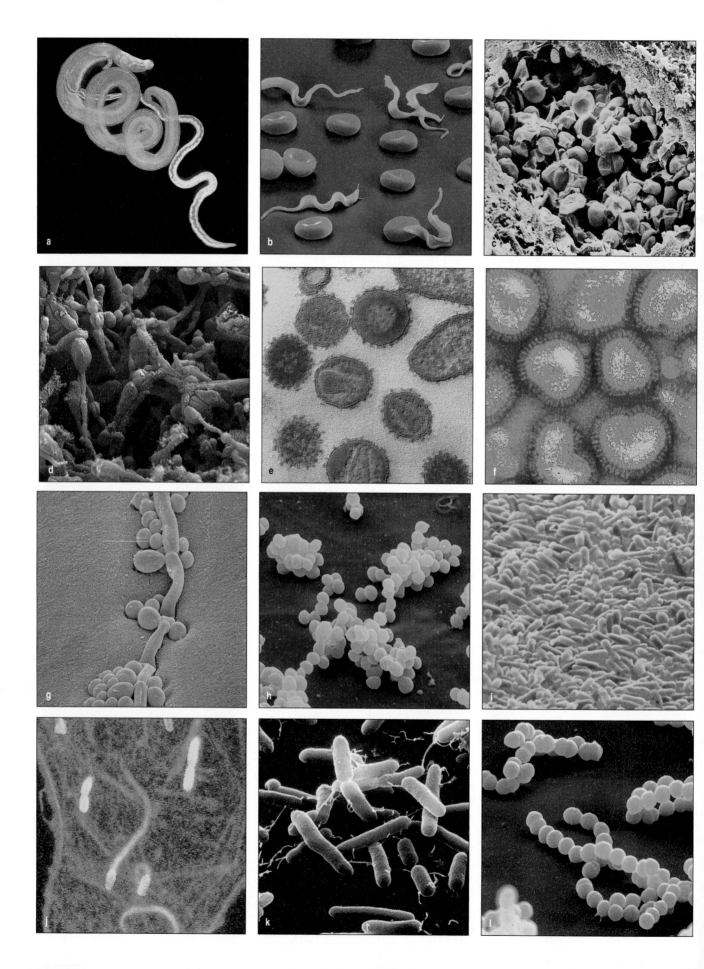

Figure 1.3 The diversity of human pathogens. Panel a: light micrograph of *Schistosoma mansoni*, the helminth worm that causes schistosomiasis. The adult intestinal blood fluke form is shown: the male is thick and bluish, the female white and thread-like. Magnification × 5. Panel b: false-color scanning electron micrograph of red blood cells and *Trypanosoma brucei*, a protozoan of the type that causes African sleeping sickness. Magnification × 1750. Panel c: false-color scanning electron micrograph of *Pneumocystis carinii*, the fungus that causes opportunistic infections in patients whose immune system is suppressed by disease or drugs. The view is of a lung alveolus from a monkey who is immunosuppressed because of infection with a simian virus that causes an acquired immune deficiency syndrome (AIDS). The fungal cells have been colored green. Magnification × 720. Panel d: scanning electron micrograph of *Epidermophyton floccosum*, the dermatophyte fungus that causes ringworm. Pear-shaped spore-producing structures (macronidia) are seen connected by filaments (hyphae). Magnification × 500. Panel e: false-color transmission electron micrograph of human immunodeficiency virus (HIV), the cause of AIDS. Magnification × 80,000. Panel f: false-color transmission electron micrograph of influenza virus, an orthomyxovirus that causes influenza. Magnification × 40,000. Panel g: false-color scanning electron micrograph of the fungus *Candida albicans*, a normal inhabitant of the human body that occasionally causes thrush and more severe systemic infections. Pseudohyphae are visible in a row. At the junctions of the hyphae, rounded yeast-like cells (blastospores) are budded off into colonies. Magnification × 1400. Panel h: false-color scanning electron micrograph of *Staphylococcus aureus*, a Gram-positive bacterium that colonizes human skin and is the common cause of pimples and boils. Some strains, however, cause food poisoning. The small spherical cells (cocci) typically form grape-like clusters. Magnification × 5000. Panel i: false-color scanning electron micrograph of *Mycobacterium tuberculosis*, the bacterium that causes tuberculosis. Magnification × 15,000. Panel j: false-color transmission electron micrograph of a human cell (colored green) and bacteria of the species *Listeria monocytogenes*, a Gram-positive coccobacillus that can contaminate processed food, causing disease (listeriosis) in immunocompromised individuals and pregnant women. Magnification × 1250. Panel k: false-color scanning electron micrograph of *Salmonella enteritidis*, a Gram-negative, rod-shaped bacterium that is a common cause of food poisoning. The hair-like flagella enable the bacteria to move. Magnification × 6500. Panel l: false-color scanning electron micrograph of *Streptococcus pyogenes*, a Gram-positive bacterium that is the principal cause of tonsilitis and scarlet fever, and can also cause ear infections. It has rounded or spherical cells that sometimes form chains as seen here. Magnification × 6500.

influenza is endemic. Warm, well-nourished, and otherwise healthy people usually recover in a couple of weeks and take it for granted that their immune system will accomplish this task. Pathogens new to the human population, in contrast, often cause high mortality in those infected—between 60% and 75% in the case of the Ebola virus.

1-2 The skin and mucosal surfaces form physical barriers against infection

The skin is the human body's first defense against infection. It forms a tough impenetrable barrier of epithelium protected by layers of keratinized cells. This barrier can be breached by physical damage, such as wounds, burns, or surgical procedures, which exposes soft tissues and renders them vulnerable to infection. Until the adoption of antiseptic procedures in the nineteenth century, surgery was a very risky business, principally because of the life-threatening infections that the procedures introduced. For the same reason,

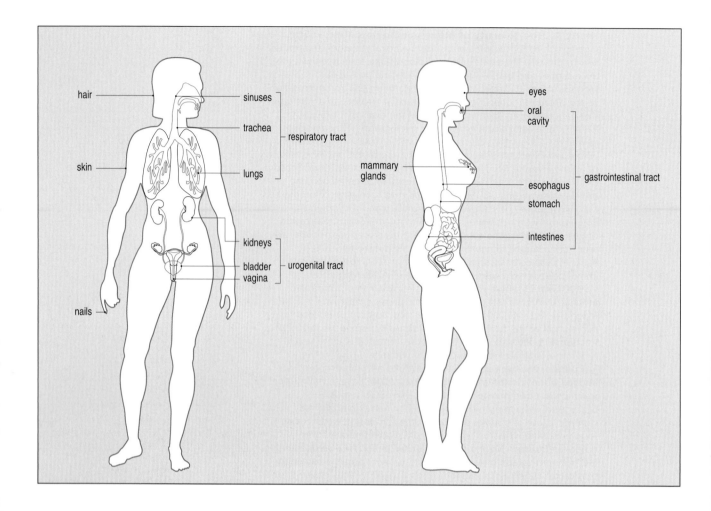

far more soldiers have died from infection acquired on the battlefield than from the direct effects of enemy action. Ironically, the need to conduct increasingly sophisticated and wide-ranging warfare has been the major force driving improvements in surgery and medicine. As an example from immunology, the burns suffered by fighter pilots during World War II stimulated studies on skin transplantation that led directly to the understanding of the cellular basis of the immune response.

Figure 1.4 The physical barriers that separate the body from its external environment. In these images of a woman, the strong barriers to infection provided by the skin, hair, and nails are colored blue and the more vulnerable mucosal membranes are colored red.

Continuous with the skin are the epithelia lining the respiratory, gastrointestinal, and urogenital tracts (Figure 1.4). On these internal surfaces, the impermeable skin gives way to tissues that are specialized for communication with their environment and are more vulnerable to microbial invasion. Such surfaces are known as **mucosal surfaces** or **mucosae** as they are continually bathed in the **mucus** that they secrete. This thick fluid layer contains glycoproteins, proteoglycans, and enzymes that protect the epithelial cells from damage and help to limit infection. The enzyme lysozyme in tears and saliva is one of a number of antibacterial substances in secretions from mucosal surfaces. In the respiratory tract, mucus is continuously removed through the action of epithelial cells bearing beating cilia and is replenished by mucus-secreting goblet cells. The respiratory mucosa is thus continually cleansed of unwanted material, including infectious microorganisms that have been breathed in. Microorganisms are also deterred by the acidic environments of the stomach, the vagina, and the skin. With such defenses, skin and mucosa provide a well-maintained physical and chemical barrier that prevents most pathogens from gaining access to the cells and tissues of the body. When that barrier is breached and pathogens gain entry to the body's soft tissues, the fixed defenses of the innate immune system are brought into play.

1-3 The innate immune response causes inflammation at sites of infection

Cuts, abrasions, bites, and wounds provide routes for pathogens to get through the skin. Touching, rubbing, picking, and poking of the eyes, nose, and mouth help pathogens to breach mucosal surfaces, as does breathing polluted air, eating contaminated food, and being around infected people. With very few exceptions, infections remain highly localized and are extinguished within a few days without illness or incapacitation. Such infections are controlled and terminated by immune mechanisms that are ready to react quickly to infection. This response consists of two parts (Figure 1.5). First is recognition that a pathogen is present. This involves soluble proteins and cell-surface receptors that bind either to the pathogen and its products or to human cells and serum proteins that become altered in the presence of the pathogen. Once the pathogen has been recognized, the second part of the response involves the recruitment of destructive **effector mechanisms** that kill and eliminate it. The effector mechanisms are provided by **effector cells** that engulf bacteria, kill virus-infected cells, or attack protozoan parasites, and a battery of serum proteins called **complement** that help the effector cells by marking pathogens with molecular flags, but also attack pathogens in their own right. Collectively, these mechanisms are called **innate immunity** or the **innate immune response**. The word 'innate' refers to the fact that all these mechanisms are determined entirely by the genes a person inherits from their parents. Many families of receptor proteins contribute to the recognition of pathogens in the innate immune response. They are of several different structural types and bind to chemically diverse ligands: peptides, proteins, glycoproteins, proteoglycans, peptidoglycan, carbohydrates, glycolipids, phospholipids and nucleic acids.

An infection that would typically be cleared by innate immunity is of the sort experienced by skateboarding adolescents when they tumble onto a San Francisco sidewalk. On returning home the graze is washed, which removes most of the dirt and the associated pathogens of human, soil, pigeon, dog, cat, raccoon, skunk, and possum origin. Of the bacteria that remain, some begin to divide and set up an infection. Cells and proteins in the damaged tissue sense the presence of the bacteria and the cells send out soluble proteins called **cytokines** that interact with other cells to trigger the innate immune response. The overall effect of the innate immune response is to induce a state of **inflammation** in the infected tissue. Inflammation is an ancient concept in medicine that has traditionally been defined by the Latin words *calor*, *dolor*, *rubor*, and *tumor*, for heat, pain, redness, and swelling, respectively. These symptoms, which are part of everyday human experience, are not due to the infection itself but to the immune system's response to it.

Figure 1.5 Immune defense involves recognition of pathogens followed by their destruction. Almost all components of the immune system contribute to mechanisms for either recognizing pathogens or destroying pathogens, or to mechanisms for communicating between these two activities. This is illustrated here by the most basic process of innate immunity, around which the entire immune system is built. Serum proteins of the complement system (turquoise) are activated in the presence of a pathogen (red) to form a covalent bond between a fragment of complement protein and the pathogen. The attached piece of complement marks the pathogen as dangerous. The soluble complement fragment summons a phagocytic white blood cell to the site of complement activation. The effector cell has a surface receptor that binds to the complement fragment attached to the pathogen. The receptor and its bound ligand are taken up into the cell by endocytosis, which delivers the pathogen to an intracellular vesicle called a phagosome, where it is destroyed.

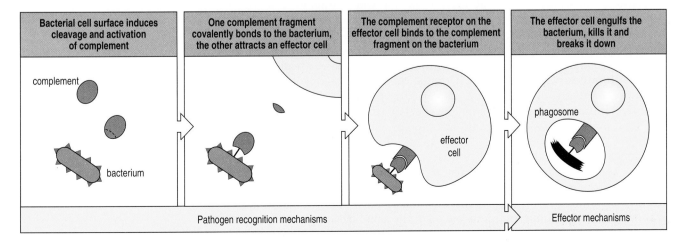

| Bacterial cell surface induces cleavage and activation of complement | One complement fragment covalently bonds to the bacterium, the other attracts an effector cell | The complement receptor on the effector cell binds to the complement fragment on the bacterium | The effector cell engulfs the bacterium, kills it and breaks it down |

complement

bacterium

effector cell

phagosome

Pathogen recognition mechanisms   Effector mechanisms

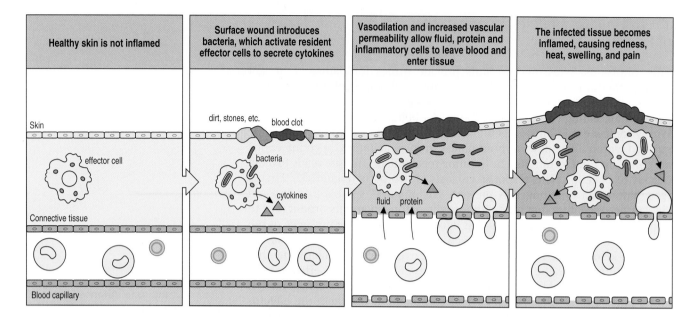

Figure 1.6 Innate immune mechanisms establish a state of inflammation at sites of infection. Illustrated here are the events following an abrasion of the skin that leads to bacteria invading the underlying connective tissue and stimulating the innate immune response.

Cytokines induce the local dilation of blood capillaries, which by increasing the blood flow causes the skin to warm and redden. Vascular dilation (vasodilation) introduces gaps between the endothelial cells, increasing the leak of blood plasma into the connective tissue. Such expansion of the local fluid volume causes edema or swelling, putting pressure on nerve endings and causing pain. Cytokines also change the adhesive properties of the vascular endothelium, inviting white blood cells to attach to it and move from the blood to the inflamed tissue (Figure 1.6). White blood cells that are usually present in inflamed tissues and release substances that contribute to the inflammation are called **inflammatory cells**. Infiltration of cells into the inflamed tissue increases the swelling, and some of the molecules they release contribute to the pain. The benefit of the discomfort and disfigurement is that inflammation enables cells and molecules of the immune system to be brought rapidly and in large numbers into the infected tissue.

1-4 The adaptive immune response adds to an ongoing innate immune response

Human beings are exposed to pathogens on a daily basis. The intensity of exposure and the diversity of the pathogens encountered increases with crowded city living and the daily input of people and pathogens from international airports. Despite this exposure, innate immunity keeps most people healthy for most of the time. Nevertheless, some infections outrun the innate immune response, an event more likely in people who are malnourished, poorly housed, deprived of sleep, or stressed in other ways. When this occurs, the innate immune response works to slow the spread of infection while it calls upon white blood cells called **lymphocytes** that increase the power and focus of the immune response. This contribution to defense is called the **adaptive immune response** because it is organized around an ongoing infection and adapts to the nuances of the infecting pathogen. Consequently, **adaptive immunity** provides a highly specialized defense against one pathogen that is of little effect against infection by a different pathogen.

The effector mechanisms used in the adaptive immune response are similar to those used in the innate immune response; the important difference is in the cell-surface receptors used by lymphocytes to recognize pathogens (Figure 1.7). In contrast to the receptor molecules of innate immunity, the receptor molecules of adaptive immunity are all of the same basic type. They are

Recognition mechanisms of innate immunity	Recognition mechanisms of adaptive immunity
Rapid response (hours)	Slow response (days to weeks)
Fixed	Variable
Limited number of specificities	Numerous highly selective specificities
Constant during response	Improve during response

Common effector mechanisms for the destruction of pathogens

Figure 1.7 The principal characteristics of innate and adaptive immunity.

not encoded by conventional genes but by genes that are cut, spliced, and modified to produce billions of variants of the basic type, each of which is borne by a different subpopulation of cells. During infection only those lymphocytes bearing receptors that recognize the pathogen are selected to participate in the adaptive response, and their numbers are expanded and matured to produce large numbers of effector cells (Figure 1.8). As this process takes time, the benefit of the adaptive immune response only begins to be felt about a week after the infection began.

The value of the adaptive immune response is well illustrated in influenza, which is caused by infection of the epithelial cells in the lower respiratory tract with influenza virus. The debilitating symptoms start 3 or 4 days after the start of infection, when the virus has begun to outrun the innate immune response. The disease persists for 5–7 days while the adaptive immune response is being organized and put to work. As the adaptive immune response gains the upper hand, fever subsides and a gradual convalescence begins in the second week after infection.

The lymphocyte populations that expand during an adaptive immune response persist in the body and provide long-term **immunological memory** of the pathogen. These memory cells allow subsequent encounters with the same pathogen to elicit a stronger and faster adaptive immune response, which terminates infection with minimal illness. The adaptive immunity provided by immunological memory is also called **acquired immunity** or **protective immunity**. For some pathogens like measles virus, one full-blown infection can provide immunity for decades, whereas for influenza the effect seems more short-lived. This is not because the immunological memory is faulty but because the influenza virus changes on a yearly basis to escape the immunity acquired by its human hosts. The first time that an adaptive

Figure 1.8 Selection of lymphocytes by a pathogen. During development, each lymphocyte is programmed to make a single species of cell-surface antigen receptor. The population of circulating lymphocytes embraces many millions of receptor species, which enables all possible pathogens to be recognized. In the panels, lymphocytes with different receptors are represented by the different colors. On infection by a particular pathogen, the small subsets of lymphocytes having receptors that bind to the pathogen or its components are stimulated to divide and differentiate, thereby producing expanded populations of effector cells from each antigen-binding lymphocyte. In the second panel the lymphocyte subpopulation selected by the pathogen is represented by that colored yellow. After activation and cell division, it gives rise to the effector lymphocytes shown in the third panel.

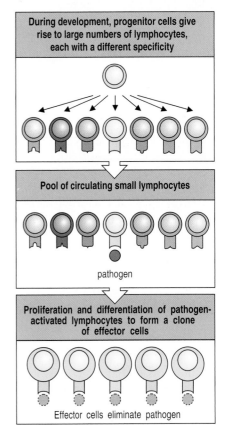

During development, progenitor cells give rise to large numbers of lymphocytes, each with a different specificity

Pool of circulating small lymphocytes

pathogen

Proliferation and differentiation of pathogen-activated lymphocytes to form a clone of effector cells

Effector cells eliminate pathogen

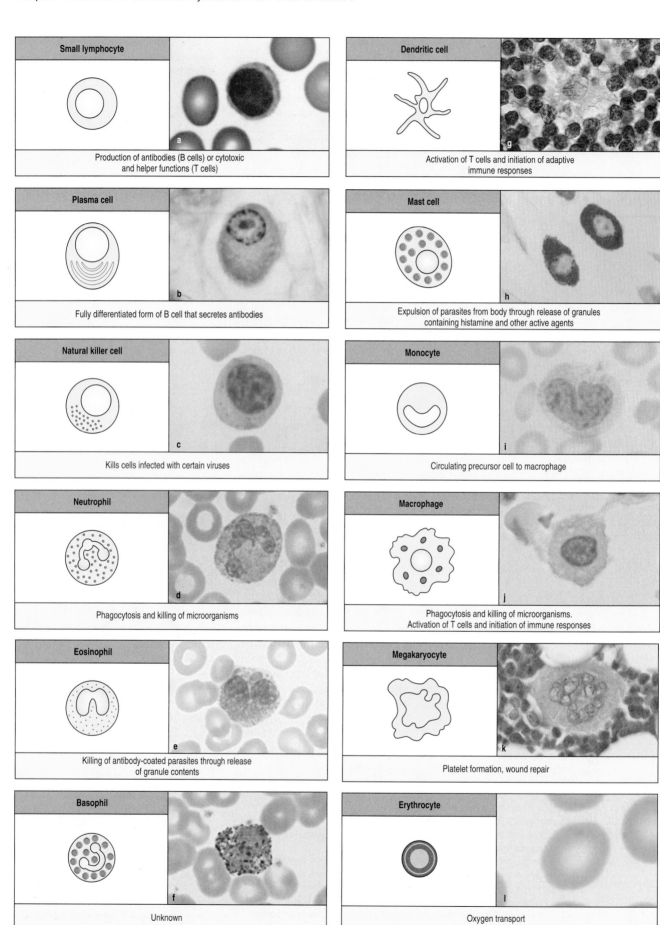

Small lymphocyte		**Dendritic cell**	
Production of antibodies (B cells) or cytotoxic and helper functions (T cells)		Activation of T cells and initiation of adaptive immune responses	
Plasma cell		**Mast cell**	
Fully differentiated form of B cell that secretes antibodies		Expulsion of parasites from body through release of granules containing histamine and other active agents	
Natural killer cell		**Monocyte**	
Kills cells infected with certain viruses		Circulating precursor cell to macrophage	
Neutrophil		**Macrophage**	
Phagocytosis and killing of microorganisms		Phagocytosis and killing of microorganisms. Activation of T cells and initiation of immune responses	
Eosinophil		**Megakaryocyte**	
Killing of antibody-coated parasites through release of granule contents		Platelet formation, wound repair	
Basophil		**Erythrocyte**	
Unknown		Oxygen transport	

Figure 1.9 Types of hematopoietic cell. The different types of hematopoietic cell are depicted in schematic diagrams, which indicate their characteristic morphological features, and in accompanying light micrographs. Their main functions are indicated. We shall use these schematic representations for these cells throughout the book. Red blood cells (erythrocytes) are seen in most of the pictures and are shown alone in panel l. They are smaller than the white blood cells and have no nucleus. Megakaryocytes (k) reside in bone marrow and release tiny non-nucleated fragments of cytoplasm, which circulate in the blood and are known as platelets. Photographs courtesy of N. Rooney (a, d, e, f, g, j), D. Friend (b, c, h, i), and L. Slomianka (k, l).

immune response is made to a given pathogen it is called the **primary response**. The second and subsequent times that an adaptive immune response is made, and when immunological memory applies, it is called a **secondary response**. The purpose of vaccination is to induce immunological memory to a pathogen so that subsequent infection with the pathogen elicits a strong fast-acting adaptive response. As all adaptive immune responses are contingent upon an innate immune response, vaccines must also induce both innate and adaptive immune responses.

1-5 Immune system cells with different functions all derive from hematopoietic stem cells

The cells of the immune system are principally the white blood cells or **leukocytes**, and the tissue cells related to them. These cells derive from a common progenitor called the **pluripotent hematopoietic stem cell**, which also gives rise to **erythrocytes**, the red blood cells, and to the cells called **megakaryocytes**. All these cell types are collectively called **hematopoietic cells** (Figure 1.9). The developmental process by which hematopoietic stem cells give rise to this variety of hematopoietic cells is called **hematopoiesis**. The anatomical site for hematopoiesis changes with age (Figure 1.10). In the early embryo, blood cells are first produced in the yolk sac and later in the fetal liver. From the third to the seventh month of fetal life the spleen is the major site of hematopoiesis. As the bones develop during the fourth and fifth months of fetal growth, hematopoiesis begins to shift to the bone marrow and by birth this is where practically all hematopoeisis takes place. In adults hematopoiesis occurs mainly in the bone marrow of the skull, ribs, sternum, vertebral column, pelvis, and femurs. Because blood cells are both vital and short-lived, hematopoiesis is active throughout life.

Hematopoietic stem cells can divide to give further hematopoietic stem cells, a process called **self renewal**; they can also become more mature stem cells that commit to one of three cell lineages: the erythroid, myeloid, and lymphoid lineages (Figure 1.11). The erythroid progenitor gives rise to the erythroid lineage of blood cells—the oxygen-carrying red blood cells (erythrocytes) and the platelet-producing megakaryocytes. Platelets are small nonnucleated cells of plate-like shape that maintain the integrity of blood vessels. A principal function of platelets is to initiate and participate in the clotting reactions that block up badly damaged blood vessels to prevent blood loss. Megakaryocytes are giant cells that arise from the fusion of multiple precursor cells and have nuclei containing multiple sets of chromosomes (megakaryocyte means cell with giant nucleus). Megakaryocytes are permanent residents of the bone marrow.

The **myeloid progenitor** gives rise to the **myeloid lineage** of cells. One group of myeloid cells are called **granulocytes** because of their prominent cytoplasmic granules, which contain reactive substances that kill microorganisms and enhance inflammation. Because granulocytes have irregularly shaped nuclei

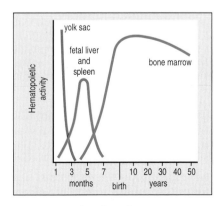

Figure 1.10 The site of hematopoiesis in humans changes during development.

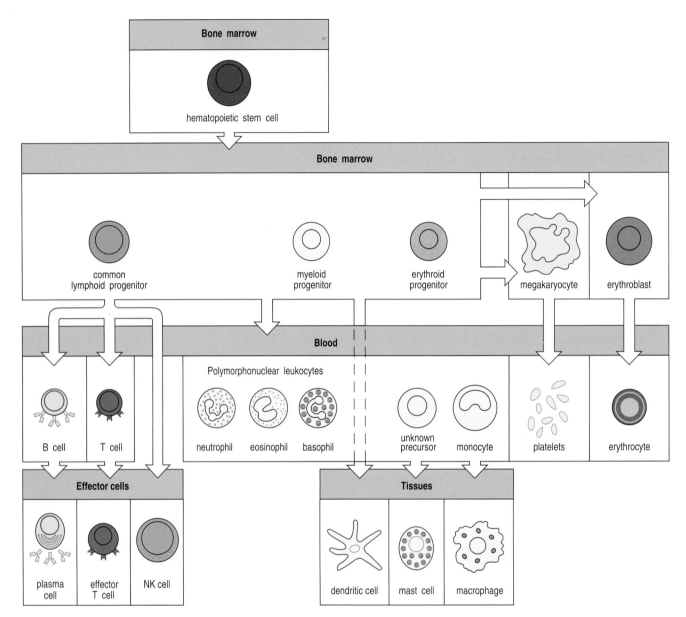

Figure 1.11 Blood cells and certain tissue cells derive from a common hematopoietic stem cell. The pluripotent stem cell (brown) divides and differentiates into more specialized progenitor cells that give rise to the lymphoid lineage, the myeloid lineage, and the erythroid lineage. The common lymphoid progenitor divides and differentiates to give B cells (yellow), T cells (blue), and NK cells (purple). On activation by infection, B cells divide and differentiate into plasma cells, whereas T cells differentiate into various types of activated effector T cell. The myeloid progenitor cell divides and differentiates to produce at least six cell types. These are: the three types of granulocyte—the neutrophil, the eosinophil, and the basophil; the mast cell, which takes up residence in connective and mucosal tissues; the circulating monocyte, which gives rise to the macrophages resident in tissues; and the dendritic cell. The word myeloid means 'of the bone marrow.' The myeloid lineage of cells was named for the neutrophil, which is the most abundant cell in the bone marrow. The erythroid progenitor gives rise to erythrocytes and megakaryocytes.

with two to five lobes, they are also called **polymorphonuclear leukocytes**. Most abundant of the granulocytes, and of all white blood cells, is the neutrophil (Figure 1.12) which is specialized in the capture, engulfment and killing of microorganisms. Cells with this function are called **phagocytes**, of which neutrophils are the most numerous and lethal. **Neutrophils** are effector cells of innate immunity that are rapidly mobilized to enter sites of infection and can work in the anaerobic conditions that often prevail in damaged tissue. Neutrophils are short-lived and die at the site of infection, forming **pus** (Figure 1.13). The second most abundant granulocyte is the **eosinophil**, which defends against helminth worms and other intestinal parasites. The least abundant granulocyte, the **basophil**, is so rare that little is known of its contribution to the immune response. The names of the granulocytes refer to the staining of their cytoplasmic granules with the commonly used histological stains: the eosinophil's granules contain basic substances that bind the acidic stain eosin, the basophil's granules contain acidic substances that bind basic stains such as hematoxylin, and the contents of the neutrophil's granules bind to neither acidic nor basic stains.

The second group of myeloid cells consists of monocytes, macrophages, and dendritic cells. **Monocytes** are leukocytes that circulate in the blood. They are distinguished from the granulocytes by being bigger, by having a distinctive indented nucleus, and by all looking the same: hence the name monocyte. Monocytes are mobile progenitors of the sedentary tissue macrophage; they travel in the blood to tissues, where they mature to **macrophages** and take up residence. The name macrophage means large phagocyte, and like the neutrophil, which was once called the microphage, the macrophage is well equipped for phagocytosis. Tissue macrophages are large irregularly shaped cells characterized by an extensive cytoplasm with numerous vacuoles, often containing engulfed materials (Figure 1.14). They are the general scavenger cells of the body, phagocytosing and disposing of dead cells and cell debris as well as invading microorganisms.

If neutrophils are the short-lived infantry of innate immunity, then macrophages are the long-lived commanders who provide intelligence to other cells and orchestrate the local response to infection. Macrophages present in the infected tissues are generally the first phagocytic cell to sense an invading microorganism. As part of their response to the pathogen, macrophages secrete the cytokines that recruit neutrophils and other leukocytes into the infected area.

Myeloid **dendritic cells** are resident in the body's tissues and have a distinctive star-shaped morphology. Although they have many properties in common with macrophages, their unique function is to act as cellular messengers that are sent to call up an adaptive immune response when it is needed. At such times, dendritic cells present in the infected tissue will leave the tissue with a cargo of degraded and intact pathogens and take it to one of several lymphoid organs that specialize in making adaptive immune responses.

The last type of myeloid cell is the mast cell, which is resident in all connective tissues. Although having granules like those of the basophil, the mast cell

Cell type	Proportion of leukocytes (%)
Neutrophil	40–75
Eosinophil	1–6
Basophil	<1
Monocyte	2–10
Lymphocyte	20–50

Figure 1.12 The abundance of leukocytes in human peripheral blood. The relative proportion of each cell type in the leukocyte population of peripheral blood taken from a vein from healthy donors is given.

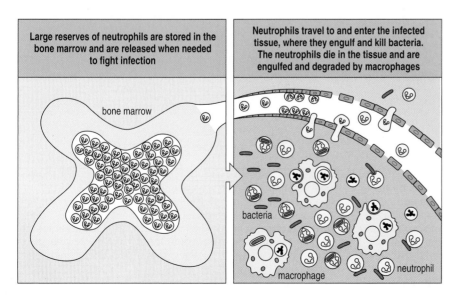

Figure 1.13 Neutrophils are stored in the bone marrow and move in large numbers to sites of infection, where they act and then die. After one round of ingestion and killing of bacteria, a neutrophil dies. The dead neutrophils are eventually mopped up by long-lived tissue macrophages, which break them down. The creamy material known as pus is composed of dead neutrophils.

is not derived from the basophil and the nature of its blood-borne progenitor is not yet known. The activation and degranulation of mast cells at sites of infection make a major contribution to inflammation.

The **lymphoid progenitor** gives rise to the **lymphoid lineage** of white blood cells. Two populations of blood lymphocytes are distinguished morphologically: large lymphocytes with a granular cytoplasm and small lymphocytes with almost no cytoplasm. The **large granular lymphocytes** are effector cells of innate immunity called **natural killer cells** or **NK cells**. NK cells are important in the defense against viral infections. They enter infected tissues, where they prevent the spread of infection by killing virus-infected cells and secreting cytokines that impede viral replication in infected cells. The **small lymphocytes** are the cells responsible for the adaptive immune response. They are small because they circulate in a quiescent and immature form that is functionally inactive. Recognition of a pathogen by small lymphocytes drives a process of lymphocyte selection, growth and differentiation that after 1–2 weeks produces a powerful response tailored to the invading organism.

The small lymphocytes, although morphologically indistinguishable from each other, comprise several sublineages that are distinguished by their cell-surface receptors and the functions that they are programmed to perform. The most important difference is between **B lymphocytes** or **B cells** and **T lymphocytes** or **T cells**. For B cells the cell-surface receptors for pathogens are **immunoglobulins**, whereas those of T cells are known as **T-cell receptors**. Immunoglobulins and T-cell receptors are structurally similar molecules that are the products of genes that are cut, spliced, and modified during lympho-

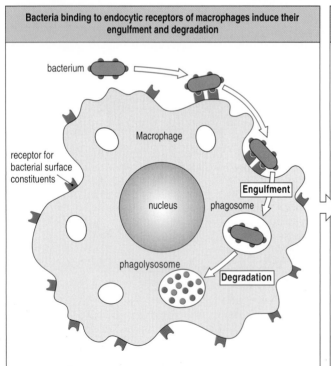
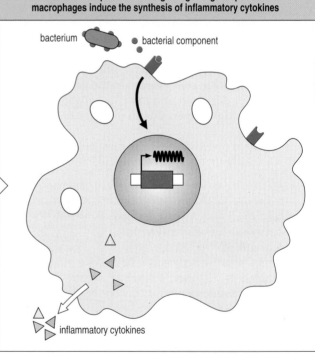

Figure 1.14 Macrophages respond to pathogens by using different receptors to stimulate phagocytosis and cytokine secretion. The left panel shows receptor-mediated phagocytosis of bacteria by a macrophage. The bacterium (red) binds to cell-surface receptors of the macrophage (blue), inducing engulfment of the bacterium into an endocytic vesicle called a phagosome within the cytoplasm of the macrophage. Fusion of the phagosome with lysosomes forms an acidic vesicle called a phagolysosome, which contains toxic small molecules and hydrolytic enzymes that kill and degrade the bacterium. The right panel shows how a bacterial component binding to a cell-surface receptor sends a signal to the macrophage's nucleus that initiates the transcription of genes for inflammatory cytokines. The cytokines are synthesized and secreted into the extracellular fluid.

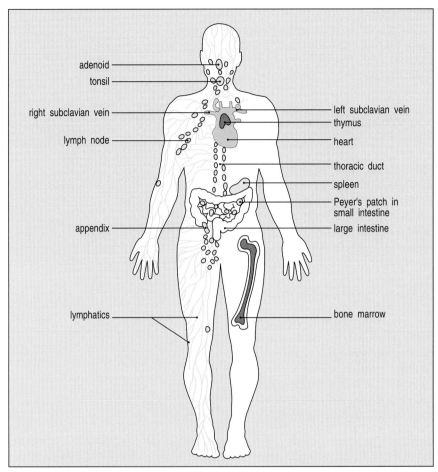

Figure 1.15 The sites of the principal lymphoid tissues within the human body. Lymphocytes arise from stem cells in the bone marrow. B cells complete their maturation in the bone marrow, whereas T cells leave at an immature stage and complete their development in the thymus. The bone marrow and the thymus are the primary lymphoid tissues and are shown in red. The secondary lymphoid tissues and the lymphatics are shown in yellow. Plasma that has leaked from the blood is collected by the lymphatics and returned to the blood supply via the thoracic duct, which empties into the left subclavian vein.

cyte development. As a consequence of these processes each B cell expresses a single type of immunoglobulin and each T cell expresses a single T-cell receptor. Within the small lymphocyte population billions of different immunoglobulins and T-cell receptors are represented.

1-6 Most lymphocytes are present in specialized lymphoid tissues

As well as circulating in the blood, lymphocytes congregate in specialized tissues known as **lymphoid tissues** or **lymphoid organs**. The major lymphoid tissues are bone marrow, thymus, spleen, adenoids, tonsils, appendix, lymph nodes, and Peyer's patches. Lymphoid tissues are also found lining the mucosal surfaces of the respiratory, gastrointestinal, and urogenital tracts. The lymphoid tissues are functionally divided into two types. **Primary** or **central lymphoid tissues** are where lymphocytes develop and mature to the stage at which they are able to respond to a pathogen. The bone marrow and the **thymus** are the primary lymphoid tissues. All other lymphoid tissues are known as **secondary** or **peripheral** lymphoid tissues; they are where mature lymphocytes become stimulated to respond to invading pathogens (Figure 1.15).

Lymph nodes lie at the junctions of an anastomosing network of **lymphatic vessels** or **lymphatics**, which originate in the connective tissues and collect the plasma that continually leaks out of blood vessels and forms the extracellular fluid. The lymphatics eventually return this fluid to the blood, chiefly via the thoracic duct, which empties into the left subclavian vein in the neck. Mature lymphocytes leave the primary lymphoid tissues and enter the bloodstream, whereupon they travel between the blood, the secondary lymphoid tissues, and the lymphatics, eventually being returned to the blood. The mixture of fluid and cells that flows through the lymphatics is called **lymph**. These movements are called **recirculation** of lymphocytes. An exception to this pattern is the spleen, which has no connections to the lymphatic system. Lymphocytes both enter and leave the spleen in the blood. At any one time only a very small fraction of lymphocytes are in the blood and lymph; the majority are in lymphoid organs and tissues.

Both B and T lymphocytes originate from lymphoid precursors in the bone marrow. But whereas B cells complete their maturation in the bone marrow before entering the circulation, T cells leave the bone marrow at an immature stage and migrate in the blood to the thymus, where they complete their maturation. Lymphocyte maturation is a highly selective process in which the vast majority of immature lymphocytes are destroyed because they fail to develop immunoglobulins or T-cell receptors that are useful to the immune system. The small fraction of immature B and T cells that successfully complete development depart from the primary lymphoid organs and enter the circulation.

1-7 Lymphocytes are activated in the secondary lymphoid tissues

Secondary lymphoid tissues provide meeting places where lymphocytes circulating in the blood encounter pathogens and their products brought from a site of infection. Precisely how this works depends on where an infection occurs. Frequent sites of infection are the connective tissues, which pathogens penetrate as a result of skin wounds. From such sites pathogens are carried by the lymphatics to the nearest **lymph node** (Figure 1.16). The

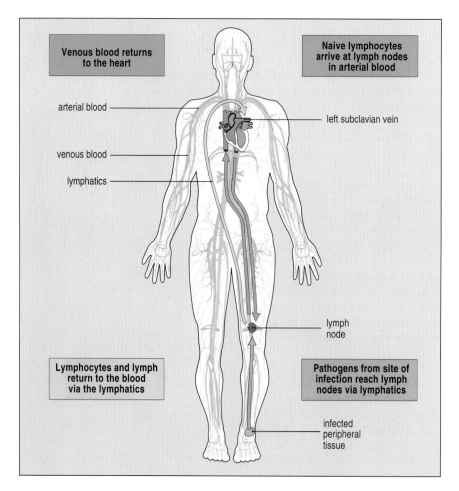

Figure 1.16 Circulating lymphocytes meet lymph-borne pathogens in draining lymph nodes. Lymphocytes leave the blood and enter lymph nodes, where they can be activated by pathogens draining in the afferent lymph from a site of infection. The circulation pertaining to a site of infection in the left foot is shown. When activated by pathogens, lymphocytes stay in the node to divide and differentiate into effector cells. If lymphocytes are not activated, they leave the node in the efferent lymph and are carried by the lymphatics to the thoracic duct (see Figure 1.15), which empties into the blood at the left subclavian vein. Lymphocytes recirculate all the time, irrespective of infection. Every minute 5×10^6 lymphocytes leave the blood and enter secondary lymphoid tissues.

lymph node receiving the fluid collected at an infected site is called the **draining lymph node**. Also traveling along this route are dendritic cells that have been activated by the infection and are carrying pathogens and their components.

Unlike the blood, the lymph is not driven by a dedicated pump and its flow is comparatively sluggish. One-way valves within lymphatic vessels and the lymph nodes placed at their junctions ensure that net movement of the lymph is always in a direction away from the peripheral tissues and towards the ducts in the upper body where the lymph empties into the blood. The flow of lymph is driven by the continual movements of one part of the body with respect to another. In the absence of such movement, as when a patient is confined to bed for a long time, lymph flow slows and fluid accumulates in tissues, causing the swelling known as **edema**.

Pathogens and dendritic cells arrive at a lymph node in **afferent lymphatic vessels**, several of which unite at the node to leave it as a single **efferent lymphatic vessel**. The anatomy of a lymph node is shown in Figure 1.17. As the lymph passes through the node, pathogens and other extraneous materials are filtered out by macrophages. This process prevents infectious organisms from reaching the blood and provides a depot of pathogen within the lymph node that can be used to activate lymphocytes. More important for this function are the dendritic cells that take up residence in parts of the lymph node

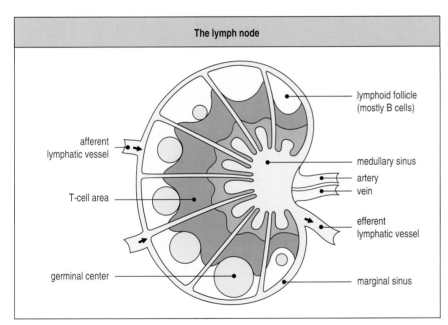

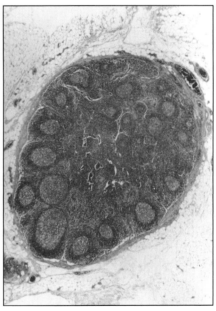

Figure 1.17 Architecture of the lymph node, the site where blood-borne lymphocytes respond to lymph-borne pathogens. Human lymph nodes are small kidney-shaped organs weighing 1 gram or less that form junctions where a number of afferent lymph vessels bringing lymph from the tissues unite to form a single, larger efferent lymph vessel. During infection, pathogens and dendritic cells carrying pathogens arrive in draining afferent lymph. The lymph node is packed with lymphocytes, macrophages and other cells of the immune system through which the lymph percolates. Dendritic cells settle in the lymph node; pathogens are picked up by the resident macrophages and then degraded. Both dendritic cells and macrophages are important stimulators of lymphocytes. Lymphocytes arrive at lymph nodes in the arterial blood. They enter the node by passing between the endothelial cells that line the fine capillaries within the lymph node (not shown). The population of lymphocytes within a node is in a continual state of flux, with new lymphocytes entering from the blood while others leave in the efferent lymph. Within the lymph node there are anatomically discrete areas where B or T cells tend to congregate. A lymph node draining a site of infection increases in size owing to the proliferation of activated lymphocytes, a phenomenon sometimes referred to as 'swollen glands.' The expansion of lymphocyte populations takes place in spherical lymphoid follicles present in the lymph node cortex. As lymphocyte division and differentiation proceed, the follicle morphology changes and it is then called a germinal center. The photograph shows a section through a lymph node in which there are prominent germinal centers. Photograph (× 7) courtesy of N. Rooney.

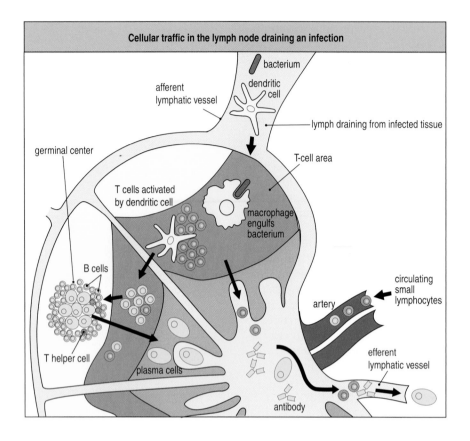

Cellular traffic in the lymph node draining an infection

bacterium

afferent lymphatic vessel

dendritic cell

lymph draining from infected tissue

germinal center

T-cell area

T cells activated by dendritic cell

macrophage engulfs bacterium

B cells

circulating small lymphocytes

artery

T helper cell

plasma cells

efferent lymphatic vessel

antibody

Figure 1.18 Cellular traffic in the lymph node draining an infection. Pathogens and dendritic cells carrying pathogens and their antigens arrive in the afferent lymph draining the site of infection. Free pathogens and debris are removed by macrophages; dendritic cells move to the T-cell areas, where they meet small lymphocytes that have entered the node from the blood. The dendritic cells stimulate the division and differentiation of pathogen-specific small lymphocytes into effector lymphocytes. Some helper T cells (blue) and cytotoxic T cells (purple) leave in the efferent lymph and travel to the infected tissue via the lymph and blood. Other helper T cells (blue) remain in the lymph node and stimulate the division and differentiation of pathogen-specific B cells (yellow) into plasma cells. Plasma cells move to the medulla of the lymph node, where they secrete pathogen-specific antibody, which is taken to the site of infection by the efferent lymph and blood. Some plasma cells leave the lymph node and travel via the efferent lymph and the blood to the bone marrow, where they continue to secrete antibody.

where the lymphocytes are predominantly T cells. These parts of the lymph node are called the **T-cell areas**.

Small lymphocytes recirculate through the body in both the blood and the lymph. When they reach the blood capillaries running through a lymph node, lymphocytes can leave the blood and enter the node. Within the node T cells move into the T-cell areas, where they form transient interactions with dendritic cells (Figure 1.18). If the receptors on the T-cell surface bind to pathogen components displayed by a dendritic cell, then the T cell is signaled to divide and differentiate into functional **effector cells**. Some T cells differentiate into **helper T cells** that stay in the lymph node and provide soluble proteins and intercellular contacts that drive the differentiation of B cells possessing immunoglobulin receptors that bind the pathogen. Effector B cells, called **plasma cells**, either stay in the lymph node or migrate to the bone marrow via lymph and blood. Plasma cells make and secrete large amounts of **antibody**, a soluble form of their cell-surface immunoglobulin. A second kind of T cell differentiates into helper T cells that leave the lymph node and travel via the efferent lymph and the blood to the infected tissue. These helper T cells interact with macrophages and secrete soluble cytokines to amplify the inflammation. A third type of effector T cell is called the **cytotoxic T cell** because these cells kill cells infected with viruses or other intracellular pathogens. Only a very small minority of the circulating small lymphocytes will be activated by any given pathogen; the rest will pass through the lymph node and leave in the efferent lymph to continue recirculation.

Pathogens can enter the blood directly, as occurs when blood-feeding insects transmit disease or when lymph nodes have failed to remove microorganisms from the lymph returned to the blood. The **spleen** is the lymphoid organ that serves as a filter for the blood. One purpose of the filtration is to remove damaged or old red cells; the second function of the spleen is that of a secondary lymphoid tissue—infectious agents are removed and used to activate lymphocytes. In this latter role, the spleen functions in a similar fashion to the lymph node, the only difference being that both pathogens and lymphocytes enter and leave the spleen in the blood (Figure 1.19).

The parts of the body that harbor the largest and most diverse populations of microorganisms are the respiratory and gastrointestinal tracts. The most heavily infested site is the oral cavity. The extensive mucosal surfaces of these tissues make them particularly vulnerable to infection and they are therefore heavily invested with secondary lymphoid tissue. The **gut-associated lymphoid tissues** (**GALT**) include the **tonsils**, **adenoids**, **appendix**, and the **Peyer's patches** which line the small intestine (see Figure 1.15). Similar but less organized aggregates of secondary lymphoid tissue line the respiratory epithelium, where they are called **bronchial-associated lymphoid tissue** (**BALT**), and other mucosal surfaces, including the gastrointestinal tract. The more diffuse mucosal lymphoid tissues are known generally as **mucosa-associated lymphoid tissue** (**MALT**).

Although different from the lymph nodes or spleen in outward appearance, the mucosal lymphoid tissues are similar to them in their microanatomy (Figure 1.20) and in their function of trapping pathogens to activate lymphocytes. The differences are chiefly in the route of pathogen entry and the migration patterns of their lymphocytes. Pathogens arrive at mucosa-associated lymphoid tissues by direct delivery across the mucosa, mediated by specialized cells of the mucosal epithelium called **M cells**. Lymphocytes first enter mucosal lymphoid tissue from the blood and, if not activated, leave via lymphatics that connect the mucosal tissues to draining lymph nodes. Lymphocytes activated in mucosal tissues tend to stay within the mucosal system, moving out from the lymphoid tissue into the lamina propria and the mucosal epithelium, where they perform their effector actions.

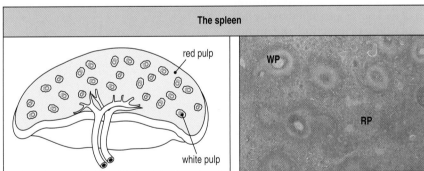

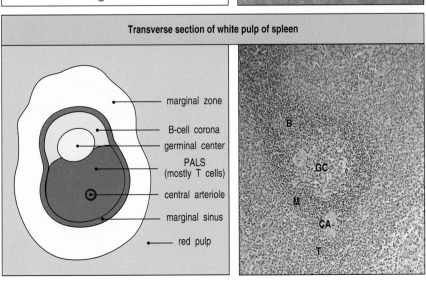

Figure 1.19 The spleen has aggregations of lymphocytes similar to those in lymph nodes. The human spleen is a large lymphoid organ in the upper left part of the abdomen, weighing about 150 grams. The upper diagram depicts a section of spleen in which nodules of white pulp are scattered within the more extensive red pulp. The red pulp (RP) is where old or damaged red cells are removed from the circulation; the white pulp (WP) is secondary lymphoid tissue, in which lymphocyte responses to blood-borne pathogens are made. The bottom diagram shows a nodule of white pulp in transverse section. It consists of a sheath of lymphocytes surrounding a central arteriole (CA). The sheath is called the periarteriolar lymphoid sheath (PALS). The lymphocytes closest to the arteriole are mostly T cells (T); B cells are placed more peripherally, forming a B-cell corona (B). Lymphoid follicles and germinal centers (GC) lie between the B- and T-cell zones. The marginal zone (M) contains differentiating B cells. Photographs courtesy of H.G. Burkitt and B. Young (top) and N. Rooney (bottom).

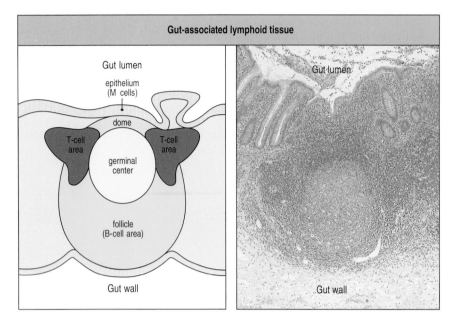

Figure 1.20 A typical region of gut-associated lymphoid tissue.
A schematic diagram (left panel) and a light micrograph (right panel) of a typical region of gut-associated lymphoid tissue. M cells of the gut epithelial wall deliver pathogens from the luminal side of the gut mucosa to the lymphoid tissue within the gut wall. These areas are organized similarly to the lymph node and the white pulp of the spleen, with distinctive B- and T-cell zones, lymphoid follicles, and germinal centers. Photograph courtesy of N. Rooney.

Summary

To restrict the nature, size, and location of microbial infection, animals have evolved a variety of defense mechanisms. The skin and contiguous mucosal surfaces provide physical and chemical barriers that confine microorganisms to the external surfaces of the body. In addition, cells and molecules of the innate and adaptive immune systems identify and eliminate microorganisms that have penetrated the physical and chemical barriers and gained entry to the soft tissues. The cells that mediate both innate and adaptive immune responses are principally the various types of leukocyte and allied tissue cells, which derive from the bone marrow. In responding to infection, the immune system uses innate mechanisms that are fast but limited, and adaptive mechanisms that are slow to start but eventually become both powerful and quick to recall. The cells and molecules of innate immunity identify common classes of pathogen and destroy them. Four key elements of innate immunity are: molecules that noncovalently bind to surface macromolecules of pathogens; molecules that covalently bond to pathogen surfaces, forming ligands for phagocyte receptors; phagocytic cells that engulf and kill pathogens; and cytotoxic cells that kill virus-infected cells. Vertebrates have evolved the additional defenses of adaptive immune responses, which involve the T and B lymphocytes. Adaptive immune responses are initiated in lymphoid tissues such as lymph nodes and spleen, in which pathogen-specific lymphocytes encounter and are activated by pathogen antigens.

Principles of adaptive immunity

In the innate immune response the diversity of pathogens that a person might encounter in a lifetime is faced by a repertoire of receptors that are always available. Some of these recognize structures shared by many different pathogens, and others recognize alterations to human cells that are commonly induced by the presence of pathogens. The adaptive immune response uses a quite different strategy in which receptors that uniquely define an infecting pathogen are first selected then mass produced on demand. Millions of different immunoglobulins and T-cell receptors are made by the B and T lymphocytes of the human immune system, and each recognizes a different molecular structure. On infection with a pathogen, only those B and T cells making recognition molecules that can bind to constituents of that pathogen are stimulated to divide, proliferate, and differentiate into effector lymphocytes.

1-8 Immunoglobulins and T-cell receptors are the highly variable recognition molecules of adaptive immunity

The molecules and mechanisms of adaptive immunity are restricted to vertebrate animals, suggesting that their evolution became feasible or advantageous only with the degree of cellular and anatomical complexity attained by the vertebrates. Immunoglobulins are expressed on the surface of B cells, where they can bind pathogens. Effector B cells, called plasma cells, secrete soluble forms of these immunoglobulins, which are known as antibodies. In contrast, T-cell receptors are only ever expressed as cell-surface recognition molecules, never as soluble proteins.

Any molecule, macromolecule, virus particle or cell that contains a structure recognized and bound by an immunoglobulin or T-cell receptor is known as its corresponding **antigen**. Surface immunoglobulins and T-cell receptors are thus also referred to as the **antigen receptors** of lymphocytes. The particular part of the antigen bound by the immunoglobulin or T-cell receptor is known as the **antigenic determinant** or **epitope**. Immunoglobulins can bind to a vast variety of different chemical structures, whereas T-cell receptors recognize a more limited range of epitopes. Immunoglobulins and T-cell receptors are said to be **specific**, or to have **specificity**, for the antigens they bind.

Immunoglobulins and T-cell receptors are structurally related molecules whose diversity of antigen-recognition sites is generated by similar genetic mechanisms. Immunoglobulins are formed from two different polypeptides called the heavy and light chains; each Y-shaped immunoglobulin molecule consists of two identical heavy chains and two identical light chains. Both types of polypeptide have an amino-terminal **variable region** that differs in amino-acid sequence from one immunoglobulin to the next, and a **constant region** that is identical in amino-acid sequence from one immunoglobulin to another (Figure 1.21). The variable regions contain the sites that bind antigens. Surface immunoglobulin is anchored in the membrane by transmembrane regions at the carboxyl ends of the heavy chains. Antibodies are a

Figure 1.21 Comparison of the basic structures of surface immunoglobulin, antibody, and the T-cell receptor. The heavy chains of surface immunoglobulin and antibody are shown in blue, the light chains in red.

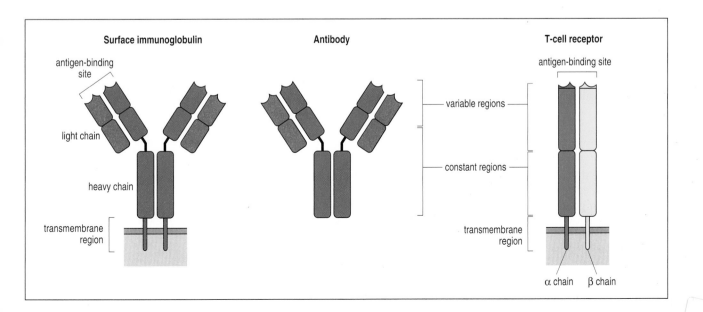

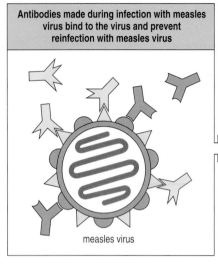

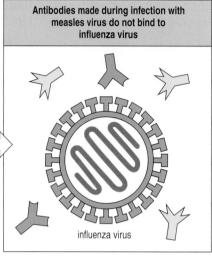

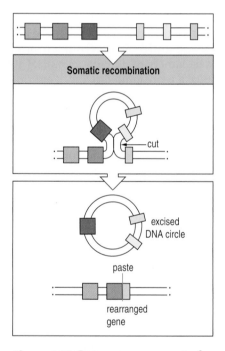

Figure 1.22 The antibodies made against a pathogen are highly specific for that pathogen. After having recovered from infection with measles virus a person's body fluids contain many different antibodies that can bind to the measles virus and prevent reinfection. None of these antibodies binds to an unrelated virus such as influenza.

secreted form of immunoglobulin that lack these transmembrane regions but are otherwise identical to surface immunoglobulins (see Figure 1.21).

A typical T-cell receptor consists of an α chain and a β chain, both anchored in the T-cell membrane. Like the light chains and heavy chains of immunoglobulins, the α and β chains of T-cell receptors each consist of a variable region and a constant region, with the variable regions forming an antigen-binding site (see Figure 1.21).

The differences in the amino-acid sequences of the variable regions of immunoglobulins and T-cell receptors create a vast variety of binding sites that are specific for different antigens, and thus for different pathogens. This is one of the hallmarks of an adaptive immune response. Antibodies secreted in response to a measles infection bind to measles virus but not to influenza virus (Figure 1.22); conversely, antibodies specific for influenza virus do not bind to measles virus.

The constant regions of antibodies contain binding sites for cell-surface receptors on phagocytes and inflammatory cells and also for complement proteins. The secreted antibody thus acts as a molecular adaptor or bridge. It binds to pathogens with one type of site and by another site to effector cells and molecules that destroy the pathogen. For immunoglobulins, but not T-cell receptors, there are several different types of constant region, known as **isotypes**, which confer different effector functions on the secreted antibody and can also target them to different anatomical sites. Every person can make antibodies of all isotypes.

1-9 The diversity of immunoglobulins and T-cell receptors is generated by gene rearrangement

The extensive diversity of variable regions in the antigen receptors of B and T cells is produced by a genetic mechanism unique to the immunoglobulin and T-cell receptor genes. In the genome, the variable regions of immunoglobulin and T-cell receptor chains are encoded in a series of separate gene segments. For a functional gene to be made, the appropriate gene segments need to be brought together by a physical rearrangement of the DNA (Figure 1.23). The immunoglobulin and T-cell receptor genes are quite unlike other human genes in this respect. Rearrangements at the heavy- and light-chain gene loci occur only in B cells, whereas rearrangements at the α- and β-chain gene loci

Figure 1.23 Gene rearrangement of the type occurring in immunoglobulin and T-cell receptor genes. In the unrearranged DNA there are three alternative 'red' segments and three alternative 'yellow' segments. A functional gene consists of one red segment joined to one yellow segment. This rearrangement is achieved by a process of 'cut and paste' in which the intervening DNA is removed as a circle. Different combinations of red and yellow segments can be brought together. In the example shown, the second red segment is brought together with the third yellow segment (counting from left to right), but other combinations of a red and a yellow segment would have been equally possible.

occur only in T cells. Before any rearrangement, the immunoglobulin and T-cell receptor genes are said to be in the **germline configuration** because that is how they are present in germ cells: eggs and sperm. The process of gene rearrangement in B and T cells is called **somatic recombination** because it recombines gene segments and occurs in the soma—those cells of the body that are not germ cells.

Many alternative forms of each gene segment are present within the unrearranged DNA and they can be rearranged in numerous different combinations. This is the basis for the diversity of the variable regions. Further diversity is introduced as a result of imprecision in the enzymatic machinery used to cut and splice the DNA during gene rearrangement. Its effect is to introduce additional nucleotides into the joints between gene segments. These processes create sequence diversity in the individual chains of the immunoglobulin and T-cell receptor polypeptides. Additional diversity in the antigen-binding sites arises from the association of different heavy and light chains in immunoglobulins and of different α and β chains in T-cell receptors.

So far we have considered the mechanisms that produce diversity in lymphocyte receptors during their development in the primary lymphoid tissues. In B cells, but not in T cells, an additional mechanism for diversification is brought into play after activation by pathogens. Activated dividing B cells in secondary lymphoid tissues initiate a process of **somatic hypermutation**, which introduces nucleotide substitutions into the immunoglobulin heavy- and light-chain genes. Some of the variant immunoglobulins produced by somatic hypermutation bind the pathogen more tightly and the cells producing them are preferentially chosen to become antibody-secreting plasma cells.

1-10 B cells recognize intact pathogens, whereas T cells recognize pathogen-derived peptides bound to proteins of the major histocompatibility complex

Antibodies made by B cells are the soluble pathogen-binding molecules of adaptive immunity. By circulating in the body fluids, antibodies can bind to bacterial cells and intact viral particles in extracellular spaces, targeting them for phagocytosis. To fulfill this function, the binding sites of antibodies interact with intact components of the pathogen surface such as glycoproteins and proteoglycans. The epitopes bound by antibodies most commonly include carbohydrate groups, or clusters of amino acids on the protein surface, or combinations of the two.

Whereas antibodies bind directly to the native structures of biological macromolecules, T-cell receptors can bind only short peptides that have been assembled into a complex with a membrane glycoprotein called a **major histocompatibility complex** (**MHC**) **molecule** (Figure 1.24). T-cell antigens are therefore peptides. They are produced within human cells by the breakdown of pathogens or their protein products, a process called **antigen processing**.

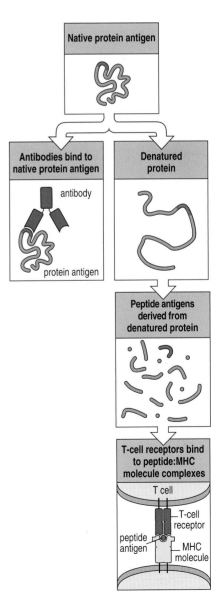

Figure 1.24 B cells recognize native proteins, whereas T cells recognize degraded proteins bound by major histocompatibility complex (MHC) molecules. The surface immunoglobulins of B cells and secreted antibodies bind to the native protein (left panel). To be recognized by a T-cell receptor, the protein must first be processed into peptide antigens by denaturation and proteolytic degradation (two center panels). Second, the peptide antigens must be bound to MHC molecules. Only then can the antigen be presented to the T cell (right panel).

Assembly of the peptide:MHC molecule complex takes place in the cell where the antigen was processed and, once formed, the complex is transported to the cell surface. Here it is accessible to T-cell receptors on the surface of T lymphocytes. MHC molecules are therefore said to **present** antigens to T cells, and cells carrying antigen:MHC complexes are known as **antigen-presenting cells**.

There are two classes of MHC molecule: class I and class II (Figure 1.25). **MHC class I molecules** present peptide antigens derived from pathogens that replicate intracellularly, such as viruses and some bacteria, and whose proteins are present in the cytosol of the infected cell (Figure 1.26). MHC class I molecules present peptides to cytotoxic T cells that function by killing infected cells and are distinguished by the presence of the CD8 glycoprotein on their surfaces. As all nucleated cells can be infected by viruses, MHC class I molecules are present on almost all cell types.

MHC class II molecules present peptides obtained from pathogens and their products that are present in the extracellular milieu and have been taken up into the endocytic vesicles of phagocytic cells (Figure 1.27). The MHC class II molecules of dendritic cells present peptides to helper T cells, which then go on to activate B cells or macrophages. Helper T cells are distinguished by the presence of the CD4 glycoprotein on their surfaces. Because helper T cells interact only with immune system cells that are specialized in the uptake and processing of pathogens, MHC class II molecules are present on only a few cell types. These **professional antigen-presenting cells** are dendritic cells, macrophages, and B cells. The CD4 T cells that interact with macrophages or B cells are distinguished by the cytokines they secrete and are called T_H1 and T_H2 (where the 'H' stands for helper) cells respectively.

In human populations there are many different genetic variants of MHC molecules; this **polymorphism** is the chief cause of rejection of tissue transplants. When donors and recipients are of different MHC types, the immune system of the recipient makes a vigorous immune response against the donor's MHC molecules, which it perceives as 'foreign.' It was this phenomenon that led to the discovery of the MHC and its original naming as a complex of genes that governed the compatibility of tissues (Greek *histo*) on transplantation.

1-11 Clonal selection of B and T lymphocytes is the guiding principle of the adaptive immune response

A direct consequence of the mechanisms of immunoglobulin and T-cell receptor gene rearrangement is that each lymphocyte expresses immunoglobulin or T-cell receptors of a single specificity. As a population, however, the lymphocytes of the human immune system make millions of different immunoglobulins and T-cell receptors. This hierarchy, whereby individual lymphocytes make a single type of recognition molecule and diversity is

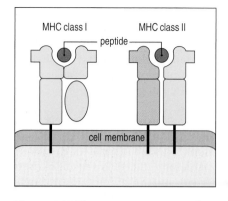

Figure 1.25 There are two types of MHC molecule, MHC class I and MHC class II.

Figure 1.26 The MHC class I pathway presents antigens derived from intracellular infections to cytotoxic T cells. In virus-infected cells, new viral proteins are made on cellular ribosomes in the cytoplasm. Some of these proteins are degraded in the cytoplasm and the resultant peptides are transported into the endoplasmic reticulum (ER). MHC class I molecules bind peptides in the ER and then transport them to the infected cell's surface, where they can be recognized by a virus-specific cytotoxic T cell. The cytotoxic T cell kills the virus-infected cell.

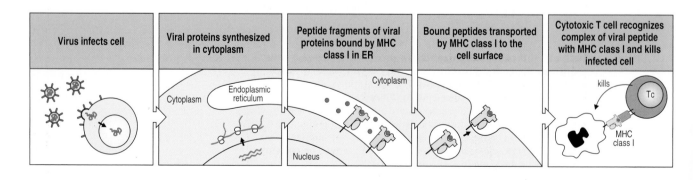

expressed at the level of lymphocyte populations, enables the lymphocyte response to be tailored toward particular pathogens. On infection, only the very small proportion of lymphocytes that have receptors that recognize the particular pathogen will be activated to divide and differentiate into effector cells. Consequently, each lymphocyte stimulated by the pathogen gives rise to a clonal population of cells all expressing the identical immunoglobulin or T-cell receptor. The process whereby pathogens select particular clones of lymphocyte for expansion is called **clonal selection** (see Figure 1.8). The use of a minute fraction of the total lymphocyte repertoire to respond to each pathogen ensures that the adaptive response is highly specific for the infection at hand.

Clonal selection is also an important influence on the development of B and T lymphocytes. During the maturation of B cells in the bone marrow, clonal selection is used to prevent the emergence of cells and antibodies that could attack the cells and tissues of the body. Clones of B cells with immunoglobulin receptors that bind strongly to the constituents of bone marrow tissue are signaled to die by programmed cell death, or **apoptosis**. During T-cell development in the thymus, clonal selection is more elaborate. First, there is positive selection, which saves those T cells that have receptors that interact well with a person's own MHC class I and class II variants (called self MHC) and discards all other T cells. Second, there is negative selection, which induces apoptosis in T cells that have receptors that bind too well to self MHC and

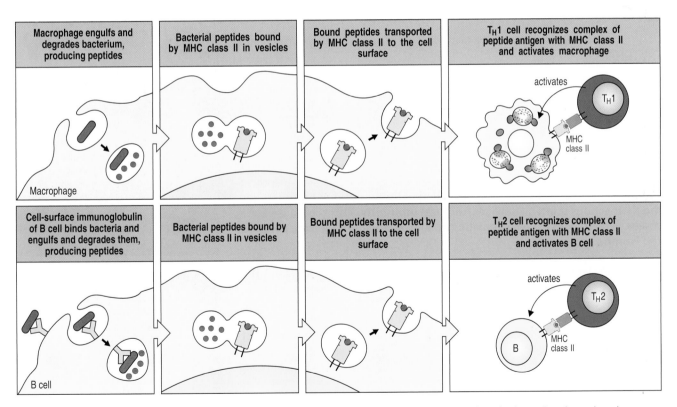

Figure 1.27 The MHC class II pathway presents antigens derived from extracellular infections to T$_H$1 cells. The upper panels show a macrophage phagocytosing extracellular bacteria and degrading their proteins in endocytic vesicles. MHC class II molecules bind peptides in endocytic vesicles and transport the peptides to the cell surface, where they are recognized by a T$_H$1 cell. Through cell contact and secretion of cytokines the T$_H$1 cell activates the macrophage. The lower panels show a B cell binding an extracellular bacterium with its cell-surface immunoglobulin. The bacterium bound to the immunoglobulin is endocytosed and its proteins are degraded into peptides. Some of these are bound by MHC class II molecules and transported to the cell surface, where they are recognized by a T$_H$2 cell. Through cell contact and secretion of cytokines the T$_H$2 cell activates the B cell. This will divide and differentiate into plasma cells secreting bacterium-specific antibody (not shown).

Figure 1.28 Stringent processes of positive and negative selection in the thymus determine the small fraction of T cells that mature and enter the peripheral circulation. Different clones of T cells are represented by different colors. Positive and negative selection are mediated by complexes of peptide with MHC class I and MHC class II molecules (the two classes of MHC molecule are not distinguished in this figure for simplicity). The epithelial cells in the cortex of the thymus that are involved in positive selection are not of hematopoietic origin; the cells of the thymic medulla that are involved in negative selection are of hematopoietic origin and include macrophages and thymocytes.

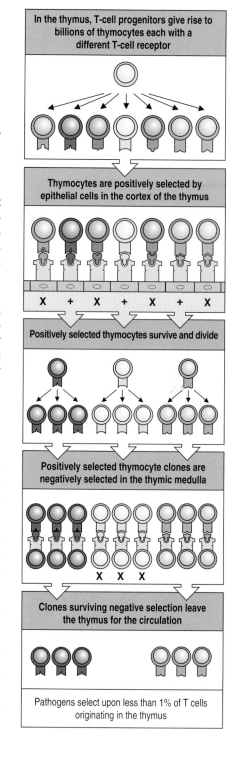

In the thymus, T-cell progenitors give rise to billions of thymocytes each with a different T-cell receptor

Thymocytes are positively selected by epithelial cells in the cortex of the thymus

Positively selected thymocytes survive and divide

Positively selected thymocyte clones are negatively selected in the thymic medulla

Clones surviving negative selection leave the thymus for the circulation

Pathogens select upon less than 1% of T cells originating in the thymus

might thus attack normal healthy tissues. These selections are so stringent that fewer than 1% of thymocytes mature to become circulating T cells (Figure 1.28). The negative selection imposed on potentially self-reactive lymphocytes helps to render the circulating pool of mature B and T cells nonresponsive to, or **tolerant** of, the normal components of the body. The lymphocyte repertoire is thus said to be **self-tolerant**.

In the immune system, apoptosis is a common feature of pathways of cell development, activation and effector function. Individual cells are induced to commit a tidy form of cell suicide, where there is no generalized and messy tissue damage of the sort that cell death by necrosis produces. Membrane changes in cells undergoing apoptosis lead to their rapid phagocytosis by macrophages.

1-12 Extracellular pathogens and their toxins are eliminated by antibodies

Immunoglobulins are divided into five **classes** or isotypes, which differ in their heavy-chain constant regions and have specialized effector functions when secreted as antibodies. The isotypes are called **IgA**, **IgD**, **IgE**, **IgG**, and **IgM**, where Ig stands for immunoglobulin. Cell-surface IgM and IgD are the antigen receptors on circulating B cells that have yet to encounter antigen. IgM is always the first antibody to be secreted in the immune response (IgD antibody is also made but in negligible amounts). As the immune response matures, antibodies of other isotypes emerge. Immunity due to antibodies and their actions is often known as **humoral immunity**. This is because antibodies were first discovered circulating in body fluids (*humors*) such as blood and lymph.

IgM, IgA, and IgG are the main antibodies present in blood, lymph, and the fluid in connective tissues. Antibodies secreted into the blood from lymph nodes, spleen, and bone marrow circulate in the bloodstream. At sites of infection and inflammation, blood vessels dilate and become permeable, increasing the flow of fluid into the infected tissue and with it the supply of antibodies for binding to extracellular pathogens (bacteria and virus particles) and the protein toxins that many bacterial pathogens secrete. IgA is also made in the lymphoid tissues underlying mucosa and then selectively transported across the mucosal epithelium to bind extracellular pathogens and their toxins on the mucosal surfaces.

One way in which antibodies reduce infection is by binding tightly to a site on a pathogen so as to inhibit pathogen growth, replication, or interaction with human cells. This mechanism is called **neutralization**. For example, certain antibodies when bound to influenza virions prevent the virus from infecting human cells. Similarly, the lethal actions of bacterial toxins can be prevented by a bound antibody (Figure 1.29).

The most important function of IgG antibodies is to facilitate the engulfment and destruction of extracellular microorganisms and toxins by phagocytes. Neutrophils and macrophages have cell-surface receptors that bind to the constant regions of the IgG heavy chains. A bacterium coated with IgG is more efficiently phagocytosed than an uncoated bacterium, a phenomenon called **opsonization** (see Figure 1.29). Opsonization can also be achieved by a coating of complement. There are no receptors on phagocytes for the constant region of IgM antibodies, but IgM bound to a pathogen's surface typically activates the complement system and the pathogen becomes coated with complement. This facilitates its uptake and phagocytosis by macrophages via another set of receptors that bind complement. A combination of IgG antibody and complement produces the greatest stimulation of phagocytosis because macrophage receptors for IgG constant regions and complement both participate in the process (see Figure 1.29).

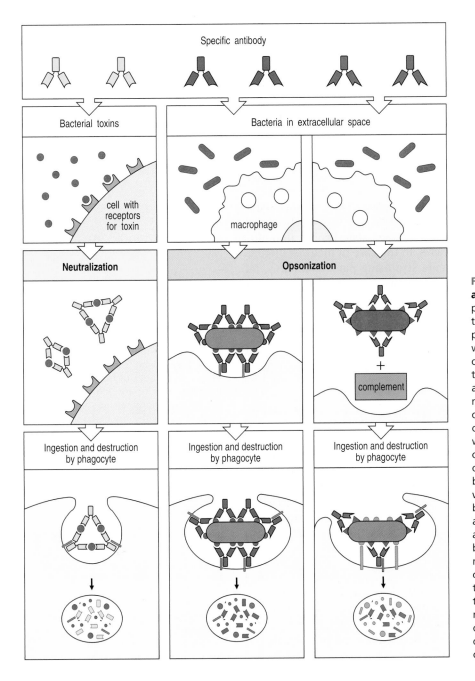

Figure 1.29 Mechanisms by which antibodies combat infection. Left panels: antibodies bind to a bacterial toxin and neutralize its toxic activity by preventing the toxin from interacting with its receptor on human cells. The complex of toxin and antibodies binds to macrophage receptors through the antibody's constant region. Finally the macrophage ingests and degrades the complex. Center panels: the opsonization of a bacterium by coating with antibody. When the bacterium is coated with IgG molecules, their constant regions point outward and can bind to the receptors on a macrophage, which then ingests and degrades the bacterium. Right panels: opsonization of a bacterium by a combination of antibody and complement. The bacterium is first coated with IgG molecules, which activate complement cleavage. Fragments of complement on the bacterial surface provide ligands for the complement receptor of macrophages. The combined interaction of macrophage receptors for complement and for the constant region of IgG makes for efficient phagocytosis.

The constant regions of IgE antibodies bind tightly to receptors on the surface of mast cells; most of the IgE antibody in the body is in this form, rather than circulating freely in blood or lymph like IgM and IgG. In response to worms and other parasitic infestations, the IgE bound to mast cells triggers strong inflammatory reactions that are thought to help expel or destroy the parasites. However, in developed countries, where parasitic infections are not a major health problem, IgE is most often encountered as the antibody involved in unwanted allergic reactions to innocuous substances.

1-13 Adaptive immune responses generally give rise to long-lived immunological memory and protective immunity

Clonal selection by pathogens is the guiding principle of adaptive immunity and explains the features of immunity that perplexed physicians in the past. The severity of a first encounter with an infectious disease arises because the **primary immune response** is developed from very few lymphocytes; the time taken to expand their number provides an opportunity for the pathogen to establish an infection to the point of causing disease. The clones of lymphocytes produced in a primary response include long-lived **memory cells**, which can respond more quickly and forcefully to subsequent encounters with the same pathogen.

The potency of such **secondary immune responses** can be sufficient to repel the pathogen before there is any detectable symptom of disease. The individual therefore appears immune to that disease. The striking differences between a primary and secondary response are illustrated in Figure 1.30. The immunity due to a secondary immune response is absolutely specific for the pathogen that provoked the primary response. It is the difference between the primary and the secondary response that has made vaccination so successful in the prevention of disease. Figure 1.31 shows how the introduction and availability of vaccines against diphtheria, polio, and measles has dramatically reduced the incidence of these diseases. In the case of measles there is also a corresponding reduction of subacute sclerosing panencephalitis, an infrequent but fatal spasticity caused by persisting measles virus that becomes manifest 7–10 years after measles infection and usually affects children who were infected as infants. When vaccination programs are successful

Figure 1.30 Comparison of a primary and secondary immune response. This diagram shows how the immune response develops during an experimental immunization of a laboratory animal. The response is measured in terms of the amount of pathogen-specific antibody present in the animal's blood serum, shown on the vertical axis, with time being shown on the horizontal axis. On the first day the animal is immunized with a vaccine against pathogen A. The levels of antibodies against pathogen A are shown in blue. The primary response reaches its maximum level 2 weeks after immunization. After the primary response has subsided, a second immunization with vaccine A on day 60 produces an immediate secondary response, which in 5 days is orders of magnitude greater than the primary response. In contrast, a vaccine against pathogen B, which was also given on day 60, produces a typical primary response to pathogen B as shown in yellow, demonstrating the specificity of the secondary response to vaccine A.

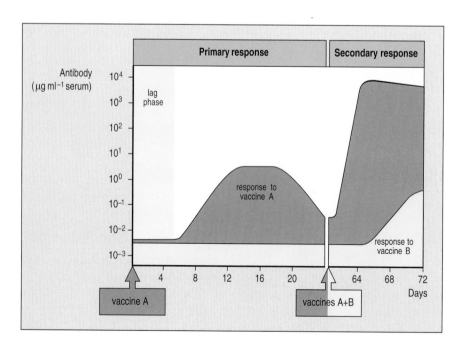

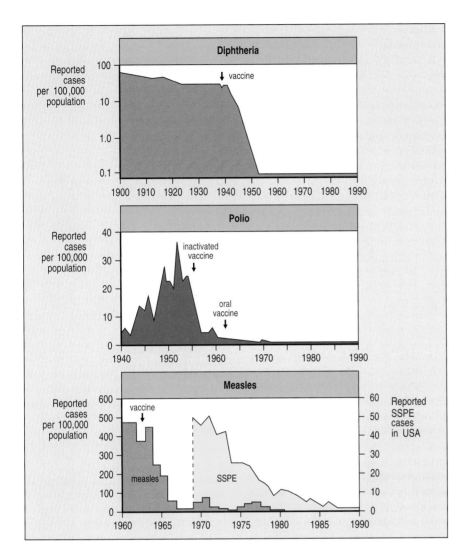

Figure 1.31 Successful vaccination campaigns. Diphtheria, poliomyelitis and measles have been virtually eliminated from the USA, as shown by these three graphs. The arrows indicate when the vaccination campaigns began. Subacute sclerosing panencephalitis (SSPE) is a brain disease that is a late consequence of measles infection for a minority of patients. Reduction of measles was paralleled by a reduction in SSPE 15 years later. Because these diseases have not been eradicated worldwide and the volume of international travel is so high, immunization must be maintained in much of the population to prevent disease recurrence.

to the point where the disease is unfamiliar to physicians and public alike, then concerns can arise with real or perceived side effects of a vaccine that affect a very small minority of those vaccinated. Such concerns lead to fewer children being vaccinated and to increasing occurrence of the disease.

1-14 The immune system can be compromised by inherited immunodeficiencies or by the actions of certain pathogens

When components of the immune system are either missing or do not work properly, this generally leads to increased susceptibility to microbial infection. One cause of defective immune responses is inherited mutations in genes encoding proteins that contribute to immunity. Most people who carry a mutant gene are healthy because their other, normal, copy of the gene provides sufficient functional protein; the small and unfortunate minority who carry two mutant copies of the gene lack the function encoded by that gene. Such deficiencies lead to varying degrees of failure of the immune system and a wide range of **immunodeficiency diseases**. In some of the immunodeficiency diseases, only one aspect of the immune response is affected, leading to susceptibility to particular kinds of infection; in others, adaptive immunity is completely absent, leading to a devastating vulnerability to all infections. These latter gene defects are rare, showing how vital is the protection normally afforded by the immune system. The discovery and study of immunodeficiency diseases has largely been the work of pediatricians, because such conditions usually show up early in life. Before the advent of antibiotics and, more recently, the possibility of bone marrow transplants and other replacement therapies, immunodeficiencies would usually have caused death in infancy.

Immunodeficiency states are caused not only by nonfunctional genes but also by pathogens that subvert the human immune system. An extreme example of an immunodeficiency due to disease is the **acquired immune deficiency syndrome** (**AIDS**), which is caused by infection with the **human immunodeficiency virus** (**HIV**). Although this disease has only been recognized by clinicians in the past 25 years, it is now at epidemic proportions, with more than 40 million people infected worldwide. HIV infects the CD4 T lymphocyte, a cell type essential for adaptive immunity. During the course of an extended infection, which can last for up to 20 years, the population of CD4 T cells gradually diminishes, eventually leading to collapse of the immune system. Patients with AIDS become increasingly susceptible to a range of infectious microorganisms, many of which rarely trouble uninfected people. Death usually results from the effects of one of these opportunistic infections rather than from the direct effects of HIV infection.

1-15 Unwanted effects of adaptive immunity cause allergy, autoimmune disease, and rejection of transplanted tissues

The sensitivity and force of the adaptive immune response, which are helpful traits in battling infection, can under some circumstances become misdirected toward innocuous materials or normal components of the human body. When this happens, various kinds of chronic and noninfectious disease emerge.

A state of **allergy** or **hypersensitivity** arises when antibodies of the IgE isotype are made against innocuous substances in the environment, for example foods, grass pollens, house dust, or dander from pets. Once such antibodies are made they circulate and become bound by their constant regions to mast cells throughout the body's tissues. On repeat exposure to the inducing antigen, called the **allergen**, it is bound by the IgE on mast cells. The combination of IgE and allergen on mast-cell surfaces triggers a response that produces symptoms (Figure 1.32) that sometimes lead to violent and life-threatening reactions such as those of asthma and systemic anaphylaxis. The damage and incapacitation caused by allergic reactions is greatly out of proportion to the threat posed by the allergen. Not all hypersensitivity diseases are caused by IgE; some are mediated by IgG, CD4 T_H1 cells or CD8 cytotoxic T cells.

Another class of noninfectious immunological disease is caused by chronic immune responses that gradually erode a target tissue. These conditions are called **autoimmune diseases** and result from **autoimmune responses** directed toward normal components of healthy human cells. In insulin-dependent diabetes, for example, the insulin-secreting β cells of the islets of Langerhans in the pancreas are the target of an autoimmune response. Because the human pancreas can produce insulin well in excess of what is normally needed, symptoms of disease do not show themselves until long after the start of the destructive response. Pancreatic β-cell destruction is believed to be due to B and T cells that respond to proteins made only in β cells. Other autoimmune diseases are rheumatoid arthritis, multiple sclerosis, myasthenia gravis, celiac disease, and Graves' disease. Autoimmune responses often seem to be by-products of a protective immune response made against an acute microbial infection. Once activated by pathogen-derived antigens, certain B or T cells cross-react with self components to which they were previously tolerant (Figure 1.33).

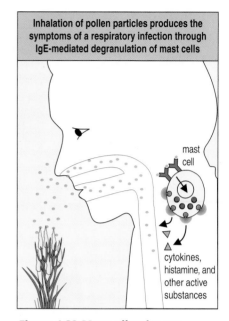

Inhalation of pollen particles produces the symptoms of a respiratory infection through IgE-mediated degranulation of mast cells

mast cell

cytokines, histamine, and other active substances

Figure 1.32 Many allergies are caused when antibodies of the IgE isotype are made against innocuous substances. The example shown here is of a pollen allergy, such as that to ragweed. For genetic and environmental reasons some members of the population make an antibody response to pollen that is of the IgE isotype. Once made, this antibody becomes bound to tissue mast cells, where it acts as a receptor. On subsequent exposure to pollen it binds to and cross-links the IgE, which is the signal for the mast cell to degranulate. The mast cell granules contain cytokines and other substances that induce inflammation, sneezing, and other symptoms that are usually associated with infectious diseases.

Over the past century, hygiene, vaccination, and antimicrobial drugs have all served to reduce mortality from infectious disease in the developed countries, particularly for children. In these populations the treatment of chronic and malignant diseases has become of growing importance in medicine. For an increasing number of these diseases, the replacement of tissues or organs by **transplantation** is now being used to restore health and prolong life. A major factor limiting the benefits of transplantation are the tissue incompatibilities caused by the extensive polymorphism of MHC class I and class II genes in the human population. MHC-compatible donors can be found within families, but only for some of the patients needing a transplant. For patients seeking a transplant from an unrelated donor, the chances of finding one who is MHC identical are low. In practice, therefore, the majority of clinical transplants are performed across MHC differences that have the potential to stimulate strong immune responses. Currently, the only means of preventing transplant rejection is the prolonged use of drugs that suppress adaptive immunity. Although facilitating graft acceptance, such drugs have toxic effects and also render the transplant patient more vulnerable to infection and cancer.

Many of the unwanted actions of adaptive immunity in hypersensitivity diseases, autoimmune diseases, and in transplant recipients are due to chronic states of inflammation that facilitate a persistent response by inflammatory cells of adaptive immunity: CD4 T_H1 cells, CD8 cytotoxic T cells and mast cells charged with IgE (Figure 1.34). The overall incidence of hypersensitivity and autoimmune diseases is increasing in the richer countries, a trend that the **hygiene hypothesis** attributes to the widespread practice of hygiene, vaccination, and antibiotic therapy. It is argued that in growing up in this environment children's immune systems become less able to deal with proper infections and more likely to charge after imaginary foes in ways that, at best, are not helpful and, at worst, are literally self-defeating.

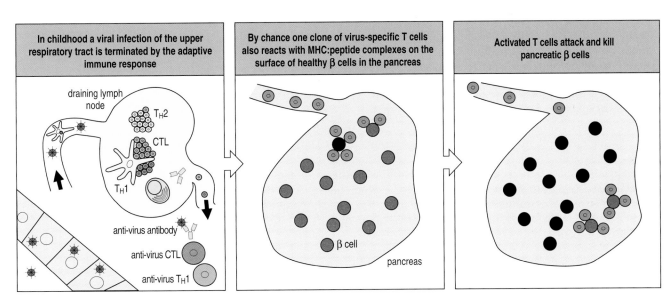

Figure 1.33 The adaptive immune response to a pathogen can sometimes trigger autoimmune reactions that eventually cause disease. Insulin-dependent diabetes mellitus (IDDM) is an autoimmune disease in which the insulin-producing β cells of the pancreas are slowly destroyed. Because the healthy pancreas can make much more insulin than is normally required, symptoms need not occur until many years after the start of the autoimmunity. A history of viral infection, for example with the Coxsackie viruses, has been correlated with IDDM and the autoimmunity may be a secondary consequence of the adaptive immune response to the virus. This could happen if a clone of T cells activated by the virus has, by chance, a T-cell receptor that reacts with an MHC:peptide complex present on the surface of pancreatic β cells. This MHC:peptide complex will then continually re-stimulate the T cells, which in turn persistently attack the β cells.

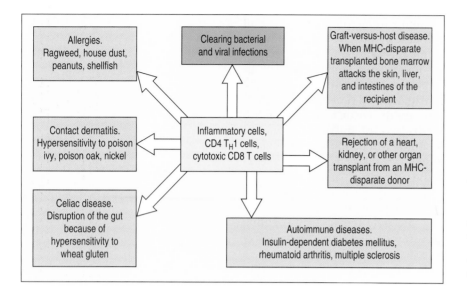

Figure 1.34 Inflammatory adaptive immune responses can be both beneficial and harmful. People in rich industrialized countries are experiencing an increase in inflammatory diseases caused by hypersensitivity to innocuous environmental antigens or by autoimmune destruction of healthy tissues of the body. Another situation in which inflammatory responses are unwanted is in transplantation of tissues and organs, which is increasingly used as therapy for inherited immuno-deficiencies, malignant disease, and organ failure in the industrialized countries. In transplantation, the major complications are of inflammatory immune responses that either reject the transplant or attack the recipient's tissue to cause severe graft-versus-host disease. Many of these unwanted immune reactions are caused by inflammatory cells of adaptive immunity: T_H1 CD4 T cells, cytotoxic CD8 T cells, or mast cells charged with IgE.

Summary

The mechanisms of adaptive immunity are ones that improve pathogen recognition rather than pathogen destruction. Adaptive immune responses are due to the lymphocytes of the immune system, which collectively have the ability to recognize a vast array of antigens. Somatic gene rearrangement and somatic mutation in the genes for antigen receptors provide the lymphocyte population with a set of highly diverse antigen receptors—immunoglobulins on B lymphocytes, and T-cell receptors on T lymphocytes. An individual lymphocyte expresses receptors of a single and unique antigen-binding specificity. A pathogen therefore stimulates only the small subset of lymphocytes that express receptors for the pathogen antigens, focusing the adaptive immune response on that pathogen.

One major difference between B and T cells is the type of antigens they recognize. Whereas the immunoglobulin receptors of B cells bind whole molecules and intact pathogens, T-cell receptors recognize only short peptide antigens that are bound to major histocompatibility complex molecules on cell surfaces. A second major difference is that all B cells have the same general function of making immunoglobulins, whereas T cells can have distinctive functions. CD8 cytotoxic T cells kill cells infected with viruses by recognizing viral peptide antigens presented by MHC class I molecules; CD4 T cells help macrophages and B cells become activated by recognizing peptides presented by MHC class II molecules. The CD4 T cells are further subdivided into T_H1 cells, which help macrophages, and T_H2 cells, which help B cells. Before they can perform these functions all T cells must themselves be activated within a secondary lymphoid tissue through recognition of their specific antigen on a professional antigen-presenting cell.

B cells differentiate into plasma cells that make antibody, a secreted form of the cell-surface immunoglobulin. The production of antibodies by a B cell usually requires help from a CD4 T_H2 cell that responds to the same antigen and activates the B cell. Antibodies bind to extracellular pathogens and their toxins and deliver them to phagocytes and other effector cells for destruction. Immunity due to antibodies and their actions is known as humoral immunity. Whereas CD4 T_H2 cells work within secondary lymphoid tissues, CD8 cytotoxic T cells and CD4 T_H1 cells must travel to the infected site to perform their functions. Immunity that is due principally to CD8 cytotoxic T cells and/or CD4 T_H1 cells is known as cell-mediated immunity.

When successful, an adaptive immune response terminates infection and provides long-lasting protective immunity against the pathogen that provoked the response. Failures to develop a successful response can arise from inherited deficiencies in the immune system or from the pathogen's ability to escape, avoid, or subvert the immune response. Such failures can lead to debilitating chronic infections or death. Another category of failure of adaptive immunity is the chronic disease caused by a misdirected response. Allergy is the consequence of a strong response to a harmless substance, whereas autoimmune disease is caused when the destructive potential of the immune system is directed at one or more of the body's own tissues. A further medical context for unwanted adaptive immunity is transplantation. Here the differences in MHC type between donor and recipient trigger an immune response that if left unchecked will reject the transplanted tissue.

Summary to Chapter 1

Throughout their evolutionary history, multicellular animals have been infected by microorganisms. In response, the animals evolved a series of defenses, which humans use today. Barriers confine microorganisms to the outer surfaces of the body, and when pathogens manage to breach the barriers they are sought out and destroyed by the immune system. In responding to microbial attack, the immune system starts with innate immunity, whose mechanisms are fast, fixed in their mode of action, and effective in stopping most infections at an early stage. The mechanisms of adaptive immunity are slow to start, but they eventually become powerful enough to terminate almost all of the infections that eluded innate immunity. The cells and molecules of innate immunity have counterparts in vertebrates and invertebrates. They identify common classes of pathogen and destroy them by using mechanisms that have stood the test of evolutionary time and are continuously useful. Mechanisms of adaptive immunity are found only in vertebrates and are distinguished by elaborate and highly specific ways of recognizing pathogens. They build on the mechanisms of innate immunity to provide a powerful response that is tailored to the pathogen at hand and can be rapidly reactivated on future challenge with that same pathogen, providing lifelong immunity to many common diseases. Adaptive immunity is an evolving process within a person's lifetime, in which each infection changes the makeup of that individual's lymphocyte population. These changes are neither inherited nor passed on but, during the course of a lifetime, they determine a person's fitness and their susceptibility to disease. The strategy of vaccination aims at circumventing the risk of a first infection, and in the twentieth century successful campaigns of vaccination were waged against several diseases that were once both familiar and feared. Through the use of vaccination and antimicrobial drugs, as well as better sanitation and nutrition, infectious disease has become a less common cause of death in many countries. In such situations, diseases due to unwanted activities of the immune system take on more importance. These include chronic conditions such as allergy and autoimmunity as well as the rejection of transplanted organs and tissues.

Questions

[handwritten: virus, bacteria, parasites FUNGI]

Question 1–1

A. Identify the four classes of pathogen that provoke immune responses in our bodies.

B. Name two pathogens for which vaccines have been developed and discuss why vaccines provide protective immunity. (Refer to Figures 1.2, 1.3, and 1.31.)

Question 1–2

Discuss the properties of skin and mucosal surfaces that make them such important components of the body's first line of defense against invading pathogens. (Refer to Figure 1.4.)

Question 1–3

Phagocytosis of microbes or microbial constituents by macrophages is followed by macrophage activation and secretion of cytokines and other bioactive molecules. What are the main effects of these molecules? (Refer to Figure 1.6.)

Question 1–4

Define the Latin words *calor, dolor, rubor,* and *tumor* in the context of inflammation. (Refer to Figure 1.6.)

Question 1–5

What are the main differences between innate immunity and adaptive immunity? (Refer to Figure 1.7.)

Question 1–6

A. Identify the two major progenitor subsets of leukocytes.

B. Where do they originate in adults?

C. Name the white blood cells that differentiate from these two progenitor lineages. (Refer to Figures 1.9–1.11 and 1.15.)

Question 1–7

A. Name the primary lymphoid tissues in mammals and the main types of secondary lymphoid tissue.

B. What is the difference in function between primary and secondary lymphoid tissues and what are the principal events that take place in each?

C. In what way is the spleen different from the other secondary lymphoid tissues? (Refer to Figures 1.15 and 1.17–1.20.)

Question 1–8

A. What molecules do (i) B cells and (ii) T cells use to recognize foreign antigens?

B. Draw simple schematics of these molecules and contrast their different antigen-recognition properties.

C. Explain how antigen is prepared for T-cell recognition. (Refer to Figures 1.21, 1.24, 1.25.)

Question 1–9

Describe three distinct mechanisms by which antibodies eradicate infection. (Refer to Figure 1.29.)

Question 1–10

A. What is meant by the term immunodeficiency disease?

B. Why is an individual's ability to combat infection compromised if they have an immunodeficiency disease?

C. What causes an immunodeficiency disease?

Question 1–11

Describe three types of unwanted and potentially harmful immune responses. (Refer to Figures 1.32, 1.33, and 1.34.)

B) measles, flu
    - creates prim. at B cells (Abs) { ~~cells~~ memory T & B cells, allows
      for both cell-mediated & humoral imm.
    - specific to them &

(-2) same parents

(-3 heat redness pain swelling
    - inc. blood flow
    - recruit other cells, sends cells

Chapter 2

Antibody Structure and the Generation of B-Cell Diversity

Antibodies are variable proteins produced by the B lymphocytes of the immune system in response to infection. They circulate as a major component of the plasma in blood and lymph. Their function is to bind to pathogenic microorganisms and their toxins when these are encountered in the extracellular spaces of the body. The molecules to which antibodies bind are, by definition, called **antigens**. All types of biological macromolecule serve as antigens, but proteins and carbohydrates are the most common. Binding of antibody to a pathogen can disable the pathogen and also renders it susceptible to destruction by other components of the immune system. Most vaccines provide their protection through stimulating the production of antibodies.

Individual antibodies are **specific**; each antibody can bind to only one or a very small number of different antigens. For this reason, immunity to the measles virus, for example, provides no protection against the influenza virus, nor does immunity to influenza protect against measles. The immune system of an individual has the potential to make antibodies against a vast number of different substances, paralleling the many different foreign antigens that a person is likely to be exposed to during their lifetime. Collectively, therefore, antibodies are diverse in their antigen-binding specificities; the total number of different specific antibodies that can be made by an individual is known as the **antibody repertoire** and it might be as high as 10^{16}.

Antibodies are the secreted form of proteins known more generally as **immunoglobulins** (**Ig**). Before it has encountered antigen, a mature B cell expresses immunoglobulin in a membrane-bound form that serves as the B cell's receptor for antigen. The development of B cells involves stages in which different components of the B-cell receptor are assembled. In this process each B cell becomes committed to expressing just one form of the B-cell receptor, but this form varies from one cell to the next. When antigen binds to this receptor, the B cell is stimulated to proliferate and to differentiate into **plasma cells**, which then secrete antibodies of the same specificity as that of the membrane-bound immunoglobulin (Figure 2.1). Antibody production is the single effector function of the B lymphocytes of the immune system.

The first part of this chapter describes the general structure of immunoglobulins, and the second part considers how immunoglobulin diversity is generated in developing B cells. After a mature B cell has encountered its specific antigen, changes are made in the specificity and effector functions of the antibodies that it produces; these are discussed in the final part of the chapter.

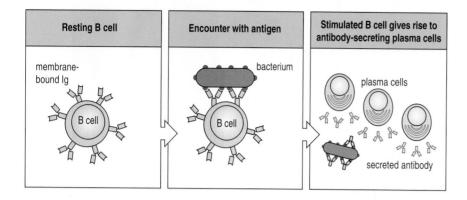

Figure 2.1 Plasma cells secrete antibody of the same antigen specificity as the membrane-bound immunoglobulin expressed by their B-cell precursor. A mature B cell expresses membrane-bound immunoglobulin (Ig) of a single antigen specificity. When a foreign antigen first binds to this immunoglobulin, the B cell is stimulated to proliferate. Its progeny differentiate into plasma cells that secrete antibody of the same specificity as the membrane-bound immunoglobulin.

The structural basis of antibody diversity

The function of an antibody molecule in host defense is to recognize and bind its corresponding antigen, and to target the bound antigen to other components of the immune system. These then destroy the antigen or clear it from the body. Antigen binding and interaction with other immune system cells and molecules are carried out by different parts of the antibody molecule. One part is highly variable in that its amino-acid sequence differs considerably from antibody to antibody. This variable part contains the site of antigen binding and confers specificity to the antibody. The rest of the molecule is far less variable in its amino-acid sequence. This constant part of the antibody interacts with other immune system components.

Immunoglobulins are split into five classes—IgA, IgD, IgE, IgG, and IgM. They are distinguished on the basis of structural differences in the constant part of the molecule and have different effector functions. We shall first consider the general structure of immunoglobulins and then look more closely at the structural features that confer antigen specificity. IgG, the most abundant antibody in blood and lymph, will be used to illustrate the common structural features of immunoglobulins.

2-1 Antibodies are composed of polypeptides with variable and constant regions

Antibodies are glycoproteins that are built from a basic unit of four polypeptide chains. This unit consists of two identical **heavy chains (H chains)** and two identical, smaller, **light chains (L chains)**, which are assembled into a structure that looks like the letter Y (Figure 2.2, top panel). An IgG molecule has a molecular weight of about 150 kDa, to which each heavy chain contributes approximately 50 kDa and each light chain approximately 25 kDa. Each arm of the Y is made up of a complete light chain paired with the amino-terminal (N-terminal) part of a heavy chain, covalently linked by a disulfide bond. The stem of the Y consists of the paired carboxy-terminal (C-terminal)

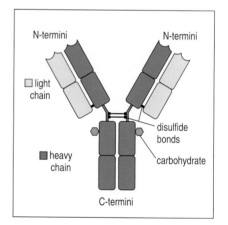

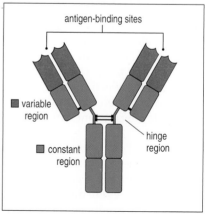

Figure 2.2 The immunoglobulin G (IgG) molecule. As shown in the top panel, each IgG molecule is made up of two identical heavy chains (green) and two identical light chains (yellow). Carbohydrate (turquoise) is attached to the heavy chains. The lower panel shows the location of the variable (V) and constant (C) regions in the IgG molecule. The amino-terminal regions (red) of the heavy and light chains are variable in sequence from one IgG molecule to another; the remaining regions are constant in sequence (blue). The carbohydrate is omitted from this panel and from most subsequent figures for simplicity. In IgG a flexible hinge region is located between the two arms and the stem of the Y.

portions of the two heavy chains. The two heavy chains are also linked to each other by disulfide bonds.

The polypeptide chains of different antibodies vary greatly in amino-acid sequence, and the sequence differences are concentrated in the amino-terminal region of each type of chain; this is known as the **variable region** or **V region**. This variability is the reason for the great diversity of antigen-binding specificities among antibodies because the paired V regions of a heavy and a light chain form the **antigen-binding site** (Figure 2.2, bottom panel). Every Y-shaped antibody molecule, therefore, has two identical antigen-binding sites, one at the end of each arm. The remaining parts of the light chain and the heavy chain have much more limited variation in amino-acid sequence between different antibodies and are, therefore, known as the **constant regions** or **C regions**.

In IgG, a relatively unstructured portion in the middle of the heavy chain forms a flexible hinge region at which the molecule can be cleaved to produce defined antibody fragments. These fragments were instrumental in providing information about antibody structure and function because they can be used for separate analysis of the antigen-binding and effector functions of the molecule. Digestion with the plant protease papain produces three fragments, corresponding to the two arms and the stem (Figure 2.3). The fragments corresponding to the arms are called **Fab** (**F**ragment **a**ntigen **b**inding) because they bind antigen. The fragment corresponding to the stem is called **Fc** (**F**ragment **c**rystallizable) because it was seen to crystallize in the first experiments

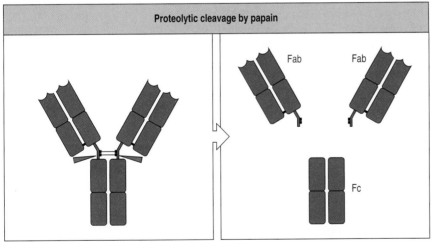

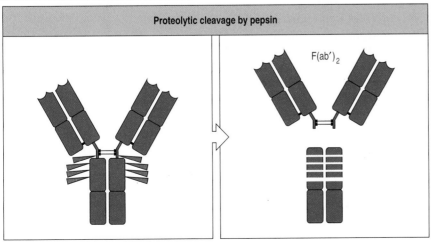

Figure 2.3 The Y-shaped immunoglobulin molecule can be dissected by using proteases. The protease papain cleaves the IgG molecule into three pieces: two Fab fragments and one Fc fragment (upper panels). The protease pepsin cleaves IgG to yield one F(ab')$_2$ fragment. The Fc fragment is not stable to pepsin degradation and is broken into a number of smaller pieces (lower panels). Red arrowheads denote the sites of protease attack.

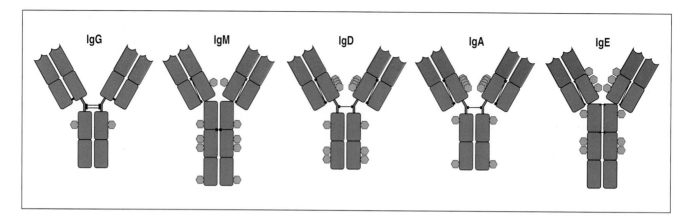

Figure 2.4 The structural organization of the human immunoglobulin isotypes. In particular, note the differences in length of the heavy-chain C regions, the locations of the disulfide bonds linking the chains, and the hinge region present in IgG, IgA, and IgD but not in IgM and IgE. The isotypes also differ in the distribution of *N*-linked carbohydrate groups, as shown in turquoise. All these immunoglobulins occur as monomers in their membrane-bound form. In their soluble, secreted form IgD, IgE and IgG are always monomers, IgA forms monomers and dimers and IgM forms only pentamers.

of this sort. The stem of a complete antibody molecule is, thus, often known as the **Fc region** or **Fc piece**, and the arms as Fab. Digestion of IgG with the gut protease pepsin produces a different fragment, **F(ab′)$_2$**, in which the two arms remain linked by disulfide bonds between the heavy chains.

Differences in the heavy-chain C regions define five main **isotypes** or **classes** of immunoglobulin, which have different functions in the immune response. They are **immunoglobulin A (IgA)**, **immunoglobulin D (IgD)**, **immunoglobulin E (IgE)**, **immunoglobulin G (IgG)**, and **immunoglobulin M (IgM)** (Figure 2.4). Their heavy chains are denoted by the corresponding lower-case Greek letter (α, δ, ε, γ, and μ, respectively).

The light chain has only two isotypes or classes, which are termed **kappa (κ)** and **lambda (λ)**. No functional difference has been found between antibodies carrying κ light chains and those carrying λ light chains; light chains of both isotypes are found associated with all the heavy-chain isotypes. Each antibody, however, contains either κ or λ light chains, not both. The relative abundances of κ and λ light chains vary with the species of animal. In humans, two-thirds of the antibody molecules contain κ chains and one-third have λ chains.

2-2 Immunoglobulin chains are folded into compact and stable protein domains

Antibodies function in extracellular environments in the presence of infection, where they can encounter variations in pH, salt concentration, proteolytic enzymes, and other potentially destabilizing factors. Their structure, however, helps them to withstand such harsh conditions. Heavy and light chains each consist of a series of similar sequence motifs; a single motif is about 100–110 amino acids long and folds up into a compact and exceptionally stable protein domain called an **immunoglobulin domain**. Each immunoglobulin chain is composed of a linear series of these domains.

The V region at the amino-terminal end of each heavy or light chain is composed of a single **variable domain (V domain)**: V_H in the heavy chain and V_L in the light chain. A V_H and a V_L domain together form an antigen-binding site. The other domains have little or no sequence diversity within a particular isotype and are termed the **constant domains (C domains)**, which make up the C regions. The constant region of a light chain is composed of a single C_L domain, whereas the constant region of a heavy chain is composed of three or four C domains, depending on the isotype. The γ heavy chains of IgG have three domains—C_H1, C_H2, and C_H3. Some other isotypes have four C domains (see Figure 2.4). In the complete IgG molecule the pairing of the four polypeptide chains produces three globular regions, corresponding to the

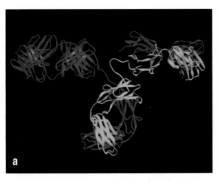

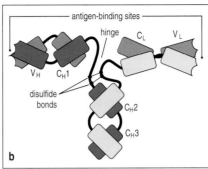

a   b

Figure 2.5 The heavy and light chains of an immunoglobulin molecule are made from a series of similar protein domains. Panel a shows a ribbon diagram tracing the course of the polypeptide backbones of chains in an IgG molecule. The heavy chains are shown in yellow and purple, the light chains in red. The molecule is composed of three similarly sized globular portions, which correspond to the Fc stem and the two Fab arms. Each of these globular portions is made up of four immunoglobulin domains. The schematic diagram in panel b shows the domain structure of an IgG molecule. The light chain (red) has one variable domain (V_L) and one constant domain (C_L). The γ heavy chain (one yellow and one purple) has a variable domain (V_H) and three constant domains (C_H1, C_H2, and C_H3). Panel a courtesy of L. Harris.

two Fab arms and the Fc stem, respectively, each of which is composed of four immunoglobulin domains (Figure 2.5).

The structure of a single immunoglobulin domain can be compared to a bulging sandwich in which two β sheets (the slices of bread) are held together by strong hydrophobic interactions between their constituent amino acid side chains (the filling). The structure is stabilized by a disulfide bond between the two β sheets. The adjacent strands within the β sheets are connected by loops of the polypeptide chain. This arrangement of β sheets provides the stable structural framework of all immunoglobulin domains, whereas the amino-acid sequence of the loops can be varied to confer different binding properties on the domain. The structural similarities and differences between the C and V domains are illustrated for the light chain in Figure 2.6.

Although immunoglobulin domains were first discovered in antibodies, very similar domains, known generally as **immunoglobulin-like domains**, have subsequently been found in many other proteins, and are particularly common in cell-surface and secreted proteins of the immune system. Such proteins collectively form the **immunoglobulin superfamily**.

2-3 An antigen-binding site is formed from the hypervariable regions of a heavy-chain and a light-chain V domain

A comparison of the V domains of heavy and light chains from different antibody molecules shows that the differences in amino-acid sequence are concentrated within particular regions called **hypervariable regions (HV)**, which

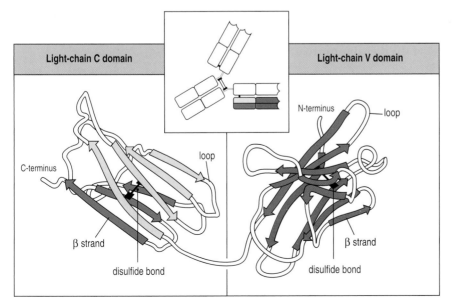

Figure 2.6 The three-dimensional structure of immunoglobulin C and V domains. An individual light chain is folded into a single C domain (left panel) and a single V domain (right panel). The inset shows the location of this light chain in an IgG molecule. The folding of the polypeptide chain backbone is depicted as ribbon diagrams in which the thick arrows correspond to the parts of the chain that form the β strands of the β sheet. The arrows point from the amino terminus of the chain to the carboxy terminus. Adjacent strands in each β sheet run in opposite directions (antiparallel), as shown by the arrows, and are connected by loops. In a C domain there are four β strands in the upper β sheet (yellow) and three in the lower sheet (green). In a V domain there are five strands in the upper sheet (blue) and four in the lower (red).

are flanked by much less variable **framework regions** (Figure 2.7, top panel). Three hypervariable regions are found in each V domain. When mapped onto the folded structure of a V domain, the hypervariable regions locate to three of the loops that are exposed at the end of the domain farthest from the constant region. The framework regions correspond to the β strands and the remaining loops. This relationship is illustrated for the light-chain V domain in Figure 2.7 (center panel). Thus, the structure of the immunoglobulin domain provides the capacity for localized variability within a structurally constant framework.

The pairing of a heavy and a light chain in an antibody molecule brings together the hypervariable loops from each V domain to create a composite hypervariable surface, which forms the antigen-binding site at the tip of each Fab arm (see Figure 2.7, bottom panel). The differences in the loops between different antibodies create both the specificity of antigen-binding sites and their diversity. The hypervariable loops are also called **complementarity-determining regions** (**CDRs**) because they provide a binding surface that is complementary to that of the antigen.

2-4 Antigen-binding sites vary in shape and physical properties

The function of antibodies is to bind to microorganisms and facilitate their destruction or ejection from the body. The antibodies that are most effective against infection are generally those that bind to the exposed and accessible molecules that make up the surface of a pathogen. The part of an antigen to which an antibody binds is called an **antigenic determinant** or an **epitope** (Figure 2.8). In nature, these structures are usually either carbohydrate or protein, or both, because the surface molecules of pathogens are commonly glycoproteins, polysaccharides, glycolipids, and peptidoglycans. Complex macromolecules such as these will usually contain several different epitopes, each of which can be bound by a different antibody. Individual epitopes are typically composed of a cluster of amino acids or part of a polysaccharide chain. Any antigen that contains more than one epitope, or more than one copy of the same epitope, is known as a **multivalent** antigen (Figure 2.9).

Antibodies can be made against a much wider range of chemical structures than proteins and carbohydrates. Such antibodies are, however, more commonly implicated in allergic reactions and autoimmune diseases than in

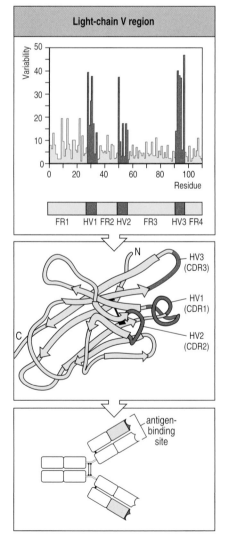

Figure 2.7 The hypervariable regions of antibody V domains lie in discrete loops at one end of the domain structure. The top panel shows the variability plot for the 110 positions within the amino-acid sequence of a light-chain V domain. It is obtained from comparison of many light-chain sequences. Variability is the ratio of the number of different amino acids found at a position and the frequency of the most common amino acid at that position. The maximum value possible for the variability is 400, the square of 20, the number of different amino acids found in antibodies. The minimum value for the variability is 1. Three hypervariable regions (HV1, HV2, and HV3) can be discerned (shown in red) flanked by four framework regions (FR1, FR2, FR3, and FR4) (shown in yellow). The center panel shows the correspondence of the hypervariable regions to three loops at the end of the V domain farthest from the C region. The location of hypervariable regions in the heavy-chain V domain is similar (not shown). The hypervariable loops contribute much of the antigen specificity of the antigen-binding site located at the tip of each arm of the antibody molecule. Hypervariable regions are also known as complementarity-determining regions: CDR1, CDR2, and CDR3.

Figure 2.8 Epitopes for antibodies are exposed on the surface of pathogens. A spherical poliovirus particle is depcited. Its coat is an array of three different proteins (indicated in yellow, blue, and pink). In the lower part of the figure one of the viral proteins, VP1 (blue), is shown separately, but folded as in the virus particle. VP1 contains several epitopes (white). They are located at the surface of the protein and are exposed on the surface of the virus particle. Photograph courtesy of D. Filman and J.M. Hogle.

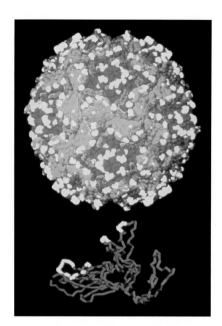

protective immunity against infection. Antibodies specific for DNA, for example, are characteristic of the autoimmune disease systemic lupus erythematosus (SLE), and allergies can be caused by antibodies against small organic molecules, such as antibiotics and other drugs.

The antigen-binding sites of antibodies vary according to the size and shape of the epitope. Antibodies that bind to the end of a polysaccharide chain, or to a small molecule such as vitamin K_1, use a deep pocket formed between the heavy- and light-chain V domains (Figure 2.10, left panel). In such cases not all the CDRs make contact with the antigen. By contrast, antibodies that bind to several of adjacent sugars within a polysaccharide, or amino acids within a polypeptide, use longer, shallower clefts formed by all the opposing CDRs of the V_H and V_L domains (see Figure 2.10, center panel). Epitopes of this kind, in which the antibody binds to parts of a molecule that are adjacent in the linear sequence, are called **linear epitopes** (Figure 2.11, left panel).

Another kind of epitope is formed by parts of a protein that are separated in the amino-acid sequence but are brought together in the folded protein. These are called **conformational** or **discontinuous epitopes** (see Figure 2.11, right panel). Antibodies that bind to complete proteins often make contact with a much larger area of the protein surface than can be accommodated within a groove between the CDRs of the V_H and V_L domains. These antibodies might have no discernible binding groove but interact with the antigen by using a surface (in area usually 700–900 $Å^2$) formed by the CDRs and which can even extend beyond them (see Figure 2.10, right panel).

The binding of antigens to antibodies is based solely on noncovalent forces—electrostatic forces, hydrogen bonds, van der Waals forces, and hydrophobic interactions. The antigen-binding sites of antibodies are unusually rich in aromatic amino acids, which can participate in many van der Waals and

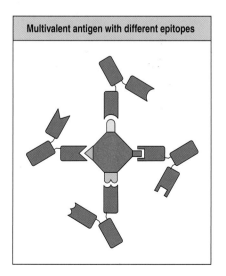

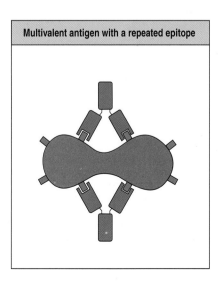

Figure 2.9 Two kinds of multivalent antigen.

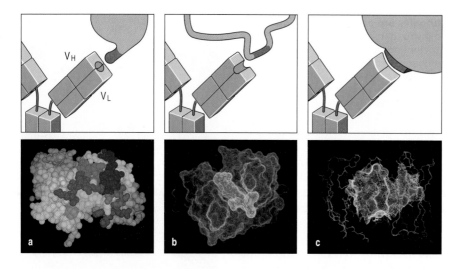

Figure 2.10 Epitopes can bind to pockets, grooves or extended surfaces in antibody-binding sites. The top row shows schematic representations of antibodies binding to three different types of epitope. The antibodies are shown in blue, the antigens in orange and the epitopes in red. On the left is a small compact epitope binding to a pocket, in the middle is an epitope consisting of a relatively unfolded part of a polypeptide chain binding to a shallow groove, and on the right is an epitope with extended surface binding to a similarly sized but complementary surface in an antibody-binding site. The crystallographic images in the bottom row are of Fab fragments binding to antigens in ways corresponding to those depicted in the top row. Panel a shows a peptide epitope (red) bound to a pocket formed by the CDR loops (colored), panel b shows a peptide epitope (red) bound to a groove formed between the two V domains (green), and panel c shows Fab binding to a surface epitope of the protein antigen, lysozyme. Here, the surface contour of the lysozyme (yellow dots) is superimposed on the antigen-binding site of the Fab. The peptide backbone of the Fab is shown in blue and the amino acids that contact lysozyme are shown in red. Photographs courtesy of I.A. Wilson, R.L. Stanfield, and S. Sheriff.

hydrophobic interactions. The better the general fit between the interacting surfaces of antigen and antibody, the stronger are the bonds formed by these short-range forces. Thus, small differences in shape and chemical properties of the binding site can give several antibodies specificity for the same epitope, but they bind to it with different binding strengths or **affinities**. Binding caused by van der Waals forces and hydrophobic interactions is complemented by the formation of electrostatic interactions (salt bridges) and hydrogen bonds between particular chemical groups on the antigen and particular amino-acid residues of the antibody.

In the immune response, the effective antibodies are those that bind tightly to an antigen and do not let go. This behavior contrasts with that of enzymes, which bind a substrate, chemically change the substrate's structure, and then release it. However, the same set of 20 amino acids is used to make enzymes and antibodies and, after intensive searching, some antibodies have been found that catalyze chemical reactions involving the antigen they bind. The application of such **catalytic antibodies** in medicine is being explored. A possible use for a catalytic antibody would be to convert a toxic chemical within the body into an innocuous product.

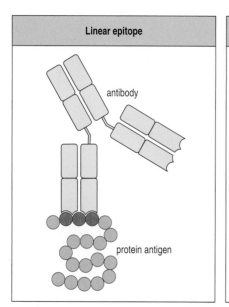

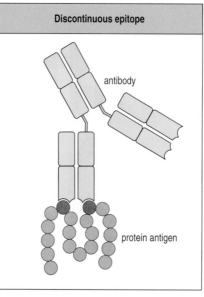

Figure 2.11 Linear and discontinuous epitopes. A linear epitope of a protein antigen is formed from contiguous amino acids. A discontinuous epitope is formed from amino acids from different parts of the polypeptide that are brought together when the chain folds.

2-5 Monoclonal antibodies are produced from a clone of antibody-producing cells

The specificity and strength of an antibody for its antigen make antibodies useful reagents for detecting and quantifying a particular antigen. They are used in this way in a range of clinical and laboratory tests and in biological research generally. The traditional method for making antibodies of desired specificity is to immunize animals with the appropriate antigen and then pre-pare **antisera** from their blood. The specificity and quality of such antisera is highly dependent upon the purity of the immunizing antigen preparation because antibodies will be made against all the foreign components it con-tains. Consequently, antisera specific for a single protein or carbohydrate can only be made when the antigen is already available in a purified form.

A more modern method for making antibodies does not require a purified antigen. In this method, B cells are isolated from the immunized animals, usually mice, and immortalized by fusion with a tumor cell to form **hybridoma** cell lines that grow and produce antibodies indefinitely. The indi-vidual hybridoma cells are then separated from each other and the ones mak-ing antibodies of the desired specificity are identified and selected for further propagation. The antibodies made by a hybridoma cell line are all identical and are thus called **monoclonal antibodies** (Figure 2.12).

Since the invention of hybridomas in the 1970s, monoclonal antibodies have replaced antisera in many clinical assays. They also have been used to iden-tify numerous previously unknown surface proteins of cells of the immune system. Monoclonal antibodies made in different laboratories are periodically compared at international workshops and assigned to different clusters of dif-ferentiation (CD), each of which represents a different molecule or a family of closely related molecules. More than 240 molecules have been numbered so far in the CD series. The reactions of monoclonal antibodies with different types of cell are measured by the technique of flow cytometry (Figure 2.13).

In medicine, flow cytometry is used to analyze the cell populations in periph-eral blood and assess for perturbations caused by disease. For example, occa-sionally some children with an inherited genetic defect have no B cells and make no antibodies. For a period of around 1 year after birth and while nurs-ing, these children are protected from infection by antibodies provided by their mothers, but this declines with age. They then become particularly sus-ceptible to infections with encapsulated bacteria such as *Streptococcus pneu-moniae* or *Haemophilus influenzae*, which, if left untreated with antibiotics, would eventually lead to death. Knowledge that a person is B-cell deficient allows infections to be prevented by regular injections of antibodies purified from the pooled sera of healthy blood donors.

Monoclonal antibodies are also used as therapeutic agents, for example, in preventing the imminent rejection of a transplanted kidney. Unfortunately,

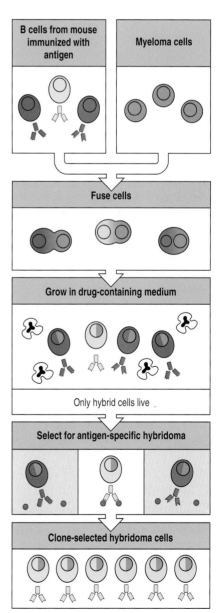

Figure 2.12 Production of a monoclonal antibody. Lymphocytes from a mouse immunized with the antigen are fused with myeloma cells using polyethyleneglycol. The cells are then grown in the presence of drugs that kill myeloma cells but permit the growth of hybridoma cells. Individual cultures of hybridomas are tested to determine if they make the desired antibody. The cells are then cloned to produce a homogeneous culture of cells making a monoclonal antibody. Myelomas are tumors of plasma cells; those used to make hybridomas were selected not to express heavy and light chains. Thus, hybridomas only express the antibody made by the B-cell fusion partner.

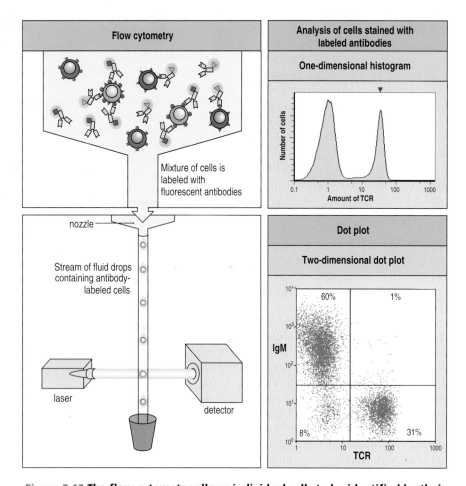

Figure 2.13 The flow cytometer allows individual cells to be identified by their cell-surface molecules. The left-hand panels illustrate the principles of flow cytometry. Human cells are labeled with mouse monoclonal antibodies specific for human cell-surface proteins, antibodies of different specificity being tagged with fluorescent dyes of different color. The labeled cells pass through a nozzle, forming a stream of droplets each containing one cell. The stream of cells passes through a laser beam, which causes the fluorescent dyes to emit light of different wavelengths. The emitted signals are analyzed by a detector: cells with particular characteristics are counted and the abundances of the labeled cell-surface molecules measured. The right-hand panels show how the data obtained can be represented, as exemplified by the presence of IgM and T-cell receptor (TCR) on lymphocytes from the human spleen. The expression of IgM and TCR distinguishes B cells and T cells, respectively. When the presence of one type of molecule is analyzed, the data is usually displayed as a histogram, as shown in the upper right panel for TCR. Two populations of cells are distinguished in the histogram. The larger peak to the left are lymphocytes (mostly B cells) that do not bind the anti-TCR monoclonal antibody and comprises 58% of the total cell number. The smaller peak to the right corresponds to the T cells that bind the anti-TCR antibody and comprises 32% of the total. Two-dimensional plots (lower right panel) are used to compare the expression of two cell-surface molecules. The horizontal axis represents the amount of fluorescent anti-TCR antibody bound by a cell, the vertical axis represents the amount of fluorescent anti-IgM antibody bound. Each dot represents the values obtained for a single cell. Many thousands of cells are usually analyzed, and the dots can blend into each other in those parts of the plot which are heavily populated. The dot plot is divided into four quadrants that roughly correspond to the four cell populations distinguished by analysis with two antibodies. In the top left quadrant are B cells; these bind anti-IgM antibody but not anti-TCR antibody and comprise 60% of the cells. In the bottom right quadrant are the T cells; these bind anti-TCR antibody but not anti-IgM antibody and comprise 31% of the cells. In the bottom left quadrant are cells that bind neither anti-IgM nor anti-TCR antibody; these comprise 8% of the cells and probably include NK cells and some contaminating non-lymphoid leukocytes. Theoretically, cells in the upper right quadrant would correspond to lymphocytes that bind both anti-IgM and anti-TCR. As no lymphocytes express both IgM and TCR at their surface, these double reactions (1% of the total) arise from imprecision in using quadrants to separate the cell population and from experimental artifact, for example the nonspecific sticking of antibody molecules to cells that do not express their specific antigen.

many of these applications have limited use because after the initial administration the human patient makes antibodies specific for the C regions of the mouse antibodies used. Reaction of these antibodies with any subsequent doses of the mouse monoclonal antibodies diminishes their therapeutic effect and increases the likelihood of complications. To reduce this problem, genetic engineering can be used to **humanize** a useful mouse monoclonal antibody. In this procedure, sequences encoding the CDR loops of the mouse monoclonal antibody's heavy and light chains are used to replace the sequences encoding the corresponding CDR loops of a human immunoglobulin with irrelevant specificity. Thus, cells expressing the humanized heavy and light chains make an antibody in which only the CDR loops are of mouse origin and the rest is human.

Summary

IgG antibodies are made of four polypeptide chains—two identical heavy chains and two identical light chains. Each chain has a V region that contributes to the antigen-binding site and a C region that, in the heavy chain, determines the antibody isotype and its specialized effector functions. The IgG molecule is Y-shaped, in which the stem and the arms are of comparable sizes and can flex with respect to each other. Each arm contains an antigen-binding site, whereas the stem contains binding sites for cells and effector molecules that enable bound antigen to be cleared from the body. Immunoglobulin chains are built from a series of structurally related immunoglobulin domains. Within the V domain, sequence variability is localized to three hypervariable regions corresponding to the three loops that are clustered at one end of the domain. In the antibody molecule the hypervariable loops of the heavy and light chains form a variable surface that binds antigen. The type of antigen bound by an antibody depends on the shape of the antigen-binding site: antigens that are small molecules can be bound within deep pockets; linear epitopes from proteins or carbohydrates can be bound within clefts or grooves; and the binding of conformational epitopes of folded proteins takes place over an extended surface area. Monoclonal antibodies are antibodies of a single specificity that are derived from a clone of identical antibody-producing cells, which are used in diagnostic tests and as therapeutic agents.

Generation of immunoglobulin diversity in B cells before encounter with antigen

The number of different antibodies that can be produced by the human body seems to be virtually limitless. To achieve this, the immunoglobulin genes are organized differently from other genes. In all cells, except B cells, the immunoglobulin genes are in a fragmented form that cannot be expressed; instead of containing a single complete gene, immunoglobulin heavy-chain and light-chain loci consist of families of **gene segments**, sequentially arrayed along the chromosome, with each set of segments containing alternative versions of parts of the immunoglobulin V region. The immunoglobulin genes are inherited in this form through the germ line (egg and sperm); this arrangement is therefore called the **germline form** or **germline configuration**.

For an immunoglobulin gene to be expressed, individual gene segments must first be rearranged to assemble a functional gene, a process that occurs only in developing B cells. Immunoglobulin-gene rearrangements occur during the development of B cells from B-cell precursors in the bone marrow. When gene rearrangements are complete, heavy and light chains can be produced and membrane-bound immunoglobulin appears at the B-cell surface. The B cell can now recognize and respond to an antigen through this receptor. Much of the diversity within the mature antibody repertoire is generated during this process of gene rearrangement, as we shall see in this part of the chapter.

2-6 The DNA sequence encoding a V region is assembled from two or three gene segments

In humans, the immunoglobulin genes are found at three chromosomal locations: the heavy-chain locus on chromosome 14, the κ light-chain locus on chromosome 2 and the λ light-chain locus on chromosome 22. Different gene segments encode the leader peptide (L), the V region (V) and the constant region (C) of the heavy and light chains. Gene segments encoding C regions are commonly called C genes (Figure 2.14). Within the heavy-chain locus are C genes for all the different heavy-chain isotypes.

The gene segments encoding the leader peptide and the C region consist of exons and introns, like those found in other human genes, and they are ready to be transcribed. In contrast, the V regions are encoded by two (V_L) or three (V_H) gene segments that require rearrangement in order to produce an exon that can be transcribed. The two types of gene segment that encode the light-chain V region are called **variable (V)** and **joining (J) gene segments**. The heavy-chain locus includes an additional set of **diversity (D) gene segments** that lies between the arrays of V and J gene segments (see Figure 2.14).

The V region of a light chain is encoded by the combination of one V and one J segment, whereas the C region is encoded by a single C gene. The main differences between V gene segments are in the sequences that encode the first and second hypervariable regions of the V domain (see Section 2-3); the third hypervariable region is determined by the junction between the V and J segments. In an individual B cell, only one of the light-chain loci (κ or λ) gives rise to a functional light-chain gene. The V region of the heavy chain is encoded

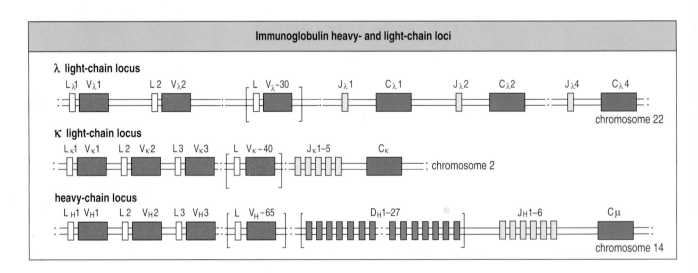

Figure 2.14 The germline organization of the human immunoglobulin heavy-chain and light-chain loci. The upper row shows the λ light-chain locus, which has about 30 functional V_λ gene segments and four pairs of functional J_λ gene segments and C_λ gene segments. The κ locus (center row) is organized in a similar way, with about 40 functional V_κ gene segments accompanied by a cluster of five J_κ gene segments but with a single C_κ gene segment. In approximately half of the human population, the entire cluster of V_κ gene segments is duplicated (not shown for simplicity). The heavy-chain locus (bottom row) has about 65 functional V_H gene segments, a cluster of about 27 D segments and six J_H gene segments. For simplicity, only a single C_H gene ($C\mu$) is shown in this diagram. This diagram is not to scale: the total length of the heavy-chain locus is over 2 megabases (2 million bases), whereas some of the D segments are only six bases long. L, leader sequence.

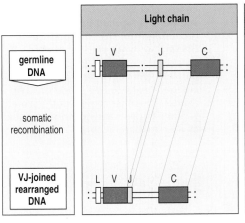
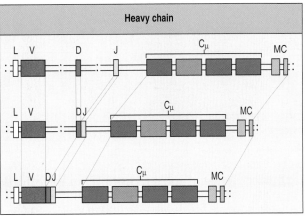
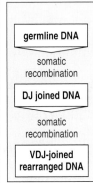

by one V, one D, and one J segment. The first two hypervariable regions are caused by differences in the sequences of the heavy-chain V gene segments; the third hypervariable region is determined by differences in the D gene segments and the junctions they make with the V and J gene segments. The C region of a heavy chain is encoded by one of the C genes.

2-7 Random recombination of gene segments produces diversity in the antigen-binding sites of immunoglobulins

During the development of B cells the arrays of V, D, and J segments are cut and spliced by DNA recombination. This process is called **somatic recombination** because it occurs in cells of the soma, a term that embraces all of the body's cells except the germ cells. A single gene segment of each type is brought together to form a DNA sequence encoding the V region of an immunoglobulin chain. For light chains, a single recombination occurs, between a V_L and a J_L segment (Figure 2.15, left panels), whereas for heavy chains, two recombinations are needed, the first to join a D and a J_H segment, and the second to join the combined DJ segment to a V_H segment (see Figure 2.15, right panels). In each case, the particular V, D, and J gene segments that are joined together are selected at random. Because of the multiple gene segments of each type, numerous different combinations of V, D, and J gene segments are possible. Thus, the gene rearrangement process generates many different V-region sequences in the B-cell population. This is one of the factors contributing to the diversity of immunoglobulin V regions.

For the human κ light chain, approximately $40\,V_\kappa$ gene segments and $5\,J_\kappa$ gene segments (Figure 2.16) can be recombined in 200 (40 × 5) different ways. Similarly, some $30\,V_\lambda$ gene segments and $4\,J_\lambda$ gene segments can be recombined in 120 (30 × 4) different ways. At the heavy-chain locus, $65\,V_H$ segments, some 27 D segments and $6\,J_H$ segments can produce 10,530 (65 × 27 × 6) combinations. If recombination between different gene segments were the only mechanism producing V-region diversity, then a maximum of 320 different light chains and 10,530 different heavy chains could be made.

Somatic recombination is carried out by enzymes that cut and rejoin the DNA, and it exploits some of the mechanisms more ubiquitously used by cells for DNA recombination and repair. The recombination of V, J, and D

Figure 2.15 V-region sequences are constructed from gene segments. Light-chain V-region genes are constructed from two segments (left panel). A variable (V) and a joining (J) gene segment in the genomic DNA are joined to form a complete light-chain V-region (V_L) exon. After rearrangement the light-chain gene consists of three exons, encoding the leader (L) peptide, the V region and the C region, that are separated by introns. Heavy-chain V regions are constructed from three gene segments (right panels). First the diversity (D) and J gene segments join, then the V gene segment joins to the combined DJ sequence, forming a complete heavy-chain V-region (V_H) exon. For simplicity, only one of the heavy-chain genes, C_μ, is shown here. Each immunoglobulin domain is encoded by a separate exon and two additional membrane-coding exons (MC, colored light blue) specify the hydrophobic sequence that will anchor the heavy chain to the B-cell membrane.

Figure 2.16 The numbers of functional gene segments used to construct the variable regions of human immunoglobulin heavy chains and light chains.

	Number of gene segments		
Segment	Light chains		Heavy chain
	κ	λ	H
Variable (V)	40	30	65
Diversity (D)	0	0	27
Joining (J)	5	4	6

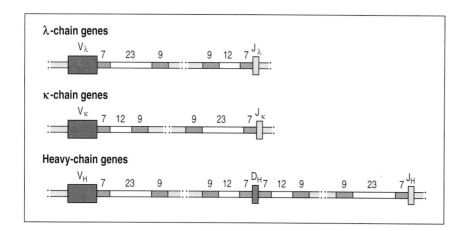

Figure 2.17 Each V, D, or J gene segment is flanked by recombination signal sequences (RSSs). Two types of RSS exist. One consists of a nonamer (9 bp, shown in purple) and a heptamer (7 bp, shown in orange) separated by a spacer of 12 bp (white). The other consists of the same 9- and 7-nucleotide sequences separated by a 23-bp spacer (white).

gene segments is directed by sequences called **recombination signal sequences (RSSs)**, which flank the 3′ side of the V segment, both sides of the D segment, and the 5′ side of the J segment (Figure 2.17). There are two types of RSS and recombination can occur only between different types. One type of RSS comprises a defined heptamer [7 base pairs (bp); CACAGTG] sequence and a defined nonamer (9 bp; ACAAAAACC) separated by a 12-bp spacer, the other comprises the heptamer and the nonamer sequences separated by a 23-bp spacer. As well as providing recognition sites for the enzymes that cut and rejoin the DNA, RSSs ensure that the gene segments are joined in the correct order, as shown for light-chain VJ recombination in Figure 2.18. Because of

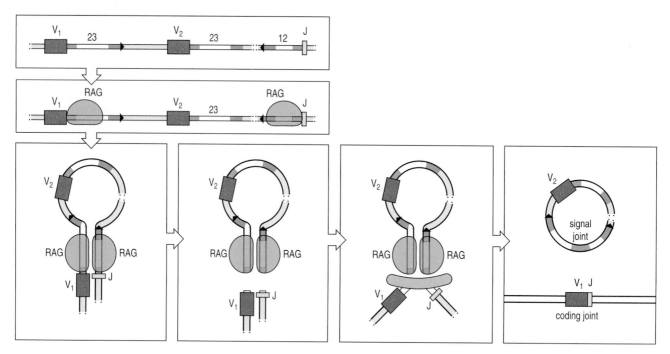

Figure 2.18 Gene segments encoding the variable region are joined by recombination at recombination signal sequences. The recombination between a V (red) and a J (yellow) segment of a light-chain gene is shown here. A RAG complex binds to the 23-bp spacer and another to the 12-bp spacer, so that a recombination signal sequence (RSS) containing a 12-bp spacer is brought together with that containing a 23-bp spacer. This is known as the 12/23 rule and ensures that gene segments are joined in the correct order. The DNA molecules are broken at the ends of the heptamer sequences (orange) and are then joined together with different topologies. The region of DNA that was originally between the V and J segments to be joined is excised as a small circle of DNA that has no function. The joint made in forming this circle is called the signal joint. Within the chromosomal DNA, the V and J segments are joined to form the coding joint. Formation of this joint involves opening up of hairpins that were formed at the original point of cleavage at one end of each V and J segment and then repairing the DNA in a way that introduces additional variability into the nucleotide sequence around the joint. The additional enzymes involved in these processes are represented in blue. Nonamers are shown in purple, heptamers in orange, spacers in white.

this strict requirement, recombination in the heavy-chain DNA cannot join V_H directly to J_H without the involvement of D_H, because the V_H and J_H segments are flanked by the same type of RSS (see Figure 2.17).

2-8 Recombination enzymes produce additional diversity in the antigen-binding sites of immunoglobulins

The set of enzymes needed to recombine V, D, and J segments is called the **V(D)J recombinase**. Two of the component proteins are made only in lymphocytes; they are specified by the **recombination activating genes** (**RAG-1** and **RAG-2**). The other components are present in all nucleated cells and have activities that repair double-stranded DNA, bend DNA, or modify the ends of broken DNA strands. They include the enzymes DNA ligase IV, DNA-dependent protein kinase (DNA-PK), and the Ku protein associated with DNA-PK. The RAG-1 and RAG-2 proteins interact with each other and with other proteins, known as the high-mobility group of proteins, to form a RAG complex. The first step in recombination is when one RAG complex binds to one type of RSS and another binds the second type of RSS (see Figure 2.18). Interaction of the two RAG complexes aligns the two RSSs and then cleaves the DNA within the ends of the immunoglobulin gene segments. This process creates a single-stranded DNA "hairpin" at the end of each immunoglobulin segment and a clean break at the ends of the two heptamer sequences. The DNA molecules are held in place by the RAG complexes, whereas the broken ends are rejoined by DNA repair enzymes in a process called nonhomologous-end joining to form both the signal joint and the coding joint.

The enzymes that open the hairpins and form the coding joint introduce additional sequence diversity into the third hypervariable region (CDR3) of immunoglobulin heavy and light chains that is not determined by germline DNA (Figure 2.19). Firstly, the nick made to open the hairpin can occur at several different positions. When the nick occurs at a different position from where the hairpin was originally formed, nucleotides from one DNA strand are transferred to the other. This creates a new sequence that is palindromic because base pairs that were complementary in the two different DNA strands are now side-by-side on the same strand (a palindromic sequence is identical when read from either end). Because of this relationship, the added nucleotides are called **P nucleotides** (for palindromic nucleotides). The opened hairpins can then be variably modified by exonucleases that remove the germline-encoded nucleotides and by the enzyme **terminal deoxynucleotidyl transferase** (**TdT**), which randomly adds back nucleotides. The nucleotide sequence added by this enzyme need not correspond to the germline sequence in either sequence or length. When the single-stranded

Figure 2.19 The generation of junctional diversity during gene rearrangement. The process is illustrated for a D to J rearrangement. The RSSs are brought together and endonuclease cleavage occurs within the heptamer sequences, as shown in the top panel. This leads to excision of the DNA that separates the D and J segments and modification of the D and J segments by extra nucleotides derived from the heptamers. The two DNA strands at the end of the modified D and J segments join together to form hairpins. Further cleavage on one DNA strand at the boundaries between the heptamers and the D and J segments releases the hairpins and generates short palindromic sequences. These extra nucleotides are known as P nucleotides. Terminal deoxynucleotidyl transferase (TdT) adds nucleotides randomly to the ends of the single DNA strands. These nucleotides, which are nongermline, are known as N nucleotides. The single strands pair and through the action of exonuclease, DNA polymerase, and DNA ligase, the double-stranded DNA molecule is repaired to give the coding joint.

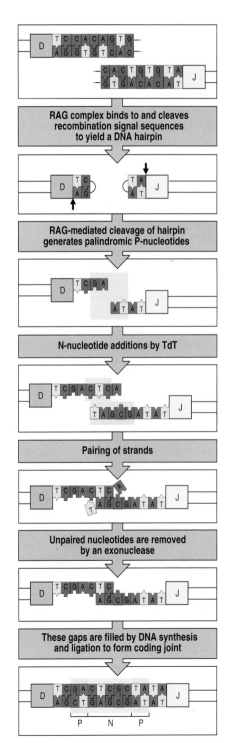

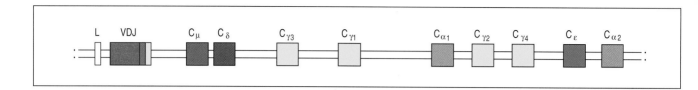

tails of the two gene segments are able to pair, the sequences are filled in with complementary nucleotides to complete the coding joint. The new nucleotides inserted by TdT are called **N nucleotides** because they are non-templated (not encoded) in germline DNA.

The contribution of P nucleotides and N nucleotides to the resulting amino-acid sequence diversity in the third hypervariable region is called **junctional diversity**. This is an important source of immunoglobulin variability, adding a factor of 3×10^7 to the overall diversity. The third hypervariable region of the light-chain V domain is encoded by the junction formed between the V and J segments (see Figure 2.7), whereas the third hypervariable region of the heavy-chain V domain is formed by the D segment and its junctions with the rearranged V and J segments.

2-9 Naive B cells use alternative mRNA splicing to make both IgM and IgD

Rearrangement of the V, D, and J segments of the heavy-chain locus brings the gene's promoter and enhancer into a closer juxtaposition that enables tran-scription to proceed. The resulting mRNA is then spliced and translated to give a heavy-chain protein. The μ and δ heavy-chain genes are the first to be transcribed, so the first immunoglobulins that a B cell expresses on its surface are IgM and IgD. These are the only immunoglobulin isotypes that can be produced simultaneously by a B cell. They are also the only isotypes that B cells produce before they encounter antigen, which is at an early stage in their life when they are called **naive B cells**. Simultaneous expression of both μ and δ chains from the same heavy-chain locus is accomplished by differential splicing of the same primary RNA transcript, a process that involves no rearrangement of genomic DNA.

In the rearranged heavy-chain locus the exons encoding the leader peptide and the V region are on the 5′ side of the C genes, of which the μ and δ C genes are the nearest (Figure 2.20). In each C gene, separate exons encode each of the domains, as is shown for the μ and δ genes in Figure 2.21. Transcription starts upstream of the exons encoding the leader peptide and the V region, continues through the μ and δ C genes and terminates downstream of the δ gene, before the γ3 C gene. This long primary RNA transcript is then spliced and processed in two different ways: one that yields mRNA for the μ heavy chain (Figure 2.21, left panel) and one that yields mRNA for the δ heavy chain (Figure 2.21, right panel). In making μ chain mRNA from the primary tran-script, the entire δ gene is removed as well as the introns from the μ gene. Conversely, in making δ chain mRNA the entire μ gene is removed as well as the introns from the δ gene.

2-10 Each B cell produces immunoglobulin of a single antigen specificity

In a developing B cell, the process of immunoglobulin-gene rearrangement is tightly controlled so that only one heavy chain and one light chain are finally expressed, a phenomenon known as **allelic exclusion**. This ensures that each B cell produces monoclonal immunoglobulin of a single antigen specificity.

Figure 2.20 Rearrangement of V, D, and J segments produces a functional heavy-chain gene. The assembled VDJ sequence lies some distance from the cluster of C genes. Only functional C genes are shown here. The four different γ genes specify four different subtypes of the γ heavy chain, whereas the two α genes specify two subtypes of the α heavy chain. For simplicity, individual exons in the C genes are not shown. The diagram is not to scale.

Thus, although every B cell has two copies, or alleles, of the heavy-chain locus and two copies of each light-chain locus, only one heavy-chain locus and one light-chain locus are rearranged to produce functional genes.

In a population of B cells the same functional light-chain gene rearrangement is found associated with different functional heavy-chain gene rearrangements. Conversely, the same functional heavy-chain gene rearrangement is found associated with different functional light-chain gene rearrangements. This combinatorial association of heavy and light chains within individual B cells, and in the immunoglobulins they produce, makes an important contribution to the overall diversity in the antigen-binding sites of immunoglobulins. B cells are free to produce any combination of light and heavy chains, and, thus, the potential number of antibodies of different specificities that can be made is the product of the total numbers of different heavy and light chains.

B-cell monospecificity has the consequence that an encounter with a given pathogen engages only the subset of B cells that make antibodies which bind to the pathogen. This ensures the specificity of the antibody response against an infection. For this reason, vaccination against diphtheria, for example, provides no protection against the influenza virus, neither does vaccination against the influenza virus protect against diphtheria.

The fact that the DNA sequence of the expressed immunoglobulin genes varies from one clone of B cells to the next can be used to detect the large clonal populations of cancer cells in patients with B-cell lymphoma or leukemia. Because the cancer cells are derived from a single clone of B cells, they can be readily distinguished from healthy B cells by comparing the immunoglobulin-gene rearrangements by DNA analysis (Figure 2.22). In the clinic, such analyses are used to determine the presence of cancer cells in blood samples or tissue biopsies and to monitor the response to therapy.

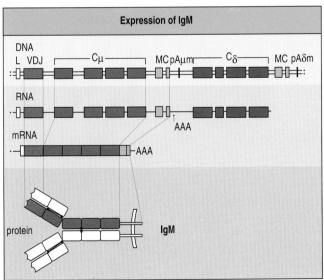
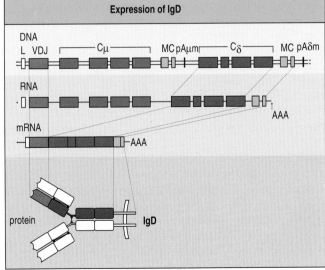

Figure 2.21 Co-expression of IgD and IgM is regulated by RNA processing. In mature B cells, transcription initiated at the V_H promoter extends through both the C_μ and C_δ genes. For simplicity we have not shown all the individual C-gene exons but only those of relevance to the differential production of IgM and IgD. The long primary transcript is then processed by cleavage, polyadenylation, and splicing. Cleavage and polyadenylation at the μ site (pAμ) and splicing between C_μ exons yields an mRNA encoding the μ heavy chain (left panel). Cleavage and polyadenylation at the δ site (pAδ) and a different pattern of splicing that removes the C_μ exons yields mRNA encoding the δ heavy chain (right panel). AAA designates the poly(A) tail. MC, membrane-coding.

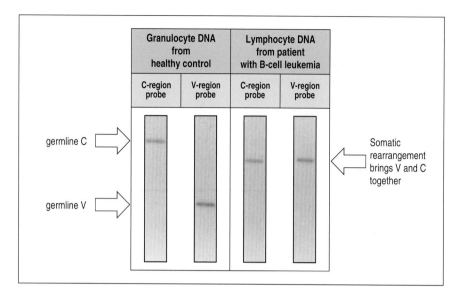

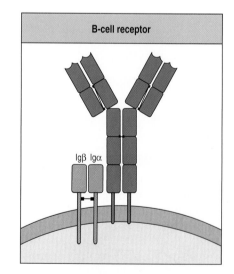

Figure 2.22 **DNA analysis of immunoglobulin genes can help to diagnose B-cell tumors.** The left two photographs (healthy control) show agarose gel electrophoresis of a restriction enzyme digest of DNA from peripheral blood granulocytes of a healthy person. In these cells, mostly neutrophils, the immunoglobulin genes are in the germline configuration. The state of immunoglobulin DNA sequences is examined by hybridization with V-region and C-region probes. The V and C regions are found in quite separate DNA fragments. The two right photographs (patient with B-cell leukemia) are of a similar restriction digest of DNA from peripheral blood lymphocytes from a patient with chronic lymphocytic leukemia, in which a particular clone of B cells is greatly expanded. For this reason, a unique rearrangement is visible. In this DNA, the V and C regions are found in the same fragment. Normal B lymphocytes in the patient's blood each have a different gene rearrangement, none of which is sufficiently well represented to be visible as a band. Photographs courtesy of S. Wagner and L. Luzzatto.

2-11 Immunoglobulin is first made in a membrane-bound form that is present on the B-cell surface

When a B cell first makes immunoglobulin the heavy chain (but not the light chain) has a hydrophobic sequence near the carboxy terminus by which it associates with cell membranes. Like all proteins destined for the cell surface, immunoglobulin chains enter the endoplasmic reticulum as soon as they are synthesized. There they associate with each other to form immunoglobulin molecules attached to the endoplasmic reticulum membrane. By themselves these immunoglobulin molecules cannot be transported to the cell surface. For that to happen they must associate with two additional transmembrane proteins called Igα and Igβ (Figure 2.23). These proteins are invariant in sequence, unlike the immunoglobulins, and travel to the B-cell surface with the immunoglobulins. At the cell surface the complex of immunoglobulin with Igα and Igβ forms the B-cell receptor for antigen.

In responding to specific antigen the immunoglobulin component of the B-cell receptor is responsible for binding to the specific antigen. However, this interaction alone cannot give the signal that tells the cell's interior when an antigen has bound. The cytoplasmic portions of immunoglobulin heavy chains are very short and do not interact with the intracellular proteins that signal B cells to divide and differentiate. That function is provided by the longer cytoplasmic tails of the Igα and Igβ components of the B-cell receptor.

Summary

In the human genome, the immunoglobulin heavy-chain and light-chain genes are in a form that is incapable of being expressed. In developing B cells, however, the immunoglobulin genes undergo structural rearrangements that permit their expression. The V domains of immunoglobulin light and heavy chains are encoded in two (V and J) or three (V, D, and J) different kinds of gene segment respectively that are brought into juxtaposition by recombination reactions. One mechanism contributing to the diversity in V-region sequences is the random combination of different V and J segments in light-chain genes and of different V, D, and J segments in rearranged heavy-chain genes. A second mechanism is the introduction of additional nucleotides (P and N nucleotides) at the junctions between gene segments during the process of recombination. A third mechanism that creates diversity in the antigen-binding sites of antibodies is the association of heavy and light

Figure 2.23 **Membrane-bound immunoglobulins are associated with two other proteins, Igα and Igβ.** Igα and Igβ are disulfide-linked. They have long cytoplasmic tails that can interact with intracellular signaling proteins. The immunoglobulin shown here is IgM, but all isotypes can serve as B-cell receptors.

chains in different combinations. Gene rearrangement in an individual B cell is strictly controlled so that only one type of heavy chain and one type of light chain are expressed, resulting in an individual B cell and its progeny expressing only immunoglobulin of a single antigen specificity. A naive B cell that has never encountered antigen expresses only membrane-bound immunoglobulin of the IgM and IgD classes. The μ and δ heavy chains are produced from a single transcriptional unit that undergoes alternative RNA processing and splicing.

Diversification of antibodies after B cells encounter antigen

A mature B cell's first encounter with antigen marks a watershed in its development. Binding of antigen to the surface immunoglobulin of a mature naive B cell triggers the cell's proliferation and differentiation, and ultimately the secretion of antibodies. As the immune response progresses, antibodies with different properties appear. In this part of the chapter we will look at the structural and functional changes that occur in immunoglobulins after antigen has been encountered and how antibodies with different functions are produced.

2-12 Secreted antibodies are produced by an alternative pattern of heavy-chain RNA processing

Gene rearrangement in an immature B cell leads to the expression of functional heavy and light chains and to the production of membrane-bound IgM and IgD on the mature B cell. After an encounter with antigen, these isotypes are produced as secreted antibodies. IgM antibodies are produced in large amounts and are important in protective immunity, whereas IgD antibodies are produced only in small amounts and have no known effector function.

All the isotypes, or classes, of immunoglobulin (IgA, IgD, IgE, IgG, and IgM) can be made in two forms: one that is bound to the cell membrane and serves as the B-cell receptor for antigen, and one, the antibody, that is secreted in order to bind to antigen and aid its destruction. During differentiation to antibody-secreting plasma cells, B cells change from making the membrane-bound form to making the secreted form; plasma cells only make secreted antibody. The difference between membrane-bound and secreted immunoglobulin lies at the carboxy terminus of the heavy chain: here, membrane-associated immunoglobulin has a hydrophobic anchor sequence that is inserted into the membrane, whereas antibody has a hydrophilic sequence. This difference is determined by different patterns of RNA splicing and processing of the same primary RNA transcript, and involves no rearrangement of the underlying genomic DNA. The alternative patterns of splicing for IgM are compared in Figure 2.24.

The hydrophilic carboxy terminus of the secreted μ chain is encoded at the 3′ end of the exon encoding the fourth C-region domain, whereas the hydrophobic membrane anchor of the membrane-associated μ chain is encoded by two small, separate exons downstream. The splicing to give secreted μ chain is the simpler pattern: the sequence encoding the hydrophilic carboxy terminus is retained and the sequences 3′ of that, including the exons encoding the hydrophobic membrane anchor, are discarded (Figure 2.24, right panel). To produce mRNA encoding membrane-associated μ chain, alternative splicing in the exon encoding the fourth C-region domain removes the sequence encoding the hydrophilic anchor, whereas the exons encoding the hydrophobic carboxy terminus are retained and incorporated into the mRNA when the introns are spliced out (Figure 2.24, left panel).

2-13 Rearranged V-region sequences are further diversified by somatic hypermutation

The diversity generated during gene rearrangement is concentrated in the third CDR of the V_H and V_L domains. Once a B cell has been activated by antigen, however, further diversification of the whole of the V-domain coding sequences occurs through a process of **somatic hypermutation**. This almost randomly introduces single-nucleotide substitutions (point mutations) at a high rate throughout the rearranged V regions of heavy- and light-chain genes (Figure 2.25). The immunoglobulin constant regions are not affected, and neither are other B-cell genes. The mutations occur at a rate of about one mutation per V-region sequence per cell division, which is more than a million times greater than the ordinary mutation rate for a gene. Mutation occurs at both copies of each immunoglobulin locus, even though one copy is not being expressed. Hypermutation appears to act selectively at certain types of DNA sequence motif that are more common than usual within the sequences encoding the hypervariable CDRs. The mechanism involves activation-induced cytidine deaminase (AID), an enzyme that converts cytidine residues in the DNA to uracil. In turn, the uracil residues are excised by the enzyme uracil-DNA glycosylase (UNG) and, on subsequent replication of the B-cell DNA, nontemplated nucleotides are introduced at these sites.

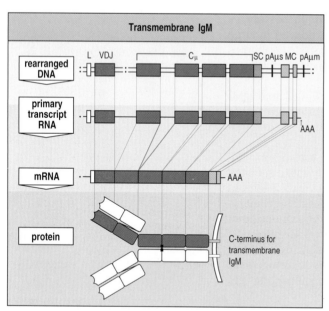
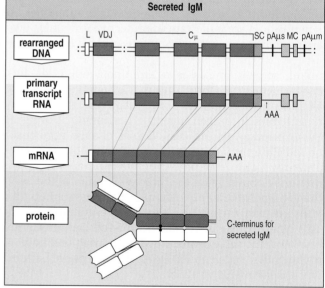

Figure 2.24 The surface and secreted forms of an immunoglobulin are derived from the same heavy-chain gene by alternative RNA processing. Each heavy-chain C gene has two exons [membrane-coding (MC), light blue] encoding the transmembrane region and cytoplasmic tail of the surface form of that isotype, and a secretion-coding (SC) sequence (orange) encoding the carboxy terminus of the secreted form. The events that dictate whether a heavy-chain RNA will result in a secreted or transmembrane immunoglobulin occur during processing of the initial transcript and are shown here for IgM. Each heavy-chain C gene has two potential polyadenylation sites (shown as pAμs and pAμm). In the left panel, the transcript is cleaved and polyadenylated at the second site (pAμm). Splicing between a site located between the fourth C_μ exon and the SC sequence, and a second site at the 5′ end of the MC exons, removes the SC sequence and joins the MC exons to the fourth C_μ exon. This generates the transmembrane form of the heavy chain. In the right panel, the primary transcript is cleaved and polyadenylated at the first site (pAμs), eliminating the MC exons and giving rise to the secreted form of the heavy chain. AAA designates the poly(A) tail.

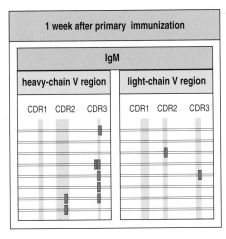

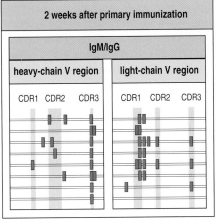

Figure 2.25 Somatic hypermutation introduces diversity into expressed immunoglobulin genes. Antibodies raised by immunization with the same antigen were collected 1 and 2 weeks after immunization and their amino-acid sequences were determined. Each line represents one antibody and the red bars represent amino-acid positions that differ from the prototypic sequence. One week after primary immunization, most of the antibodies were IgM and showed very little sequence variation in the V region. Two weeks after immunization, both IgG and IgM were present and all six CDRs were affected.

Somatic hypermutation gives rise to B cells bearing mutant immunoglobulin molecules on their surface. Some of these mutant immunoglobulin molecules have a higher affinity for the antigen; they will, therefore, be more likely to bind the antigen, and so B cells bearing them will be preferentially selected to mature into antibody-secreting cells. Therefore, as the immune response proceeds antibodies of progressively higher affinity for the immunizing antigen are produced. This phenomenon is called **affinity maturation**.

2-14 Isotype switching produces immunoglobulins with different C regions but identical antigen specificities

IgM is the first antibody produced in an immune response. Unlike IgG, IgM is secreted as a circular pentamer of Y-shaped immunoglobulin monomers (Figure 2.26). Because of its 10 antigen-binding sites, IgM binds strongly to the surface of pathogens with multiple repetitive epitopes, but it is limited in the effector mechanisms that it uses to clear antigen from the body. Antibodies with other effector functions are produced by the process of **isotype switching** or **class switching**, in which a further DNA recombination event enables the rearranged V-region coding sequence to be used with other heavy-chain C genes.

Isotype switching is accomplished by a recombination within the cluster of C genes that excises the previously expressed C gene and brings a different one into juxtaposition with the assembled V-region sequence. Thus, the antigen

Figure 2.26 IgM is secreted as a pentamer of immunoglobulin monomers. The left two panels show schematic diagrams of the IgM monomer and pentamer. The IgM pentamer is held together by a polypeptide called the J chain, for joining chain (not to be confused with a J segment). The monomers are cross-linked by disulfide bonds to each other and to the J chain. The right panel shows an electron micrograph of an IgM pentamer, showing the arrangement of the monomers in a flat disc. The lack of a hinge region in the IgM monomer makes the molecule less flexible than, say, IgG, but this is compensated for by the large number of binding sites. Given a pathogen with multiple identical epitopes on its surface, IgM is likely to be able to bind to it with several of its binding sites simultaneously. Photograph (× 900,000) courtesy of K.H. Roux and J.M. Schiff.

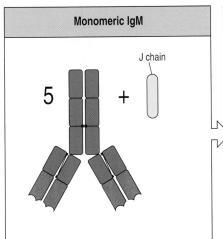

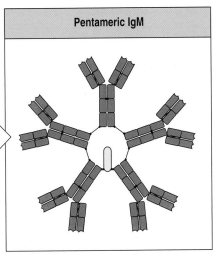

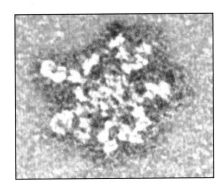

specificity of the antibody remains unchanged, even though its isotype changes. Flanking the 5′ side of each C gene, with the exception of the δ gene, are highly repetitive sequences that mediate recombination. They are called **switch sequences** or **switch regions** (Figure 2.27). In switching the heavy-chain isotype, the switch region that flanks the μ gene interacts with a switch region flanking one of the other C genes. This interaction permits recombination in which the μ, δ, and any other intervening C genes are excised as a circular DNA molecule, bringing the V region into juxtaposition with the new C gene. The mRNA for the new immunoglobulin is then produced as described in Section 2-9. Recombination can take place between the μ switch region and that of any other isotype. Sequential switching can also occur, for example from μ to γ3 to α1. Isotype switching only occurs during an active immune response and the patterns of isotype switching are regulated by cytokines secreted by antigen-activated T cells. The enzyme AID is important for class switch recombination as well as for somatic hypermutation, suggesting that the initial stages of these rather different reactions may be similar.

2-15 Antibodies with different C regions have different effector functions

In humans and other mammals the five classes of immunoglobulin are IgA, IgD, IgE, IgG, and IgM (see Figure 2.4). Certain classes are divided further into subclasses, which differ in both nomenclature and properties between species. In humans, IgA is divided into two subclasses (IgA1 and IgA2), whereas IgG is divided into four subclasses (IgG1, IgG2, IgG3, and IgG4), which are numbered according to their relative abundance in plasma, IgG1 being the most abundant. The heavy chains of the human IgA subclasses are designated α1 and α2 and the heavy chains of the human IgG subclasses by γ1, γ2, γ3, and γ4. The α, δ, and γ heavy-chain C regions are made up of three C domains, whereas the μ and ε heavy chains have four (see Figure 2.4). Each C domain is encoded by a separate exon in the relevant C gene. Additional

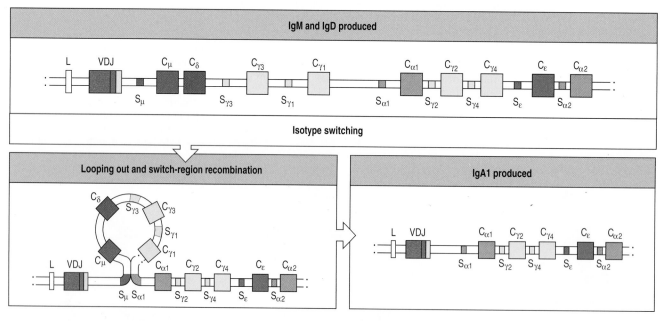

Figure 2.27 Isotype switching involves recombination between specific switch regions. Repetitive DNA sequences are found to the 5′ side of each of the heavy-chain C genes, with the exception of the δ gene. Switching occurs by recombination between these switch regions (S), with deletion of the intervening DNA. The initial switching event takes place from the μ switch region; switching from other isotypes can take place subsequently.

Immunoglobulin class or subclass								
	IgM	IgD	IgG1	IgG2	IgG3	IgG4	IgA1	IgA2	IgE
Heavy chain	μ	δ	γ_1	γ_2	γ_3	γ_4	α_1	α_2	ε
Molecular weight (kDa)	970	184	146	146	165	146	160	160	188
Serum level (mean adult mg ml^{-1})	1.5	0.03	9	3	1	0.5	2.0	0.5	5×10^{-5}
Half-life in serum (days)	10	3	21	20	7	21	6	6	2

Figure 2.28 The physical properties of the human immunoglobulin isotypes. The molecular weight given for IgM is that of the pentamer (see Figure 2.26) the predominant form in serum. The molecular weight given for IgA is that of the monomer. Large amounts of IgA are also produced in the form of dimers, which is the form found in secretions.

exons are used to encode the hinge region and the carboxy terminus, depending on the isotype. The physical properties of the human immunoglobulin isotypes are shown in Figure 2.28.

Antibodies aid the clearance of pathogens from the body in various ways. **Neutralizing** antibodies directly inactivate a pathogen or a toxin and prevent it from interacting with human cells. Neutralizing antibodies against viruses, for example, bind to a site on the virus that is normally used to gain entry to cells. Another function of antibodies is **opsonization**, a term used to describe the coating of pathogens with an immune-system protein (see Figure 1.29, p. 27). The common **opsonins** are antibodies and complement proteins. Opsonized pathogens are more efficiently ingested by phagocytes, which have receptors for the Fc region of some antibodies and for certain complement proteins. Complement activation by antibodies bound to a pathogen's surface can also lead to the direct lysis of the bacterium by complement.

IgM is the first antibody produced in an immune response against a pathogen. It is made principally by plasma cells resident in lymph nodes, spleen, and bone marrow and circulates in blood and lymph. On initiation of an immune response, most of the antibodies that bind the antigen will be of low affinity and the multiple antigen-binding sites of IgM are needed if enough antibody is to bind sufficiently strongly to a microorganism to be of any use. When bound to antigen, sites exposed in the constant region of IgM initiate reactions with complement, which can kill microorganisms directly or facilitate their phagocytosis. Because somatic hypermutation leads to antibodies of increased affinity for the antigen, two antigen-binding sites are then sufficient to produce strong binding; by switching isotype, different effector functions can be brought into play while preserving antigen specificity. Synthesis of IgM then gives way to synthesis of IgG.

IgG is the most abundant antibody in the internal body fluids, including blood and lymph. Like IgM, it is made principally in the lymph nodes, spleen, and bone marrow and circulates in lymph and blood. IgG is smaller and more flexible than IgM, properties that give it easier access to antigens in the extracellular spaces of damaged and infected tissues. The flexibility of the hinge region in IgG enables the two Fab arms to move relative to each other. This enables both the antigen-binding sites to bind to repeated epitopes on the surfaces of pathogens. IgG antibodies can also implement more effector functions than IgM (Figure 2.29). Once they have bound an antigen, the IgG1 and IgG3 subclasses can directly recruit phagocytic cells to ingest the antigen:antibody complex, as well as activating the complement system. During pregnancy, IgG antibodies can be transferred across the placenta,

Function	IgM	IgD	IgG1	IgG2	IgG3	IgG4	IgA	IgE
Neutralization	+	–	++	++	++	++	++	–
Opsonization	–	–	+++	*	++	+	+	–
Sensitization for killing by NK cells	–	–	++	–	++	–	–	–
Sensitization of mast cells	–	–	+	–	+	–	–	+++
Activation of complement system	+++	–	++	+	+++	–	+	–

Property	IgM	IgD	IgG1	IgG2	IgG3	IgG4	IgA	IgE
Transport across epithelium	+	–	–	–	–	–	+++ (dimer)	–
Transport across placenta	–	–	+++	+	++	+/–	–	–
Diffusion into extravascular sites	+/–	–	+++	+++	+++	+++	++ (monomer)	+
Mean serum level (mg ml⁻¹)	1.5	0.03	9	3	1	0.5	2.1	5×10^{-5}

Figure 2.29 Each human immunoglobulin isotype has specialized functions and distinct properties. The major effector functions of each isotype (+++) are shaded in dark red; lesser functions (++) are shown in dark pink, and very minor functions (+) in pale pink. Other properties are similarly marked, with mean serum concentrations shown in the bottom row. Opsonization refers to the ability of the antibody itself to facilitate phagocytosis. Antibodies that activate the complement system indirectly cause opsonization via complement. *IgG2 acts as an opsonin in the presence of a genetic variant of its phagocyte Fc receptor, which is found in about 50% of Caucasians. NK cells, natural killer cells.

providing the fetus with protective antibodies from the mother in advance of possible infection.

Monomeric IgA is made by plasma cells in lymph nodes, spleen, and bone marrow and is secreted into the bloodstream. Dimeric IgA is made in the lymphoid tissues underlying mucosal surfaces and is the antibody that is secreted into the lumen of the gut; it is also the principal antibody of other secretions, including milk, saliva, sweat, and tears. The mucosal surface of the gastrointestinal tract provides an extensive surface of contact between the human body and the environment, and the transport processes involved in the uptake of food make it vulnerable to infection. More IgA is made than any other isotype. Some of this is against the resident microorganisms that colonize mucosal surfaces, keeping their population in check. In contrast to monomeric IgA, dimeric IgA has a J chain identical to that present in pentameric IgM (Figure 2.30).

The IgE class of antibodies is highly specialized towards the activation of mast cells, which are present in epithelial tissues. IgE bound tightly to mast cells triggers strong inflammatory reactions in the presence of its antigen, and is thought to be involved in the expulsion of worms and other parasites. In medical practice, however, the major impact of IgE is in the allergies that result when it is produced against otherwise harmless antigens.

IgE binds tightly to a high-affinity IgE receptor carried by tissue mast cells and the basophils of the blood. Even though IgE is present in the body at minute

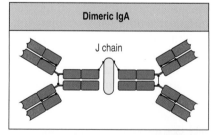

Dimeric IgA

J chain

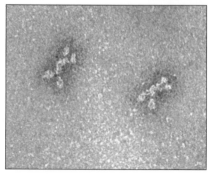

Figure 2.30 IgA molecules can form dimers. In mucosal lymphoid tissue IgA is synthesized as a dimer in association with the same J chain found in pentameric IgM. In dimeric IgA, the monomers have disulfide bonds to the J chain but not to each other. The bottom panel shows an electron micrograph of dimeric IgA. Photograph (× 900,000) courtesy of K.H. Roux and J.M. Schiff.

concentrations in both normal and allergic individuals, the affinity of this receptor for IgE is so high that it is constantly occupied with IgE. Binding of antigen to receptor-bound IgE induces mast cells to release stored histamine and other activators, which recruit cells and molecules of the immune system to local sites of trauma, causing inflammation.

Summary

The C region of the heavy chain defines classes of immunoglobulin that serve different functions in immunity. On binding of antigen to the B-cell receptor, the cell receives signals that cause both cell division and proliferation and the initiation of mechanisms that diversify the antigen specificity and effector function of the antibody. Antigen specificity is diversified by somatic hypermutation, which introduces single-nucleotide substitutions randomly throughout the V-region sequence. This leads to the selection of B cells that make higher-affinity antibodies. With higher-affinity antibodies, the dependence on pentameric IgM for strong binding to antigen is relaxed and the switching of immunoglobulin isotype from IgM/IgD to IgG, IgE, or IgA by further somatic recombination events produces antibodies of different isotype, tissue distribution and effector function while preserving antigen specificity. B cells control the extent to which its immunoglobulin is a cell-surface receptor or a secreted effector antibody by alternative splicing of the primary RNA transcript made from the same rearranged heavy-chain gene.

Summary to Chapter 2

The principal function of B lymphocytes is to produce antibodies, which are secreted immunoglobulins that bind tightly to infectious agents and tag them for destruction or elimination. Each antibody is highly specific for its corresponding antigen; the antibody repertoire of each person is enormous because it is composed of many millions of different antibodies that can bind a wide variety of different antigens. Antibodies can also be divided into five different effector classes—IgM, IgG, IgD, IgA, and IgE—that have different functions in the immune response. This chapter has provided an overview of the structure and function of the antibody molecule and of the unusual genetic mechanisms that create this diversity in specificity and effector function. Within an antibody molecule, the V regions that bind antigen are physically separated from the C region that interacts with effector molecules and cells of the immune system, such as complement, phagocytes, and other leukocytes. Antigen binding is the property of the paired V domains of the heavy and light chains, which can form an almost unlimited number of different binding sites with structural complementarity to a vast range of molecules. The immunoglobulin genes (heavy chain, κ light chain, and λ light chain) are expressed only in B cells, and their expression involves an unusual process of DNA rearrangement in which somatic DNA recombination assembles a V-region coding sequence from sets of gene segments that are present in the unrearranged gene. The random selection of gene segments for assembly creates much of the collective diversity of antigen-binding sites. The lymphocyte-specific proteins and ubiquitous enzymes of DNA repair and recombination are involved in the recombination machinery. Imprecision is inherent in some of their reactions, which creates additional diversity at the junctions between gene segments. In an individual B cell, only one rearranged heavy-chain gene and one rearranged light-chain gene become functional, ensuring that each B cell expresses immunoglobulin of a single specificity. The series of gene rearrangements that result in production of membrane-bound IgM, the first immunoglobulin produced, is summarized in Figure 2.31.

On mature B cells, the membrane-bound immunoglobulin functions as the specific receptor for antigen; when encountering antigen the B cell is stimulated to proliferate and differentiate into plasma cells that secrete antibody of the same specificity as the membrane-bound immunoglobulin. This ensures that an immune response is directed only against the invading pathogen or immunizing antigen. On stimulation of a B cell with its specific antigen, a mutational mechanism is also switched on that introduces point mutations into the rearranged V-region DNA. Thus, the diversity of antibodies is in part due to inherited variation that is encoded in the genome and in part to non-inherited diversity that develops in B cells during an individual's lifetime. The first antibody produced after an encounter with antigen is always IgM. As the immune response proceeds, the process of isotype switching then transfers the antigen specificity onto different C regions to produce antibody molecules with different effector functions—IgG, IgA, and IgE. The changes in the immunoglobulin genes that occur throughout the life of a B cell are summarized in Figure 2.32.

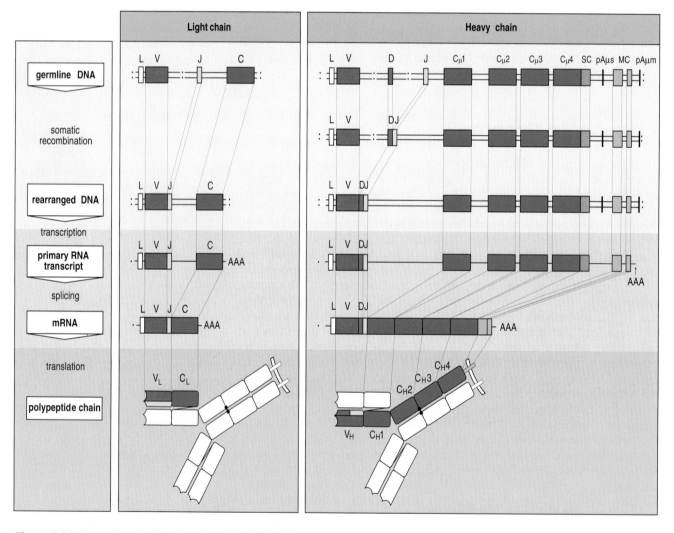

Figure 2.31 Biosynthesis of cell-surface IgM in B cells. Before immunoglobulin light- (center panel) and heavy- (right panel) chain genes can be expressed, rearrangements of gene segments are needed to produce exons encoding the V regions. Once this has been achieved the genes are transcribed to give primary transcripts containing both exons and introns. The latter are spliced out to produce mRNAs that are translated to give κ or λ light chains and μ heavy chains that assemble inside the cell and are expressed as membrane IgM at the cell surface. The main stages in the biosynthesis of the heavy and light chains are shown in the panel on the left.

Changes in immunoglobulin genes during a B cell's life		
Event	Mechanism	Permanence of change to the B cell's genome
1 V-region assembly from gene fragments	Somatic recombination of genomic DNA	Irreversible
2 Generation of junctional diversity	Imprecision in joining rearranged DNA segments adds nongermline nucleotides (P and N) and deletes germline nucleotides	Irreversible
3 Assembly of transcriptional controlling elements	Promoter and enhancer are brought closer together by V-region assembly	Irreversible
4 Transcription activated with coexpression of surface IgM and IgD	Two patterns of splicing and processing RNA are used	Reversible and regulated
5 Synthesis changes from membrane Ig to secreted antibody	Two patterns of splicing and processing RNA are used	Reversible and regulated
6 Somatic hypermutation	Point mutation of genomic DNA	Irreversible
7 Isotype switch	Somatic recombination of genomic DNA	Irreversible

Figure 2.32 The changes in the immunoglobulin genes that occur over a B cell's lifetime.

Questions

Question 2–1
A. What is the difference between antibodies and immunoglobulins?
B. Which cell types produce each type?

Question 2–2
Describe the structure of an antibody molecule and how this structure enables it to bind to a specific antigen. Include the following terms in your description: heavy chain (H chain), light chain (L chain), variable region, constant region, Fab, Fc, antigen-binding site, hypervariable region, and framework region. (Refer to Figures 2.2, 2.3, and 2.7.)

Question 2–3
A. What is an epitope?
B. Define the term multivalent antigen.
C. How does a linear epitope differ from a conformational epitope?
D. Do antibodies bind their antigens via noncovalent or covalent bonding? (Refer to Figures 2.9–2.11.)

Question 2–4
A. What is the difference between polyclonal antibodies and monoclonal antibodies?
B. How is each produced? (Refer to Figure 2.12.)

Question 2–5
A. Explain briefly how a vast number of immunoglobulins of different antigen specificities can be produced from the relatively small number of immuno-

globulin genes present in the genome. Include the following terms in your explanation: somatic recombination, germline configuration, V, D, and J segments. (Refer to Figures 2.14 and 2.15.)
B. What is the final arrangement of gene segments in the rearranged immunoglobulin heavy-chain gene V region, and in what order do these gene segment rearrangements occur?
C. In what order do the various immunoglobulin loci rearrange?

Question 2–6
How do recombination signal sequences ensure that gene segment rearrangement occurs in the right order?

Question 2–7
How is additional diversity introduced into the variable region by the molecular mechanism of somatic recombination? Include the following terms in your answer: junctional diversity, P nucleotides, N nucleotides, terminal deoxynucleotidyl transferase (TdT). (Refer to Figures 2.17–2.19.)

Question 2–8
Explain how mature, naive B cells co-express IgM and IgD. (Refer to Figure 2.21.)

Question 2–9
Describe the process responsible for altering the expression of membrane-bound immunoglobulin to secreted antibody. (Refer to Figures 2.24 and 2.31.)

Question 2–10

A. What is affinity maturation and what molecular process enables it to occur?

B. Describe this process and its consequences. (Refer to Figure 2.25.)

Question 2–11

A. What is isotype switching?

B. Explain the molecular mechanism of isotype switching.

C. Why is isotype switching important? (Refer to Figure 2.27.)

Question 2–12

Which immunoglobulin isotypes (out of IgM, IgG1, IgG2, IgG3, IgG4, IgA, IgE, and IgD) participate in (a) neutralization; (b) opsonization; (c) sensitization for killing by NK cells; (d) sensitization of mast cells; (e) activation of complement? Which isotypes (f) are transported across epithelium; (g) are transported across the placenta; (h) diffuse into extravascular sites? (Refer to Figure 2.29.)

Question 2–13

Isotype switching and immunoglobulin gene rearrangement by somatic recombination are both recombinational processes but have very different outcomes. Give five ways in which they differ from each other.

Question 2–14

Monoclonal antibodies are used for a wide range of applications including serological assays and diagnostics probes in the laboratory, and as therapeutic reagents in the clinic. Discuss why "humanizing" monoclonal antibodies is necessary for use as therapeutic reagents but not necessary when monoclonal antibodies are used as serological or diagnostic reagents.

Question 2–15

What would be the effect of a genetic defect that resulted in a lack of recombination between the switch regions in the immunoglobulin C-region genes?

Chapter 3

Antigen Recognition
by T Lymphocytes

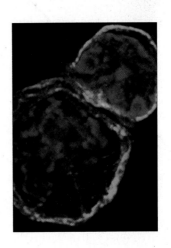

In Chapter 2 we saw that the function of B lymphocytes is to make immunoglobulins, a family of proteins with variable binding sites for antigen. T lymphocytes are a related lineage of antigen-specific cells that perform complementary functions to those of B cells in the adaptive immune response (see Chapter 1). Whereas the sole function of B cells is to produce secreted antibodies, T cells have more diverse roles, all of which involve interactions with other cells. B cells and T cells use similar means of recognizing antigen, but because of the distinct functions of B and T cells there are important differences between them, particularly in the type of antigen that they recognize.

In this chapter we shall consider the antigens that stimulate T cells and the receptors that T cells use to recognize them. In the first part of the chapter we will describe the structure of the T-cell antigen receptor and the mechanisms that generate its antigen specificity and diversity. The antigen receptor on T cells is commonly referred to simply as the **T-cell receptor**. T-cell receptors have much in common with the immunoglobulins. They have a similar structure, are produced as a result of gene rearrangement, and are highly variable and diverse in their antigen specificity. Like B cells, each clone of T cells expresses a single species of antigen receptor and thus different clones possess different and unique antigen specificities. This clonal distribution of diverse T-cell receptors is produced by genetic mechanisms that are similar to those that generate immunoglobulins during B-cell development.

Despite the fundamental similarity in immunoglobulin and T-cell receptor structures, the antigens recognized by T cells are quite distinct from those recognized by immunoglobulins. Immunoglobulins bind epitopes on a wide range of intact molecules, such as proteins, carbohydrates, and lipids. These kinds of molecule are present on the surfaces of bacteria, viruses, and parasites and also on soluble toxins. T-cell receptors, in contrast, recognize and bind to mainly one type of antigen, which must be presented to them on the surface of another human cell.

T-cell receptors recognize peptide antigens that are bound to specialized antigen-presenting glycoproteins called **MHC molecules**, which are expressed on almost all the cells of the body. There are an exceptionally large number of genetically determined variants of the MHC molecules in the human population, and differences between individuals in the MHC molecules that they possess are the primary cause of graft rejection and graft-versus-host disease in clinical transplantation. Because of this clinical effect, these antigen-presenting molecules were first discovered as determinants of tissue incompatibility in transplantation long before their function in antigen presentation was known. The genes that encode them are genetically linked in a chromosomal region called the **major histocompatibility complex** or **MHC**; the glycoproteins have, therefore, become known as MHC molecules.

In the second part of the chapter we shall consider the complex of peptide antigen and MHC molecule that is recognized by the T-cell receptor and we will trace the pathways by which peptides are derived from the proteins of infecting microorganisms and become bound to MHC molecules. The third part of the chapter will deal with the MHC itself and the immunological implications of the exceptional genetic polymorphism of the MHC genes.

T-cell receptor diversity

The T-cell receptor is a membrane-bound glycoprotein that closely resembles a single antigen-binding arm of an immunoglobulin molecule. It is composed of two different polypeptide chains and has one antigen-binding site. T-cell receptors are always membrane bound and there is no secreted form as there is for immunoglobulins. Like immunoglobulins, each chain has a variable region, which binds antigen, and a constant region. During T-cell development, gene rearrangement produces sequence variability in the variable regions of the T-cell receptor by the same mechanisms that are used by B cells to produce the variable regions of immunoglobulins. However, after the T cell is stimulated with antigen, there is no further mutation in the antigen-binding site and there is no switching of constant-region isotype as occurs for immunoglobulins. These differences correlate with the fact that T-cell receptors are used only as receptors to recognize antigen, whereas immunoglobulins serve as both recognition and effector molecules.

3-1 The T-cell receptor resembles a membrane-associated Fab fragment of immunoglobulin

A T-cell receptor consists of two different polypeptide chains, termed the **T-cell receptor α chain** (**TCRα**) and the **T-cell receptor β chain** (**TCRβ**). The genes encoding the α and β chains have similar germline organization to the genes encoding immunoglobulin heavy- and light-chain genes in that they are made up of sets of gene segments that must be rearranged to form a functional gene. As a consequence of gene rearrangements that occur as part of T-cell development, each mature T cell expresses one functional α chain and one functional β chain, which together define a unique T-cell receptor molecule. Within the population of T cells possessed by a healthy human being there are many millions of different T-cell receptors, each of which defines a clone of T cells and a single antigen-binding specificity.

Comparison of the amino-acid sequences of the T-cell receptor α and β chains from different T-cell clones shows that they are organized into variable regions (V regions) and constant regions (C regions), like those found in immunoglobulin chains. The α and β chains are folded into discrete protein domains resembling those in immunoglobulin chains. Each chain consists of an amino-terminal V domain, followed by a C domain, and then a membrane-anchoring domain. The antigen-recognition site of T-cell receptors is formed from the $V_α$ and $V_β$ domains and is the most variable part of the molecule, as in the immunoglobulins. The three-dimensional structure of the four extracellular domains of the T-cell receptor is very similar to that of the antigen-binding Fab fragment of IgG (Figure 3.1).

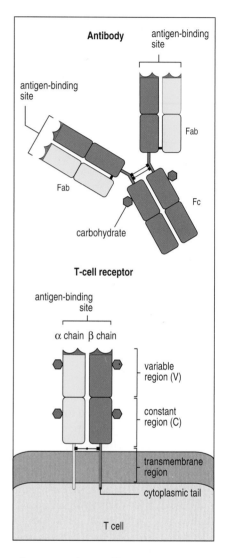

Figure 3.1 The T-cell receptor resembles a membrane-bound Fab fragment. Comparison of the T-cell receptor and an IgG antibody molecule. The T-cell receptor is a heterodimer composed of an α chain of 40–50 kDa and a β chain of 35–46 kDa in size. The extracellular portion of each chain consists of two immunoglobulin-like domains: the domain nearest to the membrane is a C region and the domain farthest from the membrane is a V region. Both the α and β chains span the cell membrane and have very short cytoplasmic tails. The three-dimensional structure formed by the four immunoglobulin-like domains of the T-cell receptor resembles that of the antigen-binding Fab fragment of antibody.

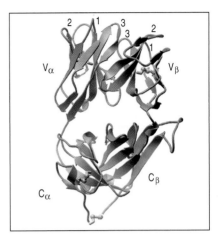

Figure 3.2 Three-dimensional structure of the T-cell receptor showing the antigen-binding CDR loops. The ribbon diagram shows the α chain (in magenta) and the β chain (in blue). The receptor is viewed from the side as it would sit on a cell surface with the highly variable CDR loops, which bind the peptide:MHC molecule ligand, arrayed across its relatively flat top surface. The CDR loops are numbered 1–3 for each chain. Courtesy of I.A. Wilson.

Comparison of the amino-acid sequences of V domains from different clones of T cells shows that sequence variation in the α and β chains is clustered into regions of hypervariability, which correspond to loops of the polypeptide chain at the end of the domain farthest from the T-cell membrane. These loops form the binding site for antigen and are termed complementarity-determining regions (CDRs), as in the immunoglobulins. The T-cell receptor α- and β-chain V domains each have three CDR loops, called CDR1, CDR2, and CDR3 (Figure 3.2).

Immunoglobulins possess two or more binding sites for antigen; this strengthens the interactions of soluble antibody with the repetitive antigens found on the surfaces of microorganisms. T-cell receptors possess a single binding site for antigen and are used only as cell-surface receptors for antigen, never as soluble antigen-binding molecules. Antigen binding to T-cell receptors occurs always in the context of two opposing cell surfaces, where multiple copies of the T-cell receptor bind to multiple copies of the antigen:MHC complex on the opposing cell, thus achieving multipoint attachment.

3-2 T-cell receptor diversity is generated by gene rearrangement

In Chapter 2 we divided the mechanisms that generate immunoglobulin diversity into two categories: those operating before the B cell is stimulated with specific antigen and those operating afterwards. In the first category were the gene rearrangements that generate the V-region sequence, whereas in the second category were changes in mRNA splicing that produce a secreted immunoglobulin, C-region DNA rearrangements that switch the heavy-chain isotype, and somatic hypermutation of the V-region gene to produce antibodies of higher affinity. In T cells the mechanisms that generate diversity before antigen stimulation are essentially the same as those in B cells, but after antigen stimulation the picture is quite different: whereas immunoglobulin genes continue to diversify, the genes encoding T-cell receptors remain unchanged.

This fundamental difference reflects the fact that the T-cell receptor is used only for the recognition of antigen and not for the generation of effector functions, which is handled by other T-cell molecules. In contrast, the effector functions of B cells are solely dependent on secreted antibodies, whose different C-region isotypes trigger different effector mechanisms. The diversification of an antibody that occurs subsequently to antigen stimulation is directed towards optimizing both its antigen-binding and effector functions.

The human T-cell α-chain locus is on chromosome 14 and the β-chain locus is on chromosome 7. The organization of the gene segments encoding T-cell receptor α and β chains is essentially like that of the immunoglobulin gene segments (Figure 3.3). The main difference is the simplicity of the T-cell receptor C region: there is only one C_α gene, and, although there are two C_β genes, no functional distinction between them is known. The T-cell receptor

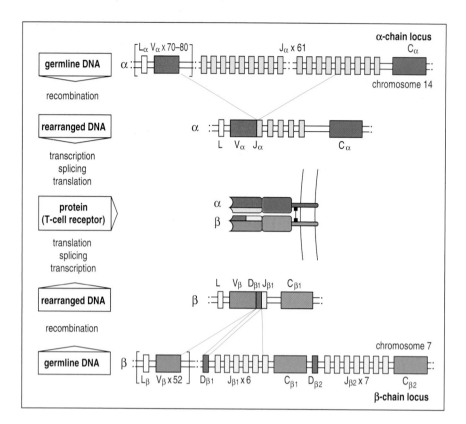

Figure 3.3 Organization and rearrangement of the T-cell receptor genes. The top and bottom rows of the figure show the germline arrangement of the variable (V), diversity (D), joining (J), and constant (C) gene segments at the T-cell receptor α- and β-chain loci. During T-cell development, a V-region sequence for each chain is assembled by DNA recombination. For the α chain (top), a V_α gene segment rearranges to a J_α gene segment to create a functional exon encoding the V domain. For the β chain (bottom), rearrangement of a V_β, a D_β, and a J_β gene segment creates the functional V-domain exon. The assembled genes are transcribed and spliced to produce mRNA (not shown) encoding the α and β chains. Exons encoding the membrane-spanning regions are not shown. L, leader sequence.

α-chain locus is otherwise more similar to an immunoglobulin light-chain locus, containing sets of V and J gene segments only; the β-chain locus is similar to an immunoglobulin heavy-chain locus, containing D gene segments in addition to V and J gene segments. The V domain of the T-cell receptor α chain is thus encoded by a V gene segment and a J gene segment; that of the β chain is encoded by a diversity (D) gene segment in addition to a V and a J gene segment.

T-cell receptor gene rearrangement occurs during T-cell development in the thymus, and the mechanisms involved are similar to those outlined in Chapter 2 for the immunoglobulin genes. In the α-chain gene, a V gene segment is joined to a J gene segment by somatic DNA recombination to make the V-region sequence; in the β-chain gene, recombination first joins a D and a J gene segment, which are then joined to a V gene segment (see Figure 3.3). The T-cell receptor gene segments are flanked by recombination signal sequences similar to those found in immunoglobulin genes and the same RAG complex and other DNA-modifying enzymes are involved in the recombination process in both cell types (see Section 2-8, p. 51). During recombination, additional, nontemplated P and N nucleotides (see Section 2-8, p. 51) are inserted in the junctions between the V, D, and J gene segments of the T-cell receptor β-chain coding sequence and between the V and J gene segments of the α-chain sequence. These mechanisms contribute junctional diversity in the CDR3 to T-cell receptor α and β chains.

Rare genetic defects that cause one or other of the RAG genes not to work are one of the causes of a syndrome called severe combined immunodeficiency disease (SCID). The disease is called combined because functional B and T lymphocytes are equally absent and the disease is severe compared to immunodeficiencies in which only B cells are lacking. Without a bone marrow transplant or other medical intervention, children with SCID die in infancy from common infections (Figure 3.4). In the case of the RAG genes, missense mutations that produce RAG proteins with partial enzymatic activity have also been found; these cause a rapidly fatal immunodeficiency that differs from SCID in some of its symptoms and is known as Omenn syndrome (Figure 3.5).

After gene rearrangement functional α- and β-chain genes consist of exons encoding the leader peptide, V region, and C region, as well as the membrane-spanning region. The exons are separated by introns, which in the case of the intron between the V-domain and the C-domain exon may contain unre-arranged gene segments (see the α-chain gene in Figure 3.3). Upon transcription, the primary RNA transcript is spliced to remove the introns and is processed to give mRNA. Translation of the α-chain and β-chain mRNA produces α and β chains, respectively. Like all proteins destined for the cell membrane, newly synthesized α and β chains enter the endoplasmic reticulum. There they pair to form the **α:β T-cell receptor** (see Figure 3.3).

3-3 Expression of the T-cell receptor on the cell surface requires association with additional proteins

T-cell receptors are diverse and specific receptors for antigen. By themselves, however, heterodimers of α and β chains are unable to leave the endoplasmic reticulum and be expressed on the T-cell surface. In this respect the T-cell receptor resembles immunoglobulin. Before leaving the endoplasmic reticulum, an α:β heterodimer associates with four invariant membrane proteins. Three of the proteins are encoded by closely linked genes of human chromosome 11 and are homologous to each other: these proteins are collectively termed the **CD3 complex** and individually called CD3γ, CD3δ, and CD3ε. The fourth protein is known as the ζ chain and is encoded by a gene on human chromosome 1.

At the cell surface the CD3 proteins and the ζ chain remain in stable association with the T-cell receptor and form the functional **T-cell receptor complex** (Figure 3.6). In this complex the CD3 proteins and the ζ chain transduce signals to the cell's interior after antigen has been recognized by the α and β chain heterodimer. Correlating with these functional differences the cytoplasmic domains of the CD3 proteins and the ζ chain contain sequences that associate with intracellular signaling molecules. In contrast, the T-cell receptor α and β chains have very short cytoplasmic tails that lack signaling function, as is also true for immunoglobulin chains. In people lacking functional CD3δ or CD3ε chains, transport of T-cell receptors to the cell surface is inefficient and, thus, their T cells express abnormally low numbers of receptors. As a consequence of both low T-cell receptor expression and impaired signal transduction these people suffer from immunodeficiency.

3-4 γ and δ chains form a second class of T-cell receptor expressed by a distinct population of T cells

There is a second type of T-cell receptor that is similar in overall structure to the α:β receptor but is formed of two different protein chains termed γ (not to be confused with CD3γ) and δ. The γ chain resembles the α chain, and the δ chain resembles the β chain (Figure 3.7). T cells express either α:β receptors or γ:δ receptors but never both. Those expressing α:β T-cell receptors are called **α:β T cells**, whereas those expressing γ:δ receptors are called **γ:δ T cells**. Cells with γ:δ receptors form a small subset of all T cells. Much more is known of the functions of α:β T cells than of γ:δ T cells. Consequently, in the rest of this book, T cells will generally refer to α:β T cells and T-cell receptor will refer to the α:β T-cell receptor, unless specified otherwise.

suffering from *Candida albicans* **infection in the mouth.** Courtesy of Fred Rosen.

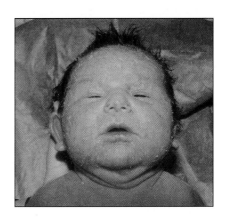

Figure 3.5 Bright red rash on the face and shoulders of an infant with Omenn syndrome. Photograph kindly provided by Luigi Notarangelo.

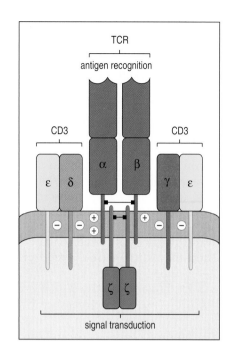

3.6 Polypeptide composition of the T-cell receptor complex. The functional antigen receptor on the surface of T cells is composed of eight polypeptides and is called the T-cell receptor complex. The α and β chains bind antigen and form the core T-cell receptor (TCR). They associate with one copy each of CD3γ and CD3δ and two copies each of CD3ε and the ζ chain. These associated invariant polypeptides are necessary for transport of newly synthesized TCR to the cell surface and for transduction of signals to the cell's interior after the TCR has bound antigen. The transmembrane domains of the α and β chains contain positively charged amino acids (+), which form strong electrostatic interactions with negatively charged amino acids (−) in the transmembrane regions of the CD3γ, δ, and ε chains.

The organization of the γ and δ loci resembles that of the β and α loci, but there are some important differences (Figure 3.8). The δ gene segments are situated within the α-chain locus on chromosome 14, between the V_α and J_α gene segments. This location means that DNA rearrangement within the α-chain locus inevitably results in the deletion and inactivation of the δ-chain locus. The human γ-chain locus is on chromosome 7. The γ- and δ-chain loci contain fewer V gene segments than the α- or β-chain loci, and so, in theory, might produce less diverse receptors, but for the γ chain this limitation is compensated for by an increase in junctional diversity.

Rearrangement at the γ and δ loci proceeds as for the α and β loci, with the exception that during δ-gene rearrangement two D segments can be incorporated into the final gene sequence. This increases the variability of the δ chain in two ways. First, the potential number of combinations of gene segments is increased. Second, extra N nucleotides can be added at the junction between the two D segments, as well as at the VD and DJ junctions.

T cells bearing γ:δ receptors comprise about 1–5% of the T cells found in the circulation, but they can be the dominant T-cell population in epithelial tissue. The immune function of γ:δ T cells is less well defined than that of α:β T cells, as are the antigens to which these cells respond and the ligands that their receptors engage. Unlike α:β T cells, the γ:δ T cells are not restricted to the recognition of antigens associated with MHC molecules.

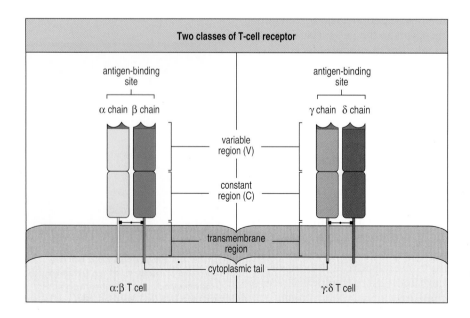

Figure 3.7 There are two classes of T-cell receptors. The α:β T-cell receptor (left panel) and the γ:δ T-cell receptor (right panel) have similar structures, but they are encoded by different sets of rearranging gene segments and have different functions.

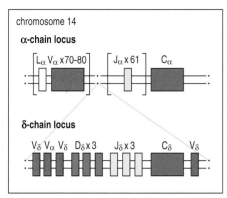

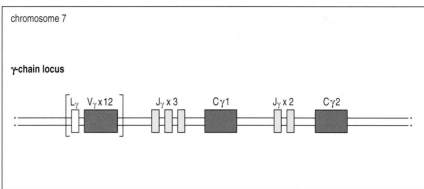

Figure 3.8 The organization of the human T-cell receptor γ- and δ-chain loci. The γ and δ loci, like the α and β loci, contain sets of variable (V), diversity (D), joining (J), and constant (C) gene segments. The δ locus is located within the α-chain locus on chromosome 14, lying between the clusters of V_α and J_α gene segments. There are at least three V_δ gene segments, three D_δ gene segments, three J_δ gene segments, and a single C_δ gene segment. V_δ segments are interspersed amongst V_α and other gene segments. The γ locus, on chromosome 7, resembles the β locus, with two C gene segments each with its own set of J segments.

Summary

T cells recognize antigen through a cell-surface receptor known as the T-cell receptor, which has structural and functional similarities to the membrane-bound immunoglobulin that serves as the B-cell receptor for antigen. T-cell receptors are heterodimeric glycoproteins in which each polypeptide chain consists of a V domain and a C domain, similar to those found in immunoglobulin chains, and a membrane-spanning region. The three-dimensional structure of the extracellular domains of the T-cell receptor resembles that of the antigen-binding Fab fragment of IgG. There are two types of T-cell receptor: one made up of α and β chains and expressed by T cells whose function is understood, and another made up of γ and δ chains and carried by T cells whose function remains elusive. All four types of T-cell receptor chain are encoded by genes that resemble the immunoglobulin genes and require similar DNA rearrangements in order to be expressed. The critical difference between T-cell receptors and immunoglobulins is that T-cell receptors serve only as cell-surface receptors and are not secreted as soluble proteins with effector function; T cells use other molecules for effector function. This explains why the T-cell receptor has only a single binding site for antigen that does not change its affinity on encountering antigen and a simple constant region that does not switch isotype. Expression of the T-cell receptor at the T-cell surface requires association with proteins of the CD3 complex. These proteins transmit signals to the interior of the cell when the T-cell receptor binds antigen.

Antigen processing and presentation

Unlike the immunoglobulins of B cells, which can recognize a wide range of molecules in their native form, a T-cell receptor can recognize antigen only in the form of a peptide bound to an MHC molecule on the surface of a human cell. This means that pathogen-derived proteins must be degraded into peptides to be recognized by T cells. This **antigen processing** occurs inside the body's own cells; the peptides are then assembled into peptide:MHC molecule complexes for display on cell surfaces, where they are recognized by T cells. The binding of a peptide antigen by an MHC molecule and its display at the cell surface is termed **antigen presentation** (Figure 3.9).

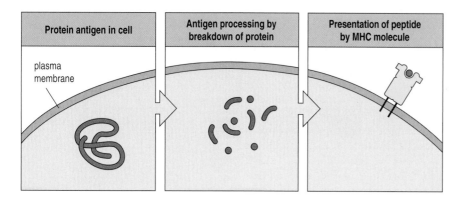

| Protein antigen in cell | Antigen processing by breakdown of protein | Presentation of peptide by MHC molecule |

plasma membrane

Figure 3.9 Antigen processing and presentation. The antigens recognized by T cells are peptides that arise from the breakdown of macromolecular structures, the unfolding of individual proteins, and their cleavage into short fragments. These events constitute antigen processing. For a T-cell receptor to recognize a peptide antigen, the peptide must be bound by an MHC molecule and displayed at the cell surface, a process called antigen presentation.

The microorganisms that infect the human body can be broadly divided into those that propagate within cells, such as viruses, and those, such as most bacteria, that live in the extracellular spaces. The T-cell population that fights infection is composed of two subpopulations: one specialized to fight intracellular infections, the other to fight extracellular sources of infection. The latter include bacterial species that live in extracellular spaces and virus particles present in the extracellular fluid after release from infected cells. In this part of the chapter we shall consider how these two sources of antigen are distinguished by the immune system and processed by intracellular pathways, and how the appropriate type of T cell is activated.

3-5 Two classes of T cell are specialized to respond to intracellular and extracellular sources of infection

Circulating $\alpha{:}\beta$ T cells fall into one of two mutually exclusive classes: one is defined by expression of the **CD4** glycoprotein on the cell surface and the other by expression of the **CD8** glycoprotein (Figure 3.10). These two classes of T cell have different functions and deal with different types of pathogen. **CD8 T cells** are cytotoxic and their main function is to kill cells that have become infected with a virus or some other intracellular pathogen. This response prevents the multiplication of the pathogen and further infection of healthy cells. The general function of **CD4 T cells** is to help other cells of the immune system to respond to extracellular sources of infection. Different aspects of this response are carried out by two subclasses of CD4 T cells—T_H1 and T_H2. The H in their names stands for 'helper' and CD4 T cells are often also known as **helper T cells**. T_H2 **cells** are involved mainly in stimulating B cells to make antibodies, which bind to extracellular bacteria and virus particles, whereas a function of T_H1 **cells** is to activate tissue macrophages to phagocytose and kill extracellular pathogens, and to secrete cytokines and other biologically active molecules that affect the course of the immune response (Figure 3.11).

The human immunodeficiency virus (HIV), which causes acquired immunodeficiency syndrome (AIDS), selectively infects CD4 T cells by exploiting the CD4 molecule as its receptor. On binding to CD4 on a T-cell surface the virus gains entry to the cell.

Figure 3.10 The structures of the CD4 and CD8 glycoproteins. CD4 has four extracellular immunoglobulin-like domains (D_1–D_4) with a hinge between the D_2 and D_3 domains. CD8 consists of an α and a β chain, which both have an immunoglobulin-like domain that is connected to the membrane-spanning region by an extended stalk. C denotes the carboxy terminus.

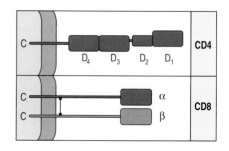

3-6 Two classes of MHC molecule present antigen to CD8 and CD4 T cells respectively

The MHC molecules are crucial in ensuring that the appropriate class of T cells is activated in response to a particular source of infection. There are two different classes of MHC molecule—**MHC class I** and **MHC class II**—and each presents peptides from one kind of antigen source to one type of T cell. MHC class I molecules present antigens of intracellular origin to CD8 T cells, whereas MHC class II molecules present antigens of extracellular origin to CD4 T cells. The basis for this correlation is specific molecular interactions between the CD8 glycoprotein and MHC class I molecules and between CD4 and MHC class II molecules (Figure 3.12). These occur when a T-cell receptor recognizes its specific peptide:MHC molecule ligand. Because of their close involvement in antigen recognition, the CD4 and CD8 molecules are called T-cell **co-receptors**.

Before we consider how the appropriate MHC molecules become associated with peptides from different sources, we shall look at the structure and general peptide-binding properties of MHC molecules.

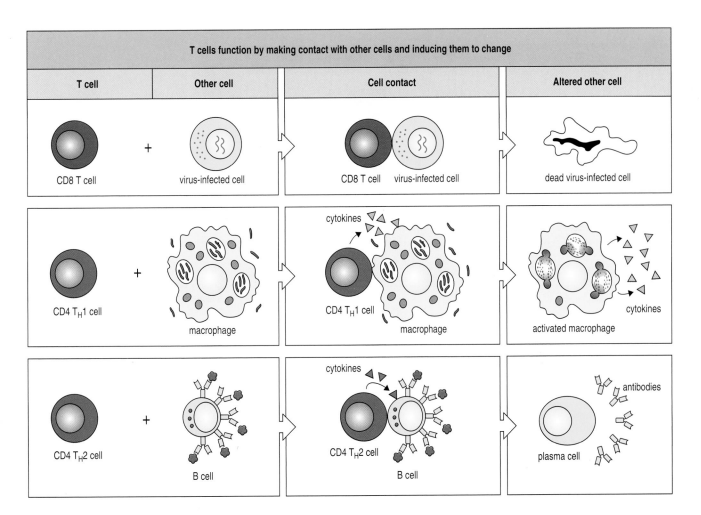

Figure 3.11 T cells function by making contact with other cells. The functions of three types of T cell are shown. Top panels: the cytotoxic CD8 T cell makes contact with a virus-infected cell, recognizes that it is infected, and kills it. Middle panels: the CD4 T$_H$1 helper cell contacts a macrophage that is engaged in the phagocytosis of bacteria and secretes cytokines that increase the microbicidal powers of the macrophage and its secretion of inflammatory cytokines. Bottom panels: the CD4 T$_H$2 helper cell contacts a B cell that is binding its specific antigen and secretes cytokines that cause the B cell to differentiate into an antibody-secreting plasma cell.

3-7 The two classes of MHC molecule have similar three-dimensional structures

MHC class I and class II molecules are membrane glycoproteins whose function is to bind peptide antigens and present them to T cells. Underlying this common function is a similar three-dimensional structure, which is formed in different ways in the two types of molecule.

The MHC class I molecule is made up of a transmembrane heavy chain, or α chain, which is noncovalently complexed with the small protein **β_2-microglobulin** (Figure 3.13). The heavy chain has three extracellular domains (α_1, α_2, and α_3). The peptide-binding site is formed by the folding of α_1 and α_2, the domains farthest from the membrane, and is supported by the α_3 domain and the β_2-microglobulin. The MHC class I heavy chain is encoded by a gene in the MHC, whereas β_2-microglobulin is not.

In contrast, the MHC class II molecule consists of two transmembrane chains (α and β), each of which contributes one domain to the peptide-binding site and one immunoglobulin-like supporting domain (see Figure 3.13). Both of these chains are encoded by genes in the MHC. Thus, both classes of MHC molecule have similar three-dimensional structures consisting of two pairs of

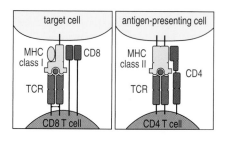

Figure 3.12 MHC class I and II molecules bind to different T-cell co-receptors. Left panel: the CD8 co-receptor of cytotoxic T cells binds to the MHC class I molecule on an infected target cell. Right panel: the CD4 co-receptor of helper T cells binds to the MHC class II molecule on an antigen-presenting cell. TCR, T-cell receptor.

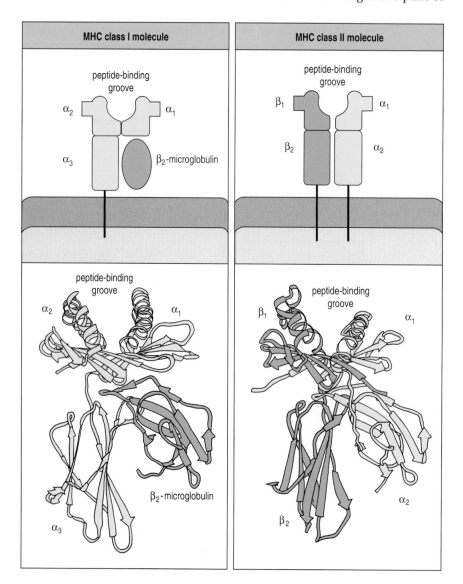

Figure 3.13 The structures of MHC class I and MHC class II molecules are variations on a theme. An MHC class I molecule (left panels) is composed of one membrane-bound heavy (or α) chain and noncovalently bonded β_2-microglobulin. The heavy chain has three extracellular domains, of which the amino-terminal α_1 and α_2 domains resemble each other in structure and form the peptide-binding site. An MHC class II molecule (right panels) is composed of two membrane-bound chains, an α chain (which is a different protein from MHC class I α) and a β chain. These have two extracellular domains each, the amino-terminal two (α_1 and β_1) resembling each other in structure and forming the peptide-binding site. The β_2 domain of MHC class II molecules should not be confused with the β_2-microglobulin of MHC class I molecules. The ribbon diagrams in the lower panels trace the paths of the polypeptide backbone chains.

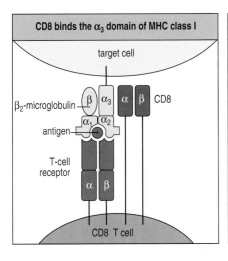
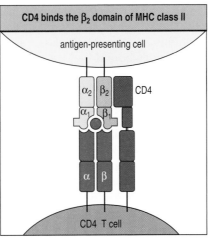

Figure 3.14 **MHC class I molecules bind to CD8, and MHC class II molecules bind to CD4.** The CD8 co-receptor binds to the α_3 domain of the MHC class I heavy chain, ensuring that MHC class I molecules present peptides only to CD8 T cells (left panel). In a complementary fashion, the CD4 co-receptor binds to the β_2 domain of MHC class II molecules, ensuring that peptides bound by MHC class II stimulate only CD4 T cells (right panel).

extracellular domains, with the paired domains farthest from the membrane resembling each other and forming the peptide-binding site. In both types of MHC molecule, the domains supporting the peptide-binding domains are immunoglobulin-like domains: α_3 and β_2-microglobulin in MHC class I molecules, and α_2 and β_2 in MHC class II molecules.

The immunoglobulin-like domains of MHC class I and II molecules are not just a support for the peptide-binding site; they also provide binding sites for the CD4 and CD8 co-receptors. Thus, the sites on the MHC molecule that interact with the T-cell receptor and the co-receptor are separated, allowing the simultaneous engagement of both T-cell receptor and co-receptor by an MHC molecule (Figure 3.14).

3-8 MHC molecules bind a variety of peptides

MHC molecules have peptide-binding sites that are capable of binding peptides of many different amino-acid sequences. This **degenerate binding specificity** contrasts with the specificity of an immunoglobulin or a T-cell receptor for a single epitope. The peptide-binding site is a deep groove on the surface of the MHC molecule (Figure 3.15), within which a single peptide is held tightly by noncovalent bonds.

Figure 3.15 **The peptide-binding groove of MHC class I and MHC class II molecules.** The T-cell receptor's view of the peptide-binding groove, with a peptide bound, is shown. In the MHC class I molecule (left panel) the groove is formed by the α_1 and α_2 domains of the MHC class I heavy chain; in the MHC class II molecule (right panel) it is formed by the α_1 domain of the class II α chain and the β_1 domain of the class II β chain. Amino-acid side-chains on the MHC molecule that are important for making interactions with the bound peptide are shown. The dotted blue lines indicate hydrogen bonds and ionic interactions made between the peptide and the MHC molecule. Peptides bind to MHC class I molecules by their ends (left panel), whereas in MHC class II molecules, the peptide extends beyond the peptide-binding groove and is held by interactions along its length (right panel).

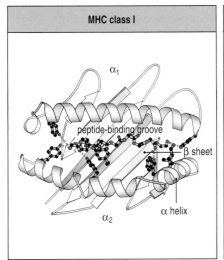
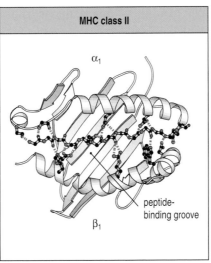

There are, however, constraints on the length and amino-acid sequence of peptides that are bound by MHC molecules. These are dictated by the structure of the peptide-binding groove, which differs between MHC class I and MHC class II molecules. The length of the peptides bound by MHC class I molecules is limited because the two ends of the peptide are grasped by pockets situated at the ends of the peptide-binding groove (see Figure 3.15, left panel). The vast majority of peptides that bind MHC class I molecules are eight, nine, or ten amino acids in length; most are nine amino acids. The differences in peptide length are accommodated by a slight kinking of the extended conformation of the bound peptide. Features that are common to all peptides—the amino terminus, the carboxy terminus, and the peptide backbone—interact with binding pockets present in all MHC class I molecules, and these form the basis for all peptide–MHC class I interactions. Most peptides bound by MHC class I molecules also have a hydrophobic or basic residue at the carboxyl terminus, which corresponds to a complementary pocket present in the binding groove of MHC class I molecules.

In MHC class II molecules, the two ends of the peptide are not pinned down into pockets at each end of the peptide-binding groove (see Figure 3.15, right panel). As a consequence, they can extend out at each end of the groove and so peptides bound by MHC class II molecules are both longer and more variable in length than peptides bound by MHC class I. Peptides that bind to MHC class II molecules are usually 13–25 amino acids in length and some are much longer.

3-9 Peptides generated in the cytosol are transported into the endoplasmic reticulum where they bind MHC class I molecules

The peptide antigens that are bound and presented by MHC molecules are generated inside cells of the body by the breakdown of larger protein antigens. Proteins derived from "intracellular" and "extracellular" antigens are present in different intracellular compartments (Figure 3.16). They are processed into peptides by two intracellular pathways of degradation, and bind to the two classes of MHC molecule in separate intracellular compartments. Peptides derived from the degradation of intracellular pathogens are formed in the cytosol and delivered to the endoplasmic reticulum. This is where MHC class I molecules bind peptides. In contrast, extracellular microorganisms and proteins are taken up by cells via phagocytosis and endocytosis and are degraded in the lysosomes and other vesicles of the endocytic pathways. It is in these cellular compartments that MHC class II molecules bind peptides. In this way, the class of the MHC molecule labels the peptide as being extracellular or intracellular in origin. In this section we shall look at the processing pathway for intracellular antigens.

When viruses infect human cells they exploit the cell's ribosomes to synthesize viral proteins, which are therefore present in the cytosol before being assembled into viral particles. In response, the infected cell uses its normal processes of breakdown and turnover of cellular proteins to degrade some of the viral proteins into peptides that can be bound by MHC class I molecules and presented to CD8 T cells.

Proteins in the cytosol are degraded by a large barrel-shaped protein complex called the **proteasome**, which has several different protease activities. It consists of 28 polypeptide subunits, each of 20–30 kDa molecular weight. Once formed, the antigenic peptides are transported out of the cytosol and into the endoplasmic reticulum (Figure 3.17). Transport of peptides across the endoplasmic reticulum membrane is accomplished by a protein, called the **transporter associated with antigen processing** (**TAP**), embedded in the membrane.

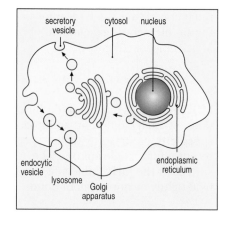

Figure 3.16 There are two major compartments within cells, separated by membranes. One compartment is the cytosol, which is contiguous with the nucleus via the pores in the nuclear membrane. The other compartment is the vesicular system, which consists of the endoplasmic reticulum, the Golgi apparatus, endocytic vesicles, lysosomes, and other intracellular vesicles. The vesicular system is effectively contiguous with the extracellular fluid. Secretory vesicles bud off from the endoplasmic reticulum and by successive fusion and budding with the Golgi membranes move vesicular contents out of the cell. In contrast, endocytic vesicles take up extracellular material into the vesicular system.

Figure 3.17 Formation and transport of peptides that bind to MHC class I molecules. In all cells, proteasomes degrade cellular proteins that are poorly folded, damaged, or unwanted. When a cell becomes infected, pathogen-derived proteins in the cytosol are also degraded by the proteasome. Peptides are transported from the cytosol into the lumen of the endoplasmic reticulum by the protein called transporter associated with antigen processing (TAP), which is in the endoplasmic reticulum membrane.

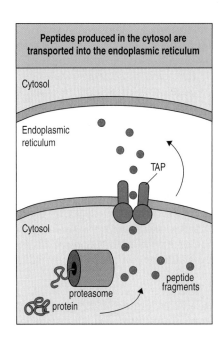

TAP is a heterodimer consisting of two structurally related polypeptide chains, TAP-1 and TAP-2. Peptide transport by TAP is dependent on the binding and hydrolysis of ATP, properties shared with the other transporters in this family. The types of peptide that are preferentially transported by TAP are similar to those that bind MHC class I molecules: they are eight or more amino acids long and have either hydrophobic or basic residues at the carboxy terminus.

Newly synthesized MHC class I heavy chains and β_2-microglobulin are also translocated into the endoplasmic reticulum where they partially complete their folding, then associate together, and finally bind peptide, at which point folding is completed. The correct folding and peptide loading of MHC class I molecules within the endoplasmic reticulum is aided by proteins known as chaperones. These are proteins that help the correct folding and subunit assembly of other proteins while keeping them out of harm's way until they are ready to enter cellular pathways and carry out their functions.

When MHC class I heavy chains first enter the endoplasmic reticulum they bind the membrane protein known as **calnexin**, which retains the partly folded heavy chain in the endoplasmic reticulum (Figure 3.18). Calnexin is a

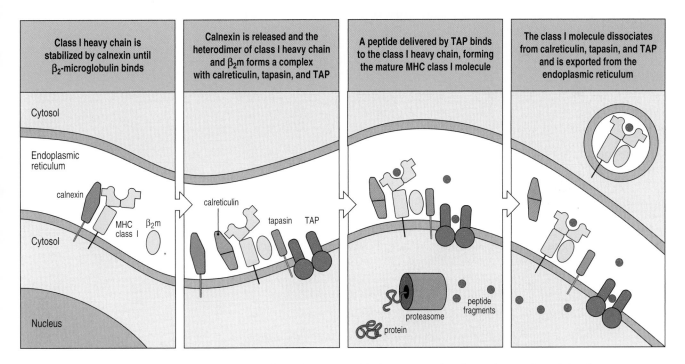

Figure 3.18 Chaperone proteins aid the assembly and peptide loading of MHC class I molecules in the endoplasmic reticulum. MHC class I heavy chains assemble in the endoplasmic reticulum with the membrane-bound protein calnexin. When this complex binds β_2-microglobulin (β_2m) the partly folded MHC class I molecule is released from calnexin and then associates with the TAP-1 subunit of TAP by interacting with the TAP-associated protein tapasin and the chaperone protein calreticulin. The MHC class I molecule is retained in the endoplasmic reticulum until it binds a peptide, which completes the folding of the molecule. The peptide:MHC class I molecule complex is then released from tapasin and calreticulin, leaves the endoplasmic reticulum, and is transported to the cell surface.

calcium-dependent lectin, a carbohydrate-binding protein that retains many multisubunit glycoproteins, including T-cell receptors and immunoglobulins, in the endoplasmic reticulum until they have folded correctly. When the MHC class I heavy chain has bound β_2-microglobulin, calnexin is released from the $\alpha:\beta_2$-microglobulin heterodimer to be replaced by a complex of proteins, one of which, **calreticulin**, resembles a soluble form of calnexin and probably has a similar chaperone function. A second component of the complex, called **tapasin**, binds to the TAP-1 subunit of the peptide transporter, positioning the partly folded heterodimer of heavy chain and β_2-microglobulin to await receipt of a suitable peptide from the cytosol.

On binding a peptide, the now completely assembled MHC class I molecule is released from all chaperones and leaves the endoplasmic reticulum in a membrane-bound vesicle. It then makes its way through the Golgi stacks to the plasma membrane. Most of the peptides transported by TAP are not successful in binding to a MHC class I molecule and are cleared out of the endoplasmic reticulum.

MHC class I molecules cannot leave the endoplasmic reticulum unless they have bound a peptide. In one form of a rare disease called **bare lymphocyte syndrome**, the TAP protein is nonfunctional and so no peptides enter the endoplasmic reticulum. Cells of patients with this defect have less than 1% of the normal level of MHC class I molecules on their surface. As a consequence of this deficiency, patients develop very poor CD8 T-cell responses to viruses and suffer from chronic respiratory infections from a young age.

Protein degradation and peptide transport occur continuously, not only when cells are infected. In the absence of infection, MHC class I molecules carry peptides derived from normal human self-proteins, as also do MHC class II molecules. These do not normally provoke an immune response because of mechanisms that act during T-cell development which eliminate or inactivate T cells reactive to these **self-peptides**. Occasionally, however, this state of T-cell tolerance to self-peptides breaks down, resulting in autoimmunity.

3-10 Peptides presented by MHC class II molecules are generated in acidified intracellular vesicles

Proteins of extracellular bacteria, extracellular virus particles, and soluble protein antigens are processed by a different intracellular pathway from that followed by cytosolic proteins, and their peptide fragments end up bound to MHC class II molecules. Most cells are continually internalizing extracellular fluid and material bound at their surface by the process of **endocytosis**. In addition, cells specialized for **phagocytosis**, the neutrophils and macrophages, engulf larger objects such as dead cells. These uptake mechanisms produce intracellular vesicles known as **endocytic vesicles** or **phagosomes** (in the case of phagocytosis), in which the vesicular membrane is derived from the plasma membrane and the lumen contains extracellular material.

These membrane-bound vesicles become part of an interconnected vesicle system that carries materials to and from the cell surface. As vesicles travel inwards from the plasma membrane, their interiors become acidified by the action of proton pumps in the vesicle membrane and they fuse with other vesicles, such as lysosomes, that contain proteases and hydrolases that are active in acid conditions. Within the **phagolysosomes** formed by this fusion, enzymes degrade the vesicle contents to produce, among other things, peptides from proteins and glycoproteins.

Microorganisms present in the extracellular environment are taken up by phagocytosis, for example by macrophages, and are then degraded within phagolysosomes. B cells also bind specific antigens via their surface immunoglobulin, then internalize these antigens by receptor-mediated endocytosis. These antigens are similarly degraded within the vesicular system. Peptides produced within phagolysosomes become bound to MHC class II molecules within the vesicular system, and the peptide:MHC class II complexes are carried to the cell surface by outward-going vesicles. Thus, the MHC class II pathway samples the extracellular environment, complementing the MHC class I pathway, which samples the intracellular environment (Figure 3.19).

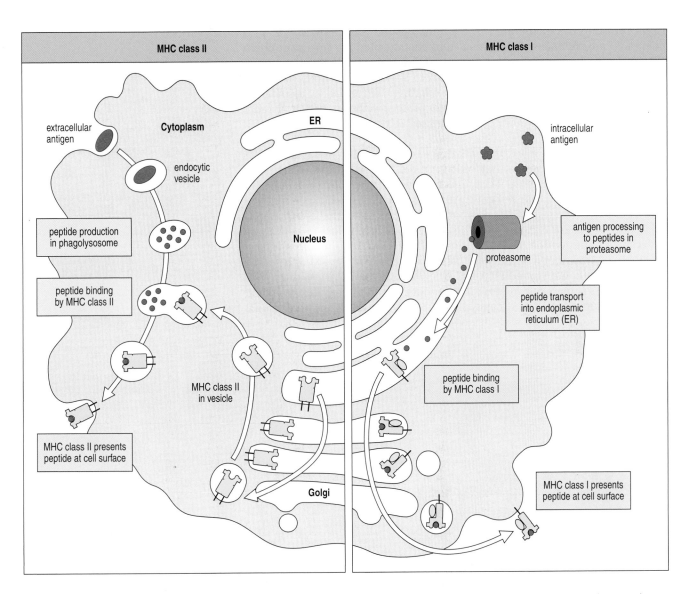

Figure 3.19 Processing of antigens presented by MHC class II and MHC class I molecules occurs in different cellular compartments. The left half of the figure shows the fate of peptides derived from extracellular antigens and pathogens. Extracellular material is taken up by endocytosis and phagocytosis into the vesicular system of the cell, in this case a macrophage. Proteases in these vesicles break down proteins to produce peptides that are bound by MHC class II molecules, which have been transported to the vesicles via the endoplasmic reticulum (ER) and the Golgi apparatus. The peptide:MHC class II complex is transported to the cell surface in outgoing vesicles. The right half of the figure shows the fate of peptides generated in the cytosol as a result of infection with viruses or intracytosolic bacteria. Proteins from such pathogens are broken down in the cytosol by the proteasome to peptides, which enter the ER. There the peptides are bound by MHC class I molecules. The peptide:MHC class I complex is transported to the cell surface via the Golgi apparatus.

Certain pathogens, for example the species of mycobacteria that cause leprosy and tuberculosis, actually exploit the vesicular system as a protected site for intracellular growth and replication. As their proteins do not enter the cytosol, they are not processed and presented to cytotoxic CD8 T cells by MHC class I molecules. They protect themselves from degradation by lysosomal enzymes by preventing fusion of the phagosome with a lysosome. This also prevents presentation of mycobacterial antigens by MHC class II molecules.

3-11 MHC class II molecules are prevented from binding peptides in the endoplasmic reticulum by the invariant chain

Newly synthesized MHC class II α and β chains are translocated from the ribosomes into the membranes of the endoplasmic reticulum. There, an α chain and a β chain associate with a third chain, called the **invariant chain** (Figure 3.20). This is so called because it is identical in all individuals, whereas the α and β chains vary from one person to another. One function of the invariant chain is to prevent the peptide-binding site formed by association of an α and a β chain from binding peptides present in the endoplasmic reticulum; these peptides are therefore targeted to MHC class I molecules only.

A second function of the invariant chain is to deliver MHC class II molecules to endocytic vesicles, where they bind peptide. These vesicles, which have been called MIIC, for MHC class II compartment, contain proteases, for example cathepsin S, that selectively attack the invariant chain. A series of cleavages leaves just a small fragment of the invariant chain that covers up the MHC class II peptide-binding site. This fragment is called **class II-associated invariant-chain peptide (CLIP)**. Removal of CLIP and binding of peptide is aided by the interaction of the MHC class II molecule with a glycoprotein in the vesicle membrane called **HLA-DM** (see Figure 3.20). HLA-DM resembles an MHC class II molecule in structure but does not bind peptides or appear on the cell surface. HLA-DM catalyzes the removal of CLIP and then allows

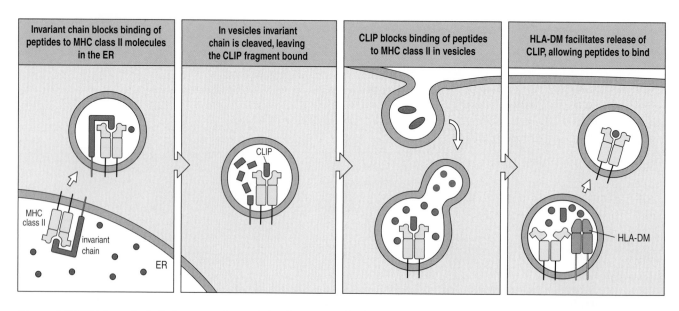

| Invariant chain blocks binding of peptides to MHC class II molecules in the ER | In vesicles invariant chain is cleaved, leaving the CLIP fragment bound | CLIP blocks binding of peptides to MHC class II in vesicles | HLA-DM facilitates release of CLIP, allowing peptides to bind |

Figure 3.20 The invariant chain prevents peptides from binding to a MHC class II molecule until it reaches the site of extracellular protein breakdown. MHC class II α and β chains are assembled with an invariant chain in the endoplasmic reticulum (ER); this complex is transported to acidified vesicles of the endocytic system. The invariant chain is broken down, leaving just a small fragment called class II-associated invariant-chain peptide (CLIP) attached in the peptide-binding site. The vesicle membrane protein HLA-DM catalyzes the release of the CLIP fragment and its replacement by a peptide derived from endocytosed antigen that has been degraded within the acidic interior of the vesicles.

the MHC class II molecule to sample other peptides until it finds one that binds strongly. Once the MHC class II molecule has lost its invariant chain and has bound peptide it is carried to the cell surface by outward-going vesicles.

3-12 The T-cell receptor specifically recognizes both peptide and MHC molecule

Once a peptide:MHC complex appears on the cell surface it can be recognized by its corresponding T-cell receptor. When the T-cell receptor binds to a peptide:MHC molecule complex it makes contacts with both the peptide and the surrounding surface of the MHC molecule. Thus, each peptide:MHC complex forms a unique ligand for its T-cell receptor.

The floor of the peptide-binding groove of both classes of MHC molecule is formed by eight strands of antiparallel β-pleated sheet, on which lie two antiparallel α helices (see Figure 3.15). The peptide lies between the helices and parallel to them, such that the top surfaces of the helices and peptide form a roughly planar surface to which the T-cell receptor binds. The residues of the peptide that bind to the MHC molecule lie deep within the peptide-binding groove and are inaccessible to the T-cell receptor; side chains of other peptide amino acids stick out of the binding site and bind to the T-cell receptor.

In its overall organization the T-cell receptor antigen-binding site resembles that of an antibody (see Figure 3.2). The interaction between T-cell receptors and ligands comprising peptides bound to MHC molecules has been visualized by X-ray crystallography. Analysis of several complexes has revealed similar interactions for peptides bound to either MHC class I or class II molecules, and the interaction is principally illustrated here for MHC class I (Figure 3.21). The T-cell receptor binds to the MHC class I:peptide complex with the long axis of its binding site oriented diagonally across the peptide-binding groove of the MHC class I molecule (Figure 3.21, panel d). The T-cell receptor binds to an MHC class II:peptide complex in a similar orientation (Figure 3.21, panel e). The CDR3 loops of the T-cell receptor α and β chains form the central part of the binding site and they grasp the side chain of one of the amino acids in the middle of the peptide. In contrast, the CDR1 and CDR2 loops form the periphery of the binding site and contact the α helices of the MHC molecule. The CDR3 loops directly contact peptide antigen and they are also the most variable part of the T-cell receptor antigen-recognition site; the α-chain CDR3 includes the joint between the V and J sequences, and the β chain CDR3 includes the joints between V and D, the whole of the D segment, and the joint between D and J.

The T-cell receptor does not interact symmetrically with the face formed by the peptide and the two α helices of the MHC molecule. Consequently, the CDR1 and CDR2 loops of the α chain make stronger contacts with the peptide:MHC complex than do the CDR1 and CDR2 loops of the β chain.

3-13 The two classes of MHC molecule are expressed differentially on cells

T-cell responses are guided in appropriate directions by differential expression of the two classes of MHC molecules on human cells (Figure 3.22). Virtually all the cells of the body express MHC class I molecules constitutively, and as all cell types are susceptible to infection by viral pathogens, this enables comprehensive surveillance by CD8 T cells. The erythrocyte is one cell type that lacks MHC class I, a property that might facilitate its persistent infection by malaria parasites.

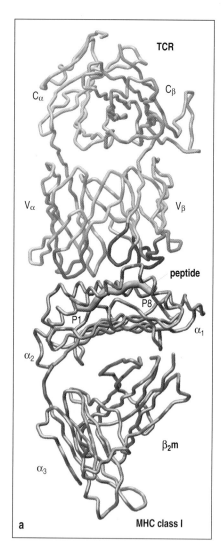

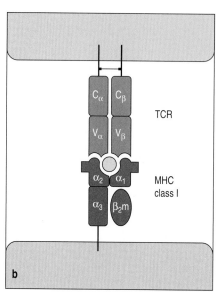

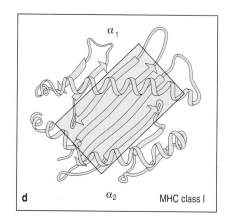

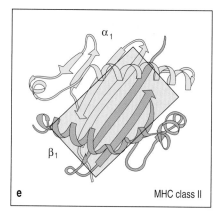

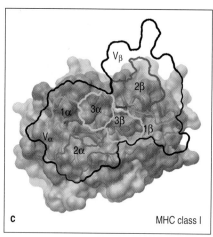

Figure 3.21 The MHC:peptide:T-cell receptor complex. Panel a shows a diagram of the polypeptide backbone of the complex of a T-cell receptor (TCR) bound to its peptide:MHC class I ligand. Panel b shows a schematic representation of this view of the receptor:ligand complex. In panel a the T-cell receptor's CDRs are colored: the α-chain CDR1 and CDR2 are light and dark blue respectively, while the β-chain CDR1 and CDR2 are light and dark purple respectively. The α chain CDR3 is yellow and the β-chain CDR3 is dark yellow. The eight amino-acid peptide is colored yellow and the positions of the first (P1) and last (P8) amino acids are indicated. Panel c is a view rotated 90° from that of panel a and shows the surface of the peptide:MHC class I ligand and the footprint made upon it by the T-cell receptor (outlined in black). Within this footprint the contributions of the CDRs are outlined in different colors and labeled. In panels d and e the diagonal orientation of the T-cell receptor with respect to the peptide-binding grooves of MHC class I and class II molecules, respectively, is shown in schematic diagrams. The T-cell receptor is represented by the black rectangle superimposed on the ribbon diagram (yellow) of the peptide-binding domains of the MHC molecules. Panels a and c courtesy of I.A. Wilson.

MHC class II molecules are, in contrast, constitutively expressed on only a few cell types, which are cells of the immune system specialized for the uptake, processing, and presentation of antigens from the extracellular environment. This distribution is consistent with MHC class II function, alerting CD4 T cells to the presence of extracellular infections. To be effective, this function need not be performed by every cell within a tissue or organ, but requires only a sufficient number of cells equipped to guard the extracellular territory. These **professional antigen-presenting cells** are: dendritic cells, which are supremely specialized for antigen presentation and the activation of T cells; macrophages, which take up antigens by phagocytosis and endocytosis; and B cells, which efficiently internalize specific antigens bound to their surface immunoglobulin.

During the course of an immune response, cells can increase the synthesis and cell-surface expression of MHC molecules beyond constitutive levels. This upregulation, which enhances antigen presentation and T-cell activation, is induced by several cytokines produced by activated immune system cells, in particular the interferons. In addition to increasing the levels of constitutively expressed MHC molecules, interferon-γ (IFN-γ) can induce the expression of MHC class II molecules on some cell types that do not normally produce them. In this manner, presentation of antigen to CD4 T cells by MHC class II molecules can be induced and increased in infected or inflamed tissues.

Tissue	MHC class I	MHC class II
Lymphoid tissues		
T cells	+++	+*
B cells	+++	+++
Macrophages	+++	++
Other antigen-presenting cells (e.g., dendritic cells)	+++	+++
Epithelial cells of the thymus	+	+++
Other nucleated cells		
Neutrophils	+++	–
Hepatocytes	+	–
Kidney	+	–
Brain	+	–[†]
Non-nucleated cells		
Red blood cells	–	–

Figure 3.22 Most human cells express MHC class I, whereas only a few cell types express MHC class II. MHC class I molecules are expressed on almost all nucleated cells, although they are most highly expressed in hematopoietic cells. MHC class II molecules are normally expressed only by a subset of hematopoietic cells and by stromal cells in the thymus, although they can be produced by other cell types on exposure to the cytokine interferon-γ. *In humans, activated T cells express MHC class II molecules, whereas resting T cells do not. [†]In the brain, most cell types are MHC class II-negative, but microglia, which are related to macrophages, are MHC class II-positive.

Summary

T cells expressing α:β T-cell receptors recognize peptides presented at cell surfaces by MHC molecules, the third type of antigen-binding molecule in the adaptive immune system. Unlike immunoglobulins and T-cell receptors, however, MHC molecules have degenerate binding sites and each MHC molecule can bind peptides of many different amino-acid sequences. The peptides are produced by the intracellular degradation of proteins of infectious agents and of self-proteins. An α:β T-cell either expresses the CD8 co-receptor and recognizes peptides presented by MHC class I molecules, or expresses the CD4 co-receptor and recognizes peptides presented by MHC class II molecules. The CD8 co-receptor interacts specifically with MHC class I molecules, and the CD4 co-receptor interacts specifically with MHC class II molecules. Protein antigens from intracellular and extracellular sources are processed into peptides by two different pathways. Peptides generated in the cytosol from viruses and other intracytosolic pathogens enter the endoplasmic reticulum, where they are bound by MHC class I molecules. Thus, these peptides are recognized by CD8 T cells, which are specialized to fight intracellular infections. As all human cells are susceptible to infection, MHC class I molecules are expressed by most cell types. Extracellular material that has been taken up by endocytosis is degraded into peptides in endocytic vesicles and these peptides are bound by MHC class II molecules within the vesicular system. The resulting complex is recognized by CD4 T cells that are specialized to fight extracellular sources of infection by mobilizing other cells of the immune system, such as B cells and macrophages. MHC class II molecules are expressed by a few cell types of the immune system that are specialized to take up extracellular antigens efficiently and activate CD4 T cells.

The major histocompatibility complex

MHC molecules and other proteins involved in antigen processing and presentation are encoded in a cluster of closely linked genes, which in humans is located on chromosome 6. This region is called the major histocompatibility complex (MHC) because it causes T cells to reject tissues transplanted from unrelated donors to recipients. For some MHC class I and class II molecules, numerous genetic variants are present in the human population. Each variant functions differently in the peptides it binds and the T cells it triggers, differences that, individually and collectively, have helped the human species survive predation by diverse and numerous pathogens. Although the magnitude of MHC diversity is much smaller than that of immunoglobulins or T-cell receptors, it has a major impact on the immune response, disease susceptibility, and the practice of medicine. In this part of the chapter we will consider the immunological and medical consequences of MHC diversity within the human population.

3-14 The diversity of MHC molecules in the human population is due to multigene families and genetic polymorphism

The human MHC is called the **human leukocyte antigen (HLA) complex** because the antibodies used to identify human MHC molecules react with the white cells of the blood—the leukocytes—but not with the red cells, which lack MHC molecules. This observation distinguished the MHC from the other known systems of cell-surface antigens, for example the ABO system matched in blood transfusion, which all involve antigens on the surface of red blood cells. Human MHC class I and II molecules are also called **HLA class I molecules** and **HLA class II molecules**, respectively.

In contrast to immunoglobulins and T-cell receptors, the MHC class I and II molecules are encoded by conventionally stable genes that neither rearrange nor undergo any other developmental or somatic process of structural change. The inherited diversity of MHC molecules has two components. The first component is provided by **gene families**, consisting of multiple similar genes encoding the MHC class I heavy chains, MHC class II α chains and MHC class II β chains. The second component is **genetic polymorphism**, which is the presence within the population of multiple alternative forms of a gene.

The products of the different molecules in a MHC class I or class II family are called isotypes. The different forms of any given gene are called **alleles** and their encoded proteins are called **allotypes**. When considering the diversity of MHC class I or II molecules that arises from the combination of multiple genes and multiple alleles, the term **isoform** can be useful to denote any particular MHC protein. The numerous alleles of certain MHC class I and II genes, and the many differences that distinguish them, make these MHC genes stand out from other polymorphic genes, and they are therefore said to be **highly polymorphic**. MHC class I and II genes that have no polymorphism are described as **monomorphic** and genes having a few alleles are described as **oligomorphic.** An important consequence of genetic polymorphism is that an individual can inherit different forms of the gene from their two parents, in which case they are said to be **heterozygous**; when they inherit the same form of the gene from both parents they are said to be **homozygous**.

In humans there are six MHC class I isotypes: **HLA-A, HLA-B, HLA-C, HLA-E, HLA-F**, and **HLA-G**, and five MHC class II isotypes: **HLA-DM, HLA-DO, HLA-DP, HLA-DQ**, and **HLA-DR** (Figure 3.23). Of the class I isotypes, HLA-A, HLA-B, and HLA-C are highly polymorphic and their function is to present antigens

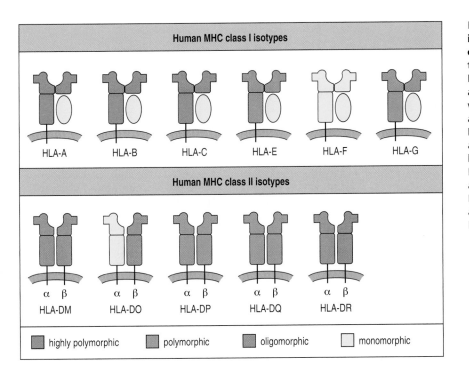

Figure 3.23 Human MHC class I and II isotypes differ in function and the extent of their polymorphism. Of the human MHC class I isotypes, HLA-A, HLA-B and HLA-C present peptide antigens to CD8 T cells and also interact with NK-cell receptors. HLA-E and HLA-G are oligomorphic and interact with NK-cell receptors. HLA-F is intracellular and of unknown function. Of the human MHC class II isotypes, HLA-DP, HLA-DQ and HLA-DR present peptide antigens to CD4 T cells, whereas HLA-DM and HLA-DO are intracellular and regulate peptide loading of HLA-DP, HLA-DQ and HLA-DR.

to CD8 T cells and to form ligands for receptors on natural killer (NK) cells. HLA-E and HLA-G are oligomorphic and form ligands for NK-cell receptors. HLA-F appears to be monomorphic and remains intracellular; its function is unknown. The human MHC class II isotypes also display a range of properties. The three highly polymorphic molecules, **HLA-DP**, **HLA-DQ**, and **HLA-DR**, are those that directly present peptide antigens to CD4 T cells, whereas the oligomorphic **HLA-DM** and **HLA-DO** molecules have functions that regulate peptide loading of HLA-DP, HLA-DQ, and HLA-DR. The numbers of alleles currently known for each HLA locus are listed in Figure 3.24.

The polymorphism of the HLA-A, B, and C molecules is the property of the heavy chain, β_2-microglobulin being monomorphic. By contrast, the polymorphic HLA class II isotypes differ in the diversity contributed by their α and β chains. In HLA-DR the α chain contributes almost no diversity and the β chain is highly polymorphic, whereas for HLA-DP and HLA-DQ class II molecules, both the α and the β chains are polymorphic. Overall there is greater diversity in HLA class I molecules than in HLA class II molecules.

3-15 The MHC class I and class II genes occupy different regions of the MHC

The HLA complex consists of about 4 million base pairs of DNA on the short arm of chromosome 6 and is divided into three regions (Figure 3.25). The **class I region**, at the end of the complex farthest from the chromosome's centromere, contains the six expressed HLA class I genes as well as several nonfunctional class I genes and gene fragments. At the other end of the complex is the **class II region**, which contains all the expressed class II genes and several nonfunctional class II genes. Separating the class I and II regions is a 1 megabase region of DNA called the **class III region** or the **central MHC**; although dense with other types of gene, it contains no class I or II genes. Notably absent from the HLA complex is the gene encoding β_2-microglobulin, the invariant light chain of HLA class I molecules, which is located on human chromosome 15.

HLA polymorphism		
MHC class	**HLA locus**	**Number of allotypes**
MHC class I	A	218
	B	439
	C	96
	E	4
	F	1
	G	6
MHC class II	DMA	4
	DMB	6
	DOA	1
	DOB	2
	DPA1	12
	DPB1	88
	DQA1	17
	DQB1	42
	DRA	2
	DRB1	269
	DRB3	30
	DRB4	7
	DRB5	12

Figure 3.24 The polymorphism of HLA class I and class II genes. The number of known functional alleles in the human population for each gene is shown.

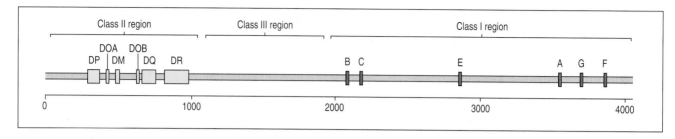

For each HLA class II isotype, the genes encoding the α and β genes are called A and B respectively, for example HLA-DMA and HLA-DMB. When there is more than one gene, including nonfunctional genes, a number in series is added, for example HLA-DQA1 and HLA-DQA2. The genes for the α and β chains of the HLA-DM, HLA-DP, HLA-DQ and HLA-DR class II isotypes cluster together in different subregions within the class II region of the MHC (see Figure 3.25). An exception is HLA-DO, for which HLA-DOA and DOB genes are separated by HLA-DM and other genes. For both HLA-DP and HLA-DQ there are two pairs of genes, one being functional (HLA-DPA1, DPB1 and HLA-DQA1,DQB1) and the other nonfunctional (HLA-DPA2, DPB2 and HLA-DQA2, DQB2). For HLA-DR there is a single HLA-DRA gene, but four different genes encoding HLA-DR β chains (DRB1, DRB3, DRB4, and DRB5) and several nonfunctional genes (DRB2, DRB6, DRB7, DRB8, and DRB9). Only the DRB1 gene is present on all chromosomes 6 and for some people this is the only DRB gene expressed. Three other types of chromosome 6 carry either DRB3, DRB4, or DRB5 in addition to DRB1 (Figure 3.26).

The particular combination of HLA alleles found on a given chromosome 6 is known as the **haplotype**. Within the HLA complex, meiotic recombination occurs at a frequency of about 2%. In the population this mechanism reassorts alleles of polymorphic HLA genes into new haplotypes, although in most families the parental HLA haplotypes are inherited intact. Although there are no more than a few hundred alleles for any one HLA gene (see Figure 3.24), in the course of human history they have been recombined into many thousands of different haplotypes. The further combination of two HLA haplotypes in each individual means that millions of different HLA isoform combinations are represented in the human population. Consequently, individuals who are homozygous for the HLA complex are rare but usually healthy. Minimally they express three class I (HLA-A, B, and C) and three class II (HLA-DP, DQ, and DR) isoforms that present antigens to their T cells. HLA heterozygous individuals can express up to six class I and eight class II isoforms, the maximum number requiring each HLA haplotypes to contain two functional DRB genes and to contribute a different allele for all of the polymorphic HLA class I and II genes.

3-16 Other proteins involved in antigen processing and presentation are encoded in the MHC class II region

The HLA complex contains a total of more than 200 genes of which the HLA class I and II genes constitute a minority. The other genes embrace a variety of functions including several that are important for the immune system. Particularly striking is the fact that the class II region of the MHC is almost entirely dedicated to genes involved in the processing of antigens and their

Figure 3.25 The MHC is divided into three regions containing different types of genes. Shown are the positions within the HLA complex (the human MHC) of the principal genes involved in antigen presentation. The HLA class I region, which contains all the class I genes, is separated from the class II region, which contains all the class II genes, by the class III region, which contains a variety of different genes (not shown), none of which contributes to processing and presentation of antigens to T cells. The three HLA class I heavy (α)-chain genes (HLA-A, HLA-B, HLA-C) are shown in red and the HLA class II gene is shown in yellow. For HLA-DM, HLA-DP, HLA-DQ and HLA-DR, the α- and β-chain genes are close together and are shown as a single yellow block; for HLA-DO the α and β genes (DOA and DOB, respectively) are separated by the DM genes and are, therefore, shown separately. Approximate distances are given in thousands of base pairs (kb).

Figure 3.26 Human MHCs differ in the number of DR genes. The MHC on every human chromosome 6 carries one gene (DRA) for the HLA class II DR α chain and one gene (DRB1) for the DR β chain. In addition, some MHCs carry either DRB3 or DRB4 or DRB5. Any DR β chain can pair with the DR α chain to form a class II molecule.

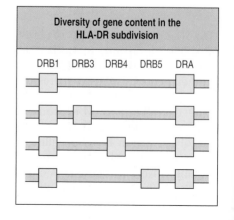

presentation to T cells. In addition to genes encoding the α and β chains of the five HLA class II isotypes, the class II region contains genes encoding the two polypeptides of the TAP peptide transporter, the gene for tapasin, and genes encoding two subunits of the proteasome called LMP2 and LMP7 (Figure 3.27). Notably absent from the class II region and the HLA complex is the gene encoding the invariant chain; it is on chromosome 5.

Genes encoding proteins that work together in antigen processing and presentation are coordinately regulated by the cytokines IFN-α, -β, and -γ, which are produced at sites of infection at an early stage in the immune response. These cytokines stimulate cells in the vicinity to increase their expression of HLA class I heavy chains, $β_2$-microglobulin, TAP, and the LMP2 and LMP7 proteasome subunits. LMP2 and LMP7 are not constitutive components of the proteasome in healthy cells, but are specifically made in response to interferon. When they replace the corresponding constitutive proteasome subunits they bias the proteasome towards production of peptides compatible with MHC class I binding requirements.

Expression of the HLA-DM, HLA-DP, HLA-DQ, HLA-DR, and invariant-chain genes is coordinated by the cytokine IFN-γ. These genes are turned on by a transcriptional activator known as **MHC class II transactivator (CIITA)**, which is itself induced by IFN-γ. Inherited impaired CIITA function leads to a form of bare lymphocyte syndrome in which HLA class II molecules are not made and CD4 T cells cannot function.

The majority of genes in the class I region are not involved in the immune system, neither do the class I genes form as compact a cluster as the class II genes (see Figure 3.25). Whereas genes encoding class II molecules are only present in the class II region of the MHC, genes encoding class I molecules and related class I-like molecules are found on several different chromosomes. A further difference is that the class II molecules all function in antigen presentation to T cells, whereas class I molecules encompass a broader range of functions, including uptake of IgG in the gut, regulation of iron metabolism, and regulation of NK cell function.

3-17 MHC polymorphism affects the binding and presentation of peptide antigens to T cells

The alleles of highly polymorphic MHC genes encode proteins that differ by 1–50 amino-acid substitutions. The substitutions are not randomly distributed within the sequence but are mainly in the domains that bind peptide

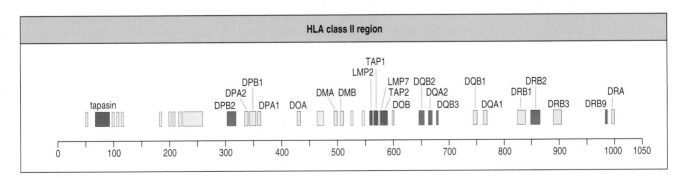

Figure 3.27 Almost all of the genes in the HLA class II region are involved in processing and presentation of antigens to T cells. A detailed map of the HLA class II region is shown. Genes shown in dark gray are pseudogenes that are related to functional genes but are not expressed. Unnamed genes in light gray are not involved in immune system function. In addition to genes encoding the MHC class II isoforms the class II region includes genes for the peptide transporter (TAP), proteasome components (LMP) and tapasin. Approximate distances are given in thousands of base pairs (kb).

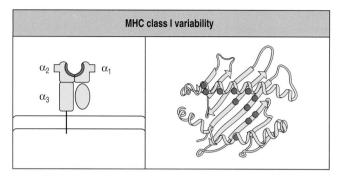

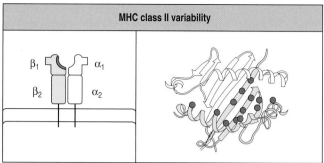

and interact with the T-cell receptor: the α_1 and α_2 domains of MHC class I and the α_1 and β_1 domains of MHC class II. More specifically, the substitutions focus at positions within those domains that contact either bound peptide or the T-cell receptor (Figure 3.28). Not all the contact residues vary as is apparent from the HLA-DR molecule in which the α_1 domain is invariant. In contrast, variability occurs in both the α_1 and β_1 domains of HLA-DP and HLA-DQ molecules.

Variation in the peptide-contact residues on the floor and sides of the peptide-binding groove determines the types of peptides that each isoform binds. At certain positions within the peptide sequence, most of the peptides that bind an MHC isoform have the same amino acid, or one of a few chemically similar amino acids. The preferences arise because the side chains of the amino acids at these positions are bound by complementary pockets within the binding groove. These amino acids are called **anchor residues** because they anchor the peptide to the MHC molecule. The combination of anchor residues that binds to a particular MHC isoform is called its **peptide-binding motif**. For MHC class I molecules, which bind mostly nonamer peptides, positions 2 and 9 are the usual anchor residues. For MHC class II the anchor residues are less clearly defined, in part because of the heterogeneity in length of the bound peptides (Figure 3.29).

The number of peptide-binding motifs is limited and so MHC allotypes that differ by only a few amino acids often bind overlapping populations of peptides. To a rough approximation, the greater the difference in sequence between two MHC allotypes the more disparate will be the populations of peptides they bind.

In the complex of peptide bound to an MHC molecule the anchor residues are buried and inaccessible to the T-cell receptor. In contrast, the peptide's other residues, which have a much greater diversity of amino acids, are available for contact with T-cell receptors. These form part of the planar surface that interacts with the T-cell receptor and includes variable residues on the upper surfaces of the MHC molecule's α helices. Any given T-cell receptor is specific for the complex of particular peptide bound to a particular MHC molecule, a requirement known as **MHC restriction** because the antigen-specific T-cell response is restricted by the MHC type. Consequently, a T cell that responds to a peptide presented by one MHC allotype will neither respond to another peptide bound by that same MHC allotype nor to the same peptide when bound to another MHC allotype (Figure 3.30).

3-18 MHC diversity results from selection by infectious disease

In the previous section we saw that the amino-acid substitutions that distinguish MHC isoforms are not randomly distributed but are concentrated at sites that affect peptide binding and presentation. Further nonrandomness is

Figure 3.28 Variation between MHC allotypes is concentrated in the sites that bind peptide and T-cell receptor. In the HLA class I molecule (left), allotype variability is clustered in specific sites (shown in red) within the α_1 and α_2 domains. These sites line the peptide-binding groove, lying either in the floor of the groove, where they influence peptide binding, or in the α helices that form the walls, which are also involved in binding the T-cell receptors. In the HLA class II molecule illustrated (right), which is a DR molecule, variability is found only in the β_1 domain because the α chain is monomorphic.

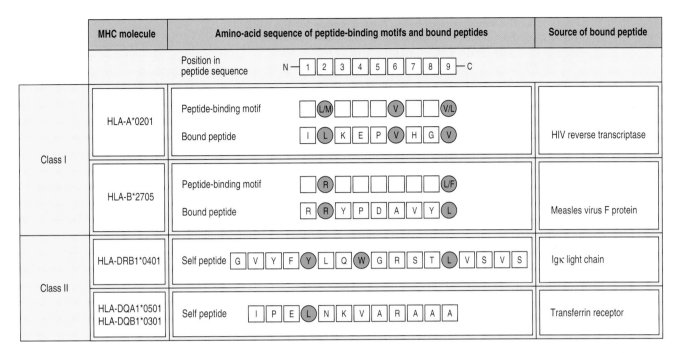

seen in the gene sequences, where the frequency of nucleotide substitutions that change an amino acid is much greater than would be generated at random. The inescapable conclusion is that MHC diversity is due to natural selection, and because of the immunological functions of MHC molecules the likely source of the selection are the infections caused by pathogens.

For an individual the advantage to having multiple MHC class I and class II genes is that they contribute different peptide-binding specificities, allowing a greater number of pathogen-derived peptides to be presented during any infection. This improves the strength of the immune response against the pathogen by increasing the number of activated pathogen-specific T cells. The same argument applies to polymorphism at any given MHC locus, the advantage to the heterozygote being that two different peptide-binding

Figure 3.29 Peptide-binding motifs and the sequences of peptides bound for some MHC isoforms. For the HLA-A and HLA-B isoforms, both the peptide-binding motif for the isoform and the complete amino-acid sequence of one peptide presented by that isoform are given. Blank boxes in the peptide-binding motifs are positions at which the identity of the amino acid can vary. For the HLA-DR and HLA-DQ isoforms, only the sequence of a self-peptide that is bound by the isoform is shown. Anchor residues are in green circles. Peptide-binding motifs for MHC class II molecules are not readily defined. The one-letter code for amino acids is used.

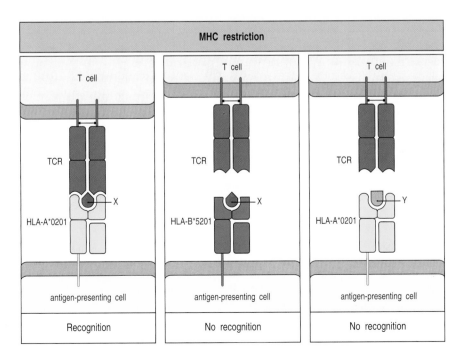

Figure 3.30 T-cell recognition of antigens is MHC restricted. The receptor of the CD8 T cell shown in the left panel is specific for the complex of peptide X with the class I molecule HLA-A*0201. Because of this co-recognition, which is called MHC restriction, the T-cell receptor (TCR) does not recognize the same peptide when it is bound to a different class I molecule, HLA-B*5201 (middle panel). Nor does the T-cell receptor recognize the complex of HLA-A*0201 with a different peptide, Y (right panel). X is HIV-1 Nef residues 190–198 AFHHVAR. Y is influenza A matrix protein residues 58–68 GILGFVFTL.

Figure 3.31 The advantage of being heterozygous for the MHC. The large circles represent the total number of antigenic peptides derived from a pathogen that can be presented by human MHC class I and II molecules. The small circles represent the subpopulation of peptides that can be presented by the MHC class I and II molecules encoded by the genes of particular MHC haplotypes. These subpopulations differ between the haplotypes. In general, heterozygous individuals will have a set of MHC class I and II molecules able to present a wider range of pathogen-derived peptides than a homozygote. However, the extent of this benefit varies. The person who has the divergent haplotypes 1 and 2 will, on average, present more peptides from any pathogen than a person who has the more related haplotypes 3 and 4.

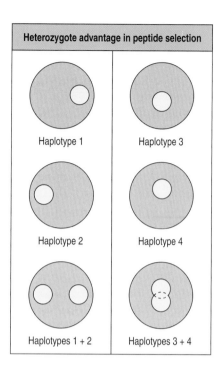

specificities can be brought into play as compared to one for the homozygote. Moreover, the high degree of polymorphism of the antigen-presenting HLA isotypes ensures that most members of the population are heterozygotes. The extent of the heterozygote advantage will vary, however, depending upon the difference in the peptide-binding specificities of the two allotypes. This type of advantage can be considered as arising from **balancing selection** because it acts to maintain a variety of MHC isoforms in the population (Figure 3.31).

A different mode of selection favors certain MHC alleles, or combinations of alleles, at the expense of others and is imposed by specific, epidemic disease. Here, presentation of particular pathogen-derived peptides by particular MHC allotypes is advantageous and in the extreme case makes the difference between life and death. As a consequence of the selection the selected alleles are driven to higher frequency while other alleles decrease in frequency and some may even be lost. Because pathogens adapt to the MHC of their host populations, it has been argued that rare, recently formed MHC alleles to which the pathogen has not adapted will more probably confer advantage to the host and be selected during disease epidemics. Instead of maintaining variety this type of selection replaces older alleles with newer variants and its characteristic outcome is change, not balance. It is therefore called **directional selection** (Figure 3.32). The numerous HLA differences between

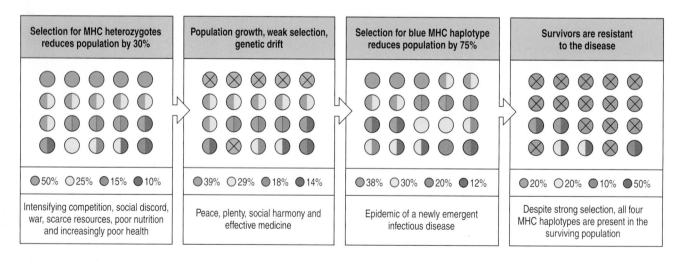

Figure 3.32 Pathogen selection on MHC polymorphism. A population is considered in which there are four different MHC haplotypes, each represented by a different color. The frequencies of different genotypes are represented by the 20 circles in each panel, the four haplotype frequencies being given in the panel below. The population first experiences a period characterized by balancing selection arising from successive epidemic infections, after which only heterozygotes survive and in which 30% of the population die, as indicated by circles containing an X. After recovery of the population during a period of relative calm and health it becomes subject to directional selection by a new and particularly nasty infection. Only individuals with the blue MHC haplotype survive and 75% of the population dies. As a result of these selections the frequencies of the MHC haplotypes change considerably, but all four MHC haplotypes are retained within the population.

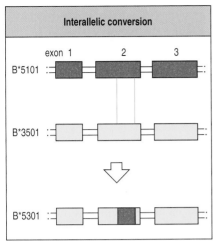

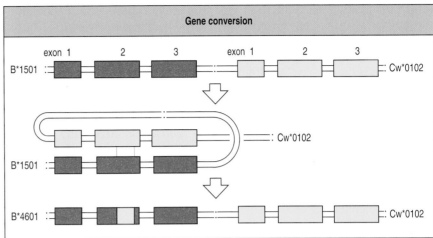

Figure 3.33 New MHC alleles are generated by interallelic conversion or gene conversion. Recombination between alleles of the same gene (HLA-B*5101 and HLA-B*3501), as shown on the left, and between alleles of different genes (HLA-B*1501 and HLA-Cw*0102), as shown on the right, can both result in the formation of a new allele in which a small block of DNA sequence has been replaced. Whereas B*5301 is characteristic of African populations and has been associated with resistance to severe malaria, B*4601 is found in southeast Asian populations and has been associated with susceptibility to nasopharyngeal carcinoma.

human populations of different ethnicity and geographical origins are evidence for directional selection. Only a minority of the HLA alleles listed in Figure 3.24 are common to all human populations, the majority being of recent origin and specific to particular ethnic groups.

New variants of HLA class I and II alleles arise through point mutation and several types of recombination, which can involve alleles of the same gene or alleles of two different genes in the same family (Figure 3.33). Particularly favored appear to be new alleles in which a small segment of one HLA allele has been replaced by the homologous section of another allele, with the introduction of several substitutions that change contact residues in the peptide-binding groove (see Figure 3.28). The recombination mechanism that produces such variants has been termed **interallelic conversion** or **segmental exchange** (Figure 3.33, left panel). Selection for new HLA-B alleles of this type has been particularly strong upon the HLA-B locus in native populations of South and Central America, because today a majority of their HLA-B alleles are 'new variants' specific to these populations.

In industrialized countries the current epidemic of HIV infection provides a unique opportunity to study the effects of HLA polymorphism on an infectious disease. In addition to a general advantage of HLA heterozygosity (Figure 3.34), certain families of alleles (HLA-B14, B27, B57, HLA-C8, C14) are associated with slow progression of the disease, whereas others are associated with rapid progression (HLA-A29, HLA-B22, 35, HLA-C16, and HLA-DR11). Almost all the correlations are with HLA class I; this is consistent with killing of virus-infected cells by cytotoxic CD8 T cells being the principal mechanism for controlling the infection.

3-19 MHC polymorphism triggers T-cell reactions that can reject transplanted organs

During T-cell development any cells having T-cell receptors that respond to complexes of peptide and MHC class I and II molecules at healthy cell surfaces are eliminated. This quality-control mechanism, which prevents a person's T cells from attacking their own healthy tissue and causing disease, only encompasses the MHC isoforms expressed by the person and not other MHC isoforms. In this context the self-MHC isoforms are described as **autologous**, and all other MHC isoforms are described as **allogeneic**. Accordingly, in every person's circulation there are T cells that respond to complexes of peptide and allogeneic MHC class I and II molecules that are present on healthy cells of other individuals. T cells with this property are called **alloreactive T cells**, and those reactive against any given allogeneic cell constitute between 1–10% of circulating T cells.

When allogeneic kidneys are transplanted to patients with renal failure a major danger is that the kidney graft will be rejected by the patient's immune system. One way this happens is when alloreactive T cells in the patient's circulation are activated by allogeneic HLA molecules expressed by the graft, leading to an **alloreaction**, a potent T-cell response that attacks the graft. To reduce the probability of kidney graft rejection, donors are selected who have identical or similar combinations of HLA alleles to the patient. The combination of HLA alleles a person has is called their **HLA type**. Immunosuppressive drugs are also used to preempt the alloreactive T-cell response and to treat rejection when it occurs.

A natural situation in which alloreactions occur is pregnancy, when the mother's immune system can be stimulated by the HLA molecules of the fetus that derive from the father but are not expressed on the mother's cells. This response leads to **alloantibodies** in the mother's circulation with specificity for paternal MHC molecules. Alloantibody is the name given to any antibody raised in one member of a species against an allotypic protein from another member of the same species. Although of no harm to the fetus, which is protected, the alloantibodies produced by pregnancy can have disastrous effects should the mother needs a kidney transplant at some future time. If the alloantibodies react with the allogeneic MHC class I of the transplanted kidney they cause a type of graft rejection that is almost impossible to treat. To avoid this outcome a patient's serum is tested for reactivity with a donor's leukocytes and transplantation is only undertaken when the reaction is acceptably low. Before molecular genetic methods were available, the HLA types of transplant recipients and donors were determined using the anti-HLA class I and class II alloantibodies present in sera obtained from multiparous women.

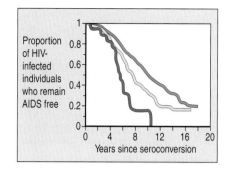

Figure 3.34 MHC heterozygosity delays the progression to AIDS in people infected with HIV-1. When people who have been infected with HIV-1 start to make detectable antibodies to the virus they are said to have undergone seroconversion. The onset of overt symptoms of AIDS occurs years after seroconversion. The rate of progress to AIDS decreases with the extent of HLA heterozygosity as compared here for individuals who are heterozygous for all the highly polymorphic HLA class I and II loci (red), to those who are homozygous for one locus (yellow) or for two or three loci (blue).

Summary

In humans, the highly polymorphic MHC class I and II genes are closely linked in the HLA region on chromosome 6, which comprises the human MHC. In contrast to immunoglobulin and T-cell receptor genes, MHC class I and II genes have a conventional organization and do not rearrange. In humans, the MHC class I genes encode the heavy (α) chains of three different class I molecules—HLA-A, HLA-B, and HLA-C—whereas the MHC class II genes encode the α and β chains of three different MHC class II molecules—HLA-DP, HLA-DQ, and HLA-DR. β_2-Microglobulin, the light chain of MHC class I molecules, is encoded outside the MHC, on chromosome 15. Certain MHC class I and class II genes are highly polymorphic; some have several hundred alleles. Genes encoding other proteins involved in antigen presentation are located in the MHC and, like the MHC class I and II genes, their expression is regulated by the interferons produced during an immune response. The strategy used by MHC molecules to bind diverse antigens contrasts with that of the T-cell receptors. MHC molecules have highly degenerate binding sites for peptides; an MHC molecule is, therefore, usually able to present a diversity of peptide antigens to a large number of T-cell receptors with highly specific binding sites. Polymorphism in families of MHC class I heavy-chain genes and MHC class II α- and β-chain genes is a secondary strategy that serves to increase the breadth and strength of T-cell immunity. It also diversifies T-cell immunity within human populations, a strategy that helps them survive epidemic disease but has also created the main immunological barrier to clinical transplantation.

Summary to Chapter 3

The general structure of T-cell antigen receptors resembles that of the membrane-bound immunoglobulins of B cells and they are encoded by similarly organized genes that undergo gene rearrangement before they are expressed. As with B cells, this gives rise to a population of T cells each expressing a unique receptor. Differences between immunoglobulins and T-cell receptors reflect the fact that the T-cell receptor is used only as a membrane-bound receptor, whereas immunoglobulins are also used as secreted effector molecules.

T-cell receptors are more limited than immunoglobulins in the antigens that they bind, recognizing only short peptides bound to MHC molecules on a cell surface. The two main classes of T cell—CD8 and CD4—are specialized to respond to intracellular and extracellular pathogens, respectively. They are activated in response to an antigen from the appropriate source by specific interactions between the MHC molecules displaying the peptide antigen and the CD4 or CD8 glycoproteins on the T-cell surface. Processing of extracellular and intracellular pathogen antigens into peptides, and their binding to MHC molecules, occurs inside the cells of the infected host. Cytotoxic CD8 T cells are activated by peptides presented by MHC class I molecules and are directed to destroy the antigen-presenting cell. Most cells express MHC class I molecules and so can present pathogen-derived peptides to CD8 T cells if infected with a virus or other pathogen that penetrates the cytosol. The function of CD4 T cells is the activation of other types of effector cell and their recruitment to sites of infection through cell–cell interactions and the production of cytokines. They are activated by peptides presented by MHC class II molecules, which are usually expressed only on specialized antigen-presenting cells that take up and process material from the extracellular environment and can activate CD4 T cells.

MHC molecules have degenerate peptide-binding sites, which enables the relatively small number of different MHC molecules present in each individual to bind peptides of many different sequences. The diversity of peptides that can be presented by the human population as a whole is further increased by the highly polymorphic nature of MHC class I and II genes. Each person differs from almost all other individuals in some or all of the MHC alleles that they possess. Thus, the population is able to respond to the pathogens it encounters with a wide diversity of individual immune responses. Because of the polymorphism of the MHC, organs or tissues transplanted between individuals of different MHC type provoke strong T-cell responses directed against the "foreign" MHC molecules.

Questions

Question 3–1
Describe (a) four ways in which T-cell receptors are similar to immunoglobulins, and (b) four ways in which they are different. (Refer to Figures 3.1–3.3, and 3.6.)

Question 3–2
Discuss several ways in which T-cell receptors differ from immunoglobulins in the way that they recognize antigen. Use the following terms in your answer: peptides, antigen-presenting cells, MHC molecules, and antigen-binding sites. (Refer to Figures 3.9 and 3.13.)

Question 3–3
Compare the organization of T-cell receptor α and β genes (the TCRα and TCRβ loci) to the organization of immunoglobulin heavy- and light-chain genes. (Refer to Figures 2.14 and 3.3.)

Question 3–4
There are considerably more J segments in the T-cell receptor loci than in the immunoglobulin gene loci. Why might this have arisen?

Question 3–5

Why do antibodies but not T-cell receptors undergo iso-type switching?

Question 3–6

What is meant by the terms (a) antigen processing and (b) antigen presentation? (c) Why are these processes required before T cells can be activated? (Refer to Figures 3.9, 3.19, and 3.21.)

Question 3–7

Pathogens that infect the human body replicate either inside cells (such as viruses) or extracellularly, in the blood or in the extracellular spaces in tissues.

A. Identify (i) the class of T cells that are stimulated by intracellular pathogens, (ii) their co-receptor, (iii) the MHC molecule used for recognition of antigen and (iv) the T-cell effector function.

B. Repeat this for the classes of T cells that are stimulated by extracellular pathogens. For the purposes of this question, count those pathogens (such as mycobacteria) that can survive and live inside intracellular vesicles after being taken up by macrophages as extracellular pathogens. (Refer to Figures 3.10, 3.12, and 3.19.)

Question 3–8

A. (i) Describe the structure of an MHC class I molecule, identifying the different polypeptide chains and domains. (ii) What are the names of the MHC class I molecules produced by humans? Which part of the molecule is encoded within the MHC region of the genome? (iii) Which domains or parts of domains participate in: antigen binding; binding the T-cell receptor; and binding the T-cell co-receptor? (iv) Which domains are the most polymorphic?

B. Repeat this for an MHC class II molecule. (Refer to Figures 3.13, 3.14, 3.21, and 3.28.)

Question 3–9

A. What is the maximum number of MHC molecules that a heterozygous individual could theoretically express? Explain your answer. (Ignore the possibility of MHC class II molecules composed of chains from different isoforms.)

B. How does this relatively small number of MHC molecules have the potential to bind the huge number of antigenic peptides encountered in the environment, and what features of a peptide determine whether it will be bound by a given MHC molecule? (Refer to Figures 3.23, 3.25, and 3.26.)

Question 3–10

A. What is the difference between MHC polygeny and MHC polymorphism?

B. How do (i) polygeny and (ii) polymorphism in the MHC genes influence the antigens a person's T cells can recognize? (Refer to Figures 3.23, 3.24, and 3.29.)

Question 3–11

A. Describe in chronological order the steps of the endogenous antigen-processing pathway for intracellular, cytosolic pathogens.

B. (i) What would be the outcome if a mutant MHC class I α chain could not associate with β_2-microglobulin, and (ii) what would happen if the TAP transporter was lacking as a result of mutation? Explain your answers. (Refer to Figures 3.17–3.19.)

Question 3–12

A. Describe in chronological order the steps of the antigen-processing pathway for extracellular pathogens.

B. What would be the outcome (i) if invariant chain were defective or missing, or (ii) if HLA-DM were not expressed? (Refer to Figures 3.19–3.20.)

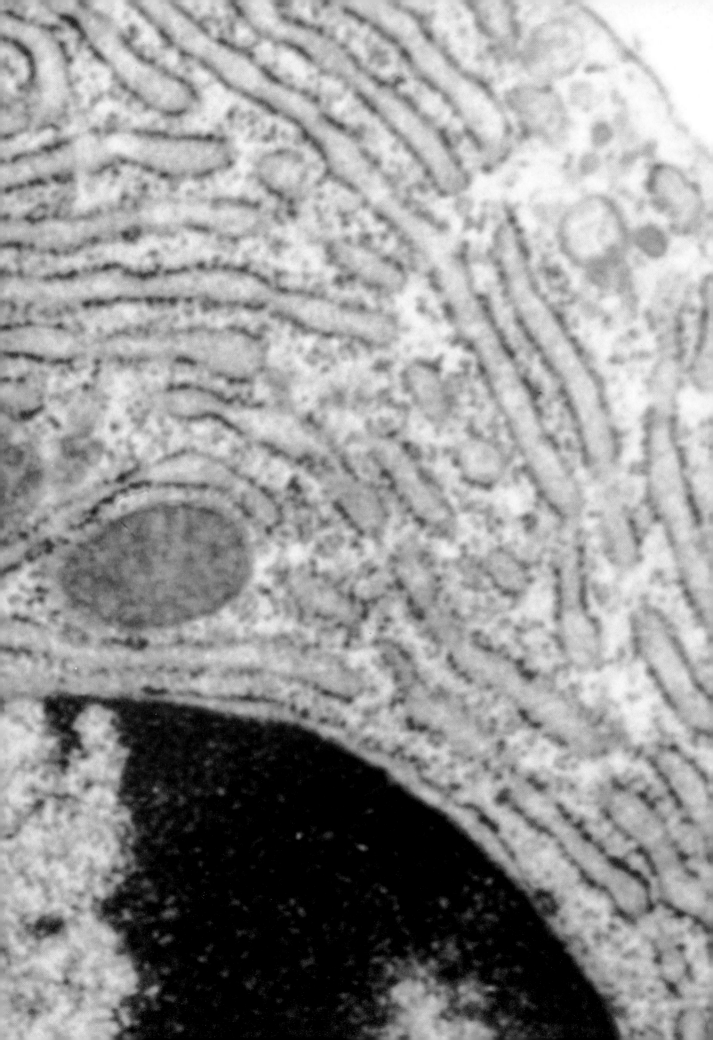

Chapter 4
The Development of B Lymphocytes

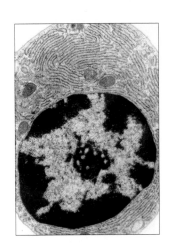

The B cells of the human immune system have the capacity to make immunoglobulins (Ig) specific for almost every nuance of chemical structure, which gives each person the potential to make antibodies against all the infectious microorganisms that could possibly be encountered in a lifetime. However, the body does not stockpile all the B cells needed to make an optimal response against all possible pathogens. That would probably mean devoting the vast majority of the body's resources to the immune system, leaving little left for the immune system to protect. Instead, the body carries a less complete inventory of B cells, but one that expands and contracts its individual clones according to need and circumstance. Fueling this system are stem cells in the bone marrow, which generate tens of billions of new B cells every day of your life.

The life cycle of B cells can be divided into four broad stages (Figure 4.1). The first stage is one of maturation and takes place in the bone marrow, which is, therefore, designated as a **primary lymphoid tissue**. During this stage, which will be described in the first part of this chapter, developing B cells acquire functional B-cell receptors through the ordered rearrangements of the immunoglobulin genes described in Chapter 2. The second stage involves testing whether a B cell's immunoglobulin receptor will bind to normal constituents of the body and, therefore, has the potential to produce autoreactivity and autoimmune disease. In the third stage, mature, naive B cells that survive this selection process leave the bone marrow, enter the blood, and from there move into **secondary lymphoid tissues** (Figure 4.2). If they do not encounter their specific antigens in the lymphoid tissues, the B cells continue to recirculate. If contact is made with antigen in a secondary lymphoid tissue, the fourth stage of a B cell's life begins. The B cell proliferates and its progeny differentiate either into plasma cells, which synthesize large quantities of antibody, or into long-lived memory B cells, which will respond more quickly than naive B cells on a subsequent encounter with the same antigen. The last three stages of the B-cell life cycle will be dealt with in the second part of this chapter.

The development of B cells in the bone marrow

The stages of B-cell development in the bone marrow are marked by successive steps in the rearrangement and expression of the immunoglobulin genes.

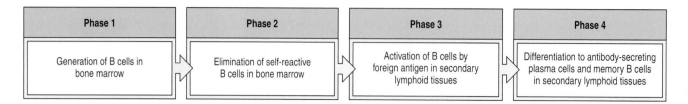

Phase 1	Phase 2	Phase 3	Phase 4
Generation of B cells in bone marrow	Elimination of self-reactive B cells in bone marrow	Activation of B cells by foreign antigen in secondary lymphoid tissues	Differentiation to antibody-secreting plasma cells and memory B cells in secondary lymphoid tissues

In this part of the chapter we shall see how gene rearrangement is controlled at each step to produce a mature but naive B cell that makes only one heavy chain and one light chain, and thus expresses immunoglobulin of a single antigen specificity on its surface.

Figure 4.1 The development of B cells can be divided into four broad phases.

4-1 B-cell development in the bone marrow proceeds through several stages

B cells derive from pluripotential hematopoietic stem cells in the bone marrow. The earliest identifiable cells of the B-cell lineage are called **pro-B cells**. These progenitor cells retain a limited capacity for self-renewal, dividing to produce both more pro-B cells and cells that will go on to develop further. Rearrangement of heavy-chain genes takes place first and this occurs in pro-B cells: D_H to J_H joining occurs at the **early pro-B cell** stage, followed by V_H to DJ_H joining at the **late pro-B cell** stage. A μ heavy chain is the first type of heavy chain to be produced. Once a B cell expresses a μ chain it is known as a **pre-B cell** (Figure 4.3). Pre-B cells represent two stages in B-cell development: the less mature **large pre-B cells** and the more mature **small pre-B cells**. Large pre-B cells are distinguished by a protein complex called the **pre-B-cell receptor**, which consists of μ heavy chains, surrogate light chains, which are only made in pre-B cells, and the Igα and Igβ polypeptides (Figure 4.4). Although some molecules of the pre-B-cell receptor are present at the cell surface, most are retained in the endoplasmic reticulum.

The presence of B-cell receptor molecules leads to intracellular signals that halt rearrangement at the immunoglobulin heavy-chain locus and synthesis of surrogate light chains. The large pre-B cell then proliferates to yield many small pre-B cells, in which the pre-B-cell receptor is no longer present, μ chains are restricted to the cytoplasm, and rearrangement of the immunoglobulin light-chain loci proceeds. Once light chains have been made, they assemble with μ chains to form IgM molecules, which are transported to the cell surface in the form of a functional B-cell receptor complex (see Section 2-11, p. 54). At this stage the B cell expresses IgM only and is defined as an **immature B cell**.

Up to this stage, the development of B cells occurs in the bone marrow and does not require interaction with specific antigen. The randomness of the gene rearrangement process leads to a proportion of the B-cell receptors being reactive against constituents of the body. Unless removed from the repertoire, B cells bearing these receptors have the potential to produce an autoimmune response, leading to autoimmune disease. The immature B cells now begin to be selected for tolerance of the normal constituents of the body, and at about this time they also start to enter the peripheral circulation. At this point in development, alternative mRNA splicing of heavy-chain gene transcripts produces IgD, as well as IgM, as membrane-bound immunoglobulin. The cell is now considered to be a **mature B cell**. Such cells are also called **naive B cells** because they have yet to be exposed to their specific antigen.

B-cell development in the bone marrow is dependent on a network of non-lymphoid **stromal cells**, which provide specialized microenvironments for

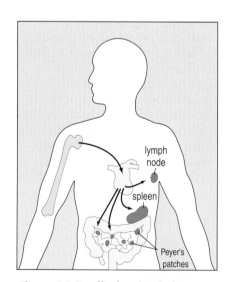

Figure 4.2 B cells develop in bone marrow and then migrate to secondary lymphoid tissues. B cells leaving the bone marrow (yellow) are carried in the blood to lymph nodes, the spleen, Peyer's patches (all shown in green), and other secondary lymphoid tissues such as those lining the respiratory tract (not shown).

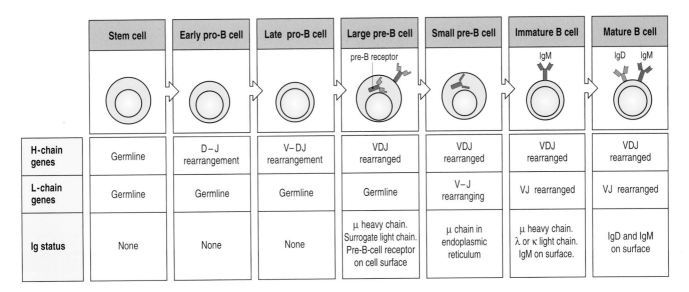

	Stem cell	Early pro-B cell	Late pro-B cell	Large pre-B cell	Small pre-B cell	Immature B cell	Mature B cell
H-chain genes	Germline	D–J rearrangement	V–DJ rearrangement	VDJ rearranged	VDJ rearranged	VDJ rearranged	VDJ rearranged
L-chain genes	Germline	Germline	Germline	Germline	V–J rearranging	VJ rearranged	VJ rearranged
Ig status	None	None	None	μ heavy chain. Surrogate light chain. Pre-B-cell receptor on cell surface	μ chain in endoplasmic reticulum	μ heavy chain. λ or κ light chain. IgM on surface.	IgD and IgM on surface

Figure 4.3 The development of B cells proceeds through stages defined by the rearrangement and expression of the immunoglobulin genes. In the stem cell, the immunoglobulin (Ig) genes are in the germline configuration. The first rearrangements are of the heavy-chain (H-chain) genes. Joining D_H to J_H defines the early pro-B cell, which becomes a late pro-B cell on joining V_H to DJ_H. Expression of a functional μ chain and its expression at the cell surface as part of the pre-B receptor defines the large pre-B cell. Large pre-B cells proliferate, producing small pre-B cells in which rearrangement of the light-chain (L-chain) gene occurs. Successful light-chain gene rearrangement and expression of IgM on the cell surface define the immature B cell. The mature B cell is defined by the use of alternative splicing of heavy-chain mRNA to place IgD on the cell surface as well as IgM.

B cells at various stages of maturation (Figure 4.5). The stromal cells perform two distinct functions. First, they make specific cell-surface contacts with the B cells through the interaction of adhesion molecules and their ligands. Second, they produce growth factors that act on the bound B cells, for example the membrane-bound, stem-cell factor (SCF), which is recognized by a receptor called Kit on immature B cells. Another important growth factor for B-cell development is interleukin-7 (IL-7), a cytokine secreted by stromal cells that acts on late pro-B and pre-B cells.

The most immature stem cells lie in a region of the bone marrow called the subendosteum, which is adjacent to the bone's interior surface. As the B cells mature, they move, while maintaining contact with stromal cells, to the central axis of the marrow cavity. Later stages of maturation are less dependent on contact with stromal cells, allowing the B cells to leave the bone marrow eventually. The final stage of development, when immature B cells become mature B cells, can occur either in the bone marrow or in secondary lymphoid organs such as the spleen.

Although B-cell development in humans and mice is quite similar, there are considerable differences in other species. For example, in chickens, immunoglobulin-gene rearrangement in the bone marrow provides little diversity. After completing heavy- and light-chain gene rearrangement, chicken B cells migrate to the bursa of Fabricius, a lymphoid organ near the cloaca of the gut. B cells were, in fact, originally named after this organ because the distinction between B cells and T cells was discovered from studying the chicken's immune response. In the bursa, B cells are signaled to

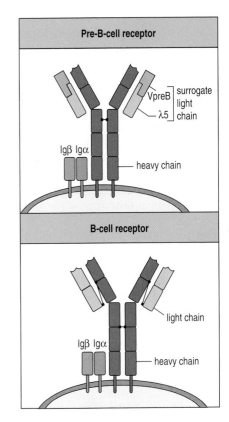

Figure 4.4 The pre-B-cell receptor resembles the B-cell receptor except for the surrogate light chain.

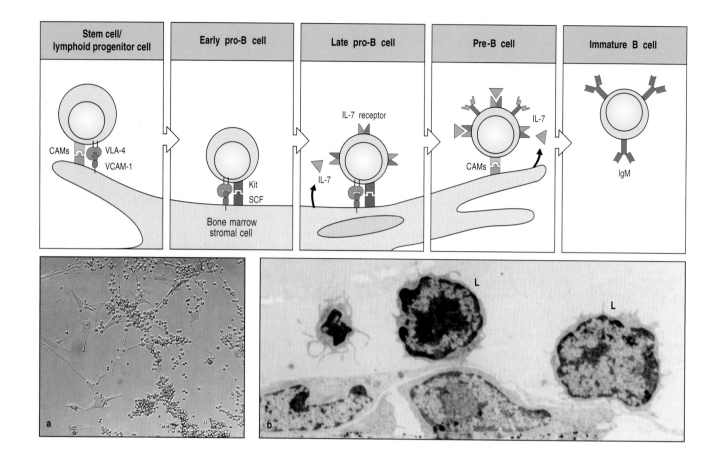

Figure 4.5 **The early stages of B-cell development are dependent on bone marrow stromal cells.** The top panels show the interactions of developing B cells with bone marrow stromal cells. Stem cells and early pro-B cells use the integrin VLA-4 to bind to the adhesion molecule VCAM-1 on stromal cells. This and interactions between other cell adhesion molecules (CAMs) promote the binding of the receptor Kit on the B cell to stem-cell factor (SCF) on the stromal cell. Activation of Kit causes the B cell to proliferate. B cells at a later stage of maturation require interleukin-7 (IL-7) to stimulate their growth and proliferation. Panel a is a light micrograph of a tissue culture showing small round B-cell progenitors in intimate contact with stromal cells, which have extended processes fastening them to the plastic dish on which they are grown. Panel b is a high-magnification electron micrograph showing two lymphoid cells (L) adhering to a flattened stromal cell. Photographs courtesy of A. Rolink (a); Paul Kincade and P.L. Witte (b).

proliferate and their rearranged heavy- and light-chain genes then diversify through gene conversion (Figure 4.6).

4-2 The survival of a developing B cell depends on the productive rearrangement of a heavy- and a light-chain gene

Because of imprecision in the gene rearrangement process, including the random addition of N and P nucleotides at joints between gene segments (see Section 2-8, p. 51), not all DNA rearrangements result in a sequence with a reading frame that can be translated into an immunoglobulin chain. Gene rearrangements that cannot be translated into a protein are called **unproductive rearrangements**. Those rearrangements that preserve a correct reading frame and give rise to a complete and functional immunoglobulin chain are known as **productive rearrangements**. At each rearrangement event, there is a one in three chance of the correct reading frame being maintained.

Every B cell has two copies of each of the immunoglobulin loci—the heavy-chain locus, the κ light-chain locus, and the λ light-chain locus. The two copies of each locus are on homologous chromosomes; one is inherited from the mother and the other from the father. In the developing B cell, gene rearrangements can be made on both homologous chromosomes. Therefore, a B cell that has made an unproductive rearrangement on one chromosome still has a chance of producing an immunoglobulin chain if it makes a productive rearrangement at the locus on the other homologous chromosome. Developing B cells are allowed to proceed to the next stage only when a productive rearrangement has been made. If all rearrangements are unproductive, the B cell does not produce immunoglobulin and dies in the bone marrow.

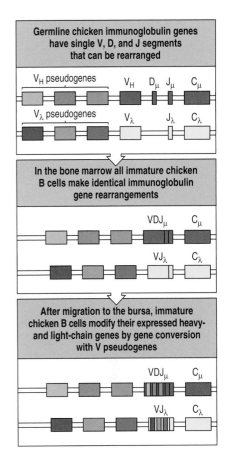

Figure 4.6 **Diversification of the immunoglobulin genes in chickens is achieved by gene conversion.** In chickens the development of mature B cells involves successive modifications of the immunoglobulin genes in two anatomical sites. In the bone marrow, the heavy- and light-chain genes are rearranged. This process produces little immunoglobulin diversity. In the bursa, diversity is introduced by gene conversion between the rearranged genes and a series of V_H and V_λ pseudogenes.

The order of gene rearrangements is outlined in Figure 4.7. The immunoglobulin heavy-chain locus rearranges before the light-chain loci. Recombination is heralded by expression of the **recombination activation genes RAG-1** and **RAG-2**. The first rearrangement event is the joining of a D_H gene segment to a J_H gene segment, which can occur concurrently in the two copies of the heavy-chain locus. B cells with unproductive D_H to J_H rearrangements on both chromosomes are prevented from developing further because, although they will undergo V_H to DJ_H rearrangement, they cannot produce a heavy chain. In humans, few B cells are lost from unproductive rearrangements at this stage because most human D segments can be translated in all three reading frames. In contrast, mouse D segments have a single productive reading frame.

The second event in heavy-chain gene rearrangement is the joining of a V_H gene segment to the rearranged DJ_H sequence. This event first occurs at only one of the heavy-chain loci. When V_H to DJ_H rearrangement on the first chromosome is unproductive, rearrangement then proceeds on the second chromosome. The two-thirds failure rate in maintaining the reading frame, and the independent chances for success offered by having two chromosomes, mean that just over half the total number of pro-B cells make a functional heavy-chain gene. They become large, dividing pre-B cells producing

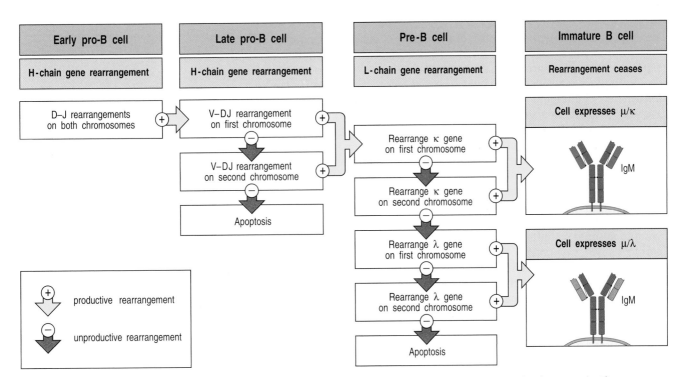

Figure 4.7 **The order of gene rearrangements leading to the expression of cell-surface immunoglobulin.** The heavy-chain genes are rearranged before the light-chain genes. Developing B cells are allowed to proceed to the next stage only when a productive rearrangement has been made. If an unproductive rearrangement is made on one chromosome of a homologous pair, then rearrangement is attempted on the second chromosome.

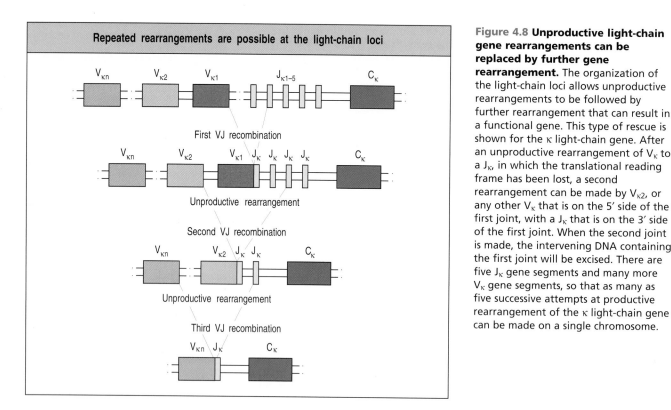

Repeated rearrangements are possible at the light-chain loci

First VJ recombination

Unproductive rearrangement

Second VJ recombination

Unproductive rearrangement

Third VJ recombination

Figure 4.8 Unproductive light-chain gene rearrangements can be replaced by further gene rearrangement. The organization of the light-chain loci allows unproductive rearrangements to be followed by further rearrangement that can result in a functional gene. This type of rescue is shown for the κ light-chain gene. After an unproductive rearrangement of V_κ to a J_κ, in which the translational reading frame has been lost, a second rearrangement can be made by $V_{\kappa2}$, or any other V_κ that is on the 5′ side of the first joint, with a J_κ that is on the 3′ side of the first joint. When the second joint is made, the intervening DNA containing the first joint will be excised. There are five J_κ gene segments and many more V_κ gene segments, so that as many as five successive attempts at productive rearrangement of the κ light-chain gene can be made on a single chromosome.

μ chains. Those pro-B cells that fail to make μ chains are programmed to die, a process called **apoptosis**. A general feature of lymphocyte development is for apoptosis to be the default pathway that is followed unless a positive signal for further differentiation is received.

The large pre-B cells go through five or six rounds of cell division, after which they become small, resting pre-B cells. At this stage, the light-chain genes begin to rearrange. Rearrangements take place at one light-chain locus at a time, with the κ locus tending to be rearranged before the λ locus. In contrast to the heavy-chain locus, several attempts to rearrange the same light-chain gene can be made by using V and J gene segments not involved in previous rearrangements (Figure 4.8). A number of attempts to rearrange the κ light-chain locus on one or other chromosome can, therefore, be made before commencing rearrangement of a λ-chain locus on a different chromosome. The possession of four light-chain loci, and the opportunity for making several attempts to rearrange each locus, means that about 85% of the pre-B cell population makes a successful rearrangement of a light-chain gene. Overall, less than half of the B-lineage cells end up producing functional immunoglobulin heavy and light chains.

Before each rearrangement there is a small amount of transcription from the individual gene segments that are about to be joined. This opens up the chromatin structure and makes the DNA accessible to the enzymes that carry out somatic recombination. It is likely that this low-level transcription is initiated by transcription factors expressed only in B cells, and is the mechanism that directs the recombination enzymes common to both B cells and T cells to the immunoglobulin genes rather than to the T-cell receptor genes (and vice versa in T cells). In addition, the gene rearrangement process brings promoter and enhancer within the locus into closer juxtaposition with each other so that the rearranged immunoglobulin gene can now be transcribed at a high rate.

4-3 Cell-surface expression of the products of rearranged immunoglobulin genes prevents further gene rearrangement

Given that a B cell can rearrange up to two heavy-chain loci and four light-chain loci, there are mechanisms for ensuring that it ends up making only one type of heavy chain and one type of light chain. The success of a gene rearrangement is signaled by the appearance of the protein product of the gene at the cell surface; once that has occurred, a signal is sent back to the interior of the cell to shut down the processes of DNA recombination and repair that are needed for gene rearrangement. The RAG genes are turned off and no further gene rearrangement is possible at this time. Once one heavy-chain gene has been successfully rearranged, all further rearrangement of heavy-chain genes is shut down, and the same happens with the light-chain genes (Figure 4.9).

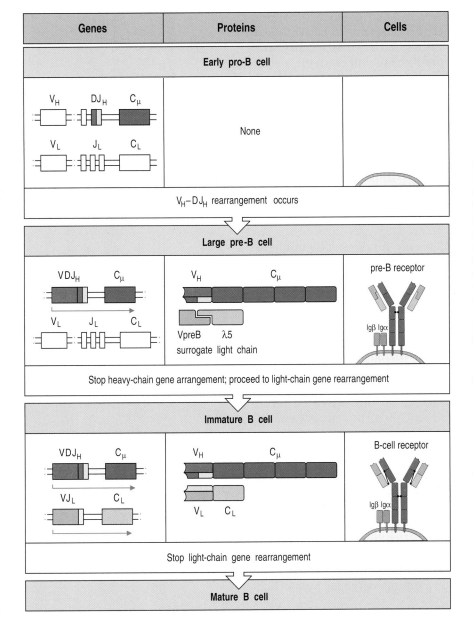

Figure 4.9 **Signaling via the functional protein product is used to terminate immunoglobulin-gene rearrangement.** First panel: in early pro-B cells, no functional μ protein is expressed and the biochemical machinery that performs gene rearrangement is active. Second panel: on completion of a productive heavy-chain gene rearrangement, μ heavy chains are synthesized and assembled into a complex with the λ5 and VpreB polypeptides that make up the surrogate light chain, and with Igα and Igβ. This complex, the pre-B-cell receptor, is transported to the cell surface, where it interacts with an unknown ligand that signals the cell to halt heavy-chain gene rearrangement and to proceed to the large pre-B-cell stage. Proliferation of the large pre-B cell yields small pre-B cells that commence immunoglobulin light-chain rearrangement. Note that λ5 and VpreB are coded by genes that are distinct from those in the immunoglobulin loci. Third panel: on completion of a productive light-chain gene rearrangement a light chain is made and assembles with μ to form IgM. This associates with Igα and Igβ and is transported to the cell surface. Its presence tells the cell to halt light-chain gene rearrangements.

Once it has formed a functional heavy-chain gene, a pro-B cell begins to synthesize μ chains. The μ chains assemble into dimers in the endoplasmic reticulum of the cell. These cannot, however, assemble with κ or λ light chains to form a regular IgM receptor because the B cell has not yet rearranged its light-chain genes. The receptor that reaches the surface at this stage contains a μ dimer, in which each μ chain is bound to an invariant **surrogate light chain**. Each surrogate light chain is formed from two proteins coded by genes away from the immunoglobulin loci; **λ5** substitutes for a light-chain constant region and **VpreB** substitutes for a variable region.

The tetramer of two μ chains and two surrogate light chains associates with the Igα and Igβ polypeptides (see Section 2-11, p. 54) to form a complex that moves to the cell surface to become the pre-B-cell receptor. The pre-B-cell receptor is present at the cell surface for only a short period of time, which represents a checkpoint in B-cell development. This checkpoint allows the B cell to confirm whether a productive rearrangement has occurred and a functional heavy chain has been made. Expression of the pre-B-cell receptor generates the positive signal that allows the B cell to avoid apoptosis and proceed to the next stage of maturation. After the checkpoint, the surrogate light chain ceases to be synthesized and the pre-B-cell receptor gradually disappears from the cell surface. The μ, Igα, and Igβ chains continue to be synthesized but are now retained in the endoplasmic reticulum. The actual mechanism by which the pre-B-cell receptor signals the pre-B cell to shut down the machinery of somatic recombination and gear up the biosynthetic machinery for DNA replication and cell division is still unknown.

The RAG genes, which were turned off in the dividing pre-B cells, are now turned on again and light-chain gene rearrangement proceeds. When a functional light-chain gene has been formed, light chains are made and assemble with μ to form IgM. This associates with Igα and Igβ to form the mature B-cell receptor, which moves to the cell surface. At this second checkpoint in B-cell maturation, the appearance of a functional B-cell receptor leads to the shutting down of light-chain gene rearrangement.

Cell division initiated by signaling through the pre-B-cell receptor results in a clone of 30–70 small pre-B cells. These all have the same rearranged heavy-chain gene but each has the potential to end up with a different light chain. The combinations of different heavy and light chains generated at this stage contribute to the diversity of the B-cell repertoire.

4-4 The proteins involved in immunoglobulin-gene rearrangement are controlled developmentally

The successive steps in immunoglobulin-gene rearrangement are accompanied by changes in the expression of the other proteins involved in the recombination reactions (Figure 4.10). The RAG-1 and RAG-2 proteins are essential components of the recombination machinery (see Section 4-2, p. 102), and their genes are specifically turned on at the stages in B-cell development when heavy- or light-chain gene rearrangement occurs. Conversely, at the large pre-B cell stage, after rearrangement of a heavy-chain gene and before rearrangement of a light-chain gene, RAG proteins are absent. Terminal deoxynucleotidyl transferase (TdT), the enzyme that adds N nucleotides at the junctions between rearranging gene segments, is expressed in pro-B cells when the heavy-chain genes begin to rearrange, and turned off in small pre-B cells when the light-chain genes begin to rearrange; this explains why N nucleotides are found in all VD and DJ joints of rearranged human heavy-chain genes but in only about half of the VJ joints of rearranged human light-chain genes.

Several proteins contribute to B-cell development by transducing signals from cell-surface receptors. Synthesis of the Igα and Igβ polypeptides is turned on at the pro-B cell stage and continues throughout the life of a B cell, which is consistent with these polypeptides being necessary for the cell-surface expression of immunoglobulin and for the transduction of signals from pre-B cell and B-cell receptors to the interior of the cell. Only when antigen-stimulated B cells differentiate into antibody-secreting plasma cells, and their antigen receptors are no longer required, are the Igα and Igβ genes turned off and immunoglobulin ceases to appear on the cell surface.

Another signal transduction molecule is the Bruton's tyrosine kinase (Btk), which is encoded by a gene on the X chromosome and is essential for B-cell maturation. Patients who lack a functional Btk gene have almost no circulating antibodies because their B cells are blocked at the pre-B-cell stage. The immune deficiency suffered by these patients is called **X-linked agamma-globulinemia** and leads to recurrent infections from common extracellular bacteria such as *Haemophilus influenzae*, *Streptococcus pneumoniae*, *Streptococcus pyogenes*, and *Staphylococcus aureus*. The infections respond to treatment with antibiotics and can be prevented by regular intravenous infusions of immunoglobulin from pooled blood from healthy donors. Because the Btk gene is located on the X chromosome and the defective gene is recessive, X-linked agammaglobulinemia is seen mostly in boys.

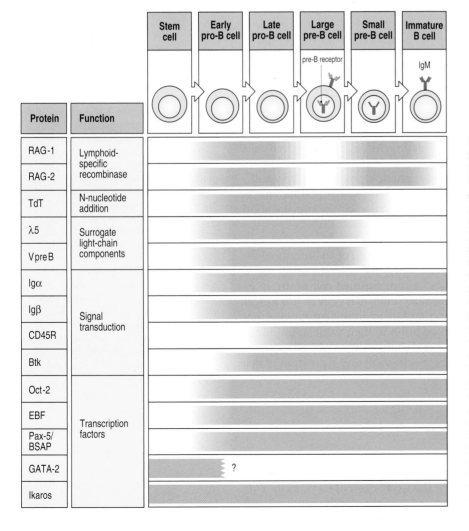

Figure 4.10 The expression of proteins involved in the rearrangement and expression of immunoglobulin genes changes during B-cell development. The rearrangement of immunoglobulin genes and the expression of the pre-B-cell receptor and IgM on the cell surface requires several categories of specialized proteins at different times during B-cell development. Examples of such proteins are listed here, with their expression during B-cell development shown with red shading. Of the proteins not discussed in the text, EBF regulates the transcription of the gene for Igα, and the B-lineage specific-activator protein (BSAP) regulates the expression of several of the other proteins listed. Oct-2 acts at the heavy-chain promoter and GATA-2 is a transcription factor active in several types of hematopoietic cells. CD45 is a cell-surface phosphotyrosine phosphatase that is restricted to hematopoietic cells. It is involved in regulating signal transduction from the pre-B-cell and B-cell receptors.

A variety of transcription factors control the genes involved in immunoglob-ulin-gene rearrangement and B-cell development, only some of which are shown in Figure 4.10. Their coordinated activity ensures that the enzymatic machinery for gene rearrangement common to B and T cells is used in B cells to rearrange and express the immunoglobulin genes, and not the T-cell recep-tor genes.

4-5 Many B-cell tumors carry chromosomal translocations that join immunoglobulin genes to genes regulating cell growth

As B cells cut, splice, and mutate their immunoglobulin genes in the normal course of events, it is hardly surprising that this process sometimes goes awry to produce a mutation that helps convert a B cell into a tumor cell. Transfor-mation of a normal cell into a tumor cell involves a series of mutations that release the cell from the normal restraints on its growth. In B-cell tumors, the disruption of regulated growth is often associated with an aberrant immunoglobulin-gene rearrangement that has joined an immunoglobulin gene to a gene on a different chromosome. Events that fuse part of a chromo-some with another are called **translocations** and, in B-cell tumors, the immunoglobulin gene has often become joined to a gene involved in the control of cellular growth. Genes that cause cancer when their function or expression is perturbed are collectively called **proto-oncogenes**. Many were discovered in the study of RNA tumor viruses that can transform cells directly. The viral genes responsible for transformation were named **oncogenes**, and it was only later realized that they had evolved from cellular genes that control cell growth, division, and differentiation.

The translocations between immunoglobulin genes and proto-oncogenes in B-cell tumors can be seen in metaphase chromosomes examined in the light microscope. Certain translocations define particular types of tumor and are valuable in diagnosis. In Burkitt's lymphoma, for example, the MYC proto-oncogene on chromosome 8 is joined by translocation to either an immunoglobulin heavy-chain gene on chromosome 14, a κ light-chain gene on chromosome 2, or a λ light-chain gene on chromosome 22 (Figure 4.11). The Myc protein is normally involved in regulating the cell cycle, but in B cells that carry these translocations its expression is abnormal, which removes some of the restraints on cell division. Single genetic changes are rarely suffi-cient to transform cells malignantly. To produce Burkitt's lymphoma, muta-tions elsewhere in the genome are required in addition to the translocation between MYC and an immunoglobulin gene.

Another translocation found in B-cell tumors is the fusion of an immunoglob-ulin gene to the proto-oncogene BCL2. Normally, the function of the Bcl-2 protein is to prevent premature apoptosis, or programmed cell death, in B-lineage cells. Production of Bcl-2 in the tumors enables B cells to live longer than is normal, during which time they accumulate additional mutations that can lead to malignant transformation. In addition to their value in immuno-logical research, B-cell tumors have provided fundamental advances in knowledge of the proteins involved in the mechanism and regulation of cell division.

4-6 B cells expressing the glycoprotein CD5 express a distinctive repertoire of receptors

Not all B cells conform exactly to the developmental pathway described above. A subset of human B cells that arises early in embryonic development is distinguished from the other B cells by the expression of CD5, a cell-sur-face glycoprotein that is otherwise considered a marker for the human T-cell lineage. This minority subset of B cells is termed **B-1 cells** because their

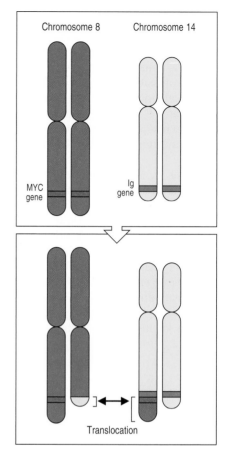

Figure 4.11 Chromosomal rearrange-ments in Burkitt's lymphoma. In this example from a Burkitt's lymphoma, parts of chromosome 8 and chromosome 14 have been exchanged. The sites of breakage and rejoining are in the proto-oncogene MYC on chromosome 8 and the immunoglobulin heavy-chain gene on chromosome 14. In these tumors it is usual for the second immunoglobulin gene to be productively rearranged and for the tumor to express cell-surface immunoglobulin. The translocation probably occurred during the first attempt to rearrange a heavy-chain gene. This counted as an unproductive rearrangement and so the other gene was rearranged. In cases where the second rearrangement is also unproductive, the cell dies and thus cannot give rise to a tumor.

development precedes that of the majority subset, whose development we have traced in previous sections and, in this context, are sometimes called **B-2 cells**. B-1 cells are also characterized by little or no IgD on the surface and a distinctive repertoire of antigen receptors. The B-1 cells are also known as **CD5 B cells**, although the CD5 molecule is not essential for their function: B-1 cells develop normally in mice lacking CD5 glycoprotein and rat B-1 cells never express CD5. B-1 cells are the dominant B cells in the pleural and peritoneal cavities. Although comprising only around 5% of the B cells in humans and mice, B-1 cells are the main type of B cells in rabbits.

B-1 cells arise from a stem cell that is most active in the prenatal period. Their immunoglobulin heavy-chain gene rearrangements are dominated by the use of the V_H gene segments that lie closest to the D gene segments in the germline. As TdT is not expressed early in the prenatal period, the rearranged heavy-chain genes of B-1 cells are characterized by a lack of N nucleotides, and their VDJ junctions are less diverse than those in rearranged heavy-chain genes of B-2 cells. Consequently, the antibodies secreted by B-1 cells tend to be of low affinity and each binds to many different antigens, which is known as **polyspecificity**. B-1 cells contribute to the antibodies made against common bacterial polysaccharides and other carbohydrate antigens but are of little importance in making antibodies against protein antigens.

Those B-1 cells that develop postnatally use a more diverse repertoire of V gene segments and their rearranged immunoglobulin genes have abundant N nucleotides. With time, B-1 cells are no longer produced by the bone marrow and, in adults, the population of B-1 cells is maintained by the division of existing B-1 cells at sites in the peripheral circulation. This self-renewal is dependent on the cytokine IL-10. The properties of B-1 and B-2 cells are compared in Figure 4.12. In the remainder of this chapter, and in the rest of

Property	B-1 cells	Conventional B-2 cells
When first produced	Fetus	After birth
N-regions in VDJ junctions	Few	Extensive
V-region repertoire	Restricted	Diverse
Primary location	Peritoneal and pleural cavities	Secondary lymphoid organs
Mode of renewal	Self-renewing	Replaced from bone marrow
Spontaneous production of immunoglobulin	High	Low
Isotypes secreted	IgM >> IgG	IgG > IgM
Response to carbohydrate antigen	Yes	Maybe
Response to protein antigen	Maybe	Yes
Requirement for T-cell help	No	Yes
Somatic hypermutation	Low–none	High
Memory development	Little or none	Yes

Figure 4.12 Comparison of the properties of B-1 cells and B-2 cells. B-1 cells develop in the omentum as well as in the liver in the fetus, and are produced by the bone marrow for only a short period around the time of birth. A pool of self-renewing B-1 cells is then established, which does not require the microenvironment of the bone marrow for its survival. The limited diversity of the antibodies made by B-1 cells and their tendency to be polyreactive and of low affinity suggests that B-1 cells produce a simpler, less adaptive, immune response than that involving B-2 cells.

this book, B cell will refer to the population of B-2 cells, unless otherwise specified.

Possibly as a direct result of their capacity for self-renewal, B-1 cells are usually the source for the common B-cell tumors causing chronic lymphocytic leukemia (CLL). A bone marrow transplant from an HLA-identical sibling is often a successful treatment for CLL. Chemotherapy and/or radiation are used to destroy the dividing cells of the patient's hematopoietic system, which includes the malignant B-1 cells. The donor bone marrow then provides the stem cells that repopulate the immune system with healthy cells.

Summary

B cells are generated throughout life from stem cells in the bone marrow. Different stages in B-cell maturation are correlated with molecular changes that accompany the gene rearrangements required to make functional immunoglobulin heavy and light chains. Many rearrangements are unproductive, an inefficiency compensated for, in part, by the presence of two heavy-chain loci and four light-chain loci in each B cell, all of which can be rearranged in the attempt to obtain a functional heavy chain and light chain. The heavy-chain genes are rearranged first, and only when this produces a functional protein is a B cell allowed to rearrange its light-chain genes. After a successful heavy-chain gene rearrangement, the μ chain and surrogate light chain form the pre-B-cell receptor, which signals the cell to halt heavy-chain gene rearrangement and to proceed toward light-chain gene rearrangement. Similarly, after a successful light-chain gene rearrangement, the assembly of IgM signals the cell to halt light-chain gene rearrangement. These feedback mechanisms, whereby protein product regulates the progression of gene rearrangement, ensure that each B cell expresses only one heavy chain and one light chain and, thus, produces immunoglobulin of a single defined antigen specificity. In this manner, the program of gene rearrangement produces the monospecificity essential for the efficient operation of clonal selection. The success rate in obtaining a functional light-chain gene rearrangement is much higher than that for the heavy-chain gene because only two gene segments are involved (compared with three for the heavy-chain gene) and the B cell has four light-chain genes (two κ and two λ) with which to work. Errors in immunoglobulin-gene rearrangement give rise to chromosomal translocations that predispose B cells carrying them to malignant transformation. A minority class of B cells, termed B-1 cells, develops very early in embryonic life. They produce antibodies that tend to bind to bacterial polysaccharides, but are polyspecific, generally of low affinity, and mainly of the IgM isotype. In aggregate, the properties of B-1 cells suggest that they represent a simpler and evolutionarily older lineage of B cell than the majority of the B-cell population.

Selection and further development of the B-cell repertoire

In the previous part of this chapter we saw how the population of immature B cells develops a repertoire of receptors of different antigen specificities. Included in this repertoire are immunoglobulins that bind to normal constituents of the body and have the potential to initiate a damaging immune response. To prevent such responses, mechanisms that either eliminate or inactivate potentially autoreactive B cells now come into play. Once the population of maturing B cells has been purged of self-reactive cells the remaining cells are released to the circulation, where they can engage specific antigen and perform their effector function of producing antibodies.

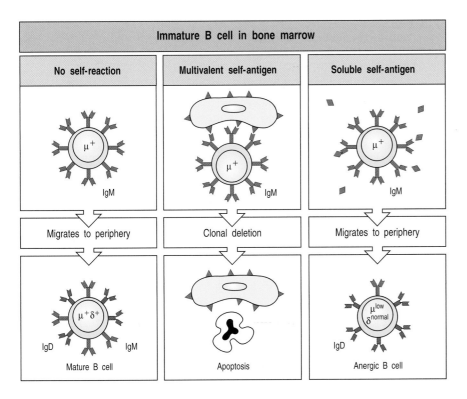

Figure 4.13 Binding to self-antigens in the bone marrow can lead to the deletion or inactivation of immature B cells. Left panels: immature B cells that do not encounter a stimulatory self-antigen leave the bone marrow and enter the peripheral circulation, expressing both IgM and IgD on their surface. Middle panels: when immature B cells express receptors that recognize common cell-surface components of human cells, they are deleted from the repertoire by the induction of apoptosis. Right panels: immature B cells that bind soluble self-antigens are rendered unresponsive or anergic to the antigen, and, as a consequence, express low levels of IgM at the cell surface. They enter the peripheral circulation, where they express IgD but remain anergic.

4-7 Self-reactive immature B cells are altered, eliminated, or inactivated by contact with self-antigens

When a B cell first expresses IgM on its surface, it is called an immature B cell. Becoming a mature B cell involves emigration from the bone marrow and the use of alternative splicing of heavy chain mRNA to place IgD, as well as IgM on the cell surface (Figure 4.13, left panels; see Section 2-9, p. 52). Quality control mechanisms prevent the maturation of B cells whose receptors bind to normal components of the human body, which in this immunological context, are called **self-antigens**. If such B cells were allowed to mature they could make potentially disease-causing antibodies. One category of self-antigens consists of the glycoproteins, proteoglycans and glycolipids that are expressed on the surfaces of human cells and are accessible for interaction with the B-cell receptors. Immature B cells that are specific for surface antigens of human cells either change the specificity of their B-cell receptor, by mechanisms that are described below, or they become programmed to die through apoptosis (Figure 4.13, center panels).

A second category of common self-antigens consists of soluble proteins and glycoproteins, some of which are present at very high concentrations in the blood and lymph through which B cells circulate. When the surface IgM of an immature B cell binds to a soluble self-antigen, the B cell is inactivated but it does not die. This state, in which the B cell matures but does not respond to subsequent exposure to antigen, is called **anergy** (Figure 4.13, right panels). In anergic cells, most of the IgM is retained inside the cell, whereas normal levels of IgD are present at the cell surface. However, binding of antigen to the surface IgD does not activate the B cell.

When an immature B cell's surface IgM molecules are cross-linked by a self-antigen on another cell, the B cell receives signals that arrest development. The amount of B-cell receptor on the cell surface is reduced and RAG protein synthesis stays high. This allows the B cell to continue rearranging the light-chain genes. Such rearrangements lead to production of a new light chain and

Figure 4.14 Replacement of light chains by receptor editing can rescue some self-reactive B cells by changing their antigen specificity. When a developing B cell produces antigen receptors that are strongly cross-linked by multivalent self-antigens, such as major histocompatibility complex molecules on cell surfaces (top panel), the B cell undergoes developmental arrest. The amount of IgM on the cell surface is reduced and the RAG genes are not turned off (second panel). Continued synthesis of RAG proteins allows the cell to continue light-chain gene rearrangement. This usually leads to a new productive rearrangement and expression of a new light chain, which combines with the previous heavy chain to form a new receptor (receptor editing; third panel). If this new receptor is not self-reactive, the cell is 'rescued' and continues normal development much like a cell that had never reacted with self (bottom right panel). If the cell remains self-reactive, it may be rescued by another cycle of rearrangement, but if it continues to react strongly with self it will undergo programmed cell death or apoptosis and be deleted from the repertoire (clonal deletion; bottom left panel).

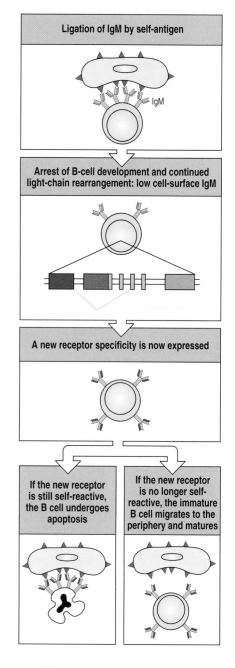

can prevent further production of the original light chain. If the immunoglobulin formed by the new light chain and the old heavy chain is not reactive with a self-antigen, then the B cell develops normally. If self-reactivity is retained, then further light-chain gene rearrangements are tried. This process of assessing the compatibility of receptors produced from successive gene rearrangements is called **receptor editing** (Figure 4.14), The multiplicity of V and J segments in the κ and λ light-chain genes provides a considerable capacity for receptor editing. Inevitably, however, some B cells fail to gain a new receptor that is not self-reactive and they succumb to apoptosis, which, in this context, is called **clonal deletion** (Figure 4.13, center panels). The apoptotic B cells are then phagocytosed by macrophages. Clonal deletion occurs within the bone marrow or shortly after the B cell enters the peripheral circulation. Some 55 billion B cells die each day in the bone marrow because they fail to make a functional immunoglobulin, or are autoreactive and subject to clonal deletion. With the presence of both IgD and IgM on the cell surface the mature B cell becomes programmed to make an immune response when it encounters its specific antigen.

4-8 Mature, naive B cells compete for access to lymphoid follicles

When mature B cells leave the bone marrow, they recirculate between the blood, the secondary lymphoid tissues, such as lymph nodes, spleen, and mucosa-associated lymphoid tissues, and the lymph (see Figure 1.15, p. 15). Within secondary lymphoid tissues, B cells congregate in organized structures called **primary lymphoid follicles** (Figure 4.15). These consist principally of B cells and a specialized stromal cell called the follicular dendritic cell. Despite their name, which comes from the long cell processes with which they touch B cells, follicular dendritic cells are unrelated to the dendritic cells that present antigens to T cells and they are not of hematopoietic origin. Mature B cells recirculate, passing from the blood into the primary lymphoid follicles and back through the lymphatic system into the blood (Figure 4.16). In the spleen, B cells enter and exit via the blood. B cells are particularly enriched in the gut-associated lymphoid tissues, which include the large lymphoid follicles known as Peyer's patches, and also the appendix and tonsils. These tissues in the gastrointestinal tract provide microenvironments in which B cells become committed to the synthesis of dimeric, secretory IgA.

Before it has made contact with its specific antigen, a mature, naive B cell must pause periodically in a primary follicle in order to survive and continue its recirculation. In the follicle it receives survival signals that may come directly from the follicular dendritic cells. Throughout a person's life, the bone marrow continues to produce new B cells that must compete with the existing population of naive B cells in the peripheral circulation. In the bone marrow of a healthy young adult, about 2.5 billion (2.5×10^9) cells per day embark on the program of B-cell development. From these progenitors, some 30 billion B cells leave the bone marrow daily to become circulating naive B cells.

There is continual competition between circulating naive B cells for passage through a limited number of follicular sites; those B cells that fail to gain regular access to a follicle die. Competition is so intense that the majority of mature, naive B cells die after only a few days in the peripheral circulation. Naive B cells that gain access to primary follicles live longer, but they too disappear from the system with a half-life of 3–8 weeks, unless they are stimulated by encounter with their specific antigen.

Anergic B cells that reach secondary lymphoid tissue are detained in the T-cell areas outside the primary lymphoid follicles and are, thus, prevented from entering the follicles. Because anergic B cells cannot be activated, they rapidly undergo apoptosis from lack of stimulation. This ensures that the population of recirculating mature B cells is continuously purged of potentially self-reactive cells.

4-9 Encounter with antigen leads to the differentiation of activated B cells into plasma cells and memory B cells

Secondary lymphoid tissues provide the sites where mature, naive B cells encounter specific antigen. When that happens, the antigen-specific B cells are detained in the T-cell areas, where they become activated by antigen-specific, CD4 helper T cells. These T cells provide signals that activate the B cells to proliferate and differentiate further. In lymph nodes and spleen some of the activated B cells proliferate and differentiate immediately into **plasma cells**, which secrete antibody (Figure 4.17). Antibody secretion is effected by a change in the processing of the heavy-chain mRNA, which

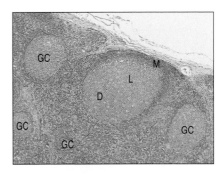

Figure 4.15 Anatomy of lymphoid follicles. This section through a human lymph node shows secondary lymphoid follicles, each of which has developed a germinal center (GC). Three zones can be recognized in a follicle, an outer mantle zone of resting B cells (M), a dark zone of proliferating blasts (D), and a light zone of differentiating cells (L). The areas between the follicles contains T cells, interdigitating cells (antigen-presenting), blood vessels, and sinuses. Courtesy of N. Rooney.

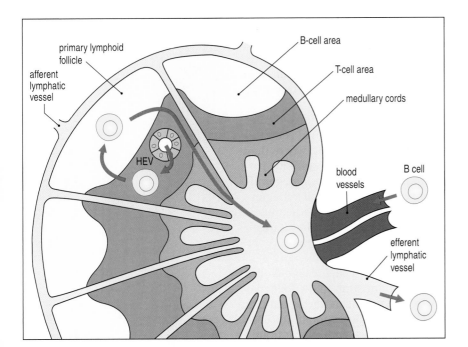

Figure 4.16 The circulation route of mature, naive B cells through a lymph node. Having matured in the bone marrow, B cells migrate in the blood to lymph nodes and other secondary lymphoid tissues. B cells leave the blood and enter the cortex of the lymph node through the walls of specialized high endothelial venules (HEV). If they do not encounter their specific antigen, the B cells pass through the primary follicles and leave the node in the efferent lymph, which eventually joins the blood at veins in the neck.

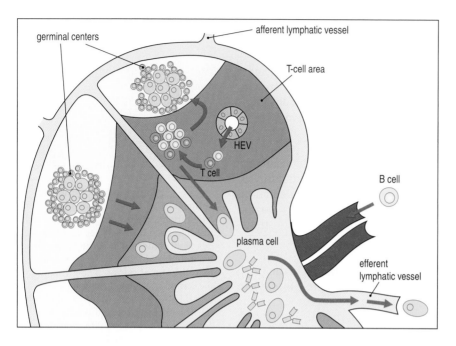

germinal centers

afferent lymphatic vessel

T-cell area

HEV

T cell

B cell

plasma cell

efferent
lymphatic vessel

Figure 4.17 B cells encountering antigen in secondary lymphoid tissues form germinal centers and undergo differentiation to plasma cells. A lymph node is illustrated here. A B cell entering the lymph node through a HEV encounters antigen in the lymph node cortex. Antigen was delivered in the afferent lymph that drained from infected tissue. The B cell is activated by CD4 helper T cells (blue) in the T-cell areas to form a primary focus of dividing cells. From this, some B cells migrate directly to the medullary cords and differentiate into antibody-secreting plasma cells. Other B cells migrate into a primary follicle to form a germinal center. B cells continue to divide and differentiate within the germinal center. Activated B cells migrate from the germinal center to the medulla of the lymph node or to the bone marrow to complete their differentiation into plasma cells.

leads to the synthesis of the secreted form of immunoglobulin rather than the membrane-bound form (see Section 2-12, p. 55).

Plasma cells are totally specialized toward the constitutive synthesis and secretion of antibody; the cellular organelles involved in protein synthesis and secretion become highly developed, and 10–20% of the total cell protein made is antibody. As part of this program of terminal differentiation, plasma cells cease to divide, have a limited life-span of about 4 weeks, and no longer express cell-surface immunoglobulin and MHC class II molecules. Thus, they become unresponsive to antigen and to interaction with T cells.

Other activated B cells migrate to a nearby primary follicle, which changes its morphology to become a **secondary lymphoid follicle** containing a **germinal center** (see Figure 4.17). Here, the activated B cells become large proliferating lymphoblasts called **centroblasts**; these mature into non-dividing B cells called **centrocytes**, which have undergone isotype switching and somatic hypermutation. Those B cells that make surface immunoglobulins with the highest affinity for the antigen are selected by the process of affinity maturation (see Section 2-13, p. 56), which occurs in germinal centers. A few weeks after a germinal center is formed, the intense cellular activity, called the germinal center reaction, dies down and the germinal center shrinks in size.

Cells that survive the selection process of affinity maturation undergo further proliferation as lymphoblasts and most migrate from the germinal center to other sites in the secondary lymphoid tissue or to bone marrow, where they complete their differentiation into plasma cells secreting high-affinity, isotype-switched antibodies. As the primary immune response subsides, germinal center B cells also develop into quiescent resting **memory B cells** possessing high-affinity, isotype-switched antigen receptors. The production of memory cells after a successful encounter with antigen establishes antigen specificities of proven usefulness permanently in the B-cell repertoire.

Memory B cells persist for long periods of time, and in their recirculation through the body they require only intermittent stimulation in the follicular environment. They are much more easily activated on encountering antigen than naive B cells. Their rapid activation and differentiation into plasma cells on a subsequent encounter with antigen enable a secondary antibody response to an antigen to develop more quickly and become stronger than the primary immune response. It also explains why IgG and antibodies of isotype other than IgM predominate in secondary responses.

When B cells become committed to differentiation into plasma cells, they migrate to particular sites in the lymphoid tissues. In the lymph nodes these are the medullary cords and in the spleen these are the red pulp. In the gut-associated lymphoid tissues, prospective plasma cells migrate to the lamina propria, which lies immediately under the gut epithelium. Prospective plasma cells also migrate from lymph nodes and spleen to the bone marrow, which becomes a major site of antibody production. So, in one sense, the life of a B cell both starts and ends in the bone marrow.

4-10 Different types of B-cell tumor reflect B cells at different stages of development

The study of B-cell tumors has provided fundamental insights into both B-cell development and the control of cell growth generally. B-cell tumors arise from both the B-1 and B-2 lineages and from B cells at different stages of maturation and differentiation. The general principle that a tumor represents the uncontrolled growth of a single transformed cell is illustrated vividly by tumors derived from B-lineage cells. In a B-cell tumor, every cell has an identical immunoglobulin-gene rearrangement, which is proof of their derivation from the same ancestral cell. Although the cells of an individual B-cell tumor are homogeneous, the B-cell tumors from different patients have different rearrangements, which reflects the multiplicity of rearrangements found in the normal B cells of a healthy person (Figure 4.18).

Tumors retain characteristics of the cell type from which they arose, especially when the tumor is relatively differentiated and slow growing. This principle is exceptionally well illustrated by the B-cell tumors. Human tumors corresponding to all the stages of B-cell development have been described, from the most immature progenitor to the highly differentiated plasma cell (Figure 4.19). Among the characteristics retained by the tumors is their location at defined sites in the lymphoid tissues. Tumors derived from mature, naive B cells grow in the follicles of lymph nodes and form **follicular center cell lymphoma**, whereas plasma-cell tumors, called **myelomas**, propagate in the bone marrow.

Although **Hodgkin's disease** was one of the first tumors to be successfully treated by radiotherapy, the tumor's origin as a germinal center B cell was only recently discovered. As a result of a somatic mutation, the tumor cells no longer have an antigen receptor. They often have a strange dendritic morphology and have been known as Reed-Sternberg cells. The disease presents clinically in several forms. In some patients it is dominated by non-malignant T cells that are stimulated by the tumor cells, whereas in others, there are both Reed-Sternberg cells and more normal-looking malignant B cells that have identical immunoglobulin-gene rearrangements.

B-cell tumors have proved invaluable in the study of the immune system. They uniquely provide cells that largely represent what happens normally, but can be obtained in a large quantity, which is highly abnormal. The very first amino-acid sequences of antibody molecules were obtained from patients with plasma-cell tumors, whose bodily fluids are dominated by a single species of antibody. B-cell tumors from mice have also been instrumental in defining the pathways of migration and recirculation of B cells. They provide defined cells that can be grown, manipulated, and modified in the laboratory and then tested for their properties *in vivo*.

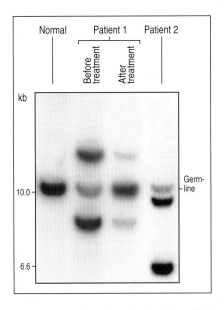

Figure 4.18 Clonal analysis of B-cell and T-cell tumors. DNA analysis of tumor cells by Southern blotting techniques can detect and monitor lymphoid malignancy. In a sample from a healthy person (left lane), immunoglobulin genes are in the germline configuration in non-B cells, so a digest of their DNA with a suitable restriction endonuclease yields a single germline DNA fragment when probed with an immunoglobulin heavy-chain, J-region probe (J_H). Normal B cells present in this sample make many different rearrangements to J_H, producing a spectrum of 'bands' each so faint that it is undetectable. By contrast, in samples from patients with B-cell malignancies (patient 1 and patient 2), in whom a single cell has given rise to all the tumor cells in the sample, two extra bands are seen with the J_H probe. These bands are characteristic of each patient's tumor and result from the rearrangement of both alleles of the J_H gene in the original tumor cells. The intensity of the bands compared with that of the germline band gives an indication of the abundance of tumor cells in the sample. After antitumor treatment (see patient 1), the intensity of the tumor-specific bands can be seen to diminish. Image courtesy of T. Vulliamy and L. Luzzatto.

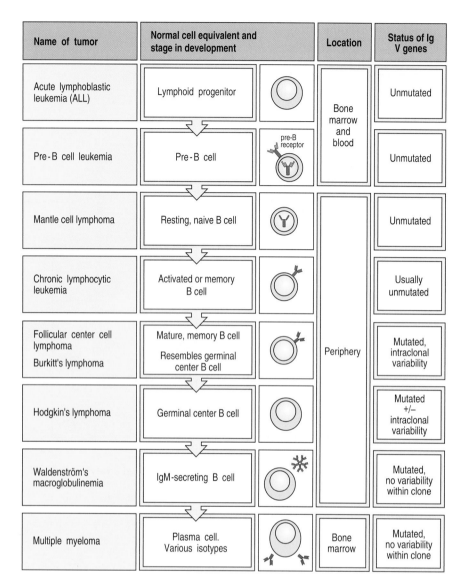

Name of tumor	Normal cell equivalent and stage in development		Location	Status of Ig V genes
Acute lymphoblastic leukemia (ALL)	Lymphoid progenitor		Bone marrow and blood	Unmutated
Pre-B cell leukemia	Pre-B cell	pre-B receptor		Unmutated
Mantle cell lymphoma	Resting, naive B cell			Unmutated
Chronic lymphocytic leukemia	Activated or memory B cell			Usually unmutated
Follicular center cell lymphoma / Burkitt's lymphoma	Mature, memory B cell / Resembles germinal center B cell		Periphery	Mutated, intraclonal variability
Hodgkin's lymphoma	Germinal center B cell			Mutated +/– intraclonal variability
Waldenström's macroglobulinemia	IgM-secreting B cell			Mutated, no variability within clone
Multiple myeloma	Plasma cell. Various isotypes		Bone marrow	Mutated, no variability within clone

Figure 4.19 The different B-cell tumors reflect the heterogeneity of developmental and differentiation states in the normal B-cell population. Each type of tumor corresponds to a normal state of B-cell development or differentiation. Tumor cells have similar properties to their normal cell equivalent, they migrate to the same sites in the lymphoid tissues, and have similar patterns of expression of cell-surface glycoproteins.

Summary

B-cell development is inherently wasteful because the vast majority of B cells die without ever contributing to an immune response. Immature B cells in the bone marrow express only IgM at the cell surface. Those cells whose IgM reacts with a self-antigen are removed at this stage by apoptosis or are rendered anergic. B cells that survive this test go on to become mature B cells expressing both IgM and IgD on their surface. Anergic B cells leaving the bone marrow are excluded from the primary lymphoid follicles and have a half-life of only a few days within the peripheral circulation. Even mature B cells that are tolerant to self-antigens and potentially responsive to microorganisms are regularly abandoned because of the intense competition for access to primary lymphoid follicles. Only after a B cell encounters specific antigen in the secondary lymphoid tissues is it activated to proliferate into a large clone of cells with identical antigen specificity. No sooner has the clone expanded than individual B cells in the clone begin to differentiate and diverge. Some become plasma cells, which service the immediate need for antigen-specific antibody, while others become memory cells providing long-term immunity for the future. Over and above these changes, B cells within the clone alter their immunoglobulins so that antibodies of higher affinity and more potent effector function can be made (Figure 4.20). The stages of antigen-dependent differentiation correspond to changes in either the structure or the expression of the immunoglobulin genes. A change in the processing of μ heavy-chain mRNA in some cells of the clone leads to the synthesis of secreted IgM antibody. In other cells of the clone, further DNA rearrangement switches the antibody isotype so that cell surface IgG, IgA, IgE, and antibodies can be produced, while somatic hypermutation of rearranged variable regions and selection for higher-affinity immunoglobulin lead to the production of antibodies of increasingly higher affinity for the antigen. Some isotype-switched B cells with high-affinity receptor immunoglobulin become memory cells, which will differentiate into plasma cells on a subsequent encounter with the same antigen. Thus, the terminal stage of B-cell development is thus the plasma cell, dedicated entirely to antibody production. B-cell tumors have been instrumental in studying B-cell development because different types of tumor correspond in cell type and location to different stages of normal B-cell development.

Summary to Chapter 4

B lymphocytes are highly specialized cells whose sole function is to recognize foreign antigens by means of cell-surface immunoglobulins and then to differentiate into plasma cells that secrete antibodies of the same antigen specificity. Each B cell expresses immunoglobulin of a single antigen specificity but, as a population, B cells express a diverse repertoire of immunoglobulins. This enables the B-cell response to any antigen to be highly specific. Immunoglobulin diversity is a result of the unusual arrangement and mode of expression of the immunoglobulin genes. In B-cell progenitors the immunoglobulin genes are in the form of arrays of different gene segments that can be rearranged in many different combinations. Gene rearrangement occurs in bone marrow and is independent of a B-cell encounter with specific antigen. The gene rearrangements needed for the expression of a functional surface immunoglobulin follow a program; the successive steps define the antigen-independent stages of B-cell development as shown in Figure 4.21.

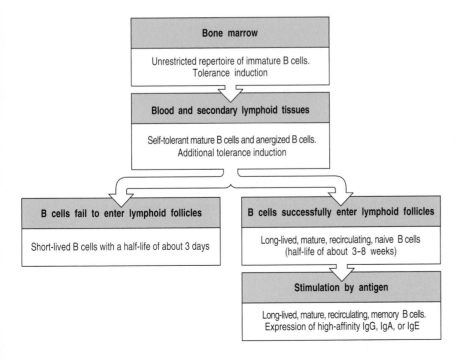

Figure 4.20 Population dynamics of B cells. The B-cell population from which antibody-producing plasma cells will eventually develop is a continually changing and heterogeneous population, consisting of B cells at different stages of development and differentiation. Immature B cells produced in the bone marrow have immunoglobulins specific for the entire range of molecules found in biological systems. Those B cells with receptors that bind to self-antigens present in the bone marrow are either eliminated by clonal deletion or inactivated. Passage through lymphoid follicles is essential for sustaining mature, naive B cells in the peripheral circulation; those that fail to enter a follicle soon die. B cells that passage through a follicle will also eventually die unless stimulated by specific antigen, whereupon they proliferate, differentiate, and give rise to memory B cells.

The success rate for individual rearrangements is far from optimal, but the use of a stepwise series of reactions allows the quality of the products to be tested at critical checkpoints after heavy- and light-chain gene rearrangement. Failure at any step leads to the withdrawal of positive signals and death of the B cell by apoptosis. Gene rearrangement is controlled so that only one functional heavy-chain gene and one functional light-chain gene are produced in each cell. Thus, individual B cells produce immunoglobulin of a single antigen specificity. For a short period after successful immunoglobulin-gene rearrangement, any interaction with specific antigen leads to the elimination or inactivation of the immature B cell, thus rendering the mature B-cell population tolerant of the normal constituents of the body.

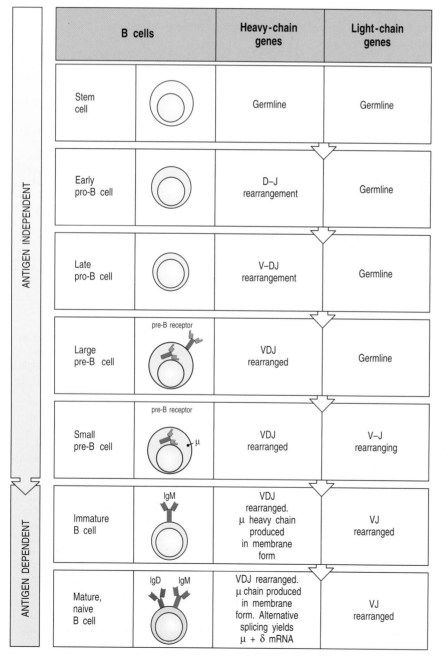

Figure 4.21 **A summary of the first two main phases of B-cell development.** This diagram shows the stages in B-cell development from the stem cell in bone marrow to the mature, naive B cell. The location of B cells at the different stages, the state of the immunoglobulin heavy- and light-chain genes, and the form of immunoglobulin expressed at each stage are indicated. This figure refers only to the development of B-2 cells.

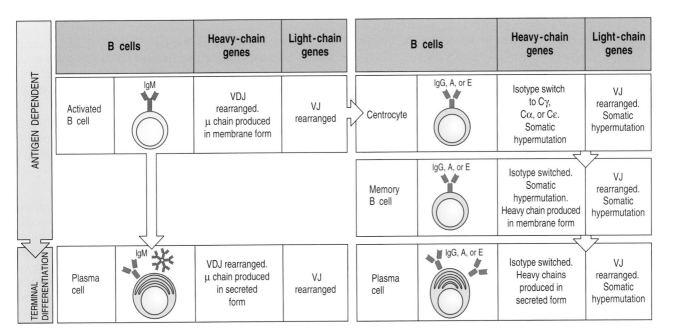

Every day the bone marrow pushes out billions of new mature, naive B cells into the peripheral circulation. There is, however, limited space for B cells in the secondary lymphoid tissues and unless naive B cells encounter specific antigen they are likely to be dead within weeks. Thus, the B-cell repertoire is never static, and newly generated specificities are continually being tested against the antigens of microorganisms causing infection. Binding of antigen to the cell-surface immunoglobulin of a B cell in secondary lymphoid tissue initiates an antigen-dependent program of cellular proliferation and development and the subsequent terminal differentiation, as shown in Figure 4.22. After encounter with antigen in secondary lymphoid tissues, B cells either differentiate directly into IgM-secreting plasma cells or undergo somatic hypermutation, isotype switching, and affinity maturation in germinal centers before differentiation into plasma cells or long-lived memory B cells. The end product of B-cell development is the plasma cell, in which surface immunoglobulin is no longer expressed and all the cell's resources are devoted to antibody secretion.

Figure 4.22 A summary of the last two main phases of B-cell development. This diagram shows the stages in B-cell development from the activated mature B cell to the terminally differentiated plasma cell. Plasma cells can differentiate directly from activated B cells, from isotype switched, somatically hypermutated centrocytes, or from memory B cells. This figure refers only to the development of B-2 cells.

Questions

Question 4–1

A. Give several properties that distinguish B-1 cells from B-2 cells.

B. Describe the chemical composition of the antigen recognized by B-1 cells and explain why these antigens are categorized as T-independent (TI-2) antigens.

C. Do you think that B-1 cells should be categorized as participants in innate immune responses or acquired immune responses? Explain your rationale. (Refer to Figure 4.12.)

Question 4–2

You are going to use flow cytometry to determine the proportion of developing B cells in the bone marrow that are immature, anergic, or mature. You have three monoclonal antibodies specific for three different B-cell surface proteins. The first has specificity for the cell-surface protein CD19 which is expressed by all developing and mature B cells; the second is specific for the Fc region of IgD; and the third is specific for the Fc region of IgM. The antibodies are conjugated to three different fluorescent tags which can be detected and distinguished by the flow cytometer.

A. Use histograms to show your analysis of CD19-positive cells and indicate which part of your histogram you would gate to analyze IgM and IgD expression. Indicate the gated population with an arrow.

B. Using a two-dimensional dot plot, compare the expression of IgD and IgM of these gated cells, and say which of these populations represents (i) immature B cells, (ii) anergic B cells, and (iii) mature B cells. (Refer to Figures 2.13 and 4.13).

Question 4–3

Explain how the inability of anergic B cells to enter secondary lymphoid follicles plays a role in eliminating B cells that have antigen receptors specific for soluble self-antigen. (Refer to Figures 4.13 and 4.16.)

Question 4–4

A. Identify properties that are shared by anergic B cells and plasma cells.

B. What key property is different?

Question 4–5

A. Explain why immunological memory is important in acquired immunity.

B. Describe how immunoglobulin expressed during a primary immune response differs qualitatively and quantitatively from the immunoglobulin expressed during a secondary immune response. (Refer to Figure 1.30 and Section 4-9.)

Question 4–6

Explain how the two checkpoints in B-cell development correlate with the process of allelic exclusion which ensures that only one heavy-chain locus and one light-chain locus produce functional gene products. (Refer to Figure 4.9.)

Question 4–7

What would be the consequence if terminal deoxynucleotidyl transferase (TdT) was expressed throughout the whole of small pre-B cell development?

Question 4–8

A. Discuss the importance of the bone marrow stroma for B-cell development.

B. What would be the effect of anti-IL-7 antibodies on the development of B cells in the bone marrow and at which stage would development be impaired? Explain your answer.

Question 4–9

B-cell tumors originate during different developmental stages of B cells during their maturation in the bone marrow or following maturation and export to the periphery.

A. Explain why B cells isolated from a particular B-cell tumor all express the same immunoglobulin.

B. How might the immunoglobulin expressed on pre-B cell leukemia cells be different from that expressed on immature B cells? (Refer to Figures 4.10, 4.18, and 4.19.)

Question 4–10

Multiple myeloma involves the unregulated proliferation of an antibody-producing plasma cell (myeloma cell) independent of antigen stimulation or T-cell help. Myeloma cells populate multiple sites in the bone marrow where they produce immense quantities of monoclonal immunoglobulin as well as suppressing normal marrow function. Myeloma cells also synthesize and secrete excessive amounts of free light chains (known as Bence-Jones protein), which, because of their low molecular weight (~25 kDa) are excreted as free light chains in the urine.

In a given patient the free light chains are both monoclonal and all of either the κ or λ type.

A. Explain both of these observations.

B. Why do you think patients with multiple myeloma are more susceptible to pyogenic infections, such as pneumonia caused by *Streptococcus pneumoniae* or *Haemophilus influenzae*?

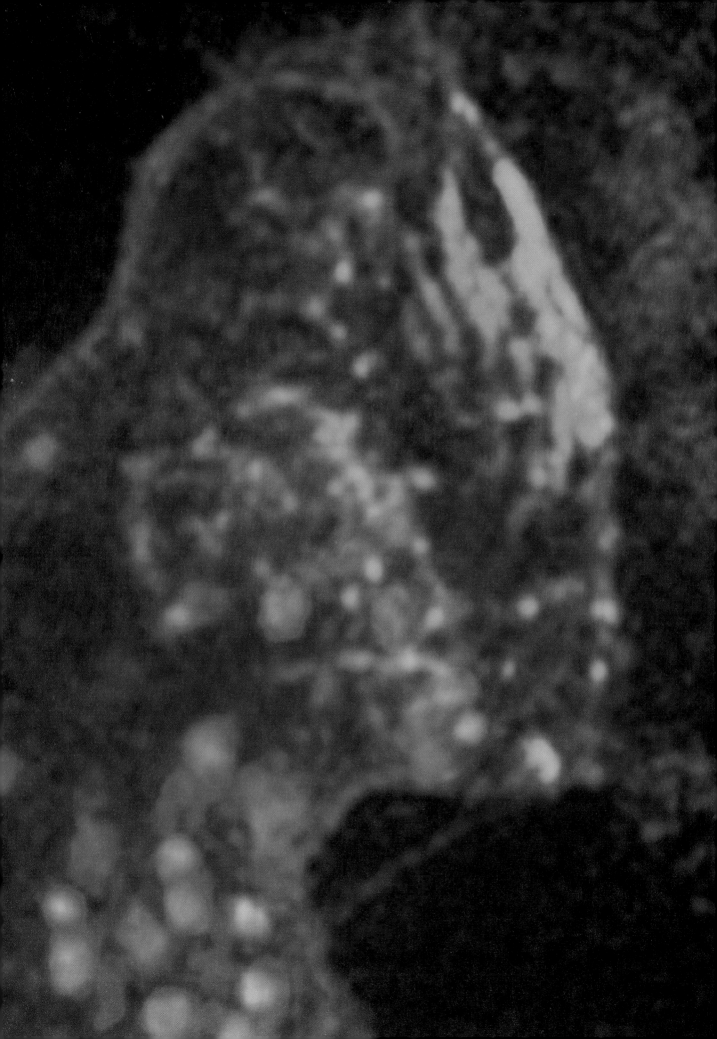

Chapter 5

The Development of T Lymphocytes

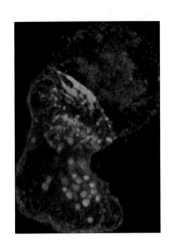

The paths of development for T and B lymphocytes have much in common: both types of cell derive from bone marrow stem cells and, during development, they must undergo gene rearrangement to produce their antigen receptors. But, whereas B cells rearrange their immunoglobulin genes while remaining in the bone marrow, the precursors of T cells have to leave the bone marrow and enter another primary lymphoid organ—the thymus—before they rearrange their T-cell receptor genes. Although gene rearrangements in developing T cells proceed in a broadly similar fashion to those in B cells, their differences cause the formation of two distinct T-cell lineages: one expressing α:β receptors and the other γ:δ receptors.

A major function of the thymus is to ensure that the mature T cells that leave the thymus and enter a person's circulation are restricted to the particular MHC class I and II isoforms expressed by that person, known in this context as **self-MHC**. The primary T-cell repertoire has no such bias; it is introduced by a process of positive selection. This results in the death by neglect of all those immature T cells that have receptors that do not interact with any of the self-MHC class I and II isoforms. The immature T cells chosen by positive selection are subsequently subjected to an additional selective process, called negative selection, which induces the death of those T cells that are potentially autoreactive because their receptors bind too strongly to a self-MHC molecule. Thus, the mature T-cell population leaving the thymus to circulate through the secondary lymphoid organs is rendered tolerant of self-antigens, responsive to foreign antigens presented by self-MHC molecules, and ready to fight infection. The α:β T-cell receptor repertoire represented in the mature T-cell population represents only a small part (some 1–2%) of the primary repertoire. A similar bias is not seen in the receptor repertoire of the γ:δ T cells, indicating that they are not subjected to such stringent selection and are, in this regard, more similar to B cells.

The first part of this chapter traces the stages in gene rearrangement that produce the primary repertoire of T-cell receptors. The second part of the chapter describes the processes of positive and negative selection that act on this repertoire in the thymus to produce the circulating population of mature naive T cells.

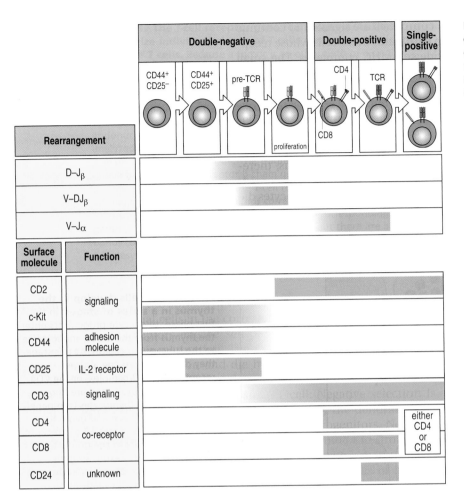

Figure 5.19 Stages of α:β T-cell development in the thymus correlate with T-cell receptor gene rearrangement and expression of cell-surface proteins by the developing T cell.

without ever performing a useful task, and only a few percent of thymocytes fulfill the stringent requirements of selection and leave the thymus to enter the circulation. Whereas the bone marrow is continually turning over the B-cell repertoire during the whole of a person's lifetime, the thymus works principally during youth, when it serves to accumulate a repertoire of T cells that can then be used throughout life. This difference might reflect the magnitude of the body's investment in the development of each useful T cell, and the savings to be made by shutting down the thymus before middle age.

Questions

Question 5–1

The surrogate light chain operating during pre-B cell development is made up of VpreB:λ5. Its expression with μ on the pre-B cell surface represents an important checkpoint in B-cell maturation. Name the T-cell analog of VpreB:λ5 and discuss how it is functionally similar. (Refer to Figure 5.6.)

Question 5–2

Successive gene rearrangements at the T-cell receptor α and β-chain loci can rescue unproductively rearranged VJ and VDJ segments, respectively (see Figures 3.3, 5.7, and 5.8). Similarly, the immunolgobulin κ and λ light-chain loci can also undergo successive gene rearrangements (see Figures 2.14 and 4.8). Interestingly, immunoglobulin heavy-chain genes are only permitted one rearrangement per locus. What aspects of the arrangement of gene segments at these loci make successive gene rearrangements possible for (a) TCRα, (b) TCRβ, (c) light chain κ and (d) light chain λ loci, but not for the (e) heavy-chain locus.

Question 5–3

MHC class II expression is restricted to a limited number of cell types.

A. What are these cell types? (Refer to Figure 3.22.)

B. Which of these cell types populate the thymus or circulate through it, and what role do they play in mediating positive and/or negative selection?

C. Can you explain why it would be detrimental for noncirculating cells that populate tissues and glands to express MHC class II?

Question 5–4

In T cells, allelic exclusion of the α-chain locus is relatively ineffective, resulting in the production of some T cells with two T-cell receptors of differing antigen specificity on their cell surface.

A. Will both these receptors have to pass positive selection for the cell to survive? Explain your answer.

B. Will both receptors have to pass negative selection for the cell to survive? Explain your answer.

C. Is there a potential problem having T cells with dual specificity surviving these selection processes and being exported to the periphery?

Question 5–5

Mature B cells undergo somatic hypermutation following activation which, following affinity maturation, results in the production of antibody with higher affinity for antigen than in the primary antibody response. Suggest some reasons why T cells have not evolved the same capacity?

Question 5–6

MHC class II deficiency is inherited as an autosomal recessive trait and involves a defect in the coordination of transcription factors involved in the regulation of all MHC class II gene expression (HLA-DP, HLA-DQ and HLA-DR).

A. What is the effect of MHC class II deficiency?

B. Explain why hypogammaglobulinemia is associated with this deficiency.

Question 5–7

Allogeneic bone marrow transplantation can be used to treat several types of T-cell tumors, including acute lymphoblastic leukemia (T-ALL).

A. Describe in general terms how an allogeneic bone marrow transplant is carried out.

B. A patient with CD3$^+$ mature T-ALL is transplanted with bone marrow that is mismatched for only one HLA class I antigen. Explain how flow cytometry could be used to monitor the progress of transplantation and the repopulation of the patient's T-cell repertoire with donor-derived T cells. For this purpose, a monoclonal antibody specific for CD3 and one specific for the mismatched HLA class I antigen from the donor are available. This antibody does not cross-react with the recipient's HLA class I antigens. Illustrate your answer by a two-dimensional dot plot. (Refer to Figures 2.13, 5.10, and 5.11.)

Question 5–8

We learned in Chapter 3 that the benefit of polygeny in the MHC class I and class II genes and the expression of multiple isotypes is the resulting increase in the number of potential peptide-binding motifs for antigen presentation to T cells. If more is better, then why hasn't natural selection favored more than three isotypes each for MHC class I (HLA-A, -B and -C) and MHC class II (HLA-DP, -DQ and -DR) and driven the expansion of polygeny further in the MHC? (Refer to Figure 5.15.)

Question 5–9

Discuss why there are T-cell tumors corresponding to early and late T-cell developmental stages but not to intermediate stages. (Refer to Figure 5.16.)

Question 5–10

A. Explain how Southern blotting could be used to identify a T-cell tumor expressing a rearranged T-cell receptor β gene.

B. What result would you expect when comparing DNA extracted from (i) spermatocytes, (ii) normal peripheral blood mononuclear cells (PBMCs) and (iii) PBMCs of a patient with T-cell leukemia? (Refer to Figure 5.17.)

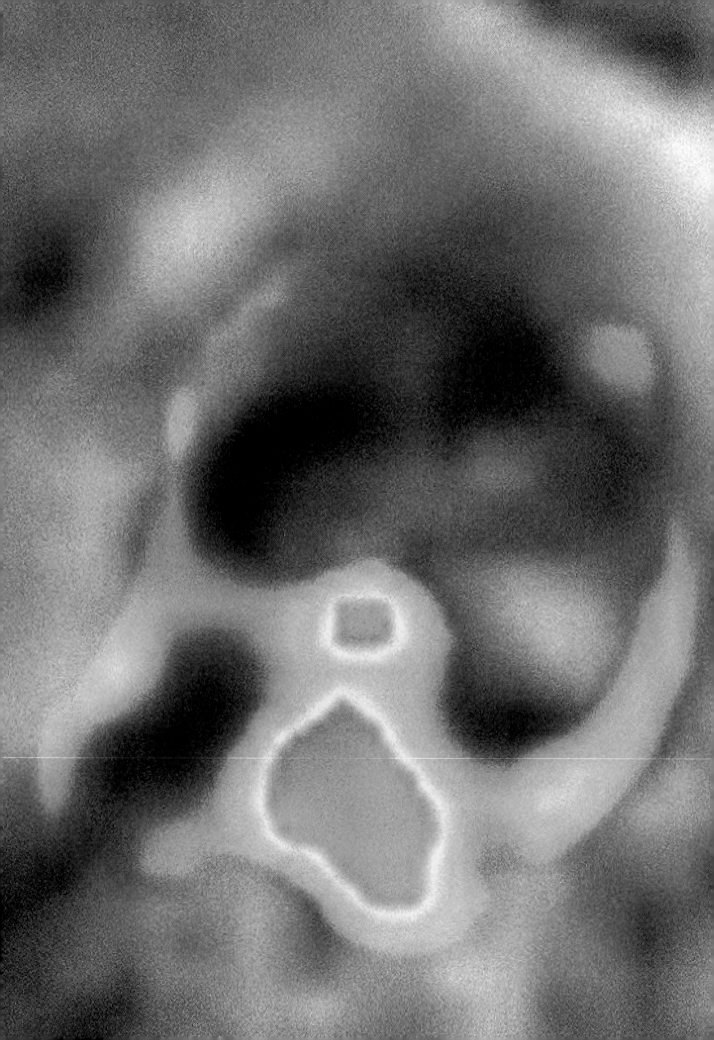

Chapter 6
T Cell-Mediated Immunity

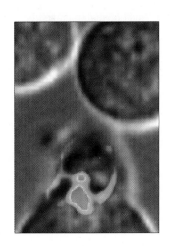

In Chapter 5 we saw how T cells develop in the thymus into a population of mature naive T cells. These T cells now circulate through the peripheral circulation, where they can be activated by specific antigen. Like B cells, naive T cells first meet specific antigen in secondary lymphoid tissues, and it is here that they are activated to undergo clonal expansion by cell proliferation followed by differentiation into effector T cells. Effector T cells can either remain in the lymphoid tissues or migrate to sites of infection, where a subsequent encounter with antigen stimulates them to perform their effector functions.

In the first part of this chapter we consider what happens when a naive T cell encounters its specific antigen for the first time and is stimulated to differentiate into an effector T cell. We shall call this process **T-cell activation**; it is sometimes also referred to as **T-cell priming**. T-cell activation is the first stage of a primary adaptive immune response against most antigens. There are three kinds of effector T cell: **cytotoxic CD8 T cells**, which kill infected cells, and two kinds of **CD4 T cell** (T_H1 and T_H2) with different functions. The general function of effector CD4 T cells is to secrete cytokines that activate other cells of the immune system. Because the function of CD4 T cells is principally to help other cells achieve their effector function, they are often called **helper T cells**. Once activated, effector T cells interact with their specific antigens that are presented by different types of antigen-presenting cell or **target cell**. This is described later in the chapter. These effector actions, which eventually lead to the removal and destruction of the pathogen, constitute the second stage of the primary immune response. The three kinds of effector T cell enable the human immune system to respond effectively to different types of infection and to different stages of the same infection.

Activation of naive T cells on encounter with antigen

Once an infection begins, the immune system faces the challenge of quickly bringing the minute fraction of naive T cells that are specific for the pathogen into contact with the pathogen's antigens. This is accomplished in the secondary lymphoid tissues, into which antigens brought from outlying tissues by the lymph meet naive T cells brought in by the blood. In this part of the chapter we shall examine the activation of naive T cells to effector T cells by professional antigen-presenting cells within secondary lymphoid tissues. Once activated, antigen-specific CD8 and T_H1 cells are then sent out to the infected sites, while T_H2 cells stay in the lymphoid tissues. We shall also see how the interaction of a naive T cell with antigen presented by cells other than professional antigen-presenting cells leads to inactivation rather than activation of the T cell. This mechanism ensures that mature T cells reactive to self-antigens are eliminated before they can become effector cells. This first phase of the primary immune response produces an expanded effector T-cell population that is ready to fight the infecting pathogen but is tolerant of self-antigens.

6-1 Dendritic cells carry antigens from sites of infection to secondary lymphoid tissues

The immune system does not attempt to initiate adaptive immune responses at the innumerable sites where a pathogen might set up an infection. Instead, it captures some of the pathogen and takes it to the organized secondary lymphoid tissues, whose purpose is the generation of adaptive immune responses. These functions are mediated by dendritic cells and macrophages, two developmentally related types of professional antigen-presenting cell. This process is much the same for infections at all sites in the body: peripheral tissues, mucosal surfaces, and blood.

Dendritic cells and macrophages are present as sentinels in all the body's tissues. Upon infection, both these cell types become active in the uptake of pathogens and in the processing and presentation of their antigens by MHC class I and II molecules. Whereas the macrophage has a range of functions in the defense and repair of damaged tissues, the only known function of the dendritic cell is to trigger T-cell responses, for which it is highly specialized and highly effective. In particular, the dendritic cell is far superior to the macrophage in stimulating naive T cells. One reason for this difference is that dendritic cells are migratory cells that can carry their load of antigen from the site of infection to the nearest secondary lymphoid tissue (Figure 6.1), which

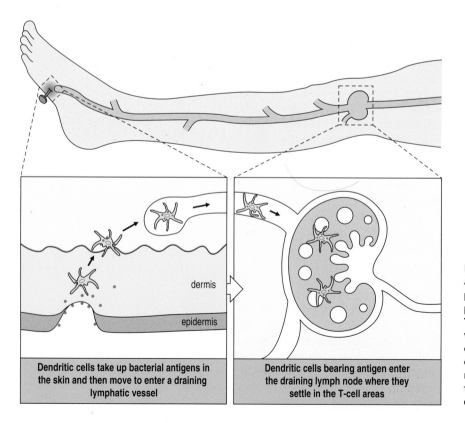

| Dendritic cells take up bacterial antigens in the skin and then move to enter a draining lymphatic vessel | Dendritic cells bearing antigen enter the draining lymph node where they settle in the T-cell areas |

Figure 6.1 Dendritic cells take up antigen in the tissues, migrate to peripheral lymphoid organs, and present foreign antigens to naive T cells. In the example illustrated, of a wound in the skin, immature dendritic cells in the skin, known as Langerhans' cells, take up antigen locally and migrate to a nearby lymph node. There they settle in the T-cell areas and differentiate into mature dendritic cells.

Figure 6.2 Dendritic cells change their functions on taking antigen from infected sites to secondary lymphoid tissues. In these images, MHC class II is stained green and a lysosomal protein is stained red. In the top panel, the cell bodies are difficult to discern but the dendrites contain endocytic vesicles that stain both for MHC class II and lysosomal protein, giving rise to a yellow fluorescence due to the combination of red and green stain. On activation and migration in the lymph to secondary lymphoid organs the morphology of the dendritic cell changes, as shown in the middle panel. The dendritic cells stop phagocytosis a change indicated in this panel by a partial separation of MHC class II (green) from the lysosomal protein (red). On reaching a lymph node (bottom panel), the now mature dendritic cells turn to antigen presentation and T-cell stimulation instead of uptake and processing of antigens. Now the lysosomal protein is quite distinct from the MHC class II molecules (green) displayed at high density on the many dendritic processes. Photographs courtesy of I. Mellman, P. Pierre, and S. Turley.

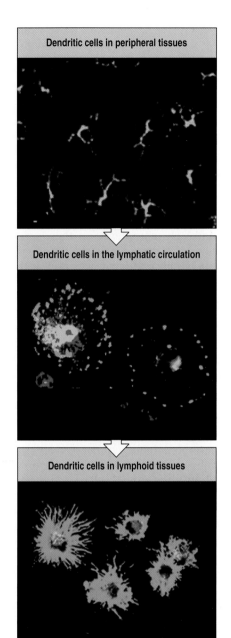

Dendritic cells in peripheral tissues

Dendritic cells in the lymphatic circulation

Dendritic cells in lymphoid tissues

is organized so that they meet naive T cells. In contrast, macrophages are resident cells. Consequently, the macrophages resident in the infected tissue, and most heavily exposed to the pathogen, have no opportunity to interact with naive T cells. Some pathogens, however, are carried passively in lymph, blood, or other fluid to secondary lymphoid tissues where their ingestion by resident macrophages can contribute to T-cell stimulation.

For infections in the skin and other peripheral tissues the T-cell response is produced against antigens carried to the draining lymph nodes; for blood infections, antigens enter the spleen. The response to infections of the respiratory mucosa is in the tonsils or other bronchial-associated lymphoid tissues, whereas the response to gastrointestinal infections is in the Peyer's patches, appendix, or other gut-associated lymphoid tissues. In each of the secondary lymphoid organs a similar sequence of events occurs, which we shall illustrate by using the lymph node.

The movement of a dendritic cell from a peripheral site of infection to a secondary lymphoid organ is accompanied by changes in the dendritic cell's surface molecules, functions, and morphology (Figure 6.2). Whereas dendritic cells in tissues are active in the capture, uptake, and processing of antigens, they lose these properties on moving to a secondary lymphoid organ while gaining the capacity to interact well with naive T cells. The dendritic cells in tissues are called **immature dendritic cells**, whereas those in lymph nodes are called **mature dendritic cells** or **activated dendritic cells**. On maturation, the finger-like processes called dendrites, after which the dendritic cell is named, become highly elaborated, which facilitates extensive interaction with T cells in the cortex of the lymph node. Whereas dendritic cells are confined to the T-cell regions of the cortex, macrophages are present in both the cortex and medulla. In addition to capturing pathogens, and other particulate matter that is carried into the lymph node with the afferent lymph, the macrophages are also responsible for eliminating the apoptotic lymphocytes produced as the inevitable by-products of the adaptive immune response.

6-2 Naive T cells first encounter antigen on antigen-presenting cells in secondary lymphoid tissues

Naive T cells enter lymphoid tissue through the blood capillaries that provide oxygen and nutrients to the tissue. In lymph nodes, for example, T cells in the blood bind to the endothelial cells of the thin-walled high endothelial venules (HEV), squeeze through the vessel wall, and enter the cortical region of the node. The naive T cell then passes through the crowded tissue, where it

Figure 6.3 Naive T cells encounter antigen during their recirculation through secondary lymphoid organs. Naive T cells (blue and green) recirculate through secondary lymphoid organs, such as the lymph node shown here. They leave the blood at high endothelial venules and enter the lymph node cortex, where they mingle with professional antigen-presenting cells (mainly dendritic cells and macrophages). T cells that do not encounter their specific antigen (green) leave the lymph node in the efferent lymph and eventually rejoin the bloodstream. T cells that encounter antigen (blue) on antigen-presenting cells are activated to proliferate and to differentiate into effector cells. These effector T cells can also leave the lymph node in the efferent lymph and enter the circulation.

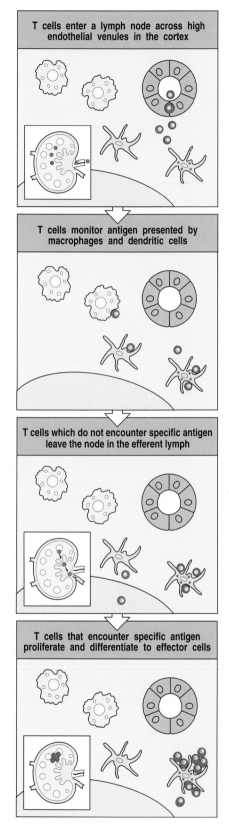

encounters dendritic cells and uses its antigen receptor to examine the peptide:MHC complexes on their surfaces. When a T cell encounters a peptide:MHC complex to which its T-cell receptor binds, the T cell is retained in the lymph node and activated. It then proliferates and differentiates into a clone of effector T cells (Figure 6.3).

For any given infection, naive T cells specific for the pathogen will represent only 1 in 10^4 to 1 in 10^6 of the total pool of circulating T cells. During most passages through a lymph node a T cell does not find its specific antigen and leaves the medulla in the efferent lymph to continue recirculation. In the absence of specific antigen, circulating naive T cells live for many years as small nondividing cells with condensed chromatin, scanty cytoplasm, and little RNA or protein synthesis. With time, the entire population of circulating T cells will pass through a lymph node. The trapping of pathogens and their antigens in the lymphoid tissue nearest to the site of infection creates a concentrated depot of processed and presented antigens. This enables the small subpopulation of T cells specific for those antigens to be efficiently pulled out of the circulating T-cell pool and activated.

Once an antigen-specific T cell has been trapped in a lymph node by an antigen-presenting cell and activated, it takes several days for the activated T cell to proliferate and for its progeny to differentiate into effector T cells. This accounts for much of the delay between the onset of an infection and the appearance of a primary adaptive immune response. Most effector T cells leave the lymph node in the efferent lymph and on reaching the blood are rapidly carried to the site of infection, where they perform their effector functions.

6-3 Homing of naive T cells to secondary lymphoid tissues is determined by cell adhesion molecules

The passage of a naive T cell out of the bloodstream and through the high endothelium to the cortex of a lymph node is controlled by contacts made between cell-surface molecules on the T cell and cell-surface molecules on the endothelial cells. These and other cell–cell contacts are initiated by **cell adhesion molecules** on the T-cell surface, which bind to complementary adhesion molecules on the surfaces of other cells. The adhesion molecules of the immune system comprise four structural classes of protein: **selectins**; mucin-like molecules called **vascular addressins**; **integrins**; and members of the immunoglobulin superfamily (Figure 6.4). Adhesion molecules of all four types are used by naive T cells.

The movement of naive T cells into secondary lymphoid tissues—which is known as **homing**—is determined by interactions between a selectin, **L-selectin**, on the naive T-cell surface, and two vascular addressins, **CD34** and **GlyCAM-1**, on the surface of the high endothelial venules (Figure 6.5). To

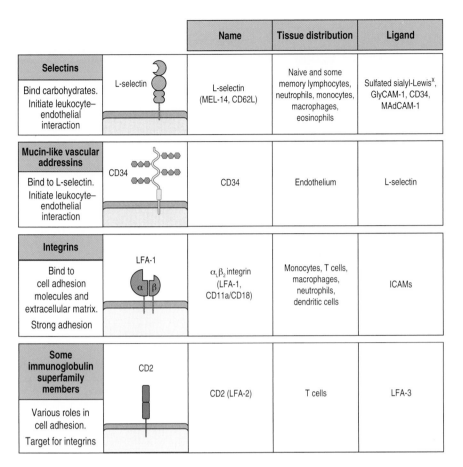

		Name	Tissue distribution	Ligand
Selectins Bind carbohydrates. Initiate leukocyte–endothelial interaction	L-selectin	L-selectin (MEL-14, CD62L)	Naive and some memory lymphocytes, neutrophils, monocytes, macrophages, eosinophils	Sulfated sialyl-Lewis^x, GlyCAM-1, CD34, MAdCAM-1
Mucin-like vascular addressins Bind to L-selectin. Initiate leukocyte–endothelial interaction	CD34	CD34	Endothelium	L-selectin
Integrins Bind to cell adhesion molecules and extracellular matrix. Strong adhesion	LFA-1 α β	$\alpha_L\beta_2$ integrin (LFA-1, CD11a/CD18)	Monocytes, T cells, macrophages, neutrophils, dendritic cells	ICAMs
Some immunoglobulin superfamily members Various roles in cell adhesion. Target for integrins	CD2	CD2 (LFA-2)	T cells	LFA-3

Figure 6.4 Leukocyte adhesion molecules. The four structural classes of adhesion molecule present on white blood cells and the cells with which they interact are: selectins, which are carbohydrate-binding lectins; mucin-like vascular addressins, which contain carbohydrate groups to which selectins bind; integrins; and some proteins in the immunoglobulin superfamily. The figure shows a schematic representation of the prototypical structure of a member of each family with a named example (alternative names in brackets), its cellular distribution, and the ligands to which it binds. The nomenclature for individual adhesion molecules has developed in a rather haphazard fashion; many names do not reflect the structural family to which the molecule belongs but are based on assays used to identify cell-surface antigens or adhesion functions. For example, lymphocyte function-associated antigen-1 (LFA-1) is an integrin, whereas LFA-2 and LFA-3 are members of the immunoglobulin superfamily. The CD nomenclature for cell-surface proteins of leukocytes gives each protein a unique number but does not reflect its structure or function in any way. ICAM, intercellular adhesion molecule.

make this interaction L-selectin binds to sulfated sialyl-Lewis^x, a carbohydrate portion of the addressin glycoprotein that is structurally related to the carbohydrates that form the Lewis series of blood-group antigens. The cooperative effect of many such interactions causes the naive T cell to slow down and attach to the high endothelial cell (Figure 6.6).

After initial attachment has been made, the contact between the naive T cell and the endothelium is strengthened by additional interactions between an integrin on the surface of the lymphocyte, known as **lymphocyte function-associated antigen-1** (**LFA-1**; α_L:β_2), and two **intercellular adhesion molecules** (**ICAMs**) expressed by vascular endothelium. These **ICAM-1** and **ICAM-2** molecules are immunoglobulin superfamily members. After making initial contact with an ICAM molecule LFA-1 is induced to undergo a conformational change that strengthens its hold. Essential for this activation is the participation of a protein called **CCL21**. This is one member of a large family of small, soluble chemoattractant proteins called chemokines. CCL21 is made by vascular endothelial cells and on binding to the CCR7 chemokine receptor on the naive T cells it activates the T cells' LFA-1. CCL21 is also made by stromal cells and dendritic cells and sets up a chemokine gradient along which the naive T cell moves and squeezes between the endothelial cells to enter the lymph-node cortex (see Figure 6.6).

Figure 6.5 Binding of L-selectin to mucin-like vascular addressins directs naive lymphocytes homing to lymphoid tissues. L-selectin on naive T cells (and naive B cells) binds to sulfated carbohydrate sialyl-Lewis^x moieties of vascular addressins CD34 and GlyCAM-1 on the high endothelial cells of lymph venules.

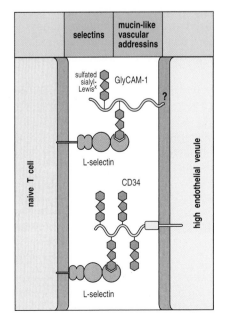

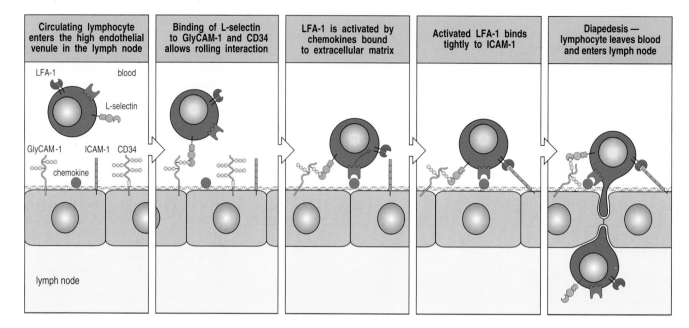

| Circulating lymphocyte enters the high endothelial venule in the lymph node | Binding of L-selectin to GlyCAM-1 and CD34 allows rolling interaction | LFA-1 is activated by chemokines bound to extracellular matrix | Activated LFA-1 binds tightly to ICAM-1 | Diapedesis — lymphocyte leaves blood and enters lymph node |

As naive T cells negotiate their way through the packed cells of the cortex, they bind transiently to antigen-presenting cells they meet. These interactions involve integrins and members of the immunoglobulin superfamily. Both T cells and professional antigen-presenting cells express the integrin LFA-1. The T cell's LFA-1 binds to ICAM-1 or ICAM-2 on the antigen-presenting cells, whereas the LFA-1 of the antigen-presenting cell binds to a third kind of ICAM—**ICAM-3**—on the T-cell surface. Adhesion is strengthened by interaction between two other members of the immunoglobulin superfamily, **CD2** on the T cell and **LFA-3** on the antigen-presenting cell, and between ICAM-3 on the T cell and the lectin DC-SIGN on the antigen-presenting cell (Figure 6.7). These transitory cell–cell interactions enable the T-cell receptor to screen the peptide:MHC complexes on the surface of the antigen-presenting cell for ones that engage the receptor and activate the T cell.

When a naive T cell encounters a specific peptide:MHC complex, a signal is delivered through the T-cell receptor. This induces a change in the conformation of the T cell's LFA-1 molecules that increases their affinity for ICAMs (Figure 6.8). The interaction of the T cell with the antigen-presenting cell is stabilized and can last for several days, during which time the T cell proliferates, and its progeny, while also remaining in contact with the antigen-presenting cell, differentiate into effector cells.

6-4 Activation of naive T cells requires a co-stimulatory signal delivered by a professional antigen-presenting cell

The intracellular signal generated by ligation of the T-cell receptor with a specific peptide:MHC complex is necessary to activate a naive T cell but is not sufficient. A second, **co-stimulatory**, signal is required. Co-stimulatory signals

Figure 6.6 Naive T and B lymphocytes circulate in the blood and enter lymph nodes by crossing high endothelial venules. Lymphocytes bind to high endothelium in the lymph node through interaction of L-selectin with vascular addressins. Chemokines, which are also bound to the endothelium, activate the integrin LFA-1 on the lymphocyte surface, enabling it to bind tightly to ICAM-1 on the endothelial cell. Establishment of tight binding allows the lymphocyte to squeeze between two endothelial cells, leaving the lumen of the blood vessel and entering the lymph node proper.

Figure 6.7 Cell-surface molecules of the immunoglobulin superfamily initiate lymphocyte adhesion to professional antigen-presenting cells. In the initial encounter of T cells with antigen-presenting dendritic cells, CD2, binding to LFA-3 on the antigen-presenting cell, synergizes with LFA-1 binding to ICAM-1 and ICAM-2. An interaction that appears to be exclusive to the interaction of naive T cells with dendritic cells is that between ICAM-3 on the naive T cell and DC-SIGN, a C-type lectin specific to dendritic cells and which binds ICAM-3 with high affinity.

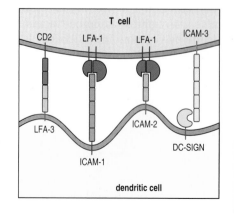

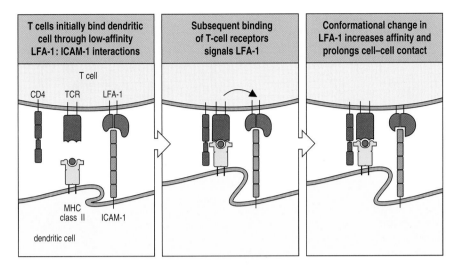

| T cells initially bind dendritic cell through low-affinity LFA-1 : ICAM-1 interactions | Subsequent binding of T-cell receptors signals LFA-1 | Conformational change in LFA-1 increases affinity and prolongs cell–cell contact |

Figure 6.8 Transient adhesive interactions between T cells and dendritic cells are stabilized by specific antigen recognition. When a T cell binds to its specific ligand on an antigen-presenting dendritic cell, intracellular signaling through the T-cell receptor (TCR) induces a conformational change in LFA-1 that causes it to bind with higher affinity to ICAMs on the antigen-presenting cell. The T cell shown here is a CD4 T cell.

are delivered only by professional antigen-presenting cells; hence, the obligatory role of these cells in activating naive T cells. A further requirement is that antigen-specific stimulation and co-stimulation must both be delivered by ligands on the same antigen-presenting cell.

The cell-surface protein on naive T cells, through which co-stimulatory signals are delivered, is called **CD28**. Its ligands are the structurally related **B7.1** (CD80) and **B7.2** (CD86) molecules, which are only expressed on professional antigen-presenting cells. B7.1 and B7.2 are collectively called **B7 molecules** and are also known as **co-stimulator molecules** or **co-stimulatory molecules**. CD28 and B7 are all members of the immunoglobulin superfamily. Activation of a naive T cell requires B7 molecules on a professional antigen-presenting to engage CD28 molecules on the T-cell surface at the same time that peptide:MHC complexes on the same antigen-presenting cell engage T-cell receptors and co-receptors (either CD4 or CD8) (Figure 6.9). The combination of intracellular signals generated by these receptor–ligand interactions is essential to induce proliferation and further differentiation of the T cell.

Although CD28 is the only B7 receptor on naive T cells, an additional receptor is expressed once they have been activated. This receptor, called **CTLA4**, is structurally similar to CD28 but binds B7 twenty times more strongly than does CD28 and functions as an antagonist. Whereas B7 binding to CD28 activates a T cell, the engagement of CTLA4 dampens down activation and limits cell proliferation. Consequently, the lymphocytes of mice engineered to lack CTLA4 undergo a massive, uncontrolled proliferation that proves fatal.

6-5 Secondary lymphoid tissues contain three kinds of professional antigen-presenting cell

The characteristic that distinguishes professional antigen-presenting cells from other antigen-presenting cells is the presence of B7 co-stimulatory molecules on their surfaces. The three kinds of professional antigen-presenting cell are the **dendritic cell**, the **macrophage**, and the **B cell**. All three cell types are present in secondary lymphoid tissues, but at different locations, as illustrated for a lymph node in Figure 6.10. Dendritic cells are present only in the cortical T-cell areas, macrophages are found throughout the cortex and medulla, and B cells are confined to the lymphoid follicles (see Section 4-8, p. 112). These distributions reflect the different functions of the three cell types and their relative importance and potency in presenting antigens to naive T cells; dendritic cells are more effective than macrophages, which are generally more effective than B cells.

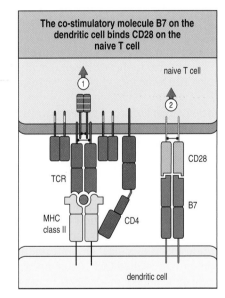

The co-stimulatory molecule B7 on the dendritic cell binds CD28 on the naive T cell

Figure 6.9 The principal co-stimulatory molecules on professional antigen-presenting cells are B7 molecules, which bind CD28 proteins on the T-cell surface. Binding of the T-cell receptor and its co-receptor CD4 to the peptide:MHC class II complex on the dendritic cell delivers a signal (arrow 1). This signal induces clonal expansion of T cells only when the co-stimulatory signal (arrow 2) is also given by the binding of CD28 to B7. Both CD28 and B7 are members of the immunoglobulin superfamily. There are two forms of B7, called B7.1 (CD80) and B7.2 (CD86), but their functional differences have yet to be understood.

Figure 6-10 Three types of professional antigen-presenting cell populate different parts of the lymph node. Dendritic cells (sometimes also known as interdigitating cells) are situated in the T-cell areas of the lymph node cortex. Although macrophages are distributed throughout the lymph node, they concentrate in the marginal sinus, where afferent lymph collects before percolating through the lymphoid tissue, and also in the medullary cords, where efferent lymph collects before leaving the node. B cells mainly populate the follicles in the cortex. These distributions reflect differences in the functions of the three types of professional antigen-presenting cell.

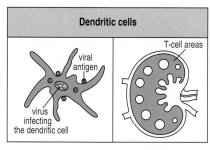

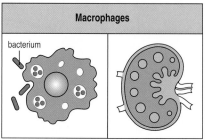

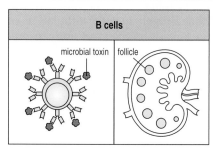

In the absence of infection neither dendritic cells, macrophages, nor B cells express co-stimulatory molecules. Thus, the capacity of professional antigen-presenting cells to activate naive T cells is only acquired during an infection. B7 expression is a direct consequence of infection, being induced by specific interaction of a potential antigen-presenting cell with microbial products through cell-surface receptors that contribute to the innate immune response.

Dendritic cells are developmentally related to macrophages and in their immature form phagocytose microbial antigens using receptors such as DEC 205, among others. Antigens can also be taken up nonspecifically by a process called macropinocytosis, in which the cell engulfs extracellular fluid. Dendritic cells are particularly important for the initiation of T-cell responses against viral infection. They can acquire viral antigens either through infection of the dendritic cells themselves or by taking up virus particles or components from the extracellular fluid or from other types of cell that have become infected.

The **Langerhans' cell** of the skin is a typical immature dendritic cell containing large granules, called Birbeck granules, which may be a form of phagosome. On infection of an area of skin, the local Langerhans' cells will take up and process microbial antigens before traveling to the T-cell areas in the cortex of the draining lymph node and maturing to become professional antigen-presenting cells. In lymph nodes, the mature dendritic cells have a characteristic morphology, which led to them being called **interdigitating reticular cells** (Figure 6.11). In addition to starting to express B7 molecules and producing more MHC molecules, activated dendritic cells also increase expression of adhesion molecules. On entering the lymph node, dendritic cells secrete a chemokine called CCL18 that attracts naive T cells toward them. An adhesion molecule called DC-SIGN is made uniquely by activated dendritic cells and, by binding tightly to ICAM-3, their interactions with naive T cells are strengthened (Figure 6.12).

Dendritic cells are found only in the T-cell areas of secondary lymphoid tissues because their sole function is to activate naive T cells, and to do this for the complete range of pathogens. Macrophages, in contrast, are found throughout the tissue of a lymph node because they have several different functions to perform. Macrophages are phagocytic cells that take up

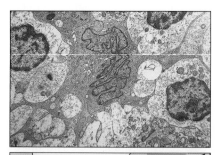

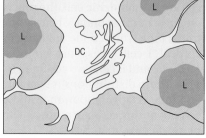

Figure 6.11 Mature dendritic cells in lymph nodes take the form of interdigitating reticular cells. Electron micrograph (top) of an interdigitating reticular cell in the T-cell area of a human lymph node, and an interpretative drawing (bottom). The interdigitating cell (DC; yellow) has a complex folded nucleus. Its cytoplasm makes a complex mesh around the surrounding T lymphocytes (L; blue). Magnification × 10,000. Photograph courtesy of N. Rooney.

microorganisms and other particulate material from the extracellular environment and degrade them in phagolysosomes loaded with hydrolytic enzymes. A major function of the macrophages in secondary lymphoid organs is to trap and degrade pathogens that arrive in the lymph from sites of infection. This enables the macrophage to process and present pathogen-derived antigens to naive T cells, and also prevents infection from reaching the blood and becoming systemic. The removal of noninfectious particulates from the lymph also prevents these entering the blood and blocking small blood vessels. A very distinct function of macrophages in secondary lymphoid tissues is to remove and degrade the numerous lymphocytes that die in these tissues as a consequence of the stimulation and proliferation of pathogen-specific B and T cells.

In the absence of infection, macrophages express no B7 molecules and few MHC molecules. Present on their surface, however, are several receptors involved in innate immune responses. These recognize carbohydrates and other components of microbial surfaces that are not present on human cells. The receptors include the mannose receptor, scavenger receptor, complement receptors, and several Toll-like receptors. When these receptors are engaged by their ligands, signals are transmitted to the macrophage to induce expression of B7 and to increase expression of MHC molecules. In this way, the presence of an infection results in macrophages becoming activated to professional antigen-presenting cell status (Figure 6.13). Because naive T cells circulate through secondary lymphoid tissues and not through peripheral sites of infection, macrophages activated in the site of infection will not serve to activate naive T cells. However, macrophages that become activated in the T-cell areas of the draining lymph node can present antigens and activate naive T cells.

Because macrophages break down pathogens in their endosomes and lysosomes, they most commonly present pathogen-derived peptides on MHC class II molecules (see Section 3-10, p. 80), which then activate naive CD4 T cells. Certain bacteria, such as *Listeria monocytogenes,* avoid destruction by leaving the endocytic vesicles and exploiting the macrophage's cytosol as a place to live. The infected cell can, however, now be detected by CD8 cytotoxic T cells through listerial antigens presented on MHC class I molecules (see Section 3-9, p. 78). Killing the cell releases the bacteria into the extracellular space, from which they can be taken up by new macrophages. In normal healthy individuals, symptoms of listerial infection are rare because the combined actions of neutrophils, macrophages, and CD8 T cells are sufficient to keep the bacterial population down. For pregnant women, the very young, or the immunosuppressed patient, however, listeriosis can be life threatening.

The third type of professional antigen-presenting cell, the B cell, binds particles and soluble protein antigens from the extracellular environment by

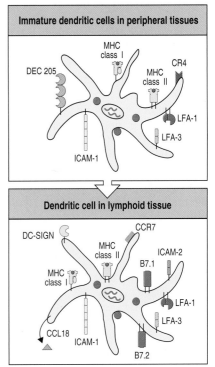

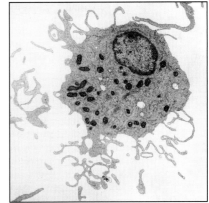

Figure 6.12 Dendritic cells mature in secondary lymphoid tissues to become professional antigen-presenting cells. Immature dendritic cells have cell-surface receptors, such as DEC 205, that facilitate the phagocytosis and pinocytosis of antigens and pathogens (top panel). When dendritic cells mature they no longer make such receptors; instead, they increase their expression of the cell-surface molecules needed to stimulate naive T cells. The genes for the B7 co-stimulators molecules are turned on and the levels of MHC and adhesion molecules are increased. Mature dendritic cells also express high levels of the dendritic-cell-specific adhesion molecule DC-SIGN, which binds ICAM-3 with high affinity. The photograph shows a mature dendritic cell. Photograph courtesy of R. Steinman.

Figure 6.13 Microbial substances induce co-stimulatory activity in macrophages. Phagocytosis of bacteria by macrophages and their breakdown in the phagolysosomes leads to the release of substances such as bacterial lipopolysaccharide, which induce the expression of co-stimulatory B7 molecules on the surface of the macrophage. Peptides derived from the degradation of bacterial proteins in the macrophage vesicular system are bound by MHC class II molecules and presented on the macrophage surface. Activation of naive T cells is accomplished by the combination of B7 binding to CD28 and peptide:MHC complexes binding to the T-cell receptor.

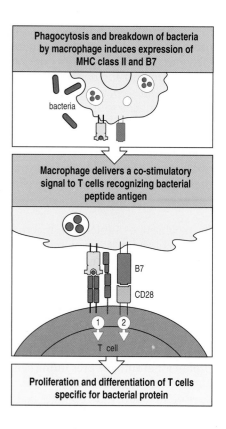

means of its surface immunoglobulin. Receptor-mediated endocytosis causes selective uptake of the antigen, followed by its processing into peptides that bind to MHC class II molecules (Figure 6.14). B cells, therefore, present peptides to CD4 T cells. Like macrophages, B cells do not express B7 co-stimulatory molecules constitutively, but are induced to do so by microbial constituents, for example LPS (bacterial lipopolysaccharide), binding to cell-surface receptors. Antigen-specific B cells that bind their specific antigen within a lymph node draining an infection will be activated there to express B7, and, through presentation of antigen-derived peptides, will activate antigen-specific naive T cells (see Figure 6.14).

Much research in immunology has involved studying the antibody and T-cell response to protein antigens in laboratory animals. Immunization with protein alone rarely led to an immune response. Reliable generation of a response required that protein antigens be mixed with certain bacteria or their breakdown products. It was discovered subsequently that the microbial components, known in this context as **adjuvants**, were inducing co-stimulatory activity in dendritic cells, macrophages, and B cells. This explains why whole microorganisms are usually more effective vaccines than highly purified antigenic macromolecules. The induction of co-stimulatory activity by common microbial constituents is believed to be a mechanism that allows the immune system to distinguish between antigens borne by infectious agents and antigens associated with innocuous proteins, including self-proteins (Figure 6.15).

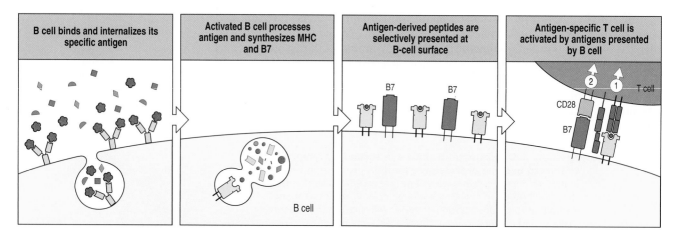

Figure 6.14 B cells can act as professional antigen-presenting cells. Immunoglobulin on the B-cell surface selectively binds specific antigen from the extracellular milieu (first panel). The antigen:immunoglobulin complexes are then internalized efficiently by receptor-mediated endocytosis. They are delivered to endocytic vesicles where they are degraded into peptides (second panel). These are bound by MHC class II molecules and the peptide:MHC class II complexes are transported to the cell surface (third panel). Naive T cells can then be activated by these B cells, which have also been induced to express co-stimulatory B7 molecules on their surface (fourth panel).

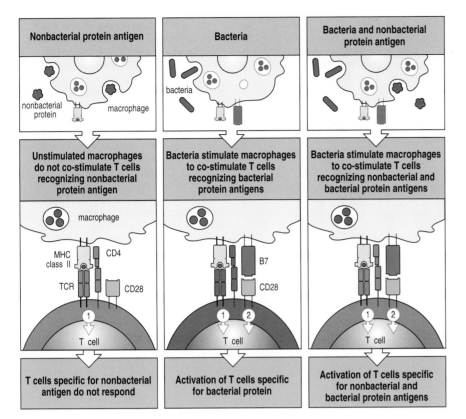

Figure 6.15 Microbial substances can induce co-stimulatory activity in macrophages. When macrophages process protein antigens in the absence of microbial components that induce co-stimulatory activity, presentation of the antigen to specific T cells causes them to become anergic (refractory to activation) (left panels). Almost all microbes can induce the expression of co-stimulators by antigen-presenting cells, and macrophages presenting peptide antigens derived by degradation of such organisms will activate naive T cells (center panels). When bacteria are mixed with protein antigens, the protein antigens are rendered immunogenic because the bacteria induce expression of co-stimulatory B7 molecules by the antigen-presenting cells (right panels). Such added bacteria act as adjuvants. The presentation of bacterial antigens to specific T cells is not shown in the right panels.

6-6 When T cells are activated by antigen, signals from T-cell receptors and co-receptors alter the pattern of gene transcription

When a T cell binds the peptide:MHC complexes on an antigen-presenting cell, the receptor–ligand interactions occur at localized and apposed areas of the two cell membranes. Within these areas, the specific MHC:peptide complexes on the antigen-presenting cell and the T-cell receptors and co-receptors cluster together, with cell adhesion molecules forming a tight seal around the area. The signal that antigen has bound to the T-cell receptor is transmitted to the interior of the T cell by the cytoplasmic tails of the CD3 proteins, which are associated with the antigen-binding α and β chains of the receptor (see Figure 3.6, p. 72). The cytoplasmic tails of all the CD3 proteins contain sequences called **immunoreceptor tyrosine-based activation motifs (ITAMs)**, which associate with cytoplasmic **protein tyrosine kinases**. These kinases are activated by receptor clustering and phosphorylate tyrosine residues in the ITAMs. Enzymes and other signaling molecules bind to the phosphorylated tyrosine residues and become activated in their turn. In this way, pathways of intracellular signaling are initiated that end with alterations in gene expression.

Signals from both the T-cell receptor and the CD4 or CD8 co-receptor combine to stimulate the T cell (Figure 6.16). The cytoplasmic tails of both CD4 and CD8 are associated with a protein tyrosine kinase called **Lck**. On formation of the T-cell receptor:MHC:co-receptor complex this kinase activates a cytoplasmic protein tyrosine kinase called **ZAP-70** (ζ chain-associated protein of 70 kDa molecular mass), which binds to the phosphorylated tyrosines on the ζ chain of the T-cell receptor complex. ZAP-70 is instrumental in initiating the intracellular signaling pathway. The importance of ZAP-70 for all

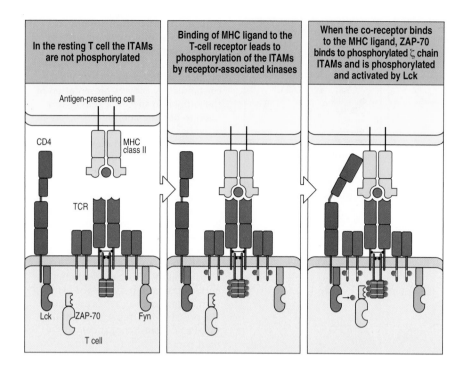

| In the resting T cell the ITAMs are not phosphorylated | Binding of MHC ligand to the T-cell receptor leads to phosphorylation of the ITAMs by receptor-associated kinases | When the co-receptor binds to the MHC ligand, ZAP-70 binds to phosphorylated ζ chain ITAMs and is phosphorylated and activated by Lck |

Figure 6.16 Clustering of the T-cell receptor and a co-receptor initiates signaling within the T cell. When T-cell receptors become clustered on binding MHC:peptide complexes on the surface of an antigen-presenting cell, activation of receptor-associated kinases, such as Fyn, leads to phosphorylation of the CD3γ, δ, and ε ITAMs (yellow, with phosphorylated tyrosines shown as small pink circles) as well as those on the ζ chain. The tyrosine kinase ZAP-70 binds to the phosphorylated ITAMs of the ζ chain but is not activated until the co-receptor binds to the MHC molecule on the antigen-presenting cell (here shown as CD4 binding to an MHC class II molecule), which brings the kinase Lck into the complex. This phosphorylates and activates ZAP-70.

subsequent signaling events was revealed by a patient with immunodeficiency due to the absence of functional ZAP-70. Although the patient had normal numbers of T cells, these were unable to develop intracellular signals on engagement of their antigen receptors.

The participation of the co-receptor is essential for effective T-cell stimulation. A target cell minimally requires about 100 specific peptide:MHC complexes to trigger a naive T cell. Human cells express between 10,000 and 100,000 MHC molecules per cell, so that 0.1–1% of the MHC molecules on a target cell must bind the same peptide for the cell to activate a T cell. In the absence of the correct co-receptor—CD8 for peptides presented by MHC class I and CD4 for peptides presented by MHC class II—stimulation of the T cell becomes highly inefficient, requiring about 10,000 specific peptide:MHC complexes on the target cell, a number almost never reached *in vivo*.

Once activated, the T-cell-specific kinase ZAP-70 triggers three signaling pathways that are common to many types of cell (Figure 6.17). In naive T cells, they lead to changes in gene expression produced by the transcriptional activator **NFAT** (**Nuclear Factor of Activated T cells**) in combination with other transcription factors. One pathway initiated by ZAP-70 leads via the second messenger inositol trisphosphate to the activation of NFAT. A second pathway leads to the activation of protein kinase C, which results in the induction of the transcription factor **NFκB**. The third signaling pathway initiated by ZAP-70 involves the activation of Ras, a GTP-binding protein, and leads to the activation of a nuclear protein called Fos, which is a component of the transcription factor **AP-1**. The co-stimulatory signals delivered through CD28, lead, among other things, to activation of the Jun protein, which, with Fos, forms the AP-1 transcription factor.

The combined actions of NFAT, AP-1, and NFκB turn on the transcription of genes that direct T-cell proliferation and the development of effector function. One of the most important of these genes is that for the cytokine interleukin-2.

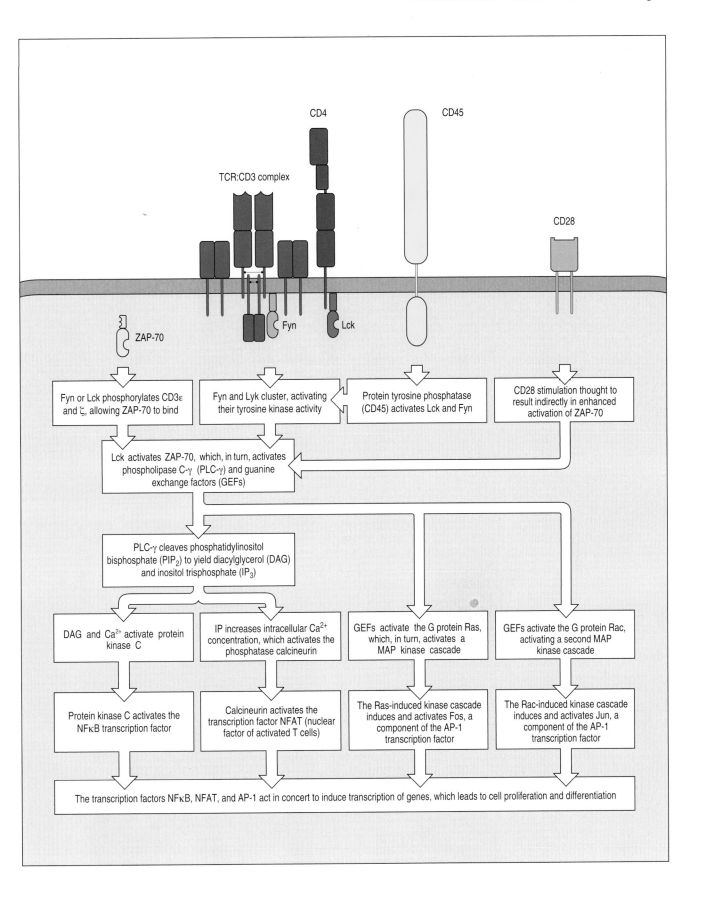

Figure 6.17 Simple outline of the intracellular signaling pathways initiated by the T-cell receptor complex, its CD4 co-receptor, and CD28. Similar pathways operate in CD8 T cells, as CD8 interacts with Lck, like CD4.

6-7 Proliferation and differentiation of activated T cells are driven by the cytokine interleukin-2

Activation by a professional antigen-presenting cell initiates a program of differentiation in the T cell that starts with a burst of cell division and then leads to the acquisition of effector function. This program is under the control of a cytokine called **interleukin-2** (**IL-2**), which is synthesized and secreted by the activated T cell itself. IL-2 binds to IL-2 receptors at the T-cell surface to drive clonal expansion of the activated cell. IL-2 is one of a number of cytokines produced by activated and effector T cells that control the development and differentiation of cells in the immune response.

The production of IL-2 requires both the signal delivered through the T-cell receptor:co-receptor complex and the co-stimulatory signal delivered through CD28 (see Section 6-4). Signals delivered through the T-cell receptor complex activate the transcription factor NFAT, which activates transcription of the IL-2 gene. IL-2 and other cytokines have powerful effects on cells of the immune system, and their production is precisely controlled in both time and space. To avoid overproduction, cytokine mRNA is inherently unstable, and sustained production of IL-2 requires the stabilization of the mRNA. Stabilization is one of the functions of the co-stimulatory signal, and it causes a 20- to 30-fold increase in IL-2 production by the T cell. A second effect of co-stimulation is the activation of additional transcription factors that increase the rate of transcription of the IL-2 gene threefold. The principal effect of co-stimulation is, therefore, to increase the synthesis of IL-2 by the T cell by some 100-fold.

IL-2 binds to a high-affinity IL-2 receptor, whose expression is also induced by T-cell activation (Figure 6.18). On binding IL-2, this receptor triggers the T cell to progress through cell division. T cells activated in this way can divide two to three times a day for about a week, enabling a single activated T cell to produce thousands of daughter cells. This proliferative phase is of crucial importance to the immune response because it produces large numbers of antigen-specific effector cells from rare antigen-specific naive T cells. In the response to certain viruses, nearly 50% of the CD8 T cells present at the peak of the response are specific for a single viral peptide:MHC complex.

The importance of IL-2 in the activation of the adaptive immune response is reflected in the mode of action of the immunosuppressive drugs cyclosporin A (cyclosporine), tacrolimus (also called FK506), and rapamycin (also called sirolimus), which are used to prevent the rejection of organ transplants. Cyclosporin A and tacrolimus inhibit IL-2 production by disrupting signals from the T-cell receptor, whereas rapamycin inhibits signaling from the IL-2 receptor. These drugs, therefore, suppress the activation and differentiation of naive T cells and all immune responses that require activated T cells.

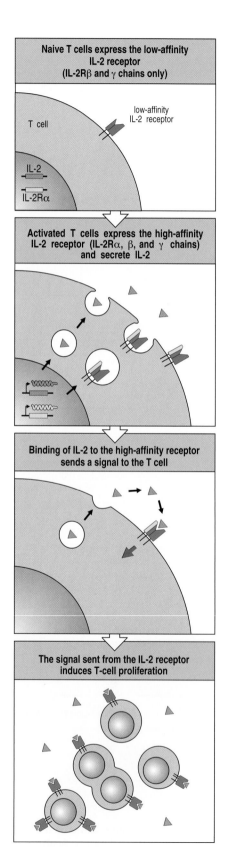

Figure 6.18 Activated T cells secrete and respond to interleukin-2 (IL-2). Naive T cells express the low-affinity receptor for IL-2, which consists of β and γ chains. Activation of a naive T cell by the recognition of a peptide:MHC complex accompanied by co-stimulation induces the synthesis and secretion of IL-2 and the synthesis of the IL-2 receptor α chain (yellow). The cell also enters the first phase (G1) of the cell-division cycle. The α chain combines with the β and γ chains to make a high-affinity receptor for IL-2. IL-2 binds to the IL-2 receptor, producing an intracellular signal that promotes T-cell proliferation.

6-8 Antigen recognition by a naive T cell in the absence of co-stimulation leads to the T cell becoming nonresponsive

Among the mature naive T cells entering the peripheral circulation from the thymus are some that are specific for self-proteins expressed by cells not encountered in the thymus. However, these T cells are unlikely to be activated because the cells presenting these self-antigens will not express co-stimulatory molecules. When the T-cell receptor on a mature naive T cell binds to a peptide:MHC complex on a cell that does not express the co-stimulatory molecule B7, the T cell becomes nonresponsive, a state described as **anergy**. In this state, the T cell cannot be activated even if it subsequently encounters its specific antigen presented by a professional antigen-presenting cell (Figure 6.19). The principal characteristic of anergic T cells is their inability to make IL-2; they are, therefore, unable to stimulate their own proliferation and differentiation. Thus, induction of tolerance in the mature T-cell repertoire seems to be based on the requirement that antigen-specific stimulation and co-stimulation of T cells are both delivered by the same antigen-presenting cell.

In Section 6-5 we saw how immunization with a protein antigen in the absence of microbes or their products does not activate T cells specific for the antigen. In this situation, the specific T cells become anergized, producing a temporary state of tolerance to the antigen.

6-9 On activation, CD4 T cells can acquire different helper functions

Toward the end of the proliferative phase, activated T cells acquire the capacity to synthesize the proteins they need to perform the specialized functions of effector T cells. For CD4 T cells these proteins comprise cell-surface molecules and soluble cytokines that activate and help other types of cell—principally macrophages and B cells—to participate in the immune response. Because of these facilitating functions, CD4 T cells are called helper cells. Upon activation, CD4 T cells can differentiate along two pathways, giving rise to either CD4 **T$_H$1 cells** or CD4 **T$_H$2 cells**. The cytokines secreted by T$_H$1 cells lead to macrophage activation, inflammation, and the production of opsonizing antibodies that enhance the phagocytosis of pathogens. The cytokines secreted by T$_H$2 cells lead mainly to B-cell differentiation and the production of neutralizing antibodies (Figure 6.20).

Figure 6.19 T-cell tolerance to antigens expressed on nonprofessional antigen-presenting cells results from antigen recognition in the absence of co-stimulation. A naive T cell can be activated only by an antigen-presenting cell carrying both a specific peptide:MHC complex and a co-stimulatory molecule on its surface. This combination results in the naive T cell's receipt of signal 1 from the T-cell receptor and signal 2 from the co-stimulator (left panel). When the antigen-presenting cell has the specific peptide:MHC complex to deliver signal 1, but no co-stimulator to deliver signal 2, the T cell enters a nonresponsive state called anergy (center panel). When the antigen-presenting cell has a co-stimulator to deliver signal 2, but no specific peptide:MHC complex to deliver signal 1, the naive T cell neither responds nor becomes anergic (right panel).

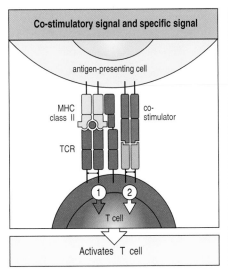

Co-stimulatory signal and specific signal

antigen-presenting cell

MHC class II / co-stimulator

TCR

1   2

T cell

Activates T cell

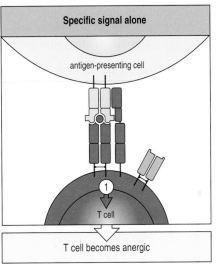

Specific signal alone

antigen-presenting cell

1

T cell

T cell becomes anergic

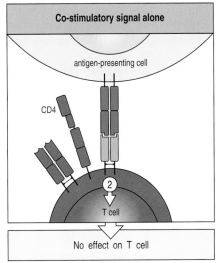

Co-stimulatory signal alone

antigen-presenting cell

CD4

2

T cell

No effect on T cell

The mechanisms that determine which differentiation pathway an activated naive T cell will take are poorly understood. Factors likely to be important include the types of cytokines already present as a result of preceding innate immune responses, the type of professional antigen-presenting cell activating the T cell, cytokines produced by the antigen-presenting cell, the abundance of specific peptide:MHC complexes on the surface of the antigen-presenting cell, and the affinity of the peptide:MHC complex for the T-cell receptor.

Although most adaptive immune responses involve contributions from both T_H1 and T_H2 cells, there are circumstances in which the response becomes biased toward either a T_H1 or a T_H2 response. A response biased toward T_H1 cells corresponds to what has traditionally been described as **cell-mediated immunity**, a response dominated by the effector cells of the immune system. On the other hand, a response biased toward T_H2 cells is dominated by antibodies, and corresponds to the traditional description of **humoral immunity**—humors being an alternative term for the body fluids in which antibodies are present.

The cytokines produced by effector T_H1 cells can suppress the differentiation of T_H2 cells, and the cytokines produced by effector T_H2 cells can suppress the differentiation of T_H1 cells. So once a CD4 T-cell response develops a bias, that bias can be reinforced. This occurs in patients with leprosy, a disease caused by infection with *Mycobacterium leprae*, a bacterium that grows within the vesicular system of macrophages. In leprosy patients, the immune response becomes strongly biased toward either a T_H1 or T_H2 response, a choice that profoundly influences disease progression. A T_H1-biased response enables the infected macrophages to suppress bacterial growth and, although skin and peripheral nerves are damaged by the chronic inflammatory response, the disease progresses slowly and patients usually survive. For patients making a T_H2-biased response the situation is quite different. Inside macrophages the mycobacteria are inaccessible to specific antibody, and by growing unchecked they cause gross tissue destruction, which is eventually fatal (Figure 6.21). The visible symptoms of disease in patients making a T_H1 or T_H2 response to *M. leprae* are so dissimilar that the conditions are given different names: tuberculoid leprosy and lepromatous leprosy, respectively.

6-10 Naive CD8 T cells can be activated in different ways to become cytotoxic effector cells

The activation of naive CD8 T cells to cytotoxic effector cells generally requires stronger co-stimulatory activity than is needed to activate CD4 T cells. Only dendritic cells, the most potent of the antigen-presenting cells, provide sufficient co-stimulation. When activated by antigen and co-stimulatory molecules on a dendritic cell, CD8 T cells are stimulated to synthesize both the cytokine IL-2 and its high-affinity receptor, which together induce the proliferation and differentiation of the CD8 T cells (Figure 6.22, left panels).

In circumstances in which the antigen-presenting cell offers suboptimal co-stimulation, CD4 T cells can help to activate naive CD8 T cells. To do this, the naive CD8 cell and the CD4 T cell recognize their specific antigens on the same antigen-presenting cell. When the CD4 T cell is already an effector cell, recognition of antigen causes it to secrete cytokines that then induce the antigen-presenting cell to increase its level of co-stimulators (see Figure 6.22, center panels). The activated antigen-presenting cell then activates the naive CD8 T cell. In this mechanism, the CD4 T cell and the CD8 T cell can either interact simultaneously (as shown in Figure 6.22 center panels) or successively with the antigen-presenting cell. When the CD4 T cell is a naive cell, it

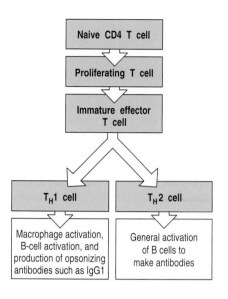

Figure 6.20 The stages of activation of CD4 T cells. Naive CD4 T cells first respond to peptide:MHC class II complexes by synthesis of IL-2 and proliferation. The progeny cells have the potential to become either T_H1 or T_H2 cells.

Infection with *Mycobacterium leprae* can result in different clinical forms of leprosy	
There are two polar forms, tuberculoid and lepromatous leprosy, but several intermediate forms also exist	
Tuberculoid leprosy	**Lepromatous leprosy**

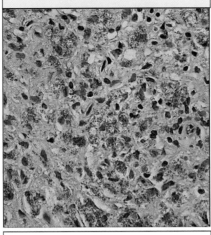

Organisms present at low to undetectable levels	Organisms show florid growth in macrophages
Low infectivity	High infectivity
Granulomas and local inflammation. Peripheral nerve damage	Disseminated infection. Bone, cartilage, and diffuse nerve damage
Normal serum immunoglobulin levels	Hypergammaglobulinemia
Normal T-cell responsiveness. Specific response to *M. leprae* antigens	Low or absent T-cell responsiveness. No response to *M. leprae* antigens

Figure 6.21 Responses to *Mycobacterium leprae* are sharply differentiated in lepromatous and tuberculoid leprosy. The photographs show sections of lesion biopsies stained with hematoxylin and eosin. Infection with *M. leprae* bacilli, which can be seen in the right-hand photograph as numerous small dark red dots inside macrophages, can lead to two very different forms of the disease. In tuberculoid leprosy (left), growth of the microorganism is well controlled by T_H1-like cells that activate infected macrophages. The tuberculoid lesion contains granulomas and is inflamed, but the inflammation is localized and causes only local peripheral nerve damage. In lepromatous leprosy (right), infection is widely disseminated and the bacilli grow uncontrolled in macrophages. In the late stages there is severe damage to connective tissues and to the peripheral nervous system. There are several intermediate stages between these two polar forms. Photographs courtesy of G. Kaplan.

is activated by the antigen-presenting cell to produce IL-2, which can then drive the proliferation and differentiation of the CD8 T cell (see Figure 6.22, right panels). This latter mechanism works because engagement of the T-cell receptor is sufficient to induce CD8 T cells to express the high-affinity IL-2 receptor, although not to make their own IL-2. This mechanism requires the two T cells to interact simultaneously with the antigen-presenting cell; the close juxtaposition of the two T cells on the surface of the antigen-presenting cell is needed to ensure that enough IL-2 is captured by the CD8 T cell to induce its activation.

The more stringent requirements for the activation of naive CD8 T cells means that they are only activated when the evidence of infection is unambiguous. Cytotoxic T cells inflict damage on any target tissue to which they are directed, and their actions will only be of benefit to the host if a pathogen is eliminated in the process. Even then, the actions of cytotoxic T cells can have deleterious effects. For example, in fighting viral airway infections, cytotoxic T cells prevent viral replication by destroying the epithelial layer, but this makes the underlying tissue vulnerable to secondary bacterial infection.

Summary

Almost all types of adaptive immune response are started by the activation of antigen-specific naive T cells. Their proliferation and differentiation to form large clones of antigen-specific effector T cells form the first stage of a primary immune response. T-cell activation takes place in the secondary

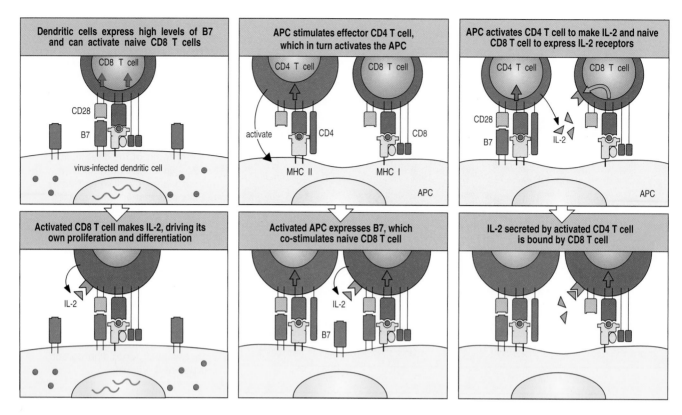

Figure 6.22 Three ways of activating a naive CD8 T cell.
The left panels show how a naive CD8 T cell can be activated directly by a virus-infected dendritic cell. The center and right panels show two ways in which antigen-presenting cells (APC) that offer suboptimal co-stimulation can interact with a CD4 T cell to stimulate a naive CD8 T cell. One way is for cytokines secreted by the CD4 T cell to improve the co-stimulation of the antigen-presenting cell, for example by the induction of B7 expression (center panels). A second way is for cytokines secreted by the CD4 T cell, for example IL-2, to act directly on a neighboring CD8 T cell (right panels).

lymphoid tissues and requires antigen to be presented to the naive T cell by a professional antigen-presenting cell. Professional antigen-presenting cells express co-stimulatory molecules that engage ligands on the T-cell surface. This interaction induces the production of the cytokine IL-2 by the T cell, which is required for clonal expansion and differentiation. Activation requires signals from the T-cell receptor, the co-receptor, and co-stimulatory molecules. When a naive T cell engages specific antigen on a nonprofessional antigen-presenting cell, the lack of co-stimulation induces anergy, a state of tolerance. The whole course of T-cell activation occurs in the immediate environment of the professional antigen-presenting cell. Throughout the process, the naive T cell and its numerous progeny maintain contact with the same antigen-presenting cell.

The three major types of professional antigen-presenting cell are the dendritic cell, macrophage, and B cell. These three cells specialize in the presentation of different types of antigen and in the induction of effector T cells with distinct functions. The same mechanisms are used to activate CD4 T cells and CD8 T cells, although CD8 T cells require stronger co-stimulation, often involving help from CD4 T cells. Whereas CD8 T cells are all destined to have cytotoxic effector function, CD4 T cells can differentiate along alternative pathways to produce effector cells that secrete different cytokines and drive the immune response in different directions. CD4 T_H1 cells secrete mainly cytokines that favor cell-mediated immune responses, whereas CD4 T_H2 cells secrete mainly cytokines that stimulate B cells to produce antibodies.

The properties and functions of effector T cells

After differentiating in the secondary lymphoid tissues, effector T cells detach themselves from the antigen-presenting cell that nursed their differentiation. CD8 cytotoxic T cells and most CD4 T_H1 cells leave the lymphoid tissues and enter the blood to seek out the sites of infection, whereas CD4 T_H2 cells mostly remain in the secondary lymphoid tissues. T-cell effector function is turned on when the T-cell receptors bind to peptide:MHC complexes on a target cell. This stimulates the T cell to release effector molecules that act on the target cell. As well as possessing specialized functions, effector T cells differ from naive T cells in ways that enable them to act more effectively as effector cells. We shall consider this first before discussing the specialized functions of CD8 T cells and CD4 T_H1 and T_H2 cells.

6-11 Effector T cells can be stimulated by antigen in the absence of co-stimulatory signals

Activated effector T cells differ from resting naive T cells in the types of molecule present at the cell surface and their abundance (Figure 6.23). One of the major changes that occurs in effector T cells enables them to respond to their specific antigen without the need for co-stimulation via B7–CD28 interaction. This means that they can respond to antigen on cells other than professional antigen-presenting cells. The benefit gained from these relaxed activation requirements is most easily understood for cytotoxic CD8 T cells. These must be able to recognize and kill all manner of cell types that become infected with viruses, even though only a small minority of cells express co-stimulators. Effector CD4 T cells, which interact mainly with B cells in lymphoid tissues and with macrophages at sites of infection, also benefit from this change. Because macrophages and B cells express varying levels of co-stimulatory activity, relaxation of the requirement for co-stimulation increases the number of antigen-presenting cells that can stimulate effector CD4 T cells, and thus strengthens the overall immune response.

Effector T cells also express two to four times as much of the cell adhesion molecules CD2 and LFA-1 as naive T cells, and so are able to interact with target cells expressing lower levels of ICAM-1 and LFA-3 than those found on the professional antigen-presenting cells (see Figure 6.23). The interaction between an effector T cell and its target is short-lived unless the T-cell receptor is engaged by specific antigen. When this happens, a conformational

Figure 6.23 Activation of T cells changes the expression of several cell-surface molecules. The example here is a CD4 T cell. Resting naive T cells express L-selectin, by which they home to lymph nodes, and relatively low levels of other adhesion molecules, such as CD2 and LFA-1. Upon activation, expression of L-selectin ceases and increased amounts of the integrin LFA-1 are made. A newly expressed integrin called VLA-4 is a homing receptor for vascular endothelium at sites of inflammation; it guides activated T cells to infected tissues. Activated T cells have more CD2 on their surface, which increases adhesion to target cells, and also a higher density of the adhesion molecule CD44. Alternative splicing of RNA made from the CD45 gene causes activated T cells to express the CD45RO isoform, which associates with the T-cell receptor and CD4. This change makes the T cell more sensitive to stimulation by lower concentrations of peptide:MHC complexes.

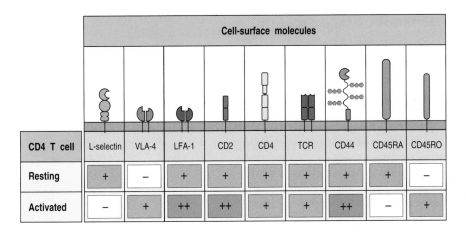

Cell-surface molecules								
CD4 T cell	L-selectin	VLA-4	LFA-1	CD2	CD4	TCR	CD44	CD45RA	CD45RO
Resting	+	−	+	+	+	+	+	+	−
Activated	−	+	++	++	+	+	++	−	+

change in LFA-1 strengthens the adhesion between the two cells. Changes in the adhesion molecules expressed by effector T cells also cause their pattern of migration to differ from that of naive T cells. Effector T cells no longer express L-selectin and thus do not recirculate through lymph nodes by leaving the blood at high endothelial venules. Instead they express the integrin **VLA-4** (Figure 6.24). This enables them to bind to adhesion molecules expressed on the endothelial cells of blood vessels in infected and inflamed tissues, and thus to enter tissues in which their effector functions are needed.

6-12 Effector T-cell functions are performed by cytokines and cytotoxins

The molecules that carry out the effector functions of T cells fall into two broad classes: **cytokines**, which alter the behavior of their target cells, and secreted cytotoxic proteins or **cytotoxins**, which are used to kill target cells. All effector T cells produce cytokines, but of different types and in different combinations. Cytotoxins, in contrast, are the specialized products of cytotoxic CD8 T cells.

Cytokines are small secreted proteins and related membrane-bound proteins that act through cell-surface receptors and generally induce changes in gene expression within their target cell. Secreted cytokines can act on the cell that made them (**autocrine** action), as with IL-2, or they can act locally on another type of cell (**paracrine** action). Many of the cytokines made by T cells are called **interleukins** and were assigned numbers according to the order of their discovery, for example IL-2. Cytokines made by lymphocytes are often called **lymphokines**. In this book, we shall use the general term cytokine for all these molecules.

Secreted cytokines generally work locally and over a short period. Membrane-bound cytokines, such as **tumor necrosis factor-α (TNF-α)**, **CD40 ligand**, and **Fas ligand**, can have an effect only on the target cell in the localized area where the T cell is bound. The secretion of soluble cytokines is similarly focused on the target cell by polarization of the T cell's intracellular secretory apparatus, which occurs on binding of the T-cell receptor to the target cell.

The cytoplasmic tails of most cytokine receptors are associated with protein kinases known as **Janus kinases (JAKs)** (Figure 6.25). Cytokine binding causes dimerization of the cytokine receptors, which, in turn, activates the kinases to phosphorylate members of a protein family called **STATs (Signal Transducers and Activators of Transcription)**. On phosphorylation, two STATs dimerize and move from the cytoplasm to the nucleus. Here they activate specific genes, which differ according to the individual cytokine receptor–JAK–STAT pathway. These are short, direct, intracellular signaling pathways that enable cells to respond rapidly to cytokine stimulation.

A common pattern is for a membrane-associated cytokine and a secreted cytokine to work synergistically within a local area. Because of these properties, clinical trials, in which individual cytokines such as IL-2 were administered systemically as a means of boosting the immune response, were often disappointing. However, some cytokines do work at a distance, for example **IL-3** and **granulocyte-macrophage colony-stimulating factor (GM-CSF)**. When released by effector CD4 T cells, these cytokines stimulate **myelopoiesis**—the production of macrophages and granulocytes in the bone marrow. The clinical use of cytokines that stimulate the production of blood cells from the bone marrow has been successful in speeding up the regeneration of the hematopoietic system in patients who have undergone destruction of their own bone marrow before bone marrow transplantation.

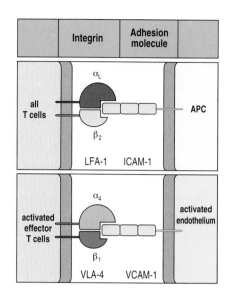

Figure 6.24 Integrin VLA-4 enables effector T cells to home to inflamed tissue. Integrins are heterodimeric proteins comprising a β chain, which defines the class of integrin, and an α chain, which defines the different integrins within a class. LFA-1 is a β2 integrin present on all leukocytes, including T cells. It binds ICAMs and is important in the adhesive interactions that mediate cell migration and in the interactions of T cells with antigen-presenting cells (APC) or target cells; its expression is increased in effector T cells. VLA-4 is a β1 integrin that increases in abundance upon T-cell activation. It binds to the cell adhesion molecule VCAM-1, which is selectively expressed on the endothelium of blood vessels in inflamed tissue, and, as discussed further in Chapter 8, is important for recruiting effector T cells into sites of infection.

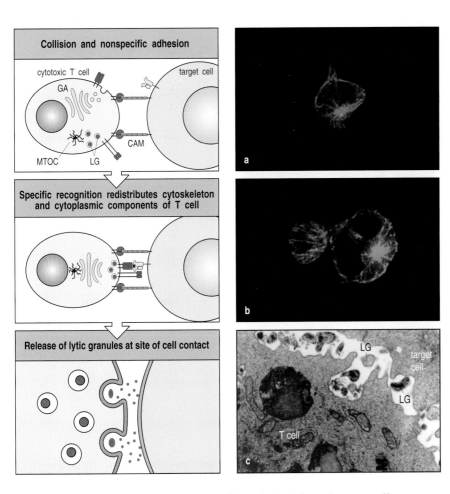

Figure 6.28 When cytotoxic T cells recognize specific antigen the delivery of cytotoxins is aimed directly at the target cell. As shown in the panels on the left, initial adhesion to a target cell has no effect on the location of the lytic granules (LG) (top panel). Engagement of the T-cell receptor causes the T cell to become polarized: the cortical actin cytoskeleton at the site of contact reorganizes, enabling the microtubule-organizing center (MTOC), the Golgi apparatus (GA), and the lytic granules to align towards the target cell (center panel). Proteins stored in lytic granules are then directed onto the target cell (bottom panel). The photomicrograph in panel a shows an unbound, isolated cytotoxic T cell. The microtubules are stained green and the lytic granules red. Note how the lytic granules are dispersed throughout the T cell. Panel b depicts a cytotoxic T cell bound to a (larger) target cell. The lytic granules are now clustered at the site of cell–cell contact in the bound T cell. The electron micrograph in panel c shows the release of granules from a cytotoxic T cell. Panels a and b courtesy of G. Griffiths. Panel c courtesy of E.R. Podack.

At sites of infection, cytotoxic CD8 T cells and the infected target cells are surrounded by healthy cells and cells of the immune system that have infiltrated the infected tissue. Because of their antigen specificity, cytotoxic T cells pick out only infected cells for attack and leave healthy cells alone. The T cell focuses granule secretion on the small localized area of the target cell where it is attached to the T cell (Figure 6.28). In this way, cytotoxic granules neither attack healthy neighbors of an infected cell nor kill the T cell itself. As the target cell starts to die, the cytotoxic T cell is released from the target cell and starts to make new granules. Once new granules have been made, the cytotoxic T cell is able to kill another target cell. In this manner, one cytotoxic T cell can kill many infected cells in succession (Figure 6.29).

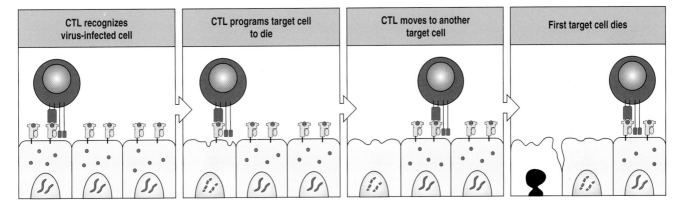

Figure 6.29 Cytotoxic CD8 T cells kill infected cells selectively. Specific recognition of peptide:MHC complexes on an infected cell by a cytotoxic CD8 T cell (CTL) programs the infected cell to die. The T cell detaches from its target cell and synthesizes a new set of lytic granules. The cytotoxic T cell then seeks out and kills another target.

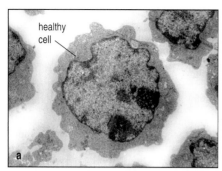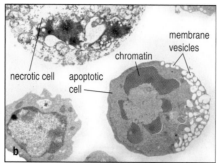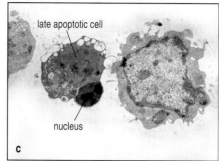

Besides their cytotoxic action, CD8 T cells also contribute to the immune response by secreting cytokines. Secretion of **IFN-γ** inhibits the replication of viruses in the infected cells and increases the processing and presentation of viral antigens by MHC class I molecules. Another effect of IFN-γ is to activate macrophages in the vicinity of the cytotoxic T cells. These macrophages get rid of the dying infected cells, thereby allowing the T cells more room for maneuver and also helping the damaged tissue to heal and regenerate.

6-14 Cytotoxic T cells kill their target cells by inducing apoptosis

Cells killed by cytotoxic CD8 T cells do not lyse or disintegrate, like cells undergoing **necrosis** due to physical or chemical injury. Instead, the cells targeted by cytotoxic T cells shrivel and shrink (Figure 6.30). This type of cell death prevents not only pathogen replication but also the release of infectious bacteria or virus particles. This cell suicide is called **apoptosis** or **programmed cell death**, and is induced in the target cell by the cytotoxins released by the cytotoxic T cell.

Soon after contact with a cytotoxic T cell, the target cell's DNA starts to be fragmented by the cell's own nucleases. These cleave between the nucleosomes to give DNA fragments that are multiples of 200 base pairs in length and are characteristic of apoptosis. Eventually the nucleus becomes disrupted and there is loss of membrane integrity and normal cell morphology. The cell destroys itself from within. It shrinks by the shedding of membrane-bound vesicles (see Figure 6.30, center panel) and the degradation of the cell contents until little is left. The changes that occur in the plasma membrane during apoptosis are recognized by phagocytes, which speed the dying cell on its way by ingesting and digesting it. The apoptotic processes that degrade the infected human cell also act upon the infecting pathogen. In particular, the breakdown of viral nucleic acids prevents the assembly of infectious virus particles that might cause further infection were they to escape from the dying cell. A five-minute contact between a cytotoxic T cell and its target cell is all it takes for the target cell to be programmed to die, even though visible evidence of death takes longer to become obvious (Figure 6.31).

Cytotoxic T cells can induce apoptosis by two different pathways. The first is initiated by cytotoxins that they release. These are **perforin**, a protein that polymerizes to form transmembrane pores of 160 Å diameter in cell membranes, another membrane-perturbing protein called **granulysin**, and the **granzymes**, a family of three serine proteases related to the pancreatic protease trypsin. A current model for the killing mechanism is that perforin and granulysin make pores in the target-cell membrane through which the granzymes can enter. Once inside, the granzymes cleave certain cell proteins, leading to the activation of nucleases and other enzymes that initiate apoptosis.

Figure 6.30 Apoptosis. Panel a shows an electron micrograph of a healthy cell with a normal nucleus. In the bottom right of panel b is an apoptotic cell at an early stage. The chromatin in the nucleus has become condensed (shown in red); the plasma membrane is well defined and is shedding vesicles. In contrast, the plasma membrane of the necrotic cell shown in the upper left part of panel b is poorly defined. The middle cell shown in panel c is at a late stage in apoptosis. It has a very condensed nucleus and no mitochondria, and the cytoplasm and cell membranes have largely been lost through vesicle shedding. Photographs courtesy of R. Windsor and E. Hirst.

A second way of inducing apoptosis is by interactions between cell-surface molecules on the cytotoxic T cell and the target cell. Activated cytotoxic T cells express the cell-surface cytokine Fas ligand, which binds to **Fas** molecules on the target-cell surface. This interaction sends signals to the target cell to undergo apoptosis. Although probably a minor pathway for killing infected cells, apoptosis induced by Fas–Fas ligand interaction is the main route by which unwanted lymphocytes are disposed of during lymphocyte development and in the course of an immune response. Individuals who lack functional Fas molecules cannot control the size of their lymphocyte population nor remove autoimmune cells. Consequently, they suffer from a disease in which the secondary lymphoid organs become swollen in the absence of infection (Figure 6.32) and autoimmune responses that attack healthy blood cells, platelets, and liver cells can be made. This disease, which is usually caused by inheritance of one nonfunctional copy of the Fas gene, is called autoimmune lymphoproliferative syndrome (ALPS).

6-15 T$_H$1 CD4 cells induce macrophages to become activated

Macrophages have receptors that bind to bacteria and other microorganisms and facilitate their phagocytosis, destruction, and intracellular degradation. As a consequence of these processes, pathogen-derived peptides are presented by MHC class II molecules on the macrophage surface, where they are able to activate naive CD4 T cells to become T$_H$1 effector cells. Some species of microorganism have adapted to the macrophage: they interfere with macrophage function by living and replicating inside the phagosome. A principal function of T$_H$1 cells is to act back on macrophages, increasing their phagocytic ability and their capacity to kill ingested microorganisms.

The enhancement of macrophage function is called **macrophage activation** and requires the interaction of peptide:MHC class II complexes on the macrophage with the T-cell receptor of the T$_H$1 cell. One effect of macrophage activation is to cause the phagosomes that contain captured microorganisms to be more efficiently fused with lysosomes, the source of hydrolytic degradative enzymes. Another effect is to increase the synthesis by macrophages of highly reactive and microbicidal molecules, such as oxygen radicals, nitric oxide (NO), and proteases, which together kill the engulfed pathogens. In patients with acquired immune deficiency syndrome (AIDS), the number of CD4 T cells decreases progressively, as does the activation of macrophages. Under these circumstances, microorganisms such as *Pneumocystis carinii* and mycobacteria, which live in the vesicles of macrophages and are normally kept in check by macrophage activation, flourish as opportunistic, and sometimes fatal, infections.

Other changes that occur on macrophage activation help to amplify the immune response. Increased expression of MHC class II molecules and B7 on

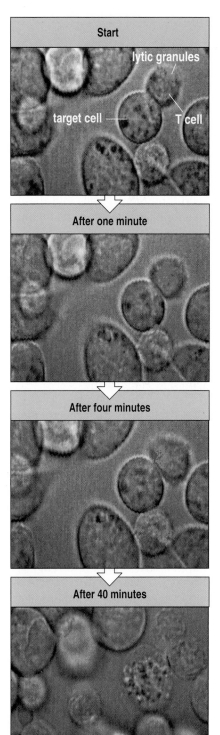

Figure 6.31 Time course of programmed cell death. The four panels show time-lapse photographs of a cytotoxic T cell killing a target cell. The lytic granules of the T cell are labeled with a red fluorescent dye. In the top panel, the T cell has just made contact with the target cell and this event is designated as the Start. At this time, the T-cell granules are distant from the point of contact with the target cell. After one minute (second panel), the granules have begun to move toward the point of target-cell attachment, a move that is essentially completed after four minutes (third panel). After 40 minutes (bottom panel), the granules have been secreted and are seen between the T cell and the target cell. The target cell has now begun to undergo apoptosis as shown by the fragmented nucleus. The T cell is ready to detach from the apoptotic cell and seek out a further target cell. Photographs courtesy of G. Griffiths.

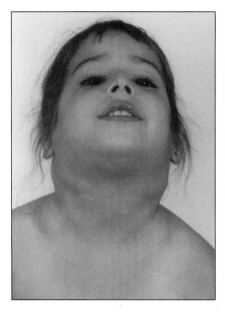
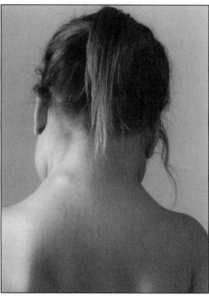

Figure 6.32 Lymphadenopathy in autoimmune lymphoproliferative syndrome (ALPS). Young girl with ALPS with very enlarged lymph nodes in her neck. Photograph courtesy of Jennifer Puck.

the macrophage surface increases antigen presentation to naive T cells, thus recruiting more T cells into the immune response. This, in turn, enhances macrophage activation and maintains increased numbers of macrophages in the activated state.

Macrophages require two signals for activation, both of which can be delivered by effector T_H1 cells. The primary signal is provided by IFN-γ, which is a characteristic cytokine produced by T_H1 cells, whereas the second signal, which makes a macrophage responsive to IFN-γ, is delivered by CD40 ligand on the surface of the T cell interacting with CD40 on the macrophage (Figure 6.33). Macrophage activation results in the increased expression of CD40 and TNF receptors, which raises the sensitivity of the macrophage to CD40 ligand and TNF-α. TNF-α, produced by the activated macrophage itself, synergizes with IFN-γ to raise the level of activation.

CD4 T cells make their effector molecules only on demand, unlike CD8 T cells. After encounter with antigen on a macrophage, an effector T_H1 cell takes several hours to synthesize the requisite effector cytokines and cell-surface molecules. During this time the T cell must maintain contact with its target cell. Newly synthesized cytokines are translocated into the endoplasmic reticulum of the T cell and delivered by secretory vesicles to the site of contact between the T cell and macrophage. Thus, they are focused on the target cell. Newly synthesized CD40 ligand is also expressed selectively at the region of contact with the macrophage. Together, this localized production of cytokines ensures the selective activation of those macrophages carrying the specific peptide:MHC complexes recognized by the T cell.

CD8 T cells are also an important source of IFN-γ and because of this they can activate macrophages. Macrophage sensitization to IFN-γ need not require the action of CD40 ligand; small amounts of bacterial polysaccharide have a similar effect and can be of particular importance when CD8 T cells, which do not express CD40 ligand, are the principal source of IFN-γ.

The microbicidal substances produced by activated macrophages are also harmful to human tissues, which inevitably suffer damage from macrophage activity. For this reason, the activation of macrophages by CD4 T_H1 cells is under strict control. Cytokines secreted by CD4 T_H2 cells, which include **transforming growth factor-β (TGF-β)**, **IL-4**, **IL-10**, and **IL-13**, inhibit

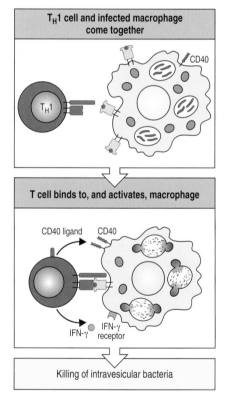

Figure 6.33 T_H1 CD4 cells activate macrophages to become highly microbicidal. When a T_H1 cell specific for a bacterial peptide contacts a macrophage that presents the peptide, the T_H1 cell is induced to secrete the macrophage-activating cytokine interferon-γ (IFN-γ) and also to express CD40 ligand at its surface. Together, these newly synthesized proteins activate the macrophage to kill the bacteria living inside its vesicles.

macrophage activation, which is an example of how cytokines secreted by T_H2 cells can control the T_H1 response. T_H1 cells stop production of IFN-γ if their antigen receptors lose contact with the peptide:MHC complexes on a macrophage, which is a further control on the T_H1 response.

6-16 T_H1 cells coordinate the host response to intravesicular pathogens

Certain microorganisms, including the mycobacteria that cause tuberculosis and leprosy, are intracellular pathogens that enjoy a protected life in the vesicular system of macrophages. The protozoan parasite *Leishmania* also survives for part of its life cycle inside vesicles in macrophages. Such microorganisms subvert the destructive mission of the macrophage for their own purposes. By sequestering themselves in this cellular compartment they cannot be reached by antibodies; neither are their peptides presented by MHC class I molecules, thus preventing the infected macrophage from being attacked by cytotoxic T cells. Mycobacteria avoid digestion by lysosomal enzymes by preventing the acidification of the phagolysosome that is required to activate the lysosomal hydrolases. Infections of this type are fought by T_H1 CD4 T cells, which help the macrophage to become activated to the point at which the intracellular pathogens are eliminated or killed.

The activation of macrophages by IFN-γ and CD40 ligand is central to the immune response against pathogens that proliferate in macrophage vesicles. In mice lacking functional IFN-γ or CD40 ligand, the ability of macrophages to kill intravesicular pathogens is impaired, and doses of mycobacteria or the parasite *Leishmania* that normal mice can withstand prove fatal. Although IFN-γ and CD40 ligand are probably the most important effector molecules of T_H1 cells, other cytokines secreted by these cells help to coordinate responses to intravesicular bacteria (Figure 6.34). Macrophages chronically infected with intravesicular bacteria can lose the capacity to be activated. Such cells can be killed by an effector T_H1 cell that uses Fas ligand or TNF to engage Fas

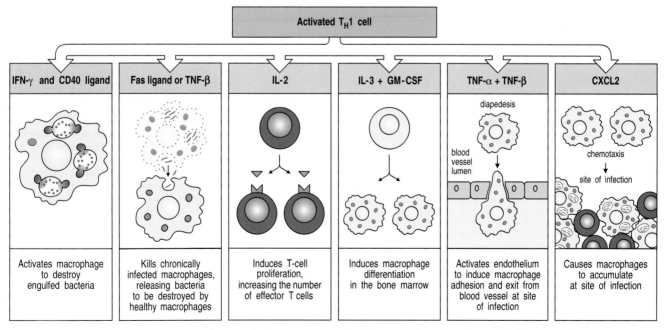

Figure 6.34 The immune response to intravesicular bacteria is coordinated by activated T_H1 cells. The activation of T_H1 cells by infected macrophages results in the synthesis of cytokines that activate the macrophage and coordinate the immune response to intravesicular pathogens.

The six panels show the effects of different cytokines and chemokines secreted by T_H1 cells. CXCL2 is the chemokine formerly known as macrophage chemoattractant protein (MCP). GM-CSF, granulocyte-macrophage colony-stimulating factor; TNF-β, tumor necrosis factor-β.

or a TNF receptor on the surface of the macrophage, inducing the macrophage to undergo apoptosis. This releases the bacteria, which are then taken up and killed by fresh macrophages.

The IL-2 produced by T_H1 cells induces T-cell proliferation and potentiates the production and release of other cytokines. Some of these recruit phagocytes—macrophages and neutrophils—to sites of infection. First, T_H1 cells secrete IL-3 and GM-CSF, which stimulate the increased production of macrophages and neutrophils in the bone marrow. Second, TNF-α and **lymphotoxin** (**LT**, also called **TNF-β**) made by T_H1 cells induces vascular endothelial cells at sites of infection to change the adhesion molecules they express so phagocytes circulating in the blood can bind to them. At this point, the chemokine **CCL2** (formerly known as **monocyte chemoattractant protein** or MCP-1), which is also produced by the T_H1 cells, guides the phagocytes between the endothelial cells and into the infected area. In total, the CD4 T_H1 cell orchestrates a multifaceted macrophage response that focuses on the destruction of pathogens taken up by macrophages.

When microbes resist the microbicidal effects of activated macrophages successfully, a chronic infection with inflammation can develop. Such areas of tissue often have a characteristic morphology, called a **granuloma**, in which a central area containing infected macrophages is surrounded by activated T cells. Giant cells resulting from the fusion of macrophages are present at the center of a granuloma, which contain the resistant pathogens. Large single macrophages, sometimes called epithelioid cells, form an epithelium-like layer around the center (Figure 6.35).

In tuberculosis, the centers of large granulomas can become cut off from the blood supply and the cells in the center die, probably from a combination of oxygen deprivation and the cytotoxic effects of the macrophages. The resemblance of the dead tissue to cheese led to this process being called **caseation necrosis**. It provides a vivid example of how CD4 T_H1 cells can produce a local pathology. Their absence, however, leads to death from disseminated infection, which is commonly seen in AIDS patients infected with opportunistic mycobacteria.

6-17 CD4 T_H2 cells activate only those B cells that recognize the same antigen as they do

During an infection, the T-cell zone of secondary lymphoid tissues contains pathogen-specific T_H2 effector cells that are the progeny of naive CD4 T cells activated by antigen-presenting dendritic cells. The main function of these T cells is to help B cells mount an antibody response against the infectious agent. Mature naive B cells passing through the lymphoid tissue pick up and present their specific antigens. As the circulating B cells pass through the T-cell zones, they make transient interactions with the T_H2 cells, whose T-cell receptors screen the peptides presented by the MHC class II molecules on the surface of the B cell. When a B cell presents the specific antigen recognized by the T_H2 cell, the adhesive interactions are strengthened and the B cell becomes trapped by the T cell. This interaction gives rise to a primary focus of activated B cells and helper T cells (see Figure 4.17, p. 114).

When the T-cell receptor of a helper T cell recognizes peptide:MHC class II complexes on the surface of a naive B cell, the T cell responds by synthesizing CD40 ligand. This molecule is involved in all T-cell interactions with B cells, which express the corresponding receptor molecule CD40. The interaction of CD40 ligand with CD40 drives the resting B cell into the cycle of cell division. The characteristic cytokine secreted by T_H2 cells on stimulation by their target cell is IL-4, which works in concert with CD40 ligand to initiate

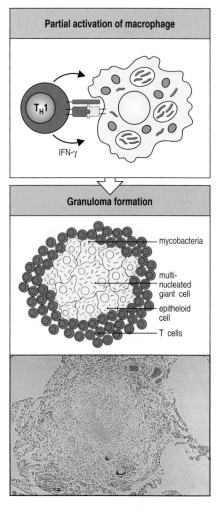

Figure 6.35 Granulomas form when an intracellular pathogen or its constituents resists elimination. In some circumstances mycobacteria (red) resist the killing effects of macrophage activation (top panel). A characteristic localized inflammatory response called a granuloma develops (second panel). The granuloma consists of a central core of infected macrophages, which can include multinucleated giant cells formed by macrophage fusions, surrounded by large single macrophages often called epithelioid cells. Mycobacteria can persist in the cells of the granuloma. The central core is surrounded by T cells, many of which are CD4 T cells. The photomicrograph (bottom panel) shows a granuloma from the lung. Photograph courtesy of J. Orrell.

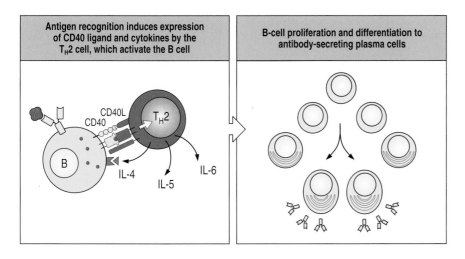

Figure 6.36 T$_H$2 cells stimulate the proliferation and differentiation of naive B cells. The specific interaction of an antigen-binding B cell with a helper T$_H$2 cell leads to the expression of CD40 ligand (CD40L) and the secretion of IL-4, IL-5, and IL-6. In concert, these T$_H$2 products drive the proliferation of B cells and their differentiation to form plasma cells dedicated to the secretion of antibody.

the proliferation and clonal expansion of B cells that precedes their differentiation into antibody-secreting plasma cells. T$_H$2 cells also produce IL-5 and IL-6, which drive further B-cell differentiation to plasma cells (Figure 6.36).

The principle governing T-cell help to B cells is that cooperation occurs only between B and T cells that are specific for the same antigen, although they usually recognize different epitopes. Such interactions are called **cognate interactions**. The peptide recognized by the T cell must be part of the same physical entity bound by the B cell's surface immunoglobulin. For example, the T cell might recognize a peptide derived from an internal protein of a virus, whereas the B cell recognizes an exposed external carbohydrate epitope of a viral capsid glycoprotein. The specialized antigen-presenting function of a B cell makes it supremely efficient at presenting peptides that derive from any protein, virus, or microorganism that specifically binds to its surface immunoglobulin. Only those B cells that selectively internalize a pathogen antigen by receptor-mediated endocytosis will present enough of the pathogen-derived peptide to engage and stimulate an antigen-specific T$_H$2 cell. It is estimated that a B cell that can use receptor-mediated endocytosis to capture a particular antigen is 10,000-fold more efficient at presenting peptides derived from that antigen than a B cell that cannot use receptor-mediated endocytosis.

Knowledge of the mechanism by which B and T cells cooperate helps in the design of vaccines. An example illustrated in Figure 6.37 is the vaccine against *Haemophilus influenzae* B, a bacterial pathogen that is life-threatening to young children when it infects the lining of the brain—the meninges—producing a meningitis that, in severe cases, causes lasting neurological damage or death. Protective immunity against *H. influenzae* is provided by antibodies specific for the capsular polysaccharides. However, a child's antibody response is weakened by the lack of associated peptide epitopes that could engage T$_H$2 cells and provide help to polysaccharide-specific B cells. To enable the immune system to make antibodies against *H. influenzae*, a vaccine was made in which the immunizing antigen was the bacterial polysaccharide covalently coupled to tetanus toxoid, a protein containing good peptide epitopes that are bound by MHC class II molecules and presented to T$_H$2 cells.

6-18 Regulatory CD4 T cells limit the activities of effector CD4 and CD8 T cells

As we saw in Section 6-9 there is a tendency for cytokines made by T$_H$1 cells to suppress the generation of T$_H$2 cells and for cytokines made by T$_H$2 cells to

Figure 6.37 Molecular complexes recognized by both B and T cells make effective vaccines. The first panel shows a naive B cell's surface immunoglobulin binding a carbohydrate epitope on a vaccine composed of a *Haemophilus* polysaccharide (blue) conjugated to tetanus toxoid (red), a protein. This results in receptor-mediated endocytosis of the conjugate and its degradation in the endosomes and lyosomes, as shown in the second panel. Peptides derived from degradation of the tetanus toxoid part of the conjugate are bound by MHC class II molecules and presented on the B cell's surface. In the third panel, the receptor of a T_H2 cell recognizes the peptide:MHC complex. This induces the T cell to secrete cytokines that activate the B cell to differentiate into plasma cells, which produce protective antibody against the *Haemophilus* polysaccharide (fourth panel).

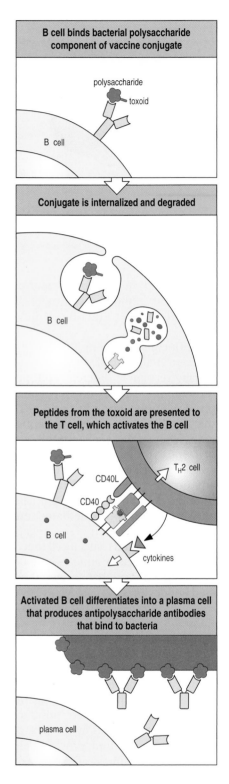

suppress the generation of T_H1 cells. Thus, each type of effector CD4 T cell can regulate the response of the other. This is an example of a more general phenomenon in which one population of lymphocytes can regulate and suppress the actions of another. The study of organ transplantation has identified populations of antigen-specific CD4 T cells whose actions can suppress the rejection response of CD4 and CD8 T cells to allogeneic MHC molecules. Similarly, in autoimmune conditions antigen-specific CD4 T cells with the capacity to suppress the response to self-antigens have been identified. Although not uniquely defined by one effector or cell-surface molecule, these **regulatory** or **suppressor CD4 T cells** usually make inhibitory cytokines, such as IL-4, IL-10, and TGF-β, and express high levels of CD25, the α chain of the IL-2 receptor. Suppression depends on physical contact between the regulatory CD4 T cell and its target T cell. Physiologically, CD4 CD25 T cells could provide a means of generating tolerance to self-antigens and preventing autoimmunity, and of limiting the proliferation of effector T cells during the immune response and the collateral damage they cause to healthy cells and tissues. In the future, manipulation or administration of donor-specific CD4 CD25 T cells might be able to reduce the likelihood of transplant rejection and, thus, reduce the doses of immunosuppressive drugs that have to be taken by patients receiving kidney and other organ transplants.

Summary

The three types of effector T cell—CD8 cytotoxic cells, CD4 T_H1 cells, and CD4 T_H2 cells—have complementary roles in the immune response. The common principle by which they function is to affect the behavior of other types of cell through intimate contact and the action of effector molecules. Naive T cells are activated to develop effector function in the secondary lymphoid tissues, whereupon cytotoxic CD8 T cells and CD4 T_H1 cells enter the blood and travel to sites of infection. In contrast, CD4 T_H2 cells remain in the secondary lymphoid tissue, where they activate naive B cells that have specificity for the same antigen as themselves. Linked recognition between T_H2 cells and B cells arises because a B cell efficiently internalizes and processes the antigen to which its surface immunoglobulin binds, and then presents peptide antigens to a T cell. When the T-cell receptor of the T_H2 cell binds to peptide:MHC complexes on the B-cell surface, the B cell becomes activated by interactions between CD40 ligand on the T cell and CD40 on the B cell, and by IL-4, the characteristic cytokine secreted by the T_H2 cell. Cytotoxic CD8 T cells induce cells overwhelmed by viral infection to die by apoptosis. This mode of death ensures that the infected cell's load of viruses is also destroyed rather than being released to infect healthy cells. Apoptosis is induced by the cytotoxic enzymes contained in secretory lytic granules that are stored by the cytotoxic T cell and are released onto the target cell membrane once contact has been established. After granule release, the T cell rapidly synthesizes new granules so it can kill several targets in succession. In contrast, the effector molecules of effector CD4 T cells are made to order

The panels of the figure are labeled:

B cell binds bacterial polysaccharide component of vaccine conjugate

polysaccharide
toxoid
B cell

Conjugate is internalized and degraded

B cell

Peptides from the toxoid are presented to the T cell, which activates the B cell

T_H2 cell
CD40L
CD40
B cell
cytokines

Activated B cell differentiates into a plasma cell that produces antipolysaccharide antibodies that bind to bacteria

plasma cell

once contact with an antigen-bearing target cell has been established. A major role of T_H1 CD4 cells is to activate macrophages, helping them to become more competent at destroying extracellular pathogens that they have taken up into their vesicular system, including those that exploit the phagocytic pathway for their own survival. IFN-γ is the characteristic cytokine of T_H1 cells and is instrumental in activating macrophages.

Summary to Chapter 6

All aspects of the adaptive immune response are initiated and controlled by effector T cells—CD4 T_H1 cells, CD4 T_H2 cells, and CD8 cytotoxic T cells. These cells differentiate from naive recirculating T cells that have been trapped and activated by antigen presented by professional antigen-presenting cells in the secondary lymphoid tissues. Inherent to the function of a professional antigen-presenting cell is the presence on the cell surface of B7 co-stimulatory proteins that interact with the CD28 protein on the naive T cell. Dendritic cells, macrophages, and B cells are the three types of professional antigen-presenting cell and each is adapted to presenting particular categories of antigen (Figure 6.38). Activation leads to cell proliferation and differentiation, all of which proceeds while the T cell remains in contact with the antigen-presenting cell. Each of the three types of effector T cell has a distinct role in the immune response, but all function through interactions with another type of cell (Figure 6.39).

Effector CD4 T_H1 cells mostly migrate from the secondary lymphoid tissues to sites of infection. There they activate tissue macrophages. This both increases the macrophages' capacity to phagocytose and kill pathogenic organisms infecting the extracellular spaces, and increases their capacity to

	Dendritic cells	Macrophages	B cells
Antigen uptake	+++ Macropinocytosis and phagocytosis by tissue dendritic cells. Viral infection	Phagocytosis +++	Antigen-specific receptor (Ig) ++++
MHC expression	Low on tissue dendritic cells. High on dendritic cells in lymphoid tissues	Inducible by bacteria and cytokines − to +++	Constitutive. Increases on activation +++ to ++++
Co-stimulator delivery	Constitutive by mature, nonphagocytic lymphoid dendritic cells ++++	Inducible − to +++	Inducible − to +++
Antigen presented	Peptides Viral antigens Allergens	Particulate antigens. Intracellular and extracellular pathogens	Soluble antigens Toxins Viruses
Location	Lymphoid tissue Connective tissue Epithelia	Lymphoid tissue Connective tissue Body cavities	Lymphoid tissue Peripheral blood

Figure 6.38 The properties of the professional antigen-presenting cells. Dendritic cells, macrophages, and B cells are the main cell types involved in the presentation of foreign antigens to naive T cells. These cells vary in their means of antigen uptake, MHC class II expression, co-stimulator expression, the type of antigen they present effectively, their locations in the body, and their surface adhesion molecules (not shown).

act as professional antigen-presenting cells. Activation also enhances the capacity of macrophages to eliminate microorganisms that deliberately parasitize macrophage vesicles. Within the secondary lymphoid tissues, effector

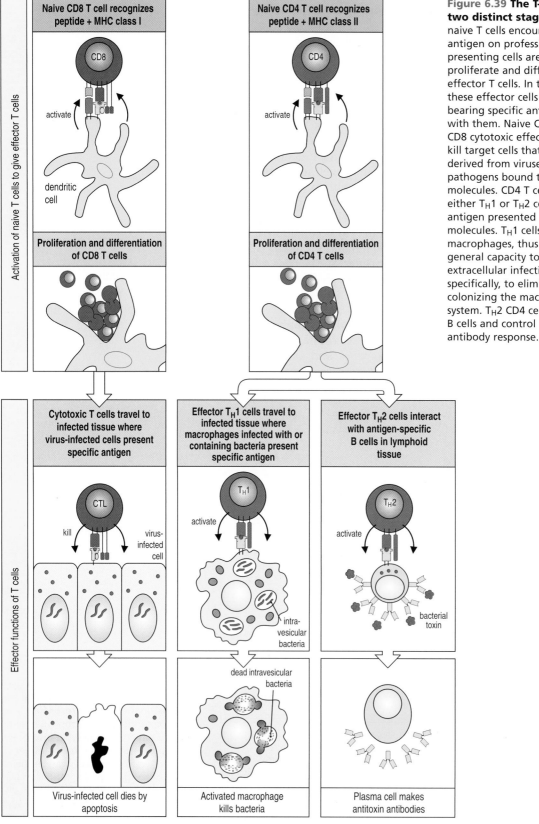

Figure 6.39 The T-cell response has two distinct stages. In the first stage, naive T cells encountering their cognate antigen on professional antigen-presenting cells are induced to proliferate and differentiate into effector T cells. In the second stage, these effector cells recognize target cells bearing specific antigen and interact with them. Naive CD8 T cells become CD8 cytotoxic effector cells (CTL), which kill target cells that present peptides derived from viruses and other cytosolic pathogens bound to MHC class I molecules. CD4 T cells differentiate into either T_H1 or T_H2 cells, which recognize antigen presented by MHC class II molecules. T_H1 cells activate macrophages, thus enhancing their general capacity to eliminate extracellular infection and, more specifically, to eliminate organisms colonizing the macrophages' vesicular system. T_H2 CD4 cells activate naive B cells and control many aspects of the antibody response.

CD4 T$_H$2 cells activate naive B cells that are specific for the same antigen as the T cell. Activated B cells divide and differentiate into antibody-secreting plasma cells, and also undergo isotype switching under the influence of effector CD4 T cells of both T$_H$2 and T$_H$1 types. The three types of effector T cell enable the human immune system to respond to different categories of infection and to different developmental stages in the course of the same infection.

Questions

Question 6–1

A. At which anatomical sites do naive T cells encounter antigen?

B. In which sites specifically would a pathogen or its antigens end up, and how, if they entered the body (i) through a small wound in the skin, (ii) from the gut, or (iii) got into the bloodstream?

C. How do T cells arrive at these sites?

D. Do all T cells leave these locations after priming, and if so, how? (See Figures 6.3 and 6.6, and Section 1-7.)

Question 6–2

Unlike innate immune responses which can begin within hours of the onset of an infection, adaptive immune responses involving T cells usually take several days. What accounts for this delay between the initiation of an infection and the engagement of an adaptive immune response?

Question 6–3

A. Which selectins, mucin-like vascular addressins and integrins play a role in the circulation of T cells between the blood and lymphoid tissues?

B. Describe in chronological order how T cells migrate across lymph node high endothelial venules (HEVs) from the blood using these molecules. (Refer to Figure 6.6.)

Question 6–4

A. Identify three types of professional antigen-presenting cells.

B. How are they distributed in secondary lymphoid tissue?

C. Which kinds of antigen do they present efficiently to T cells? (Refer to Figures 6.10 and 6.38.)

Question 6–5

A. Which cell-surface glycoprotein distinguishes professional antigen-presenting cells from other cells and is involved in co-stimulation of T cells?

B. What receptors can it bind on the T cell and what signal does it deliver in each case?

C. Explain the consequence of antigen recognition by T cells in the absence of this glycoprotein on the antigen-presenting cell. (Refer to Figure 6.9.)

Question 6–6

A. Explain the functional differences between immature and mature dendritic cells.

B. Discuss why you think these functional changes should occur.

C. Give an example of an immature and a mature dendritic cell. (Refer to Figure 6.12.)

Question 6–7

The three classes of effector T cells—cytotoxic T cells, T$_H$1 cells and T$_H$2 cells—are specialized to deal with different classes of pathogens and produce different sets of cytokines.

A. For each class, describe how antigen is recognized and the corresponding effector functions.

B. Give an example of an antigen for each class. (Refer to Figures 6.26, 6.29, 6.33, 6.36, and 6.39.)

Question 6–8

Virus-infected cells attacked and killed by effector cytotoxic T cells are often surrounded by healthy tissue which is spared from destruction.

A. Explain the mechanism that ensures that cytotoxic T cells only kill the virus-infected cells (the target cells).

B. What cytotoxins do cytotoxic T cells produce? (Refer to Figures 6.28–6.31.)

Question 6–9

What are the roles of the following molecules in the signal transduction pathway leading from the T-cell receptor: (i) the CD3 complex; (ii) protein tyrosine kinase Lck; (iii) CD45; (iv) ZAP-70; (v) the zeta chain; (vi) IP$_3$; (vii) calcineurin? (Refer to Figures 6.16 and 6.17.)

Question 6–10

A. Describe the morphology of a granuloma.

B. Which types of infection would lead to the formation of a granuloma?

C. Why is this type of pathology actually beneficial to the host? (Refer to Figure 6.35.)

Question 6–11

Cyclosporin A is an immunosuppressive drug commonly used in transplant patients to prevent graft rejection by alloreactive T cells. It acts by interfering with the signaling pathway that leads from the T-cell receptor to transcription in the nucleus of the genes for the cytokine IL-2 and the α chain of the IL-2 receptor. Why does preventing the transcription of these genes lead to immunosuppression? (Refer to Figure 6.18.)

Question 6–12

The etiological agent responsible for leprosy is *Mycobacterium leprae,* which survives and replicates within the vesicular system of macrophages. Explain the difference between tuberculoid leprosy and lepromatous leprosy in the context of T-cell differentiation and effector function. (Refer to Figures 6.20 and 6.21.)

Question 6–13

B cells are activated by CD4 T_H2 cells only if both cell types recognize the same antigen. The same epitope, however, does not need to be shared for recognition.

   A. Discuss why this characteristic is important in vaccine design.

   B. Provide an example of a conjugate vaccine used to stimulate IgG antibody synthesis to *Haemophilus influenzae* B polysaccharide. (Refer to Figure 6.37.)

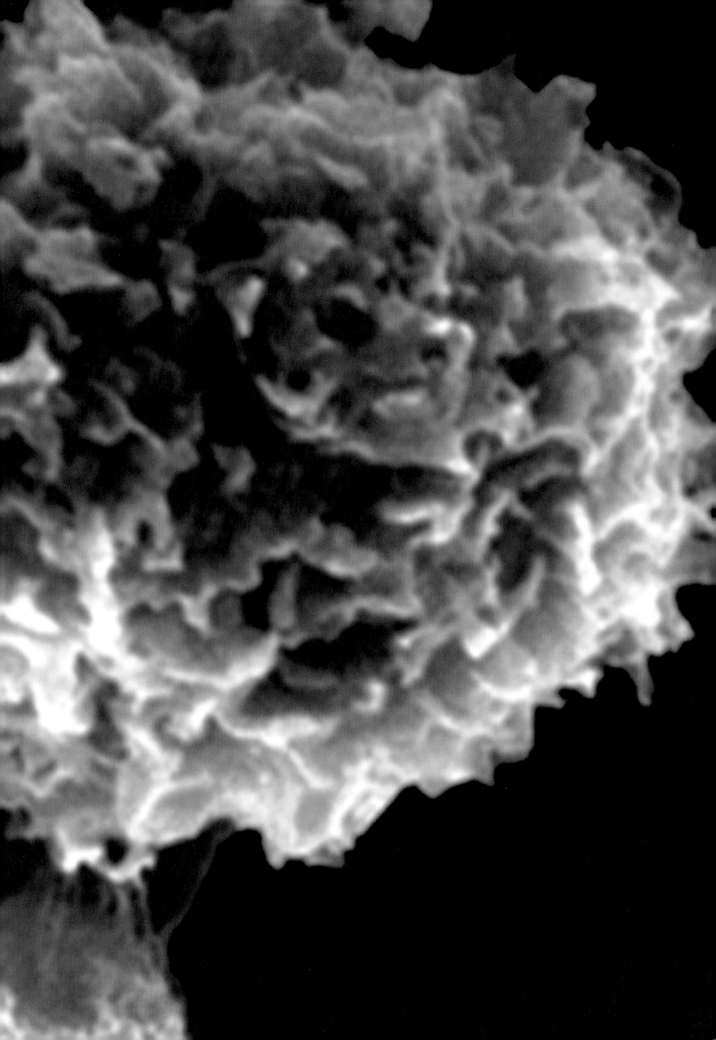

Chapter 7

Immunity Mediated by B Cells and Antibodies

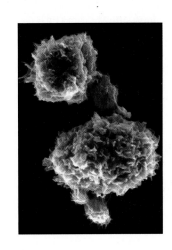

The production of antibodies is the sole function of the B-cell arm of the immune system. Antibodies are useful in the defense against any pathogen that is present in the extracellular spaces of the body's tissues. Some human pathogens, such as many species of bacteria, live and reproduce entirely within the extracellular spaces, whereas others, such as viruses, replicate inside cells but are carried through the extracellular spaces as they spread from one cell to the next. Antibodies secreted by plasma cells in secondary lymphoid tissues and bone marrow find their way into the fluids filling the extracellular spaces.

Antibodies are not in themselves toxic or destructive to pathogens; their role is simply to bind tightly to them. This can have several consequences. One way in which antibodies reduce infection is by covering up the sites on a pathogen's surface that are necessary for growth or replication, for example the viral glycoproteins that viruses use to bind to the surface of human cells and initiate infection. Such antibodies are said to **neutralize** the pathogen. In the development of a vaccine against an infectious agent or its toxic products, the gold standard that a company aims for is the induction of a neutralizing antibody. Antibodies also act as molecular adaptors that bind to pathogens with their antigen-binding arms and to receptors on phagocytic cells with their Fc regions. Thus, **opsonization**, or coating of a pathogen with antibody, promotes its phagocytosis. The antibody-directed destruction of pathogens caused by opsonization is enhanced by the actions of a set of proteins that do not discriminate between antigens and are present in blood and lymph. These proteins are collectively known as **complement** because their functions complement the antigen-binding function of the antibody.

The structure, specificity, and other properties of antibodies were discussed in Chapter 2, and the development of B cells from their origin in bone marrow to differentiation into antibody-secreting plasma cells was the subject of Chapter 4. This chapter will focus on how antibodies clear infection by targeting destructive but nonspecific components of the immune system to an infecting pathogen. In the first part of the chapter we consider the antigens that provoke a B-cell response, how the response develops, and the generation of the different antibody isotypes. The structural differences between antibody isotypes provide a variety of adaptor functions that can target antibody-bound pathogen to different types of nonspecific effector cell; these aspects of the antibody-mediated immune response will be discussed in the second part of this chapter. In the last part of the chapter, we shall look at the functions of the complement system, one of the principal mechanisms of targeting extracellular pathogens for destruction, and how it is activated by antibody.

Antibody production by B lymphocytes

The antibodies most effective at combating infection are those that are made early in an infection and bind strongly to the pathogen. On first exposure to an infectious agent, these two goals make competing demands on the immune system. As we saw in Chapters 4 and 6, B cells generally require help from activated T cells to mature into antibody-secreting plasma cells; this delays the onset of antibody production until around a week after infection. In addition, B cells take time to switch isotype and undergo affinity maturation, processes that are necessary for the production of the high-affinity antibodies that are most effective at dealing with pathogens. Thus, during the course of an infection, the effectiveness of the antibodies produced improves steadily. This experience is retained in the form of memory B cells and high-affinity antibodies, which provide long-term immunity to reinfection.

A faster primary response is made to certain bacterial antigens that are able to activate B cells without the need for T-cell help. However, the antibodies produced in such a response are predominantly of the IgM isotype and of generally low affinity. They do, however, provide an early defense, helping to keep the infection at a relatively low level until a better antibody response can be developed.

7-1 B-cell activation requires cross-linking of surface immunoglobulin

On binding to protein or carbohydrate epitopes on the surface of a microorganism, the surface IgM molecules of a naive, mature B cell become physically cross-linked to each other by the antigen and are drawn into the localized area of contact with the microbe. This clustering and aggregation of B-cell receptors sends signals from the receptor complex to the inside of the cell (Figure 7.1). Signal transduction from the B-cell receptor complex resembles, in many ways, the signaling from the T-cell receptor complex discussed in Section 6-6, p. 155. Both types of receptor are associated with cytoplasmic protein tyrosine kinases that are activated by receptor clustering, and both receptors activate similar intracellular signaling pathways.

Interaction of antigen with surface immunoglobulin is communicated to the interior of the B cell by the proteins Igα and Igβ, which are associated with IgM in the B-cell membrane to form the functional B-cell receptor. Like the CD3 polypeptides of the T-cell receptor complex, the cytoplasmic tails of Igα and Igβ each contain two immunoreceptor tyrosine-based activation motifs (ITAMs) with which the Blk, Fyn, and Lyn tyrosine kinases associate. The ITAMs become phosphorylated on tyrosine residues, which allows the Syk tyrosine kinase to bind to Igβ tails that are doubly phosphorylated. Interaction between bound Syk molecules initiates intracellular signaling pathways that lead to changes in gene expression in the nucleus (Figure 7.2).

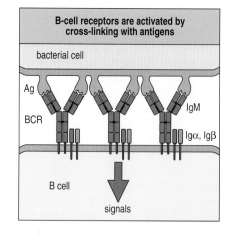

Figure 7.1 Cross-linking of antigen receptors is the first step in B-cell activation. The B-cell receptors (BCR) on B cells are physically cross-linked by the repetitive epitopes of antigens (Ag) on the surface of a bacterial cell. The B-cell receptor on a mature, naive B cell is composed of surface IgM, which binds antigen, and associated Igα and Igβ chains, which provide the signaling capacity.

Figure 7.2 Signals from the B-cell receptor initiate a cascade of intracellular signals. On clustering of the receptors, the receptor-associated tyrosine kinases Blk, Fyn, and Lyn phosphorylate the immunoreceptor tyrosine-based activation motifs (ITAMS) on the cytoplasmic tails of Igα and Igβ (shown in blue and orange, respectively). Subsequently, Syk binds to the phosphorylated ITAMs of the Igβ chain. Because there are at least two receptor complexes in each cluster, Syk molecules become bound in close proximity and can activate each other by transphosphorylation, thus initiating further signaling. Therefore, the signals produced are ultimately relayed to the B-cell nucleus where they induce changes in gene expression required for B-cell activation.

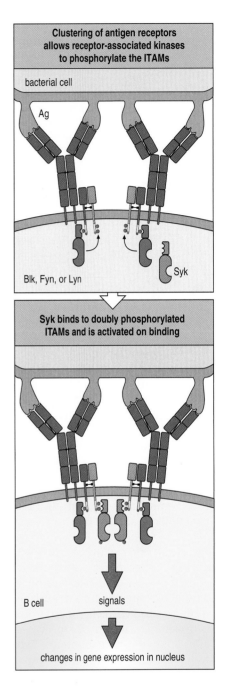

Cross-linking of the B-cell receptor by antigen generates a signal that is necessary but not sufficient to activate a naive B cell. The additional signals required are delivered in several ways. One set of signals is delivered when the B-cell receptor becomes closely associated with another protein complex on the B-cell surface known as the **B-cell co-receptor**. The B-cell co-receptor is a complex of three proteins: the first is the complement receptor 2 (CR2 or **CD21**), which binds to complement components deposited on a pathogen; the second is the protein **CD19**, which acts as the signaling chain of the receptor; and the third is the protein **CD81** (TAPA-1), whose function is not yet known. It does act, however, as a cell-surface receptor for the hepatitis C virus.

Certain antigens at pathogen surfaces catalyze a series of enymatic reactions that leads to the deposition of several fragments of complement proteins on the pathogen's surface close to the antigen molecule. One of these fragments, called C3d, is a ligand for the CR2 component of the B-cell co-receptor. When the receptor of a B cell that is specific for the antigen binds to the antigen the CR2 component of the B-cell co-receptor complex can bind to an adjacent C3d, which serves to bring the B-cell receptor and co-receptor into juxtaposition. This co-ligation of the B-cell receptor and the co-receptor brings the Igα-bound tyrosine kinase Lyn into close proximity with the CD19 cytoplasmic tail, which it phosphorylates. The phosphorylated CD19 can then bind intracellular signaling molecules that generate signals that synergize with those generated by the B-cell receptor complex (Figure 7.3). Simultaneous ligation of the B-cell receptor and co-receptor increases the signals by 1000- to 10,000-fold.

Even the combined effects of the B-cell receptor and co-receptor signals are generally insufficient to activate a naive B cell. This requires additional signals provided by CD4 helper T cells, the effector cells produced upon antigen activation of naive CD4 T cells. Whether a B cell needs T-cell help or not depends on the nature of the antigen, as we shall see in the next section.

The final outcome of B-cell activation is the proliferation and differentiation of the B cell into plasma cells and the secretion of antibodies. The morphological effects of activation are striking: the small resting B cell, which in appearance is all nucleus and no cytoplasm, gives rise to plasma cells committed to antibody secretion and whose large active cytoplasm packed with rough endoplasmic reticulum is testimony to this function (Figure 7.4).

7-2 The antibody response to certain antigens does not require T-cell help

In their cell walls and capsules, bacterial pathogens possess complex polysaccharides, lipopolysaccharides, and peptidoglycans that are chemically and antigenically distinct from those of mammalian cells and are characterized by

Figure 7.3 Signals generated from the B-cell receptor and co-receptor synergize in B-cell activation. Binding of complement receptor 2 (CR2) to complement fragments (C3d) deposited on the surface of a pathogen cross-links the B-cell co-receptor complex with the B-cell receptor. This causes them to cluster together on the B-cell surface. The cytoplasmic tail of CD19 is then phosphorylated by tyrosine kinases associated with the B-cell receptor. Phosphorylated CD19 binds intracellular signaling molecules whose signals synergize with those generated by the B-cell receptor.

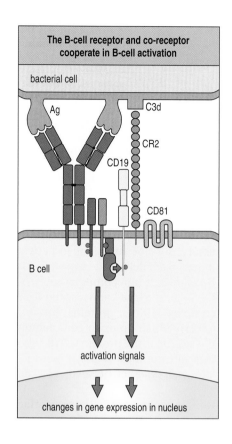

repetitive epitopes. These cell-surface molecules are a major target of the antibody response against extracellular bacterial pathogens, and some of these antigens can activate naive B cells without help from CD4 T cells. Such antigens are known as **thymus-independent antigens (TI antigens)** because immunodeficient patients born without a thymus are able to make an antibody response against them. However, they cannot respond to other antigens because that requires help from an antigen-activated helper T cell (see Section 6-17, p. 172). Antigens that need T-cell help are known as **thymus-dependent antigens (TD antigens)**. For thymus-independent antigens, the need for T-cell help is overcome in two different ways.

As well as binding to the B-cell receptor, certain thymus-independent antigens (called **TI-1 antigens**) bind to other receptors on B cells, which, in combination, induce the B cells to proliferate and differentiate. An example of a TI-1 antigen is the lipopolysaccharide (LPS) of Gram-negative bacteria. LPS binds to LPS-binding protein and CD14, which then associate with another receptor, called a Toll-like receptor, to produce activating signals. When B cells are triggered by TI-1 antigens they produce only IgM antibodies, because cytokines produced by activated T-helper cells are needed for a B cell to switch its antibody isotype. A surface-associated TI-1 antigen like LPS not only causes T-cell-independent activation of B cells specific for epitopes of LPS (Figure 7.5, left panels) but also B cells specific for other antigens of the bacterial cell surface (Figure 7.5, right panels).

The second type of thymus-independent antigen are the **TI-2 antigens**, which are typically composed of repetitive carbohydrate or protein epitopes present at high density on the surface of a microorganism. TI-2 antigens only stimulate B cells specific for the antigen and probably act by cross-linking B-cell receptors and co-receptors so extensively that the need for additional signals is overridden. Responses to TI-2 antigens are usually seen around 48 hours after antigen was encountered. Typical antigens of this kind are bacterial cell wall polysaccharides and the responding B cells are often of the B-1 subpopulation (see Section 4-6, p. 108). Human B-1 cells develop their full function only when a person is about 5 years old, perhaps explaining why infants make relatively poor antibody responses to polysaccharide antigens. IgM and IgG antibodies are both induced by TI-2 antigens and they are likely to be an important part of the early B-cell response to some common bacterial infections. Examples of TD, TI-1, and TI-2 antigens and the responses to them are presented in Figure 7.6.

Although the TI-2 antigens on some bacteria induce an early antibody response that helps to contain the infection, this response has limitations. There is little isotype switching and so the antibodies are predominantly IgM.

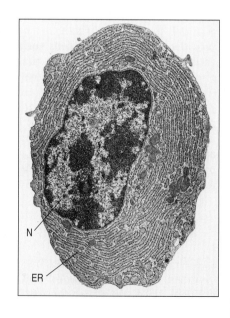

Figure 7.4 Plasma cell. Electron micrograph of a plasma cell. Note the characteristic 'clockface' pattern in the nucleus (N), which resembles the hands and face of a clock, as well as the extensive endoplasmic reticulum (ER). Photograph courtesy of C. Grossi.

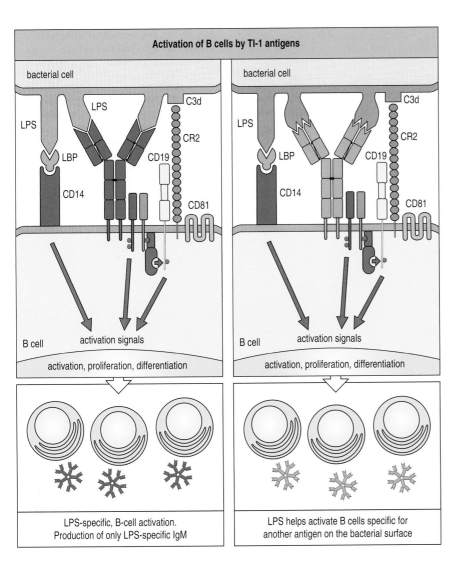

Figure 7.5 Thymus-independent (TI)-1 antigens can activate B cells without T-cell help. Certain antigens, such as the lipopolysaccharide (LPS) of Gram-negative bacteria, can on their own activate B cells to become antibody-producing plasma cells. The left hand panels depict a B cell whose B-cell receptor is specific for an epitope of LPS. In addition to binding to the B-cell receptor, LPS also forms a complex with soluble LPS-binding protein (LBP) that binds to CD14 on the B-cell surface. Signaling through both CD14, the B-cell receptor and the associated B-cell co-receptor complex is sufficient to activate the B cell, which gives rise to plasma cells producing anti-LPS antibodies. In the right-hand panels, LPS binding to CD14 provides a co-activating signal for another antigen on the bacterium that binds to its specific B-cell receptor. This B cell goes on to produce antibodies specific for the bacterial antigen, not LPS.

Within the figure:

Activation of B cells by TI-1 antigens

bacterial cell — LPS, C3d, LBP, CD14, CD19, CR2, CD81 — B cell — activation signals — activation, proliferation, differentiation — LPS-specific, B-cell activation. Production of only LPS-specific IgM

bacterial cell — LPS, C3d, LBP, CD14, CD19, CR2, CD81 — B cell — activation signals — activation, proliferation, differentiation — LPS helps activate B cells specific for another antigen on the bacterial surface

Neither is there somatic hypermutation, so there is no possibility for increasing the affinity for antigen of the antibodies produced. Lastly, TI-2 antigens do not induce long-term immunological memory and so provide no long-lasting immunity against reinfection. The development of all these attributes requires T-cell help, as we shall see next.

7-3 B cells needing T-cell help are activated in secondary lymphoid tissues where they form germinal centers

Although the antibody response to a pathogen may be initiated by thymus-independent antigens, the bulk of the pathogen-specific antibody is eventually produced by B cells stimulated by thymus-dependent antigens. Activation of these B cells occurs in the secondary lymphoid tissues where B cells, specific antigen, and helper CD4 T cells are all brought together. We will describe these processes using the lymph node as an example.

Antigens arrive at a node in the lymph draining the infected tissue, whereas antigen-specific lymphocytes enter the node from the blood (see Section 4-9, p. 113). The antigens are transported by dendritic cells that mature and take up residence in the T-cell area of the lymph node (see Section 6-1, p. 146) or are passively carried along in the lymph to be phagocytosed by macrophages resident in the lymph node. Either way, the antigens are trapped in the node

Property	TD antigen	TI-1 antigen	TI-2 antigen
Antibody response in absence of cognate T cells	No	Yes	Yes*
Antibody production in congenitally athymic individuals	No	Yes	Yes
Antibody response in infants	Yes	Yes	No
Activates T cells	Yes	No	No
Induces immunological memory	Yes	No	No
Activation of non-specific B cells	No	Yes	No
Requires repeated epitopes	No	No	Yes
	Diphtheria toxin Viral hemagglutinin Purified protein derivative (PPD) of *Mycobacterium tuberculosis*	Bacterial lipopoly-saccharide *Brucella abortus*	Pneumococcal poly-saccharide Polymerized flagellin (*Salmonella*)

Figure 7.6 Properties of different classes of antigen that elicit antibody responses. *Responses to thymus-independent (TI) -2 antigens are generally enhanced by the presence of T cells but they do not require cognate interactions. The contribution of the T cells might come from cytokines that α:β T cells produce, for example IL-5. Alternatively, the effect could be a result of γ.δ T cells, which recognize antigens other than the conventional peptide:MHC complexes and do not need a thymus for their development.

and presented there by professional antigen-presenting cells. In the T-cell area, antigen-specific CD4 T lymphocytes are activated to become effector helper T cells by engagement with dendritic cells presenting the cells' specific antigen on MHC class II molecules. B cells pass through the T-cell zone and when their specific antigen is present they are activated through cognate interactions with effector helper T cells (Figure 7.7, see also Section 6-17, p. 172).

The B-cell receptor has two distinct roles in B cell activation: binding antigen, which sends a signal to the B cell's nucleus, and internalizing antigen by receptor-mediated endocytosis, which facilitates the processing and presentation of antigen to helper T cells. When the antigen receptors of a CD4 T cell bind complexes of peptide antigen and MHC class II molecules on the B-cell surface, interaction also occurs between CD40 ligand on the T cell and CD40 on the B cell. The latter interaction signals the B cell to activate the transcription factor NFκB and increase surface expression of intercellular

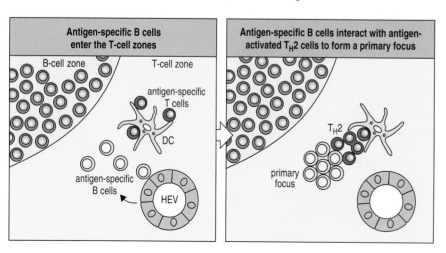

Figure 7.7 If they encounter cognate helper T cells, naive B cells become trapped in the T-cell zone of secondary lymphoid tissues. Recirculating naive B cells enter the T-cell zone of a lymph node from the blood through high endothelial venules (HEV). If they encounter helper T$_H$2 cells specific for the same antigen, they interact with them to form a primary focus of proliferating activated B cells and T$_H$2 cells.

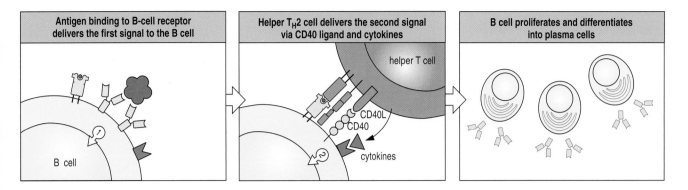

adhesion molecule 1 (ICAM-1). This strengthens the cognate interaction between the B cell and the helper CD4 T cell. A signal produced in the T cell causes it to secrete cytokines. Of these, interleukin-4 (IL-4) is characteristic of T_H2 cells and essential for B-cell proliferation. Cytokines binding to receptors on the B-cell surface drive the B cell to proliferate and differentiate into plasma cells (Figure 7.8).

B cells activated by interactions with cognate helper T cells in the T-cell areas of a lymph node form a primary focus of dividing B lymphoblasts that lasts for a few days (see Figure 7.7). Some of these B lymphoblasts move to the medullary cords and differentiate directly into plasma cells under the influence of the cytokines IL-5 and IL-6, which are secreted by T_H2 cells. They secrete predominantly IgM antibody but some isotype switching can also occur in the primary focus. For infections in which the pathogen carries no thymus-independent antigens, this will be the first antibody produced. It will start to appear several days after the onset of infection.

Other B lymphoblasts from the primary focus move into primary follicles still attached to their cognate helper T cells. Their rate of division increases to about once every 6 hours and they become large, metabolically active cells called centroblasts (see Section 4-9, p. 113). With the increase in the number of centroblasts, the morphology of a follicle changes and it becomes dominated by the **germinal center** that contains the newly formed B cells (Figure 7.9). Germinal centers appear in secondary lymphoid tissues about 1 week after the start of infection and they cause the characteristic swelling of lymph nodes draining an infection.

As they divide, the centroblasts become increasingly closely packed and form a region that is darkly staining in histological sections and is called the **dark zone** of the germinal center. The centroblasts give rise to nondividing centrocytes, which leave the close-packed lymphocytes to interact with **follicular dendritic cells** (**FDCs**) in the **light zone** of the germinal center (see Figure 7.9). Follicular dendritic cells are the characteristic stromal cell of primary lymphoid follicles. They pick up antigen but do not internalize it, and it remains bound to their surface for long periods of time. They interact with B cells through a dense network of antigen-loaded dendrites. The follicular dendritic cells are quite distinct from the dendritic cells that present antigen to naive T cells and activate them (see Section 6-5, p. 151); they do not derive from a hematopoietic stem cell and do not express MHC class II molecules.

Those helper T cells that migrated to the primary follicle together with the activated B cells also proliferate in the light zone and are intermingled with the centrocytes. With time, the vast majority of lymphocytes present in a germinal center are clones derived from one or a few founder pairs of antigen-activated B and T cells. B cells that were present in the primary follicle before entry of the activated B cell–T cell conjugates, and are not specific for

Figure 7.8 **B-cell activation in response to thymus-dependent antigens requires cognate T-cell help.** The first signal required for B-cell activation is delivered through the antigen receptor (left panel). With thymus-dependent antigens, the second signal is delivered by a cognate helper T cell that recognizes a peptide fragment of the antigen bound to MHC class II molecules on the B-cell surface (center panel). The two signals together drive B-cell proliferation and differentiation into plasma cells (right panel).

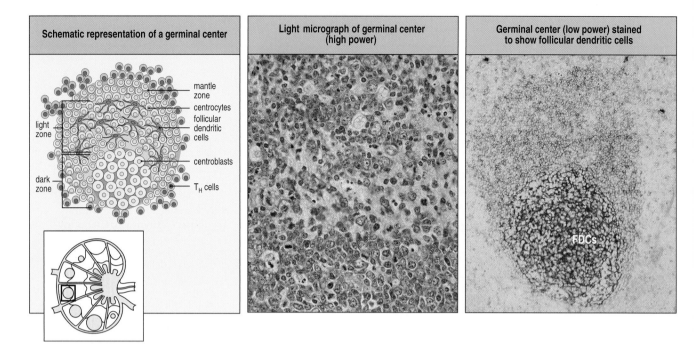

| Schematic representation of a germinal center | Light micrograph of germinal center (high power) | Germinal center (low power) stained to show follicular dendritic cells |

Figure 7.9 Germinal centers are formed when activated B cells enter lymphoid follicles. The germinal center is a specialized microenvironment in which B-cell proliferation, somatic hypermutation, and selection for antigen binding all occur. Rapidly proliferating B cells in germinal centers are called centroblasts. Closely packed centroblasts form the so-called 'dark zone' of the germinal center. This can be seen in the lower part of the center panel, which shows a light micrograph of a section through a germinal center, and in the accompanying diagram (left panel). As these cells mature, they stop dividing and become small centrocytes, moving out into an area of the germinal center called the 'light zone' (in the upper part of the center panel), where the centrocytes make contact with a dense network of follicular dendritic cell processes. The follicular dendritic cells are not stained in the center panel but can be seen clearly in the right panel, in which both follicular dendritic cells (stained blue with antibody against Bu10, a marker of follicular dendritic cells) in the germinal center, as well as mature B cells in the mantle zone (stained brown with an antibody against IgD) can be seen. The plane of this section reveals mostly the dense network of follicular dendritic cells in the light zone, although the less dense network in the dark zone can just be seen at the bottom of the figure. Photographs courtesy of I. MacLennan.

the antigen, are pushed to the outside of the germinal center, forming the **mantle zone** (see Figure 7.9).

7-4 Activated B cells undergo somatic hypermutation and affinity maturation in the specialized microenvironment of the germinal center

As we have seen in Chapters 4–6, a common theme in lymphocyte development is for a phase of activation and proliferation to be followed by one of selection. This is precisely what happens to the B cells maturing in a germinal center. Somatic hypermutation, initiated by T-cell cytokines, takes place in centroblasts dividing within the germinal center and gives rise to nondividing centrocytes with mutated surface immunoglobulin. After hypermutation, the surface immunoglobulin expressed by an individual centrocyte can have an affinity for its specific antigen that is higher, lower, or the same as that of the unmutated immunoglobulin. Thus, the population of centrocytes in a germinal center expresses immunoglobulins with a range of affinities for the specific antigen.

Centrocytes are programmed to die by apoptosis within a short period unless their surface immunoglobulin is bound by antigen and they are subsequently contacted by a helper T cell bearing CD40 ligand. To engage such a helper T cell, the centrocyte must first bind and process antigen, then present antigenic peptides at its surface in association with MHC class II molecules. The

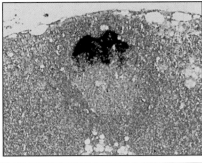

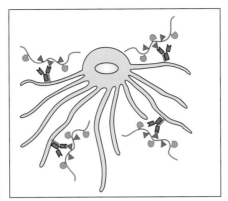

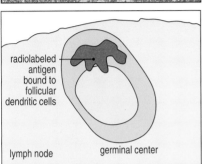

radiolabeled antigen bound to follicular dendritic cells

lymph node   germinal center

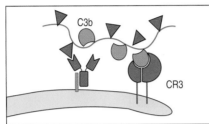

C3b

CR3

Figure 7.10 B cells recognize antigen as immune complexes bound to the surface of follicular dendritic cells. Radiolabeled antigen localizes to, and persists in, lymphoid follicles of draining lymph nodes (see light micrograph and the schematic representation below, showing a germinal center in a lymph node). Radiolabeled antigen was injected 3 days previously and its localization in the germinal center is shown by the intense dark staining. The antigen is in the form of antigen:antibody:complement complexes that can bind to both Fc receptors (top right panel) and to Fc and complement receptors (bottom right panel) on the surface of the follicular dendritic cell. These complexes are not internalized and antigen can persist in this form for long periods. Photograph courtesy of J. Tew.

mutated centrocytes now compete with each other, first for access to antigen on follicular dendritic cells and then for antigen-specific helper T cells.

Follicular dendritic cells provide a source of intact antigen. They bind antigen in the form of complexes either with antibody or with antibody and complement. Such complexes are called **immune complexes**. The first source of these complexes is the IgM produced early in the primary immune response. Later in the immune response, the immune complexes contain IgG. Follicular dendritic cells bear receptors for complement and for the Fc region of IgG (Figure 7.10). The immune complexes are not internalized and they persist for long periods at the surface of follicular dendritic cells where they can be bound by antigen-specific B cells. Bundles of membrane coated with immune complexes also bud off from the surface of follicular dendritic cells. These bundles, called **iccosomes** (**immune-complex coated bodies**) (Figure 7.11),

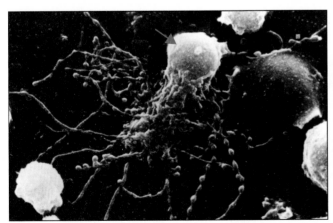

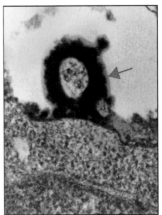

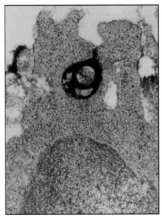

Figure 7.11 Immune complexes bound to follicular dendritic cells form iccosomes, which are released and can be taken up by B cells in the germinal center. Follicular dendritic cells have a prominent cell body and many dendritic processes. Immune complexes are bound to Fc receptors on the follicular dendritic cell surface and become clustered, forming prominent beads along the dendrites (left panel). In this scanning electron micrograph the cell body of the follicular dendritic cell is indicated by the arrow. The beads are shed from the cell as iccosomes, one of which is indicated by the pink arrow in the center panel. Iccosomes are taken up by B cells within the germinal center, as shown by the arrow in the right panel. The immune complexes in the transmission electron micrographs in the center and right panels contain horseradish peroxidase, which generates the dense staining. Photographs courtesy of A.K. Szakal.

are bound and taken up by antigen-specific B cells, which then process and present the antigen.

Newly formed centrocytes move from the dark zone of the germinal center to contact follicular dendritic cells in the light zone. If a centrocyte captures sufficient antigen from the follicular dendritic cells or iccosomes, it then moves to the outer regions of the light zone where helper T cells are concentrated. Engagement of peptide:MHC class II by the T-cell receptor complex, and of CD40 on the centrocyte by CD40 ligand on the T cell, induces the centrocyte to express the Bcl-x_L protein, which prevents its death by apoptosis (Figure 7.12).

Thus, centrocytes with the highest-affinity antigen receptors are selected for survival and further differentiation into antibody-producing plasma cells or

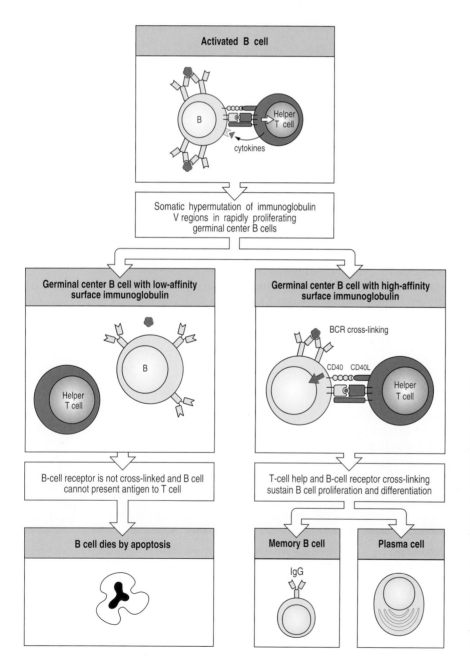

Figure 7.12 After somatic hypermutation, B cells with high-affinity receptors for antigen are rescued from apoptosis. In the germinal center, helper T cells induce B cells to undergo somatic hypermutation (top panel). B cells that have undergone somatic hypermutation interact with follicular dendritic cells (FDCs) that display immune complexes on their surface. B cells whose receptors bind antigen poorly, or do not bind antigen at all because they have mutated beyond recognition, cannot compete for access to the FDCs and die by apoptosis (left panel). B cells with receptors that bind well receive signals from the FDCs and are induced to express Bcl-x_L, which prevents apoptosis; these cells survive (right panels). BCR, B-cell receptor.

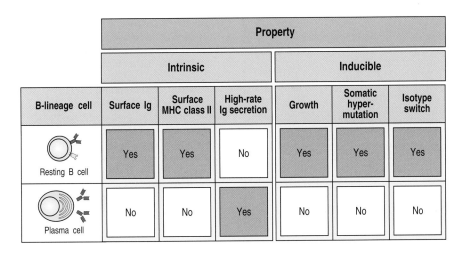

B-lineage cell	Property					
	Intrinsic			Inducible		
	Surface Ig	Surface MHC class II	High-rate Ig secretion	Growth	Somatic hyper-mutation	Isotype switch
Resting B cell	Yes	Yes	No	Yes	Yes	Yes
Plasma cell	No	No	Yes	No	No	No

Figure 7.13 Comparison of resting B cells and plasma cells. The resting B cell expresses an antigen receptor in the form of surface immunoglobulin, and can also take up protein antigen and present it as a peptide:MHC class II complex. Thus, it can activate helper T cells. Its immunoglobulin genes can also undergo somatic hypermutation, giving rise to progeny with altered immunoglobulin specificity. The plasma cell, in contrast, is a terminally differentiated B cell that is dedicated to the synthesis and secretion of soluble antibody. It no longer divides and its antibody specificity cannot be changed.

into long-lived memory cells. In this way, the affinity of antibodies for the specific antigen increases during the course of an immune response and in subsequent exposures to the same antigen. This process is known as **affinity maturation**.

Under the influence of T cells, isotype switching also takes place in B cells within the germinal center. Thus, not only does the affinity of the antibodies produced increase but also antibodies of different isotypes are made, principally IgG in the case of B cells that differentiate in the lymph nodes and spleen.

For mutated centrocytes that survive selection, the interaction with an antigen-specific helper T cell serves several purposes. The mutual engagement of ligands and receptors on the two cells generates an exchange of signals that induces the further proliferation of both B and T cells. This serves to expand the population of selected high-affinity, isotype-switched B cells. Individual B cells are also directed along pathways of differentiation leading either to plasma cells or to memory B cells. At the height of the adaptive immune response, when the main need is for large quantities of antibodies to fight infection, the centrocytes that win in this selection leave the germinal center and differentiate into antibody-producing plasma cells. The differences between resting B cells and plasma cells are summarized in Figure 7.13. In the later stages of a successful immune response, as the infection subsides, centrocytes are thought to differentiate into long-lived, memory B cells, which now possess isotype-switched, high-affinity antigen receptors. Plasma cells are the effector B cells that provide antibody for dealing with today's infection, whereas the memory B cells represent an investment in the prevention of future infection with the same pathogen should the current infection be successfully resolved.

If a centrocyte fails to obtain, internalize, and present antigen it dies by apoptosis and is phagocytosed by macrophages in the germinal center. Macrophages that have recently engulfed apoptotic centrocytes are a characteristic feature of germinal centers and, because of their contents, are called **tingible body macrophages**. Somatic hypermutation can produce centrocytes bearing immunoglobulin that reacts with a self-antigen on the surface of cells in the germinal center. When this happens, contact with helper T cells or other cells in the germinal center will render such centrocytes inactive or anergic—a mechanism similar to the one whereby self-reactive, immature B cells are inactivated in the bone marrow (see Section 4-7, p. 111).

Influence of cytokines on antibody isotype switching in mice							
Cytokine	IgM	IgG3	IgG1	IgG2b	IgG2a	IgA	IgE
IL-4	Inhibits	Inhibits	Induces		Inhibits		Induces
IL-5						Augments production	
IFN-γ	Inhibits	Induces	Inhibits		Induces		Inhibits
TGF-β	Inhibits	Inhibits		Induces		Induces	

Figure 7.14 Different cytokines induce B cells to switch to different immunoglobulin isotypes. Individual cytokines can either induce (green), augment (bright yellow), or inhibit (red) the switching of immunoglobulin synthesis to a particular isotype. The inhibitory effects are largely due to the positive effect of the cytokine on switching to another isotype. This compilation is drawn from experiments on mouse B cells. There are differences in humans, but they are not yet as well worked out. For example, switching to IgA in humans involves TGF-β and IL-10, not IL-5.

7-5 Interactions with T cells are required for isotype switching in B cells

In Chapter 2 we saw how the first immunoglobulins made by B cells are of the IgM and IgD classes, but, after activation by antigen B cells can switch their heavy-chain isotype to produce IgG, IgA, or IgE. Isotype switching takes place in activated B cells mainly within the germinal center, and the isotype to which an individual B cell switches is determined by cognate interactions with helper T cells. The particular isotype to which a switch is made depends on the cytokines secreted by the helper T cell. The roles of individual cytokines in switching the isotype of mouse immunoglobulin heavy chains are summarized in Figure 7.14. Cytokines secreted by T_H2 cells—IL-4, IL-5, and TGF-β—are the predominant players. They initiate the antibody response by activating naive B cells to differentiate into plasma cells secreting IgM, and also induce the production of other antibody isotypes including, in humans, the weakly opsonizing antibodies IgG2 and IgG4, as well as IgA and IgE. However, interferon (IFN)-γ, the characteristic cytokine produced by T_H1 cells, switches B cells to making the IgG2a and IgG3 classes of immunoglobulin (in mice) and the strongly opsonizing antibody IgG1 in humans.

T-cell cytokines induce isotype switching by stimulating transcription from the switch regions that lie 5′ to each heavy-chain C gene. For example, when activated B cells are exposed to IL-4, transcription from a site upstream of the switch regions of $C_\gamma1$ and C_ε can be detected a day or two before switching occurs. As with the low-level transcription that occurs in immunoglobulin loci before rearrangement (see Section 4-2, p. 102), this transcription could be opening up the chromatin and making the switch regions accessible to the somatic recombination machinery that will place a new C gene in juxtaposition to the V-region sequence.

The induction of isotype switching by cognate helper T cells also requires the ligation of CD40 on the B-cell surface by CD40 ligand on the T cells. The importance of helper T cells and the CD40–CD40 ligand interaction for isotype switching is apparent from the immunodeficiency of patients who lack CD40 ligand. These patients have abnormally high levels of IgM in their blood serum, which gives the name **hyper-IgM syndrome** to their condition, but almost no IgG and IgA because of the inability of their B cells to switch isotype. They cannot make antibody responses to thymus-dependent antigens and their secondary lymphoid tissues contain no germinal centers (Figure 7.15), showing the general importance of the CD40–CD40 ligand interactions in T-cell help. Aspects of cell-mediated immunity are also impaired in these patients, who are mostly male because the gene for CD40 ligand is on the X chromosome.

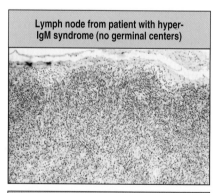
Lymph node from patient with hyper-IgM syndrome (no germinal centers)

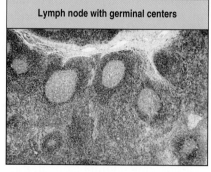
Lymph node with germinal centers

Figure 7.15 Comparison of normal and hyper-IgM syndrome lymph nodes. Bottom panel photograph courtesy of Dr Antonio Perez-Atayde.

Summary

B cells respond to specific antigen with activation, proliferation, and differentiation. They then become plasma cells that synthesize and secrete massive amounts of antibody. Activation of a mature but naive B cell requires signals delivered through its antigen receptor and most B cells also need additional signals that are delivered only on cognate interaction with an antigen-specific, helper T cell. Activating signals are also delivered through the B-cell co-receptor when this is simultaneously ligated with the antigen receptor. The signaling pathways used to activate B cells are summarized in Figure 7.16; they are similar to those used to activate T cells. The first antibodies made are always IgM; further contact with effector helper T cells is required for activated B cells to undergo isotype switching, somatic hypermutation, and affinity maturation

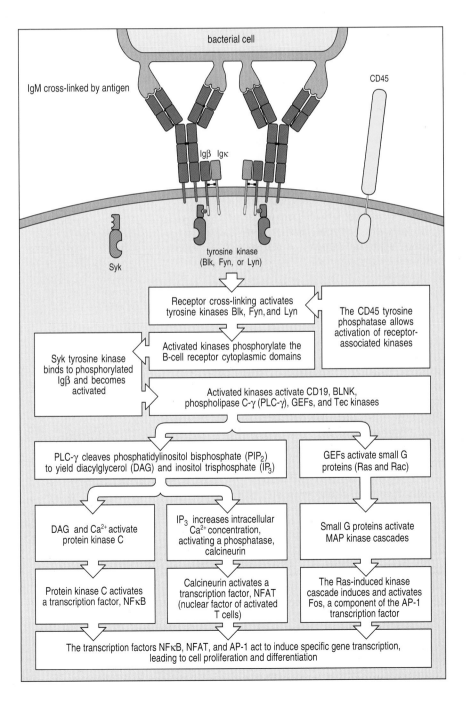

Figure 7.16 Simplified outline of the intracellular signaling pathways initiated by cross-linking of B-cell receptors by antigen. These pathways are similar to those that act in naive T cells exposed to antigen (see Figure 6.17), but some of the components, such as the kinase Blk, are specific to B cells.

within the germinal centers of secondary lymphoid organs. All this takes time, during which the pathogen can multiply, spread from the focus of infection, and cause disease. However, if the host survives, there will remain in the circulation expanded populations of high-affinity antibodies and memory B cells programmed to make them again should the need arise. Some antigens, notably certain components of bacterial cell walls and capsules, are capable of inducing a rapid antibody response that does not require T-cell help. These thymus-independent antigens are of two types. TI-1 antigens bind to a second receptor that contribute signals for mitosis and differentiation in addition to those generated through the B-cell antigen receptor. TI-2 antigens are generally microbial cell-surface macromolecules with repetitive epitopes that are present at high density on microbial surfaces and extensively cross-link the antigen receptors and co-receptors on the B-cell surface. Antibodies produced against TI antigens are predominantly IgM and the cells producing them are often of the B-1 lineage. Responses to TI antigens neither induce immunological memory nor long-lasting immunity.

Antibody effector functions

As the B-cell response to an infection gets under way, isotype switching diversifies the functional properties of the antibody Fc region, which contains binding sites for other proteins and cells of the immune system. Fc regions serve two distinct functions: they deliver antibody to anatomical sites that would otherwise be inaccessible and they link bound antigen to molecules or cells that will effect its destruction. Such cells carry receptors called **Fc receptors**, which bind to the Fc regions of antibodies of a particular class or subclass, irrespective of the antibody's antigen specificity. In this part of the chapter we shall consider how antibodies of different isotypes recruit nonspecific effector cells, such as macrophages and neutrophils, into the immune response by interaction with their Fc receptors.

7-6 IgM, IgG, and IgA antibodies protect the blood and extracellular fluids

In any antibody response, IgM is the first antibody to be produced. It is secreted as a pentamer by plasma cells in the bone marrow, spleen, and medullary cords of lymph nodes. IgM enters the blood and is carried to sites of tissue damage and infection throughout the body. The pentameric nature of IgM enables it to bind strongly to microorganisms and particulate antigens, but its large size decreases the extent to which this antibody isotype can passively leave the blood and penetrate infected tissues. There are no receptors for the IgM Fc region on phagocytic cells or other leukocytes, so IgM cannot directly recruit the destructive capabilities of these cells into the immune response. The Fc region of IgM can, however, bind complement and activate the complement system, with consequences that we shall consider in the last part of this chapter.

Later in an immune response, the dominant blood-borne antibody is the smaller IgG molecule. An important function of circulating IgM and IgG is to prevent blood-borne infection—**septicemia**—and the spread of microorganisms by neutralizing those that enter the blood. Because the blood circulation is so effective in distributing cells and molecules to all parts of the body, infections of the blood itself can have grave consequences.

IgA is synthesized by plasma cells in secondary lymphoid tissues. Monomeric IgA is made by plasma cells derived from B cells that switched their antibody isotype in the lymph nodes or spleen. In contrast, dimeric IgA is made in the secondary lymphoid tissues underlying mucosal surfaces, as we shall see in the next section. Like IgG, monomeric IgA enters the extracellular spaces and helps IgG to protect them against infection by bacteria and virus particles.

7-7 IgA and IgG are transported across epithelial barriers by specific receptor proteins

Whereas IgM, IgG, and monomeric IgA provide antigen-binding functions within the fluids and tissues of the body, dimeric IgA protects the surfaces of the epithelia that communicate with the external environment and are particularly vulnerable to infection. These epithelia include the linings of the gastrointestinal tract, the eyes, nose, throat, the respiratory, urinary, and genital tracts, and the mammary glands. Dimeric IgA is made in patches of mucosal-associated lymphoid tissues in the lamina propria, the connective tissue that underlies the basement membrane of the mucosal epithelium. In these tissues, antigen-specific B-cell and T-cell responses to local infections are developed. However, the IgA-secreting plasma cells are on one side of the epithelium and their target pathogens are on the other. To reach their targets, dimeric IgA molecules are transported individually across the epithelium by means of a receptor on the basolateral surface of the epithelial cells.

The dimeric form of IgA, but not the monomer, binds to a cell-surface receptor on the basolateral surface of epithelial cells that is called the **poly-Ig receptor** because of its specificity for IgA polymers and, to a lesser extent, for pentameric IgM (Figure 7.17). The poly-Ig receptor itself is made up of a series of immunoglobulin-like domains. On being bound, the IgA dimer is taken into the cell by receptor-mediated endocytosis and the antibody:receptor complex is carried across the cell to the apical surface in endocytic vesicles. Receptor-mediated transport of a macromolecule from one side of a cell to the other is known as **transcytosis**. Once receptor-bound IgA appears on the

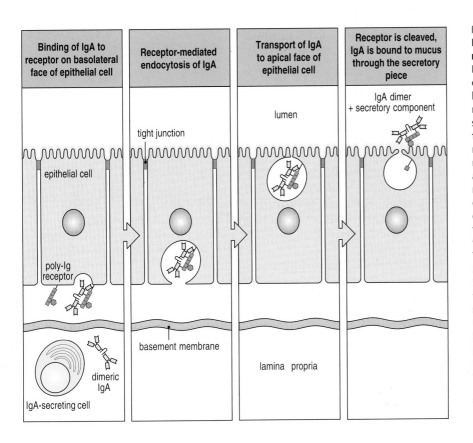

Binding of IgA to receptor on basolateral face of epithelial cell	Receptor-mediated endocytosis of IgA	Transport of IgA to apical face of epithelial cell	Receptor is cleaved, IgA is bound to mucus through the secretory piece

Figure 7.17 Transcytosis of dimeric IgA antibody across epithelia is mediated by the poly-Ig receptor. Dimeric IgA is made mostly by plasma cells lying just beneath the epithelial basement membranes of the gut, respiratory tract, tear glands, and salivary glands. The IgA dimer bound to the J chain diffuses across the basement membrane and is bound by the poly-Ig receptor on the basolateral surface of an epithelial cell. Binding to the receptor is via the C_H3 constant domains of the IgA heavy chains. The bound complex undergoes transcytosis across the cell in a membrane vesicle and is finally released onto the apical surface. There the poly-Ig receptor is cleaved, releasing the IgA from the epithelial cell membrane while still being bound to a fragment of the receptor called the secretory component or secretory piece. Carbohydrate (blue hexagon) on the poly-Ig receptor forms the secretory piece that binds to mucus at the epithelial surface, thus preventing IgA from being washed away into the gut lumen. The residual membrane-bound fragment of the poly-Ig receptor is nonfunctional and is degraded.

Figure 7.18 The Brambell receptor (FcRB) transports IgG from the bloodstream into the extracellular spaces. An IgG molecule binds to two FcRB molecules at the apical (luminal) side of the endothelial cell. After receptor-mediated endocytosis, the IgG molecule is carried in a vesicle across the endothelial cell to the basal side of the cell, where it is released into the extracellular space.

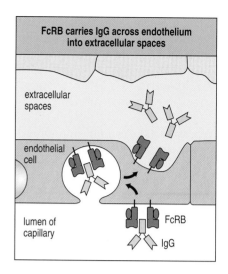

apical surface, a protease cleaves the poly-Ig receptor at sites between the membrane-anchoring region and the IgA-binding site. Dimeric IgA is released from the membrane still bound to a small fragment of the poly-Ig receptor, which is called the **secretory component**, or **secretory piece**, of IgA. The IgA is then held at the mucosal surface, being bound to mucins in mucus by the carbohydrate of the secretory piece.

IgG is actively transported from the blood into the extracellular spaces within tissues by an Fc receptor present on the endothelial cells (Figure 7.18). This receptor is sometimes called the **Brambell receptor** (**FcRB**) after the scientist who first described its function. FcRB is similar in structure to an MHC class I molecule, with the α_1 and α_2 domains forming a site that binds to the Fc region of the antibody. In the antibody:receptor complex, two molecules of FcRB bind to the Fc region of one IgG molecule. The delivery of IgG to the extracellular spaces in connective tissue helps to protect tissues against infection and also protects IgG from the degradation pathways to which serum proteins are subject. As a consequence, IgG molecules have a relatively long half-life in relation to most other plasma proteins.

During pregnancy, the fetus is physically protected by the mother from the microorganisms that inhabit the external environment. At birth, the baby is suddenly exposed to numerous pathogens. Because of their lack of actively acquired immunity, newborn infants are particularly vulnerable to infection arising from the microbial colonization of epithelia. To help the infant counter such attack, it receives IgA from its mother. The IgA is first secreted into breast milk and then transferred on breast-feeding into the baby's gut. Within this transferred IgA are antibodies against microorganisms to which the mother has previously mounted an IgA response. Within the infant's gut, the IgA molecules bind to microorganisms and their products, preventing their attachment to the gut epithelium and facilitating their expulsion in feces. The transfer of preformed IgA from mother to child in breast milk is an example of the **passive transfer of immunity**.

IgA is not the only immunoglobulin isotype that mothers donate to their children. During pregnancy, IgG from the maternal circulation is transported across the placenta and is delivered directly into the fetal bloodstream. The efficiency of this mechanism is such that at birth human babies have as high a level of IgG in their plasma as their mothers, and as wide a range of antigen specificities. Transport of IgG across the placenta is performed by FcRB (see Figure 7.18).

Mice and rats express a homologue of FcRB called **FcRn**, but its function is somewhat different. In rodents, the receptor is expressed in the intestine for a short period after birth. During this time, the newborn rodent ingests maternal IgG in colostrum, the protein-rich fluid in the postnatal mammary gland, which is then transported across the intestinal epithelium into the tissues by FcRn.

By means of these specialized transport systems mammals are supplied from birth with antibodies against common pathogens in their environment. As the young mature and make their own antibodies of all the isotypes, these are each distributed to selected sites in the body (Figure 7.19). Thus, throughout

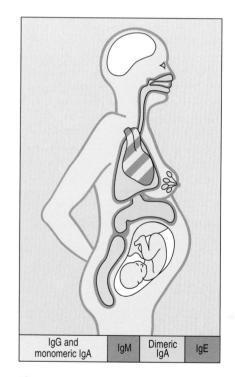

Figure 7.19 Immunoglobulin isotypes are selectively distributed in the body. IgG and IgM predominate in plasma, whereas IgG and monomeric IgA are the major isotypes in the extracellular fluid within the body. Dimeric IgA predominates in secretions across epithelia, including breast milk. The fetus receives IgG from the mother by transplacental transport. IgE is associated mainly with mast cell surfaces and is, therefore, found beneath epithelial surfaces (especially the respiratory tract, gastrointestinal tract, and skin). The brain is normally devoid of immunoglobulin.

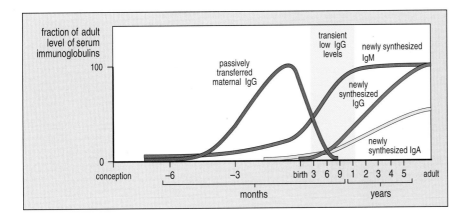

life, the production of different isotypes provides protection against infection in the extracellular spaces throughout the body.

7-8 Antibody production is deficient in very young infants

During the first year of life there is a window of time when all infants are relatively deficient in antibodies and specially vulnerable to infection. During pregnancy, maternal IgG antibodies are transported across the placenta into the fetal circulation, providing newborn infants with antibody levels comparable to those of their mothers. As the maternally derived IgG is catabolized, the antibody level gradually decreases until the infant's immune system begins to produce its own antibody at about 6 months of age (Figure 7.20).

Consequently, IgG levels are lowest in infants aged 3–12 months and this is when they are most susceptible to infection. This problem is particularly acute in babies born prematurely, who begin life with lower levels of maternal IgG and take longer to attain immunocompetence after birth than babies born at term.

7-9 High-affinity IgG and IgA antibodies are used to neutralize microbial toxins and animal venoms

Many bacteria secrete protein toxins that cause disease by disrupting the normal function of human cells (Figure 7.21). To have this effect, a bacterial toxin must first bind to a specific receptor molecule on the surface of the human cell. In some toxins, for example those of diphtheria and tetanus, the receptor-binding activity is carried by one polypeptide chain and the toxic function by another. Antibodies that bind to the receptor-binding polypeptide can be sufficient to neutralize a toxin (Figure 7.22). The vaccines for diphtheria and tetanus work on this principle. They are modified toxin molecules, called **toxoids**, in which the toxic chain has been denatured to remove its toxicity. On immunization, protective neutralizing antibodies are made against the receptor-binding chain.

Bacterial toxins are potent at low concentrations, a single molecule of diphtheria toxin is sufficient to kill a cell. To neutralize a bacterial toxin, an antibody must be of high affinity and essentially irreversible in its binding to the toxin. It must also be able to penetrate tissues and reach the sites where toxins are being released. High-affinity IgG is the main source of neutralizing antibodies for the tissues of the human body, whereas high-affinity IgA serves a similar purpose for the mucosal surfaces.

Poisonous snakes, scorpions, and other animals introduce venoms containing toxic polypeptides into humans through a bite or sting. For some venoms,

Disease	Organism	Toxin	Effects *in vivo*
Tetanus	*Clostridium tetani*	Tetanus toxin	Blocks inhibitory neuron action leading to chronic muscle contraction
Diphtheria	*Corynebacterium diphtheriae*	Diphtheria toxin	Inhibits protein synthesis leading to epithelial cell damage and myocarditis
Gas gangrene	*Clostridium perfringens*	Clostridial-α toxin	Phospholipase activation leading to cell death
Cholera	*Vibrio cholerae*	Cholera toxin	Activates adenylate cyclase, elevates cAMP in cells, leading to changes in intestinal epithelial cells that cause loss of water and electrolytes
Anthrax	*Bacillus anthracis*	Anthrax toxic complex	Increases vascular permeability, leading to edema, hemorrhage, and circulatory collapse
Botulism	*Clostridium botulinum*	Botulinum toxin	Blocks release of acetylcholine leading to paralysis
Whooping cough	*Bordetella pertussis*	Pertussis toxin	ADP-ribosylation of G proteins leading to lymphocytosis
		Tracheal cytotoxin	Inhibits ciliar movement and causes epithelial cell loss
Scarlet fever	*Streptococcus pyogenes*	Erythrogenic toxin	Causes vasodilation, leading to scarlet fever rash
		Leukocidin Streptolysins	Kill phagocytes, enabling bacteria to survive
Food poisoning	*Staphylococcus aureus*	Staphylococcal enterotoxin	Acts on intestinal neurons to induce vomiting. Also a potent T-cell mitogen (SE superantigen)
Toxic-shock syndrome	*Staphylococcus aureus*	Toxic-shock syndrome toxin	Causes hypotension and skin loss. Also a potent T-cell mitogen (TSST-1 superantigen)

Figure 7.21 Many common diseases are caused by bacterial toxins. Several examples of exotoxins, or secreted toxins, are shown here. Bacteria also make endotoxins, or nonsecreted toxins, which are usually only released when the bacterium dies. Endotoxins, such as bacterial lipopolysaccharide (LPS), are important in the pathogenesis of disease, but their interactions with the host are more complicated than those of the exotoxins and are less clearly understood.

a single exposure is sufficient to cause severe tissue damage or even death, and in such situations the primary response of the immune system is too slow to help. As exposure to such venoms is rare, protective vaccines against them have not been developed. For patients who have been bitten by poisonous snakes or other venomous creatures, the preferred therapy is to infuse them

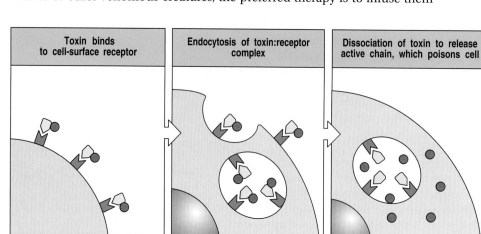
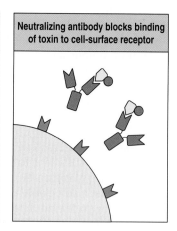

Figure 7.22 Neutralization by IgG antibodies protects cells from toxin action. Many species of bacteria cause their harmful effects by producing protein toxins. These toxins are usually of modular construction. One part of the toxin binds to a cellular receptor, which allows the toxin to be internalized, whereupon the second part poisons the cell. Antibodies that bind to the toxin and prevent it from attaching to its receptor and poisoning the cell are called neutralizing antibodies.

with antibodies specific for the venom. These antibodies are produced by immunizing large domestic animals—such as horses—with the venom. Transfer of protective antibodies in this manner is known as **passive immunization** and is analogous to the way in which newborn babies acquire passive immunity from their mothers.

7-10 High-affinity neutralizing antibodies prevent viruses and bacteria from infecting cells

The first step in the infection of a human cell by a virus is its attachment to the cell by means of a cell-surface protein, which is used as the virus receptor. The influenza virus, for example, binds to oligosaccharides on cell-surface glycoproteins on epithelial cells of the respiratory tract. The virus binds through a protein in its outer envelope, which is known as the **influenza hemagglutinin** because the protein can **agglutinate**, or clump together, red blood cells by binding to oligosaccharides on the red cell surface. Neutralizing antibodies that have been developed during primary immune responses to influenza and other viruses are the most important aspect of subsequent immunity to these viruses. Such antibodies coat the virus, inhibit its attachment to human cells, and prevent infection (Figure 7.23, upper panels).

Some bacteria that exploit mucosal surfaces maintain their populations by binding to and colonizing the surface of the epithelial cells, as does the

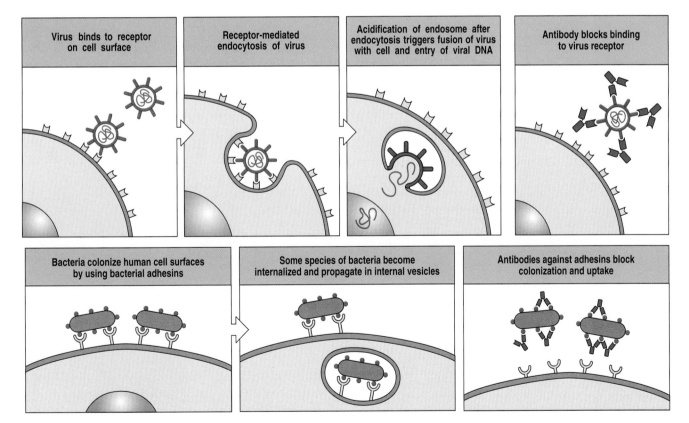

Figure 7.23 Viral and bacterial infection of cells can be blocked by neutralizing antibodies. Upper panels: for a virus to infect a cell, it must gain entry to the cytoplasm. This requires binding of the virus to the cell surface, internalization in an endosome, and fusion of viral and cell membranes to release viral nucleic acid into the cytoplasm. Antibodies binding to viral surface proteins can inhibit either the initial binding of virus or its subsequent entry into the cell. Lower panels: many bacterial infections require an interaction between the bacterium and the surface of a human cell. This is particularly true for infections of mucosal surfaces. The attachment process involves very specific molecular interactions between bacterial adhesins and their ligands on human cells. Antibodies specific for epitopes of the bacterial adhesins can, therefore, block infection.

bacterium *Neisseria gonorrhoeae*, which causes gonorrhea. Others enter epithelial cells, as do the species of *Salmonella* that cause food-borne gastrointestinal infections. IgA antibodies against the adhesion proteins (adhesins) responsible for binding to epithelial cells limit bacterial populations within the gastrointestinal, respiratory, urinary, and reproductive tracts and prevent disease-causing infections in these tissues (see Figure 7.23, lower panels).

7-11 The Fc receptors of hematopoietic cells are signaling receptors that bind the Fc regions of antibodies

Although the binding of a neutralizing antibody to a pathogen or toxin prevents further infection, it does not in itself remove the antigen from the body. This is accomplished by phagocytic effector cells, principally neutrophils, blood monocytes, and tissue macrophages. These cells express various receptors that bind to the Fc regions of antibodies of different isotypes and are known generally as Fc receptors.

The Fc receptors of phagocytes and other hematopoietic cells are functionally and structurally distinct from the FcRB of endothelial cells. They each consist of several polypeptide chains, the most important of which is an α chain made up of immunoglobulin-like domains (Figure 7.24). This chain binds to the Fc region of the antibody and determines the isotype specificity of the receptor. Associated with the α chain are other polypeptide chains that

Receptor	Fcγ RI (CD64)	Fcγ RII-A (CD32)	Fcγ RII-B2 (CD32)	Fcγ RII-B1 (CD32)	Fcγ RIII (CD16)	FcεRI	FcαRI (CD89)
Structure	α 72 kDa / γ	α 40 kDa / γ-like domain	ITIM	ITIM	α 50–70 kDa / γ or ζ	α 45 kDa / β 33 kDa / γ 9 kDa	α 55–75 kDa / γ 9 kDa
Relative binding strength	IgG1 / 200	IgG1 / 4	IgG1 / 4	IgG1 / 4	IgG1 / 1	IgE / 20,000	IgA1, IgA2 / 20
Cell type	Macrophages Neutrophils* Eosinophils* Dendritic cells	Macrophages Neutrophils Eosinophils Platelets Langerhans' cells	Macrophages Neutrophils Eosinophils	B cells Mast cells	NK cells Eosinophils Macrophages Neutrophils Mast cells FDCs	Mast cells Eosinophils* Basophils FDCs	Macrophages Neutrophils Eosinophils†
Effect of ligation	Uptake Stimulation Activation of respiratory burst Induction of killing	Uptake Granule release (eosinophils)	Uptake Inhibition of stimulation	No uptake Inhibition of stimulation	Induction of killing (NK cells)	Secretion of granules	Uptake Induction of killing

Figure 7.24 Receptors for the Fc regions of immuno-globulins are present on a variety of immune-system cells. The subunit structure, relative binding strength, and cellular distribution of the Fc receptors are shown. The complete multimolecular structure of most receptors is not yet known but they may all be multichain molecular complexes similar to the Fcε receptor I (FcεRI). Receptor structure can vary slightly from one cell type to another. For example, FcγRIII in neutrophils is expressed as a protein with a glycophosphatidylinositol membrane anchor and has no associated γ chains, whereas in natural killer (NK) cells it is a transmembrane protein associated with γ chains as shown. The information in the figure is based on the Fc receptors of mouse cells, with the exception of the relative binding strengths that pertain to the human receptors. The Fcγ receptors also bind the other subclasses of IgG. FcγRIII binds IgG1 and IgG3 with equal strength. For the other FcγRs, IgG1 binds most strongly, IgG2 least strongly, and IgG3 and IgG4 with intermediate affinity. FDCs, follicular dendritic cells. *In these cases, Fc receptor expression is inducible rather than constitutive. †In eosinophils, the molecular weight of CD89α is 70–100 kDa.

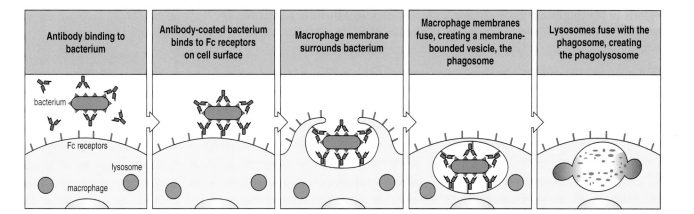

| Antibody binding to bacterium | Antibody-coated bacterium binds to Fc receptors on cell surface | Macrophage membrane surrounds bacterium | Macrophage membranes fuse, creating a membrane-bounded vesicle, the phagosome | Lysosomes fuse with the phagosome, creating the phagolysosome |

bacterium

Fc receptors

lysosome

macrophage

function either in the folding of the Fc receptor and its movement to the cell surface or signal the cell once the receptor has bound its ligand. One of the signaling components, the γ chain, is closely related in amino-acid sequence to the ζ chain of the T-cell receptor complex.

FcγRII-B1 and -B2 are inhibitory receptors that help to control the activation of naive B cells, mast cells, macrophages, and neutrophils. These receptors bear **immunoreceptor tyrosine-based inhibition motifs** (**ITIMs**) in their cytoplasmic tails, which associate with intracellular proteins that develop inhibitory signals.

7-12 Phagocyte Fc receptors facilitate the recognition, uptake, and destruction of antibody-coated pathogens

As we saw in Chapter 6, phagocytic cells can recognize, ingest, and destroy bacteria in the absence of specific antibody. This capacity is of paramount importance in containing infection during the period before an antigen-specific immune response has been made and in enabling macrophages to take up, process, and present antigen to T cells in the early phases of an adaptive immune response. However, the speed with which pathogens can be bound and engulfed by phagocytes greatly increases when the pathogens are coated with antibodies, or opsonized. This is because the principal phagocytic cells of the body—the macrophages and neutrophils—express Fc receptors, called **Fcγ receptors**, which are specific for the Fc regions of IgG antibodies, particularly that of IgG1 (see Figure 7.24).

When IgG molecules specific for the surface components of a pathogen bind to the pathogen with their Fab arms, the Fc regions are left exposed on the outside of the antibody-coated particle. The pathogen becomes coated with many IgG molecules, presenting multiple Fc regions to the Fc receptors on a phagocyte. On contact with a phagocyte, multiple ligand–receptor interactions are made, producing a stable and strong binding from interactions that are individually of low affinity and short-lived. The low affinity of Fcγ receptors for individual IgG molecules means that they bind only transiently to free IgG molecules in the absence of antigen. This property enables high concentrations of IgG of diverse antigenic specificities to circulate in the body's fluids and not clog up the Fc receptors of phagocytes in the absence of antigen.

After the pathogen has been bound to the phagocyte, interactions between antibody Fc regions and their receptors facilitate the engulfment of the antibody-coated pathogen (Figure 7.25). The surface of the phagocyte gradually extends around the surface of the opsonized pathogen through cycles of binding and release between the Fc receptors of the phagocyte and the Fc

Figure 7.25 Fc receptors on phagocytes trigger the uptake and breakdown of antibody-coated pathogens. Specific IgG molecules coat the pathogen surface, here a bacterium, and tether the bacterium to the surface of the phagocyte by binding to the Fc receptors. Signals from the Fc receptors enhance phagocytosis of the bacterium and the fusion of lysosomes containing degradative enzymes with the phagosome.

ojecting from the pathogen surface. The engulfment is an active process triggered by signals from the Fc receptors, which is similar, in some ways, to walking.

A coating of antibodies makes different sorts of microorganisms appear similar to the macrophage, and, thus, enables it to deal with them all by using a single effector mechanism. Encapsulated bacteria such as *Streptococcus pneumoniae* have evolved cell-surface structures that are resistant to direct phagocytosis; for these species a coating with antibody that masks their surface is essential if they are to be phagocytosed.

Once an opsonized bacterium has been endocytosed, it becomes enclosed in an acidified vesicle called a phagolysosome, formed from the fusion of the phagosome with lysosomes and neutrophil granules that contain hydrolytic enzymes and microbicidal peptides. Activated neutrophils and macrophages also produce oxygen radicals, nitric oxide, and other oxidizing agents with powerful microbicidal actions. The engulfed bacteria are killed by the combined effects of these substances.

As well as destroying microorganisms intracellularly, activated macrophages also attack larger antibody-coated parasites, such as worms, that they have bound via their Fc receptors but are too big for them to engulf. In this case, the toxic contents of the lysosomes and diffusible metabolites, such as nitric oxide, are secreted by the macrophage and poured onto the parasite.

7-13 IgE binds to high-affinity Fc receptors on mast cells, basophils, and activated eosinophils

IgE antibodies against a wide variety of different antigens are normally present in small amounts in all humans. They are produced in responses dominated by CD4 T_H2 cells, in which the cytokines produced favor switching to the IgE isotype. A consequence of the low affinity of Fc receptors for IgG is that free IgG molecules do not form stable interactions with cells expressing these receptors. The Fc receptor for IgE on **mast cells**, **basophils**, and activated **eosinophils** has quite the opposite properties. This receptor, called FcεRI, has such a high affinity ($\sim 10^{10}\,M^{-1}$) for the Fc region of IgE that IgE molecules are tightly bound in the absence of antigen and the cells are almost always coated with antibody. In the absence of allergy or parasitic infection, a single mast cell carries IgE molecules specific for many different antigens.

Mast cells are sentinels posted throughout the body's tissues, particularly in the connective tissues lying in the mucosa of the gastrointestinal and respiratory tracts and in connective tissues along blood vessels—especially those in the dermis of the skin. The cytoplasm of the resting mast cell is filled with large granules containing **histamine** and other molecules that contribute to inflammation, which are known generally as **inflammatory mediators**. Mast cells become activated to release their granules when antigen binds to the IgE molecules bound to FcεRI on the mast-cell surface (Figure 7.26). To activate the cell, the antigen must cross-link at least two IgE molecules and their associated receptors, which means that the antigen must have at least two topographically separate epitopes recognized by the cell-bound IgE. Cross-linking of FcεRI generates the signal that initiates the release of the mast cell granules. After degranulation the mast cell synthesizes and packages a new set of granules.

Inflammatory mediators secreted into the tissues by activated mast cells, basophils, and eosinophils increase the permeability of the local blood vessels, enabling other cells and molecules of the immune system to move out of the bloodstream and into tissues. This causes local accumulation of fluid and

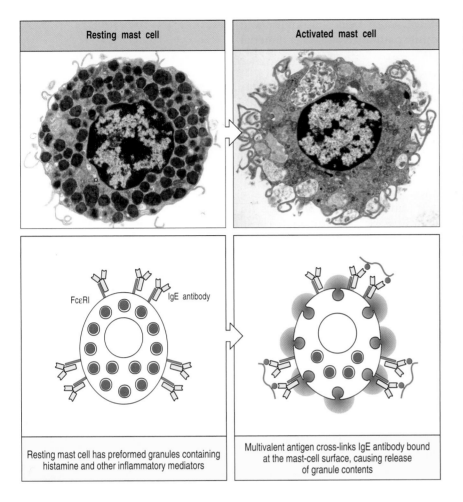

Resting mast cell	Activated mast cell

Resting mast cell has preformed granules containing histamine and other inflammatory mediators

Multivalent antigen cross-links IgE antibody bound at the mast-cell surface, causing release of granule contents

Figure 7.26 IgE cross-linking on mast-cell surfaces leads to the rapid release of mast-cell granules containing inflammatory mediators. Resting mast cells contain numerous granules containing inflammatory mediators such as histamine and serotonin. The cells have high-affinity Fc receptors (FcεRI) on their surface that are occupied by IgE molecules (left panels). Antigen cross-linking of bound IgE cross-links the FcεRI molecules, triggering the degranulation of the mast cell, and the release of inflammatory mediators into the surrounding tissue, as shown in the right panels. Photographs courtesy of A.M. Dvorak.

the swelling, reddening, and pain that characterize inflammation. Inflammation in response to an infection is beneficial because it recruits cells and proteins required for host defense into the sites of infection.

The prepackaged granules and the high-affinity FcεRI receptor already armed with IgE make the mast cell's response to antigen impressively fast. The infections that are the 'natural' targets of IgE-activated mast cells and eosinophils are thought to be those caused by parasites.

Parasites are a heterogeneous set of organisms that include the unicellular protozoa and multicellular invertebrates, notably the helminths—intestinal worms and the blood, liver, and lung flukes—and ectoparasitic arthropods such as ticks and mites. As a group, parasites establish long-lasting, persistent infections in human hosts and are well practised in the avoidance and subversion of the human immune system. Most parasites are much larger than any microbial pathogen. The largest human parasite is the tapeworm *Diphyllobothrium latum*, which can reach 9 meters in length and lives in the small intestine, causing vitamin B_{12} deficiency and, in some patients, megaloblastic anemia. Multicellular parasites cannot be controlled by the cellular and molecular mechanisms of destruction that work for microorganisms, so a different strategy based on IgE has evolved.

Inflammatory mediators released by mast cells, basophils, and eosinophils cause the contraction of smooth muscle surrounding the airways and the gut. In addition to violent muscular contractions that can expel parasites from the airways or gut, the increased permeability of local blood vessels supplies an outflow of fluid across the epithelium, which can help to flush out parasites.

In summary, the combined actions of IgE, mast cells, basophils, and eosinophils serve to physically remove parasite pathogens and other material from the body.

Eosinophils can also use their Fcε receptors to act directly against multicellular parasites. Such organisms, even small ones such as the blood fluke *Schistosoma mansoni*, which causes schistosomiasis, cannot be ingested by phagocytes. However, if the parasite induces an antibody response and becomes coated with IgE, activated eosinophils will bind to it through FcεRI and then pour the toxic contents of their granules directly onto its surface (Figure 7.27).

For human populations in developed countries where parasite infections are rare, the mast cell's response is most frequently seen as a detriment because its actions are the cause of allergy and asthma. People with these conditions make IgE in response to relatively innocuous substances, for example grass pollens or shellfish, which are often either airborne or eaten. Such substances are known as **allergens**. Having made specific IgE, any subsequent encounter with the allergen leads to massive mast-cell degranulation and a damaging response that is quite inappropriate to the threat posed by the antigen or its source. In extreme cases, the ingestion of an allergen can lead to a systemic life-threatening inflammatory response called anaphylaxis.

Parasite infections that invoke a protective IgE response are not major health problems in the developed world, whereas allergy and asthma do not seem to be prevalent in the developing countries where infection with parasites is endemic. This is one of various pieces of circumstantial evidence suggesting that if elements of the immune system are left unstimulated by infection, they can respond in ways that are frankly unhelpful.

7-14 Fc receptors activate natural killer cells to destroy antibody-coated human cells

Natural killer cells (NK cells) are large effector lymphocytes that circulate in the blood (see Figure 1.9, p. 10) and whose chief role is in innate immunity. However, they also express an Fc receptor called FcγRIII, or CD16, which is specific for IgG1 and IgG3. In experimental situations, NK cells have been shown to recognize and kill human cells coated with antibody against cell-surface components (Figure 7.28). This **antibody-dependent cell-mediated cytotoxicity (ADCC)** requires the presence of preformed antibody. This is not available during a primary immune response, but NK-cell cytotoxicity might

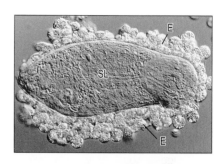

Figure 7.27 Eosinophils (E) attacking a schistosome larva (SL) in the presence of serum from an infected patient. Large parasites, such as worms, cannot be ingested by phagocytes; however, when the worm is coated with antibody, especially IgE, eosinophils can attack it by using their high-affinity Fc receptors (FcεRI). Similar attacks can be mounted by other Fc-receptor-bearing cells on various large targets. Photograph courtesy of A. Butterworth.

Figure 7.28 Antibody-coated target cells can be killed by natural killer cells (NK cells) in antibody-dependent cell-mediated cytotoxicity (ADCC). NK cells are large granular lymphocytes that are distinct from B and T cells and have FcγRIII receptors (CD16) on their surface. When these cells encounter cells coated with IgG antibody, they rapidly kill the target cell. The importance of ADCC in host defense or tissue damage is uncertain.

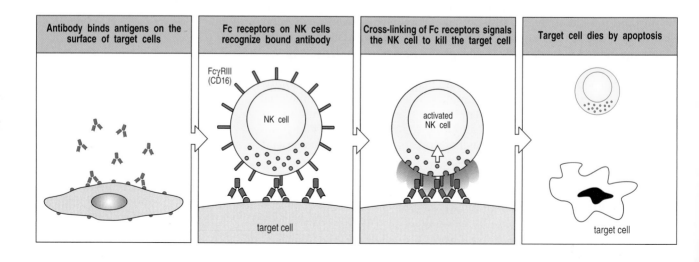

Antibody binds antigens on the surface of target cells

Fc receptors on NK cells recognize bound antibody

FcγRIII (CD16)

NK cell

Cross-linking of Fc receptors signals the NK cell to kill the target cell

activated NK cell

Target cell dies by apoptosis

target cell

target cell

have a role in secondary responses where antibody is already present. Another situation in which preformed antibody is present is in newborn infants, who have passively acquired IgG against many pathogens to which they have yet to be exposed.

Summary

Secreted antibodies are the only effector molecules produced by B cells; their principal function is as adaptor molecules that neutralize the pathogen and bring together pathogens or their products with the effector cells that destroy them. Antibodies can become bound through their Fc regions to Fc receptors on various types of effector cell. These interactions with Fc receptors are specific for immunoglobulin isotype and are used by the immune system for two main purposes. The first is to deliver antibodies to sites where they would not be carried by the circulation of the blood and lymph. To this end, the poly-Ig receptor of epithelium provides the lumen of the intestines and other mucosal surfaces with a continual supply of IgA that binds to the microorganisms that inhabit and infect those tissues. FcRB delivers IgG from plasma into the extracellular fluid in tissues and also, during pregnancy, delivers maternal IgG to the fetal circulation, which is useful after birth against many types of infection. The second purpose of the Fc receptors is to attach pathogens or antigens, which have bound to specific antibody, to effector cells that will respond in ways that eliminate infection. Fc receptors for IgG can deliver antibody-bound bacteria to phagocytes. In these cases, antibody binds first to the antigen and then to the Fc receptor. In contrast, the IgE antibody binds first to the Fc receptor of mast cells, basophils, and activated eosinophils and then awaits its antigen.

The antigen–antibody mediated pathway of complement activation

The binding of antibodies to antigens involves noncovalent bonds. Interactions between them are potentially reversible, depending on local concentrations of antibody and antigen, and are, therefore, susceptible to disruption, especially where low-affinity antibodies are concerned. Once a pathogen or antigen has been identified as foreign, it becomes advantageous to mark it for destruction in a more permanent manner. This is accomplished by a system of blood proteins known collectively as the complement system, or just complement. The name complement was given because the effector functions provided by these proteins 'complement' the antigen-binding function of antibodies in the defense against pathogens. Activation of the complement system initiates a series of enzymatic reactions in which the proteolytic cleavage and activation of successive complement components leads to the covalent bonding, or 'fixation', of particular complement fragments to the pathogen surface. Phagocytes bear surface receptors that recognize these fragments and this recognition facilitates the uptake and destruction of complement-coated microbes by neutrophils and macrophages.

Complement fixed on bacterial surfaces also nucleates a complex of proteins that attacks pathogens by poking holes in their cell membranes. Antibody bound to a pathogen triggers **complement activation** that proceeds by a series of enzymatic reactions called the **classical pathway of complement activation**. Pathogens also trigger complement activation by two other pathways that do not involve antibody and are considered part of innate immunity. They are the **lectin pathway of complement activation**, which is activated by binding of a plasma protein to mannose-containing peptidoglycans

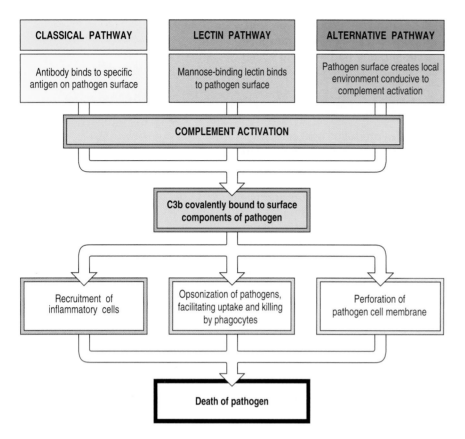

Figure 7.29 The three pathways of complement activation. The classical pathway is initiated by the binding of either IgM or IgG antibodies to a microbial surface, whereas the lectin-mediated pathway is initiated by the mannose-binding lectin of plasma, which binds to carbohydrates found on bacterial cells. The alternative pathway is triggered by the local physicochemical environment created by the constituents of some bacterial surfaces.

on microbial surfaces, and the **alternative pathway of complement activation,** which is triggered by direct environmental influence of the microbial surface. The antibody-mediated classical pathway will be discussed here; the other two pathways are covered in Chapter 8.

7-15 Complement components are plasma proteins with various functions

All complement components are made in the liver and circulate in the plasma. They comprise more than 30 proteins with a variety of biochemical functions. Many are enzymes and these are secreted and circulate in an inactive form known as a **zymogen**. Activation of complement takes place in the tissues, into which plasma leaks from the blood, and also in the blood itself. The complement components can be grouped on the basis of their functions; the members of each group are often structurally similar.

The three pathways of complement activation differ in the way they are triggered and in their first few reactions (Figure 7.29). However, all three pathways converge on the same reaction, the cleavage of complement component C3 into fragments C3b and C3a and the covalent binding of C3b to the pathogen's surface. This binding of C3b to the surface of pathogens is the most important function of the complement system and is called complement fixation. The bound C3b tags the pathogen for destruction by phagocytosis and also nucleates protein complexes that damage the pathogen's membrane. In addition, the soluble C3a fragment recruits inflammatory cells to the site of infection. These three effector mechanisms of complement are common to all three pathways of complement activation. Thus, the classical pathway of complement activation involves some components that are specific to the classical pathway and others that also function in the other pathways.

Early-acting components of the classical pathway		
Function	Protein	Concentration in serum µg ml^{-1}
Noncovalent binding to antigen:antibody complexes	C1q	70
Serine proteases that activate themselves or other complement proteins by cleaving a peptide bond	C1r C1s C2	34 31 25
Covalent attachment to pathogen surface and to antigen:antibody complexes	C4 C3	600 1200

Figure 7.30 Components of the early part of the classical pathway of complement activation.

Six complement components contribute to the early reactions of the classical pathway, those leading to the deposition of C3b fragments on pathogen surfaces (Figure 7.30). Of these, five are specific to the classical pathway, the exception being C3. Inherited deficiency of each of the complement components has been described in humans, and these deficiencies vary considerably in their effects on immunity. Most severe is the increased susceptibility to bacterial infections of patients deficient in C3, affirming the importance of C3 in complement function. C3 is also the most abundant complement component in plasma.

Binding of antibody to antigen on the pathogen's surface is an essential prerequisite for activation of the classical pathway. The first stage in activation involves the binding of the complement component C1q to the antigen:antibody complex. In the second stage, C1q serves as a scaffold for the binding and activation of the proteases C1r and C1s to form the active C1 molecule. This is the first of a series, or cascade, of proteases in which each enzyme cleaves and activates the next enzyme in the pathway. Each protease is highly specific for the complement component it cleaves, and cleavage is usually at a single site. In the third stage, cleavage of complement component C4, followed by cleavage of C3, allows fragments of these proteins to bind covalently to the antigen:antibody complex and to the surrounding area on the pathogen surface. These first three stages of complement activation by the classical pathway are each elaborated in the following three sections.

7-16 C1 uses different polypeptides to bind antibody and to activate complement components

The classical pathway of complement activation is triggered when the complement component C1 binds to the Fc region of an antibody that is part of an antibody:antigen complex. C1 is a complex of three proteins, one of which—C1q—specifically recognizes and binds to the Fc region of the antibody, whereas the other two—C1r and C1s—are inactive proteases. The structure of C1 is dominated by C1q, a large assembly of 18 polypeptides that resembles a bunch of six tulips when viewed in the electron microscope (Figure 7.31). Each tulip is composed of three similar polypeptides. The amino-terminal two-thirds of the polypeptides form the stalk, whereas the carboxy-terminal one-third of the polypeptides form the globular flower, which contains the binding site for antibody. The stalks flex so that the antibody-binding sites can move with respect to each other to allow multipoint attachment to an antibody-coated surface.

The complement cascade is initiated when antibody is bound to multiple sites on a cell surface, normally that of a pathogen. IgM is the isotype that is

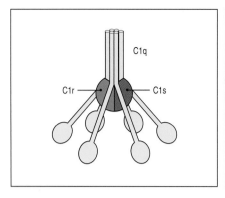

Figure 7.31 The complement component C1. The C1 molecule consists of a complex of C1q, C1r, and C1s. The C1q component consists of six identical subunits, each with one binding site for the Fc region of IgM or IgG and extended amino-terminal stalk regions that interact with each other and with two molecules each of the proteases C1r and C1s. The electron micrograph on the right contains images of three C1q molecules. Photograph courtesy of K.B.M. Reid.

most efficient at activating complement. The other human isotypes that activate the complement system are IgG1 and IgG3 and, to a lesser extent, IgG2. Pentameric IgM has five Fc regions, each of which can provide a binding site for one of the six binding sites of C1q. Multipoint attachment of C1q to IgM is required for a stable interaction; this can readily be satisfied by a single molecule of each type. IgG can also bind C1q, but it has only one Fc region; consequently, at least two molecules of IgG bound to a microbial surface within 30–40 nm of each other are required to bind one molecule of C1q (Figure 7.32). This difference explains why IgM activates complement much more effectively than IgG, and why free IgG cannot activate complement in solution. Free IgM cannot activate complement because it has a conformation that does not bind C1q. On binding to specific antigen, IgM changes its conformation to one that can bind C1q.

C1r and C1s are serine proteases that are activated when C1q binds to an antibody Fc region. Serine proteases are proteolytic enzymes that have a serine residue at the active site. They are typically synthesized in an inactive form, the zymogen, and become enzymatically active only after proteolytic cleavage by another protease. On binding to antibody, one molecule of C1r is induced to cut itself, thereby becoming enzymatically active. It then cuts and

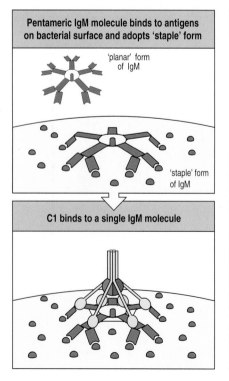

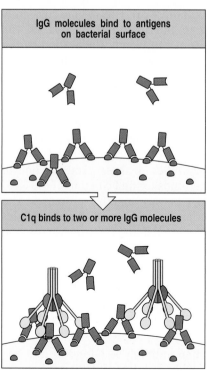

Figure 7.32 The classical pathway of complement activation is initiated by binding of C1q to antibody on a bacterial surface. The binding of C1q to a molecule of pentameric IgM is shown in the left panels. On establishing multipoint binding to bacterial cell-surface antigens, the IgM molecule adopts a less planar conformation, the so-called staple conformation (upper panel). This distortion allows the C1q molecule to establish multipoint attachment to the Fc regions of a single IgM molecule, using the hinges in the C1q stalks to position the globular Fc-binding sites (lower panel). It also exposes binding sites for the C1q heads. The binding of C1q to IgG is shown in the right panels. The C1q molecule needs to find pathogen-bound IgG molecules that are close enough to each other for the C1q molecule to span between them. As a consequence, the activation of complement by IgG depends more on the amount and density of antibodies bound to a pathogen surface than does complement activation by IgM.

activates the second C1r molecule and both C1s molecules. Activated C1s is the protease that binds, cleaves, and activates the next two components of the classical pathway, C4 and C2.

7-17 Fragments of C2 and C4 associate on the pathogen surface to form the classical C3 convertase

When a C4 molecule interacts with the activated C1s protease, it is cleaved into a large fragment called C4b and a small fragment called C4a. This cleavage exposes a high-energy thioester bond, which, in uncut C4, is sequestered and stabilized within the hydrophobic interior of the protein. On exposure to plasma the thioester is rapidly subjected to nucleophilic attack. The vast majority of C4b thioesters are hydrolyzed and remain in solution, but some react with the amino and hydroxyl groups of proteins and carbohydrates at the pathogen surface, and when this happens C4b becomes covalently bonded to the pathogen. The overall result is that C4b fragments are covalently attached to the pathogen's surface in the vicinity of the antigen:antibody complex that started the complement activation. During this reaction, C4b also becomes covalently attached to the antibodies bound to the pathogen and to any associated complement components (Figure 7.33).

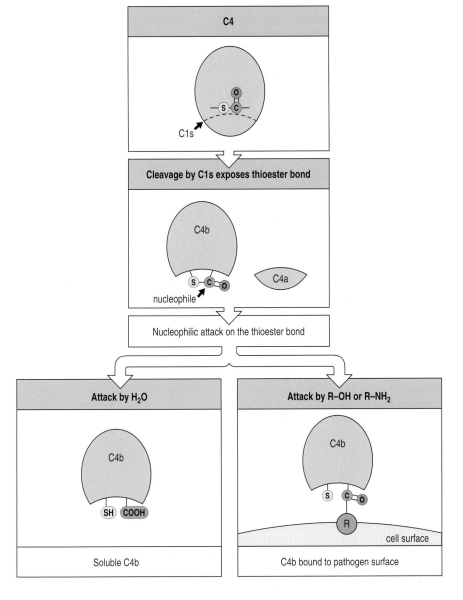

Figure 7.33 Cleavage of C4 exposes a reactive thioester bond that covalently attaches the C4b fragment to the pathogen surface. Circulating C4 is an inactive serine protease consisting of α, β, and γ polypeptide chains in which a thioester bond in the α chain is protected from hydrolysis within the hydrophobic interior of the protein. The thioester bond is denoted in the top two panels by the circled letters S, C, and O. The C4 molecule is activated by cleavage of the α chain by C1s to give fragments C4a and C4b. This exposes the thioester bond of C4b to the environment. The thioester bonds of most of the C4b fragments will be spontaneously hydrolyzed by water as shown in the bottom left panel, but a minority will react with the hydroxyl and amino groups on molecules on the pathogen's surface, bonding C4b to the pathogen surface as shown in the bottom right panel.

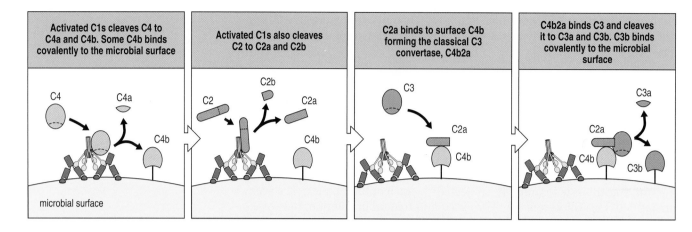

Figure 7.34 **Activated C1s cleaves C4 and C2 to produce C4b and C2a, which associate to form the classical C3 convertase.** The steps in the reaction are outlined here and detailed in Section 7-17.

Complement component C2 is the second substrate for the activated C1s protease. C2 is a serine protease zymogen that upon cleavage by C1s, produces a large C2a fragment with protease activity and a small C2b fragment. (For historical reasons, the small cleavage product of C2 is called C2b and the larger product is called C2a, whereas for other complement proteins, the larger fragment is called 'b' and the smaller 'a'.) On release from C1, C2a binds to a C4b fragment bonded to the pathogen surface. This C4b2a complex, called the **classical C3 convertase**, is a surface-associated serine protease whose function is to cleave complement component C3 and 'convert' it from an inactive to an active form.

7-18 Cleavage of C3 yields C3b covalently bound to pathogen surfaces

C3 is structurally very similar to C4 and, like C4, it contains a sequestered thioester bond that becomes activated on cleavage of the molecule. So the reaction that occurs when C3 binds to and is cleaved by C4b2a is just like the C1s-mediated cleavage of C4. Upon cleavage, the smaller C3a fragment is removed and the thioester bond of the larger C3b fragment becomes exposed and susceptible to attack. Consequently, some of the C3b fragments become covalently bonded to the pathogen's surface around each site where a C4b2a complex is active (Figure 7.34). Although each C4b2a molecule is only active for a few minutes it can cleave up to 1000 molecules of C3, many of which become bonded to the microbial surface. So the number of C3b molecules deposited on the surface at a site of complement fixation is much larger than the number of C4b molecules.

Although the complement system has many components, and can seem horribly complicated, its workings are based on a few simple principles that are elaborated in different ways. Of these principles the most important is the use of thioester bonds in C3 and C4 to covalently tag microbial surfaces. Although many enzymes transiently form thioester bonds during catalysis, for a stable protein to contain a thioester bond is truly exceptional. Apart from C3 and C4, only one other protein has been found to have this property, the structurally related protease inhibitor of plasma, α_2-macroglobulin.

7-19 Partial lack of C4 is the most common immune protein deficiency in humans

The two types of C4—C4A and C4B— have different properties. The thioester of the C4A form is preferentially attacked by the amino groups of macromolecules, whereas that of C4B is preferentially attacked by the hydroxyl groups. This increases the efficiency of C4 deposition and coverage of the whole

Figure 7.35 Humans differ in the number and type of genes for complement component C4. C4A and C4B exhibit differences in the way they bond to pathogen surfaces. The genes for C4A and C4B are located in the central part of the MHC, between the class I region and the class II region (see Figure 3.25, p. 88). Although a majority of MHC haplotypes have one gene for C4A and one for C4B, a considerable minority have other arrangements involving loss or duplication of one of the genes. These differences lead to variation in C4 function within the population and to immunodeficiency in some individuals.

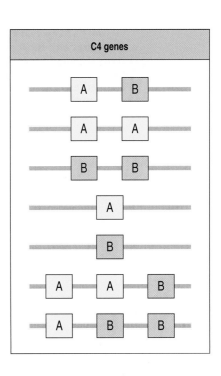

pathogen surface. The two genes encoding C4A and C4B are closely linked and situated in the class III region of the MHC, where, through gene duplication and deletion, further diversification of the C4 genes has evolved (Figure 7.35). In humans, 13% of chromosomes lack a functional C4A gene and 18% of chromosomes lack a functional C4B gene; thus, more than 30% of the human population is deficient for one or other form of C4, and a partial lack of C4 is the most common human immunodeficiency. Reflecting the complementary functions of the two forms of C4, deficiency in C4A is associated with susceptibility to the autoimmune disease **systemic lupus erythematosus (SLE)**, whereas deficiency in C4B is associated with lowered resistance to infection. As well as the simple presence or absence of the C4A and C4B genes, there are more than 40 different alleles of the C4 genes, which could be associated with further differences in C4 function.

7-20 C3b produced by the classical C3 convertase permits the formation of a more powerful alternative C3 convertase

The cleavage of C3 by the classical C3 convertase can be seen as an amplification step that increases the number of complement fragments attached to the pathogen surface over that accomplished by cleavage of C4 alone. This process of amplification is now taken one step further. C4 and C3 are closely related proteins and C3b, like C4b, can assemble a C3 convertase. The other component of this alternative convertase is derived from the plasma protein factor B that is closely related in structure and function to C2, which furnishes the second subunit of the classical C3 convertase (Figure 7.36). When factor B binds to C3b on the pathogen surface it becomes susceptible to cleavage by factor D, another serine protease in plasma, that cleaves off a small Ba fragment leaving the larger, and now proteolytically active, Bb fragment associated with C3b. The C3bBb convertase is homologous in both structure and function to C4b2a (Figure 7.37): it cleaves C3 molecules, exposing their thioester bonds to attack by water, plasma proteins, and components of the microbial surface. The process by which the alternative convertase cleaves and activates C3 is homologous to that followed by the classical convertase

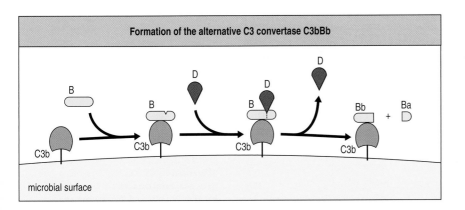

Formation of the alternative C3 convertase C3bBb

Figure 7.36 Formation of the alternative C3 convertase. The C3 convertase of the alternative pathway is assembled from C3b and the active Bb fragment of factor B. Because C3b, the product of the alternative C3 convertase, actually makes more convertase, this means that the alternative convertase is inherently more active than the classical C3 convertase in depositing C3b on pathogen surfaces.

Figure 7.37 The two types of C3 convertase have similar structures and functions. In the C3 convertase produced by the classical pathway, C4bC2a, the activated protease C2a cleaves C3 to C3b and C3a (not shown). In the analogous C3 convertase of the alternative pathway, C3bBb, the activated protease Bb carries out exactly the same reaction.

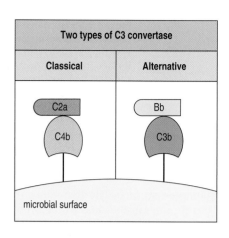

(Figure 7.38). What distinguishes the convertase C3bBb from C4b2a is that the product of the enzymatic reaction, C3b, can assemble more enzyme. This distinctive property of C3bBb produces exponential amplification of the C3 conversion reaction started by C4b2a, an accelerating reaction that would rapidly consume the supplies of C3 were there not control mechanisms that dampen the response, as we shall see later. By the end of the complement-fixing reaction, most of the C3b fragments that cover the pathogen surface around the initiating antigen:antibody complex are as a result of the action of the C3bBb convertase (Figure 7.39).

In the alternative pathway of complement activation, where antibody is not involved, C3bBb is the only participating C3 convertase. For this, and historical reasons, C3bBb is called the C3 convertase of the alternative pathway or the **alternative C3 convertase**. Likewise, factor B and factor D are designated as components of the alternative pathway of complement activation. However, these designations can be misleading because formation and catalysis of the C3bBb convertase makes an essential contribution to the antibody-driven classical pathway of complement activation, in which C3bBb serves in addition to the classical convertase rather than as an alternative.

7-21 Fragments of C3 and C4 on pathogen surfaces are recognized by receptors on various cell types

Several types of immune-system cell have surface receptors that bind to the C3 and C4 fragments deposited on the pathogen surface and stimulate a cellular response to the pathogen. There are four types of **complement receptors**, falling into two structural groups: CR1 and CR2 in one group and CR3

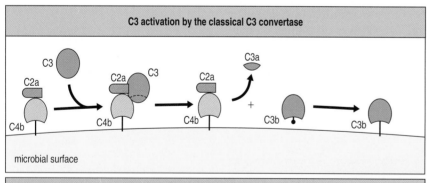

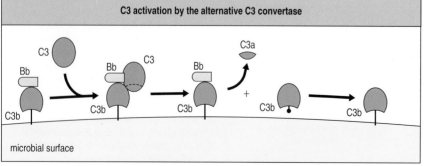

Figure 7.38 C3 activation by the alternative C3 convertase is a process analogous to C3 activation by the classical C3 convertase.

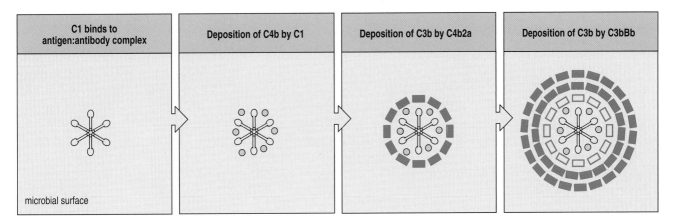

| C1 binds to antigen:antibody complex | Deposition of C4b by C1 | Deposition of C3b by C4b2a | Deposition of C3b by C3bBb |

microbial surface

Figure 7.39 A bird's-eye view of the fixation of C4b and C3b fragments on a pathogen surface around an antigen:antibody complex. Antibody bound to an antigen on a microbial surface binds C1 (first panel), which leads to the deposition of C4b (pink circles) around the antigen:antibody complex (second panel). When C4b binds C2a to form the classical C3 convertase, a limited number of C3b molecules (green rectangles) are produced (third panel). These can bind Bb to form the alternative convertase, C3bBb (yellow rectangles), which leads to deposition of many more C3b fragments on the microbial surface (green rectangles, fourth panel).

and CR4 in the other. All four receptors differ in their cellular distribution (Figure 7.40). Macrophages and neutrophils (polymorphonuclear leukocytes) express **complement receptor 1 (CR1)**, which on binding to C3b or C4b on a pathogen surface facilitates uptake and destruction of the pathogen by these phagocytic cells (Figure 7.41). In this case, C3b, and to a lesser extent C4b, are acting as opsonins in what is the principal function of complement, to destroy microorganisms. By itself, the interaction of C3b with CR1 does not stimulate phagocytosis and the intracellular killing of pathogens, but it enhances these functions once they have been initiated by the binding of IgG to an Fcγ receptor or by the cytokine IFN-γ, which is produced by activated T cells. The interaction of C3b with CR1 is more significant than with C4b because complement activation deposits much more C3b than C4b on pathogen surfaces.

Complement receptor 2 (CR2) has a different tissue distribution from CR1, CR3, and CR4 because it is expressed on B cells and follicular dendritic cells. As we saw in Section 7-1, CR2 (also called CD21) is a component of the B-cell co-receptor. It binds to complement fragments iC3b, C3d, and C3dg, which are formed by degradation of C3b at the pathogen surface. When a B-cell

Receptor	Ligand	Functions	Cell types
CR1	C3b, C4b	Promotes C3b and C4b decay Stimulates phagocytosis Erythrocyte transport of immune complexes	Erythrocytes, macrophages, monocytes, polymorphonuclear leukocytes, B cells, FDCs, podocytes in kidney glomeruli
CR2	C3d, C3dg, iC3b	Part of B-cell co-receptor	B cells, FDCs
CR3	iC3b	Stimulates phagocytosis	Macrophages, monocytes, polymorphonuclear leukocytes, FDCs
CR4	iC3b	Stimulates phagocytosis	Macrophages, monocytes, polymorphonuclear leukocytes

Figure 7.40 Distribution and function of receptors for complement proteins. There are several different complement receptors that are specific for different complement components or their fragments. CR1 and CR3 are especially important in the phagocytosis of complement-coated bacteria by macrophages and neutrophils. CR1 on erythrocytes clears immune complexes from the circulation and CR2 is mainly present on B cells, where it is also part of the B-cell co-receptor complex. C3d, C3dg, and iC3b are cleavage products of C3b. FDC, follicular dendritic cell.

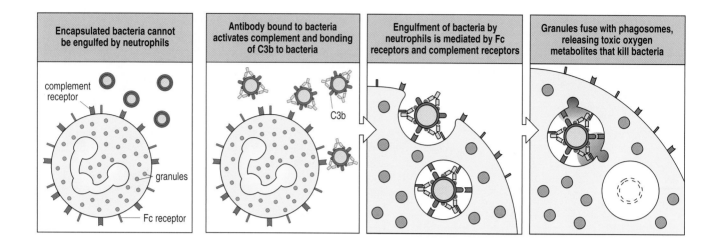

| Encapsulated bacteria cannot be engulfed by neutrophils | Antibody bound to bacteria activates complement and bonding of C3b to bacteria | Engulfment of bacteria by neutrophils is mediated by Fc receptors and complement receptors | Granules fuse with phagosomes, releasing toxic oxygen metabolites that kill bacteria |

receptor interacts with its specific antigen on a pathogen, the signal delivered to the B cell is strengthened if the CR2 component of the B-cell co-receptor also interacts with a nearby iC3b, C3d, or C3dg fragment on the pathogen surface (see Figure 7.3). CR2 on follicular dendritic cells in the lymphoid follicles enables these cells to bind antigens that have been tagged with C3b or its products (see Figure 7.10) and to retain them for long-term stimulation of B cells. The Epstein–Barr virus (EBV), which causes infectious mononucleosis and certain lymphomas, also binds to CR2 and exploits this interaction to infect B cells. In this context, CR2 is known as the EBV receptor of human B cells. CR1 and CR2 are both elongated molecules consisting of a string of small compact structural modules.

In contrast, **complement receptors 3** and **4** (**CR3** and **CR4**) are β-integrins. They bind to iC3b on pathogen surfaces. CR3 and CR4 are expressed on phagocytes, where they augment the activities of Fc receptors and CR1 in activating phagocytosis. Unlike the interaction between C3b and CR1, the binding of iC3b to CR3 is sufficient in itself to stimulate phagocytosis. β-Integrins also function as cell adhesion molecules, and CR3 and CR4 are involved in the adherence of leukocytes to endothelial cells in sites of inflammation.

7-22 Complement receptors remove immune complexes from the circulation

In the previous sections we have been principally concerned with the binding of antibody and complement to pathogenic microorganisms, which are large particles. High-affinity antibodies also bind to soluble protein antigens such as bacterial toxins, forming complexes that cannot be engulfed by phagocytes because they contain too few molecules of IgG to form a stable interaction with Fcγ receptors. Such soluble immune complexes are present in the circulation after the immune response to most infections and they are removed through the action of complement. The number of IgG molecules in an immune complex is sufficient to bind C1 and activate the enzymes that cleave first C4 and then C3, so that the antigen and antibody molecules within the complex become covalently tagged with C4b and C3b. Having been tagged in this way, the complex can now be bound by circulating cells that express CR1. Of these, the most numerous is the erythrocyte, and the vast majority of immune complexes become bound to the surface of red blood cells. During their circulation in the blood, erythrocytes pass through areas of the liver and the spleen where tissue macrophages remove and degrade the complexes of complement, antibody, and antigen from the erythrocyte surface while leaving the erythrocyte unscathed (Figure 7.42).

Figure 7.41 Encapsulated bacteria are more efficiently engulfed by phagocytes when the bacteria are coated with antibody and C3b. Encapsulated bacteria are naturally resistant to uptake by phagocytes, here represented by a neutrophil. When such bacteria are coated with antibody and C3b they become susceptible to phagocytosis mediated by Fc receptors and C3b receptors. Fc receptors bind IgG, whereas complement receptors bind C3b, inducing efficient phagocytosis and also the activation of the neutrophil. When the neutrophil is activated, its granules fuse with the phagosome containing the bacteria, releasing bactericidal metabolites. Macrophages also phagocytose and kill encapsulated bacteria in the same manner.

Figure 7.42 Erythrocyte CR1 helps to clear immune complexes from the circulation. Immune complexes bind to CR1 on erythrocytes, which transport them to the liver and spleen. Here they are removed by macrophages expressing receptors for Fc regions and for bound complement components.

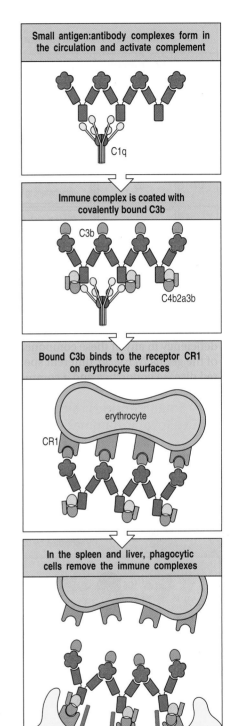

If immune complexes are not removed, they have a tendency to enlarge by aggregation and to precipitate at the basement membrane of small blood vessels, most notably those of the kidney glomeruli, where blood is filtered to form urine and is under particularly high pressure. Immune complexes that pass through the basement membrane bind to CR1 receptors expressed by podocytes, specialized epithelial cells that cover the capillaries. Deposition of immune complexes within the kidney probably occurs at some level all the time, and mesangial cells within the glomerulus are specialized in the elimination of immune complexes and in stimulating the repair of the tissue damage they cause.

A feature of the autoimmune disease **SLE** is a level of immune complexes in the blood sufficient to cause massive deposition of antigen, antibody, and complement on the renal podocytes. These deposits damage the glomeruli, and kidney failure is the principal danger for patients with this disease. A similar deposition of immune complexes can also be a major problem for patients who have inherited deficiencies in the early components of the complement pathway and cannot tag their immune complexes with C4b or C3b. Such patients cannot clear immune complexes; these accumulate with successive antibody responses to infection and inflict increasing damage on the kidneys.

7-23 The terminal complement proteins lyse pathogens by forming a membrane pore

As we have seen, the most important product of complement activation is C3b bonded to pathogen surfaces and soluble antigens. However, the cascade of complement reactions does extend beyond this stage, involving five additional complement components (Figure 7.43). C3b can bind to either of the C3 convertases to produce enzymes that act on the C5 component of complement and are called **C5 convertases**. The C5 convertase of the classical pathway, the **classical C5 convertase**, consists of C4b, C2a, and C3b and is designated C4b2a3b, whereas the C5 convertase of the alternative pathway, the **alternative C5 convertase**, consists of Bb plus two C3b fragments and is designated $C3b_2Bb$ (Figure 7.44).

Complement component C5 is structurally similar to C3 and C4 but lacks the thioester bond and has a different function. It is cleaved by one or other C5 convertase into a smaller C5a fragment and a larger C5b fragment. The function of C5b is to initiate the formation of a **membrane-attack complex**, which can make holes in the membranes of bacterial pathogens and eukaryotic cells. In succession, C6 and C7 bind to C5b—interactions that expose a hydrophobic site in C7, which inserts into the lipid bilayer. When C8 binds to C5b a hydrophobic site in C8 is exposed, and on insertion into the membrane this part of C8 initiates polymerization of C9, the component that forms the transmembrane pores. C9 is structurally similar to perforin, present in the lytic granules of cytotoxic T cells, but forms pores of about 100 Å diameter compared to the 160 Å pores formed by perforin (Figure 7.45). The components of the membrane-attack complex are listed and their activities summarized in Figure 7.43.

Although in the laboratory the perforation of membranes by the membrane-attack complex appears dramatic, clinical evidence demonstrating the

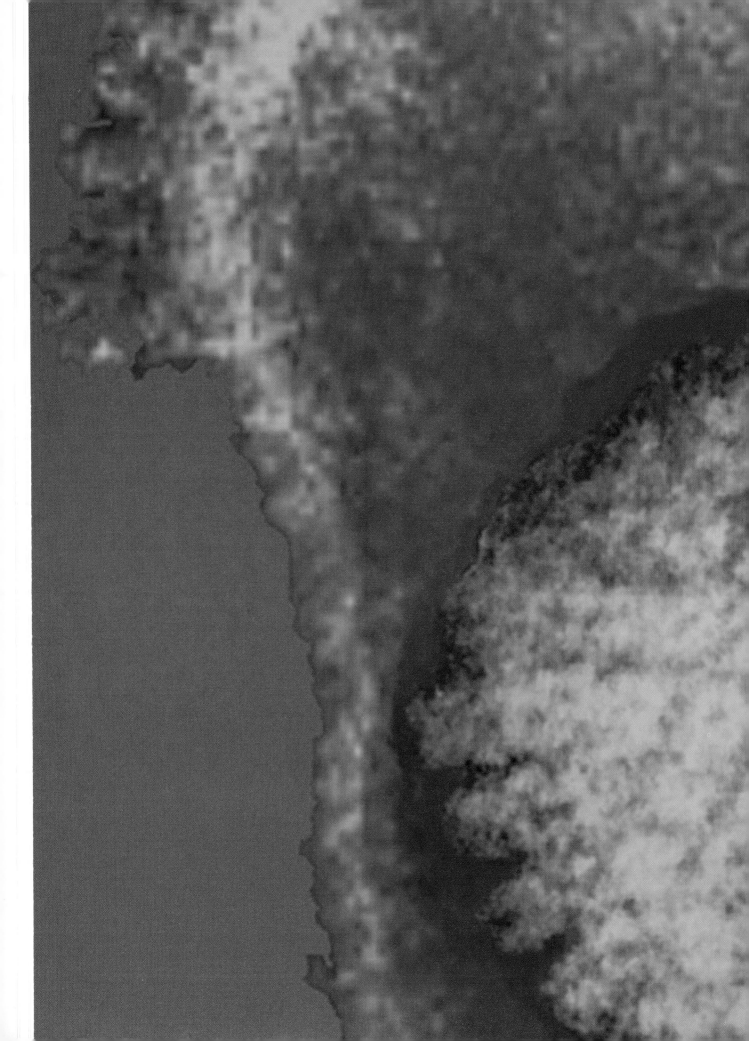

Chapter 8

The Body's Defenses Against Infection

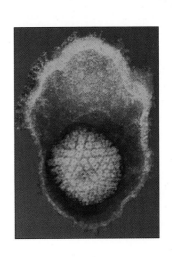

Most infectious diseases suffered by the human species are caused by pathogens that are smaller than a single human cell. For these agents, the human body constitutes a vast resource-rich environment in which to live and reproduce. In the face of such threats, the body deploys multifarious defense mechanisms that have accumulated over hundreds of millions of years of invertebrate and vertebrate evolution. In Chapter 1 we took a broad look at the nature of infections and of the body's defenses against them. We saw how these defenses fall into three categories: physical barriers that prevent pathogens from entering the body's tissues; the fixed or 'hard-wired' mechanisms of **innate immunity**, which attack an infection from its very beginning; and the mechanisms of the **adaptive immune response**, which respond to the infection in a very specific fashion. Much of medical practice is concerned with the diseases that result from the small proportion of infections that innate immunity fails to terminate, and in which the spread of the pathogen to lymphoid tissues stimulates an adaptive immune response. In such situations the attending physicians and the adaptive immune response work together to effect a cure, a partnership that has historically favored the scientific investigation of adaptive immunity over innate immunity.

The adaptive immune response is the component of the immune defenses that is enhanced by vaccination and provides long-term protection against many infectious diseases. Chapters 2–7 of this book focus on the cellular and molecular basis of the adaptive immune response. They describe how lymphocytes recognize pathogens through antigen-specific receptors, how B and T cells develop, and how they fulfill their effector functions. In this chapter we see how, in practice, the adaptive immune response cooperates with innate immunity to fight infections. In the first part of the chapter we examine the mechanisms of innate immunity and why they are an essential prerequisite for any adaptive immune response. The second part shows how the various arms of the immune system cooperate in producing an adaptive immune response. In the last part of the chapter we see how a primary adaptive immune response produces immunological memory—**immunity**—that lessens the impact of subsequent encounters with the same pathogen.

Innate immunity

Although we become conscious of infectious agents only when we are suffering from the diseases they cause, microorganisms are always with us. Fortunately, the vast majority of microorganisms we come into contact with are prevented from ever causing an infection by barriers at the body's surface.

The outer epithelial surfaces of the body provide an effective physical barrier against most organisms. They are also colonized by nonpathogenic resident microorganisms that compete with the invading pathogen for nutrients and living space. Furthermore, almost all infectious agents that penetrate those barriers and start an infection are eliminated quickly by the innate immune response before causing any obvious symptoms of disease. The physical barriers and the mechanisms of innate immunity can be considered as fixed defenses, completely determined by genes inherited from our parents and ready for action at all times. This part of the chapter introduces the agents that cause disease and describes how the fixed defenses cope with infections by pathogens to which the immune system has had no previous exposure.

The proportion of infections that are successfully eliminated by innate immunity is difficult to assess, mainly because such infections are overcome before they have caused symptoms severe enough to command the attention of those infected or their physicians. Intuitively, it seems likely to be a high proportion, especially in light of the human body's capacity to sustain vast populations of resident microorganisms without causing symptoms of disease. The importance of innate immunity is also implied by the rarity of inherited deficiencies in nonadaptive immune mechanisms and the considerable impairment of protection when they do occur (Figure 8.1).

8-1 Infectious diseases are caused by pathogens of diverse types that live and replicate in the human body

Human pathogens are of four kinds: viruses, bacteria, fungi, and parasites. The first three categories each correspond to a single taxonomic group of microorganisms, whereas the term 'parasites' refers to various disease-causing eukaryotic organisms that are mostly unicellular protozoa and multicellular worms (Figure 8.2). Pathogens are diverse in their structure, in the manner in which

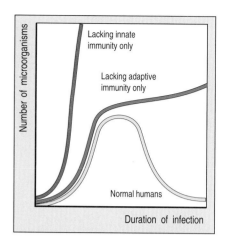

Figure 8.1 The benefits of innate and adaptive immunity. Shown here is the time course of a primary infection that, in normal individuals, is cleared from the body by the combined effects of innate and adaptive immunity (yellow line). In a person who lacks innate immunity, uncontrolled infection occurs because the adaptive immune response cannot be deployed without the preceding innate response (red line). In a person who lacks adaptive immune responses, the infection is initially contained by innate immunity but cannot be cleared from the body (green line).

Type	Disease	Pathogen	General classification*	Route of infection
Viruses	Severe acute respiratory syndrome	SARS virus	Coronaviruses	Oral/respiratory/ocular mucosa
	West Nile encephalitis	West Nile virus	Flaviviruses	Bite of an infected mosquito
	Yellow fever	Yellow fever virus	Flaviviruses	Bite of infected mosquito (*Aedes aegypti*)
	Hepatitis B	Hepatitis B virus	Hepadnaviruses	Sexual transmission; infected blood
	Chickenpox	Varicella-zoster	Herpes viruses	Oral/respiratory
	Mononucleosis	Epstein-Barr virus	Herpes viruses	Oral/respiratory
	Influenza	Influenza virus	Orthomyxoviruses	Oral/respiratory
	Measles	Measles virus	Paramyxoviruses	Oral/respiratory
	Mumps	Mumps virus	Paramyxoviruses	Oral/respiratory
	Poliomyelitis	Polio virus	Picornaviruses	Oral
	Jaundice	Hepatitis A virus	Picornaviruses	Oral
	Smallpox	Variola	Pox viruses	Oral/respiratory
	AIDS	Human immunodeficiency virus	Retroviruses	Sexual transmission, infected blood
	Rabies	Rabies virus	Rhabdoviruses	Bite of an infected animal
	Common cold	Rhinoviruses	Rhinoviruses	Nasal
	Diarrhea	Rotavirus	Rotaviruses	Oral
	Rubella	Rubella	Togaviruses	Oral/respiratory

Type	Disease	Pathogen	General classification*	Route of infection
Bacteria	Trachoma	*Chlamydia trachomatis*	Chlamydias	Oral/respiratory/ocular mucosa
	Bacillary dysentery	*Shigella flexneri*	Gram-negative bacilli	Oral
	Food poisoning	*Salmonella enteritidis, S. typhimurium*	Gram-negative bacilli	Oral
	Plague	*Yersinia pestis*	Gram-negative bacilli	Infected flea bite, respiratory
	Tularemia	*Pasteurella tulaensis*	Gram-negative bacilli	Handling infected animals
	Typhoid fever	*Salmonella typhi*	Gram-negative bacilli	Oral
	Gonorrhea	*Neisseria gonorrhoeae*	Gram-negative cocci	Sexually transmitted
	Meningococcal meningitis	*Neisseria meningitidis*	Gram-negative cocci	Oral/respiratory
	Meningitis, pneumonia	*Haemophilus influenzae*	Gram-negative coccobacilli	Oral/respiratory
	Legionnaire's disease	*Legionella pneumophila*	Gram-negative coccobacilli	Inhalation of contaminated aerosol
	Whooping cough	*Bordetella pertussis*	Gram-negative coccobacilli	Oral/respiratory
	Cholera	*Vibrio cholerae*	Gram-negative vibrios	Oral
	Anthrax	*Bacillus anthracis*	Gram-positive bacilli	Oral/respiratory by contact with spores
	Diphtheria	*Corynebacterium diphtheriae*	Gram-positive bacilli	Oral/respiratory
	Tetanus	*Clostridium tetani*	Gram-positive bacilli (anaerobic)	Infected wound
	Boils, wound infections	*Staphylococcus aureus*	Gram-positive cocci	Wounds; oral/respiratory
	Pneumonia, scarlet fever	*Streptococcus pneumoniae*	Gram-positive cocci	Oral/respiratory
	Tonsilitis	*Streptococcus pyogenes*	Gram-positive cocci	Oral/respiratory
	Leprosy	*Mycobacterium leprae*	Mycobacteria	Infected respiratory droplets
	Tuberculosis	*Mycobacterium tuberculosis*	Mycobacteria	Oral/respiratory
	Respiratory disease	*Mycoplasma pneumoniae*	Mycoplasmas	Oral/respiratory
	Typhus	*Rickettsia prowazeckii*	Rickettsias	Bite of infected tick
	Lyme disease	*Borrelia burgdorferi*	Spirochetes	Bite of infected deer tick
	Syphilis	*Treponema pallidum*	Spirochetes	Sexual transmission
Fungi	Aspergillosis	*Aspergillus* species	Ascomycetes	Opportunistic pathogen, inhalation of spores
	Athlete's foot	*Tinea pedis*	Ascomycetes	Physical contact
	Candidiasis, thrush	*Candida albicans*	Ascomycetes (yeasts)	Opportunistic pathogen, resident flora
	Pneumonia	Pneumocystis carinii	Ascomycetes	Opportunistic pathogen, resident lung flora
Protozoan parasites	Leishmaniasis	*Leishmania major*	Protozoa	Bite of an infected sand fly
	Malaria	*Plasmodium falciparum*	Protozoa	Bite of an infected mosquito
	Toxoplasmosis	*Toxoplasma gondii*	Protozoa	Oral, from infected material
	Trypanosomiasis	*Trypanosoma brucei*	Protozoa	Bite of an infected tsetse fly
Helminth parasites (worms)	Common roundworm	*Ascaris lumbricoides*	Nematodes (roundworms)	Oral, from infected material
	Schistosomiasis	*Schistosoma mansoni*	Trematodes	Through skin by bathing in infected water

Figure 8.2 (opposite page and above) Diverse microorganisms cause human disease. Pathogenic organisms are of four main types—viruses, bacteria, fungi, and parasites, which are mostly protozoa or worms. Some important pathogens in each category are listed along with the diseases they cause. *The classifications given are intended as a guide only and are not taxonomically consistent; families are given in the case of the viruses; general groupings often used in medical bacteriology for the bacteria; and higher taxonomic divisions for the fungi and parasites.

	Direct mechanisms of tissue damage by pathogens			Indirect mechanisms of tissue damage by pathogens		
	Exotoxin release	Endotoxin release	Direct cytopathic effect	Immune complexes	Anti-host antibody	Cell-mediated immunity
Pathogenic mechanism						
Infectious agent	*Vibrio cholerae*	*Yersinia pestis*	Influenza virus	*Treponema pallidum*	*Streptococcus pyogenes*	*Mycobacterium tuberculosis*
Disease	Cholera	Plague	Influenza	Kidney damage in secondary syphilis	Rheumatic fever	Tuberculosis

Figure 8.3 Pathogens damage tissues in different ways. Pathogens can directly kill cells and damage tissues in three ways. Exotoxins released by microorganisms act at the surfaces of host cells, usually via a cell-surface receptor (first column). When phagocytes degrade certain microorganisms, endotoxins are released that induce the phagocytes to secrete cytokines, causing local or systemic symptoms (second column). Cells infected by pathogens are usually killed or damaged in the process (third column). Tissue damage can also be caused in an indirect fashion. Antibodies produced by B cells in response to pathogens can form large antigen:antibody complexes that damage blood vessels and kidneys (fourth column). Antibodies against a pathogen can react with a human cell, causing it to be treated by the immune system as though it were a foreign invader (fifth column). In clearing pathogens from an infected site, inflammatory cells of the immune system damage tissue and leave debris (sixth column). A representative example of a pathogen and the disease it causes is given under each type of damage.

they exploit the human body, and in the type of damage they cause. Upon infection, tissue damage and disease symptoms can be caused directly by the pathogen, or indirectly as a consequence of the response of the host's immune system to the pathogen (Figure 8.3). For purposes of host defense, a distinction can be made between pathogens that replicate in the spaces between human cells to produce extracellular infections and pathogens that replicate inside human cells to produce intracellular infections. Both extracellular and intracellular spaces can be further subdivided (Figure 8.4). These different sites of replication affect the types of immune mechanism that can be used to eliminate the pathogen. Extracellular forms of pathogens are accessible to soluble molecules of the immune system, whereas intracellular forms are not. Those intracellular pathogens that live in the nucleus or cytosol

	Extracellular		Intracellular	
	Interstitial spaces, blood, lymph	Epithelial surfaces	Cytoplasmic	Vesicular
Site of infection				
Organisms	Viruses Bacteria Protozoa Fungi Worms	*Neisseria gonorrhoeae* *Candida albicans* Worms	Viruses *Chlamydia* species Protozoa	Mycobacteria Trypanosomes *Cryptococcus neoformans*
Protective immunity	Antibodies Complement Phagocytosis	IgA antibodies Antimicrobial peptides	Cytotoxic T cells NK cells	Activated macrophages

Figure 8.4 Pathogens exploit different compartments of the body that are defended in different ways. Virtually all pathogens have an extracellular stage in their life cycle. For the other compartments, a representative example of each type of pathogen that exploits the compartment is given. For some pathogens, all stages of their life cycle are extracellular, whereas others exploit intracellular sites as places to grow and replicate. Different components of the immune system contribute to protective immunity against different types of microorganism in different locations. NK cells, natural killer cells.

are attacked by killing the infected cell, which interferes with the pathogen's life cycle and exposes pathogens that are released from the killed cells to the soluble molecules of the immune system. Those pathogens that live in intracellular vesicles, in contrast, can be attacked by activating the infected cell to intensify its antimicrobial activity. However, virtually all pathogens, whether viruses, bacteria, fungi or parasites, are present at some time in the extracellular spaces and can be attacked by antibodies.

Most pathogens infect only a few related host species and for this reason humans are only infrequently infected through transmission from another vertebrate species, such as the domesticated animals with which humans are often in contact, or wild animals that are hunted, butchered, and eaten. The vast majority of human infections result from transmission of the pathogen either directly or indirectly from another person who is already infected. Transmission can be directly from one person to another, or, as in the case of many parasites, it requires an intermediate passage through a distantly related organism, for example an insect or mollusc, that is necessary for completing the pathogen's life cycle.

The ability of different pathogens to persist outside the body varies considerably and determines the ease with which a particular disease is spread. The bacterial disease anthrax is spread by spores that are resistant to heat and desiccation and can, therefore, be passed over long distances from one person to another. It is these properties that make anthrax a 'hot topic' in discussions of germ warfare. In contrast, the human immunodeficiency virus (HIV) is very sensitive to changes in its environment and can be passed between individuals only by intimate contact and the exchange of infected body fluids and cells.

Experiments involving deliberate infection show that a large initial dose of pathogenic microorganisms is usually necessary to cause disease. To establish an infection a microorganism must colonize a tissue in sufficient numbers to overwhelm the cells and molecules of innate immunity that are rapidly recruited to the site of invasion. Even in these circumstances the effects will be minor unless the infection can spread within the body. Extracellular pathogens usually spread to other tissues by carriage in lymph or blood, which leads to the initiation of an adaptive immune response in secondary lymphoid tissue. Intracellular infections spread from one cell to its neighbors within a tissue, or by the release of infectious microorganisms that can be spread over greater distances in the extracellular fluids, again leading to the activation of adaptive immunity (Figure 8.5).

Recovery from an infectious disease involves clearance of infectious organisms from the body and repair of the damage caused by both infection and

Figure 8.5 Stages in infection and host defense. To enter the body through the skin, a pathogen must first adhere to it (first panel). A wound allows the pathogen to penetrate the connective tissue underlying the protective epithelium (second panel). The pathogen replicates while the defenses of innate immunity develop a state of inflammation at the infected site (third panel). Pathogens are carried to the draining lymph node, where they are trapped and stimulate antigen-specific T and B cells to mount an adaptive immune response (fourth panel). After lymphocyte division and differentiation, antibodies and effector T cells arrive at the infected site and work with the cells and molecules of innate immunity to clear the pathogen (last panel).

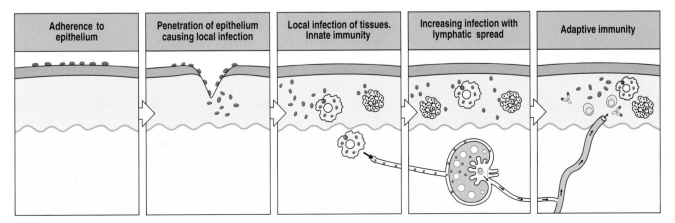

| Adherence to epithelium | Penetration of epithelium causing local infection | Local infection of tissues. Innate immunity | Increasing infection with lymphatic spread | Adaptive immunity |

the immune response. A cure is not always possible. Infection can overwhelm the immune system with death as the consequence, as was once common for smallpox infections. In intermediate situations the infection persists but its pathological effects are controlled by the adaptive immune response, as usually occurs on infection with herpes viruses.

8-2 Surface epithelia present a formidable barrier to infection

The surface of the human body is protected by a continuous covering of epithelial cells, which forms a barrier between the internal milieu of the body's organs and tissues and the external world containing pathogens. These epithelia comprise the skin and the mucosal epithelial linings of the respiratory, gastrointestinal, and urogenital tracts (Figure 8.6). Infections occur only

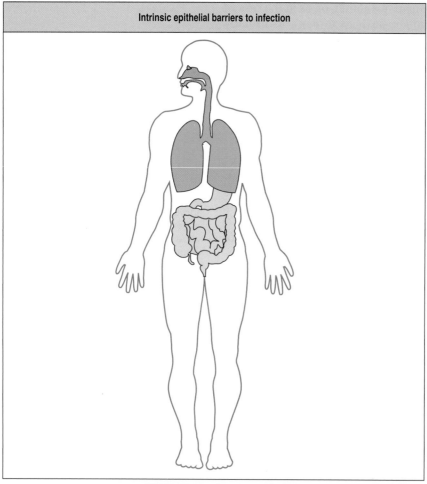

Intrinsic epithelial barriers to infection

	Skin	Gut	Lungs	Eyes/nose
Mechanical	Epithelial cells joined by tight junctions			
	Longitudinal flow of air or fluid		Movement of mucus by cilia	
Chemical	Fatty acids	Low pH		Salivary enzymes (lysozyme)
		Enzymes (pepsin)		
	Antibacterial peptides (defensins)			
Microbiological	Normal flora			

Figure 8.6 Many barriers prevent bacteria from crossing epithelia and colonizing tissues. Surface epithelia provide mechanical, chemical, and microbiological barriers to infection.

when these barriers are breached or when pathogens colonize the surfaces or interiors of the epithelial cells themselves. The skin principally provides a physical barrier, whereas the mucosal tissues permit various forms of communication between the body and its environment, including interactions with other members of the human species. Because the skin provides almost all of the outer covering of the body, it is much more frequently breached by wounds or burns than the mucosa and it is thus more susceptible to the passive entry of pathogens. In contrast, the relatively protected mucosa are more susceptible to pathogens that actively exploit the communication functions of these tissues to gain entry to the body.

Surface epithelia are more than just a physical barrier to infection: they are also loaded with chemical weapons that inhibit the attachment and growth of microorganisms. Lysozyme in tears and saliva degrades bacterial cell walls, while the acid and hydrolytic enzymes secreted by the lining of the stomach create an environment unfriendly to bacterial growth (see Figure 8.6). All epithelial surfaces secrete antimicrobial peptides called defensins. These comprise the α defensins, which include neutrophil peptides and the cryptdins made by Paneth cells of the small intestine, and the β defensins, which are secreted by epithelial cells of the epidermis, respiratory tract, and gastrointestinal tract. Defensins kill bacteria, fungi, and enveloped viruses by perturbing their membranes. To prevent defensins from disrupting human cells they are synthesized as part of longer, inactive polypeptides and function poorly under physiological conditions of ionic strength, needing the lower ionic concentrations of sweat, tears, or the lumen of the gut to become active.

Most epithelia are also coated with a flora of nonpathogenic microorganisms that compete with pathogens for nutrients and for attachment sites on epithelial cells (see Figure 8.6). More than 500 microbial species live in the healthy human gut; they are called commensal species, meaning that they 'eat at the same table'. As well as inhibiting colonization by pathogens they enhance human nutrition by further processing digested food and making some vitamins. *Escherichia coli*, a major bacterial component of the normal gut flora, secretes antibacterial proteins called colicins, that prevent the colonization of the gut by other bacteria. When a patient takes a course of antibiotic drugs the nonpathogenic flora is killed together with the pathogens that caused the disease. After such treatment the body is recolonized by microorganisms; in this situation other pathogenic organisms can establish themselves, causing further disease (Figure 8.7).

8-3 Complement activation by the alternative pathway tags microorganisms for destruction

As soon as a pathogen penetrates an epithelial barrier and starts living in the tissues the defense mechanisms of innate immunity are brought into play. One of the first components of innate immunity to be activated is complement. Complement proteins are ubiquitous in blood and lymph, making complement activation a molecular defense mechanism that can be used immediately an infection begins. We saw in Chapter 7 how complement activated through the classical and alternative pathways works together with antibodies to opsonize pathogens so that they can be phagocytosed more easily. Complement activation through the **alternative pathway** occurs at the very beginning of a primary infection, long before any pathogen-specific antibody is made.

The first step in the alternative pathway involves the spontaneous hydrolysis and activation of complement component C3, a process that occurs continually at low rate in blood, lymph, and extracellular fluids. The rate increases in

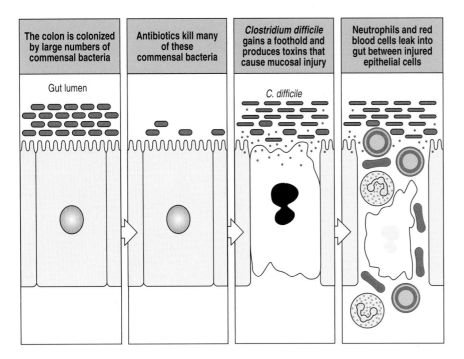

| The colon is colonized by large numbers of commensal bacteria | Antibiotics kill many of these commensal bacteria | *Clostridium difficile* gains a foothold and produces toxins that cause mucosal injury | Neutrophils and red blood cells leak into gut between injured epithelial cells |

Figure 8.7 Antibiotic treatments disrupt the natural ecology of the colon. When antibiotics are taken orally to counter a bacterial infection, beneficial populations of commensal bacteria in the colon are also decimated. This provides an opportunity for pathogenic strains of bacteria to populate the colon and cause further disease. The pathogen shown here is *Clostridium difficile*; it produces a toxin that can cause severe bloody diarrhea in patients treated with antibiotics.

the vicinity of certain pathogens. The product of spontaneous C3 hydrolysis, iC3, binds to **factor B** in the blood or extracellular fluid, making factor B susceptible to cleavage by **factor D**. This reaction produces iC3Bb, a soluble form of C3 convertase that then cleaves C3 into C3a and C3b. Although most of the C3b fragments produced are either hydrolyzed by water or become attached to serum proteins, some covalently bond to the pathogen's surface (Figure 8.8). When factor B binds to these C3b fragments and is cleaved by factor D, the complex of C3bBb is formed on the pathogen's surface. This complex is the **C3 convertase of the alternative pathway** (Figure 8.9).

Once some C3 convertase molecules have been assembled, they cleave more C3 and fix more C3b at the pathogen's surface, leading to the assembly of yet more convertase. This process is one of progressive amplification that, from the initial deposition of a few molecules, can rapidly coat the pathogen with C3b. When C3b fragments are bound at high density to a pathogen's surface they form effective ligands for the complement receptors (CR1, for example)

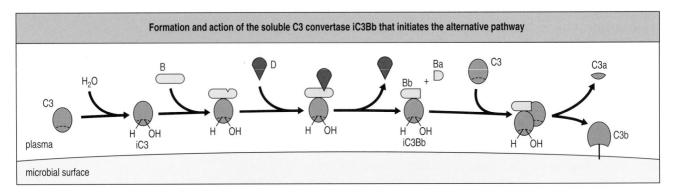

Formation and action of the soluble C3 convertase iC3Bb that initiates the alternative pathway

Figure 8.8 Formation and action of the soluble C3 convertase that initiates the alternative pathway of complement activation. In the plasma close to a microbial surface the thioester bond of C3 spontaneously hydrolyzes at low frequency. This activates the C3 which then binds factor B.

Cleavage of B by factor D produces a soluble C3 convertase, called iC3Bb, which then activates C3 molecules by cleavage into C3b and C3a. Some of the C3b fragments become covalently attached to the microbial surface.

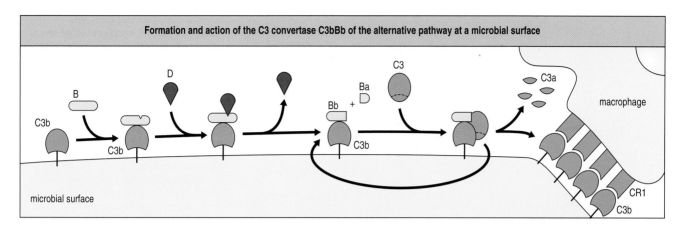

Formation and action of the C3 convertase C3bBb of the alternative pathway at a microbial surface

of macrophages resident in the infected tissue, as described in Section 7-21, p. 212. The covalently coupled C3b fragments thereby target the pathogen for destruction by phagocytosis (see Figure 8.9).

The sequence of complement-activating reactions can also be continued by binding of C3b to existing C3bBb complexes at the pathogen's surface to form the alternative C5 convertase, C3b$_2$Bb, the counterpart of the C5 convertase of the classical pathway C4b2a3b (see Section 7-23, p. 215, and Figure 7.44, p. 216). This enzyme activates C5 and the terminal components of complement (C6–C9), forming pores in the microbial membrane that are sufficient to lyse certain pathogens, notably the *Neisseria* bacteria that cause meningitis and gonorrhea. The activation of C5 also releases C5a, a small soluble peptide that facilitates pathogen destruction by the recruitment of neutrophils into the site of infection. C5a is the most potent of the anaphylatoxins (see Section 7-24, p. 216).

The actions of several proteins that control complement activation ensure that C3b fragments are densely deposited on microbial surfaces but not on the surfaces of human cells. The plasma protein **properdin (factor P)** binds to the C3 convertase C3bBb on microbial surfaces and protects it from inhibition by factor H, a plasma protein that dampens down complement reactions. In contrast, initiation of the alternative pathway on the surface of human cells is stopped at an early stage by the human cell-surface proteins CR1, decay-accelerating factor (DAF), and membrane co-factor protein (MCP), as well as by the plasma protein factor H. As described in Section 7-23, p. 215, these control proteins all act to destroy C3 convertase activity by binding to C3b and displacing Bb, and/or by rendering C3b susceptible to cleavage by factor I (Figure 8.10).

Because complement activation limits bacterial infections, some bacteria have evolved mechanisms to evade the actions of complement. Bacteria with an abundance of sialic acid on their surface, such as *Streptococcus pyogenes* and *Staphylococcus aureus*, mimic human cells. Consequently, when C3b becomes bound to their surface it is readily inactivated by factor H. Because antibodies coat the bacterial surface and mask the sialic acids before complement is bound, these bacteria are only resistant to the effect of complement when no specific antibacterial antibody is present.

The combined effects of the reactions that promote and regulate C3 activation is to deposit C3b on the surface of human cells. This mechanism provides a simple and effective way of distinguishing self from nonself, and for guiding mechanisms of death and destruction towards invading pathogens and away from human cells.

Figure 8.9 Formation and action of the C3 convertase, C3bBb, of the alternative pathway at a microbial surface. Through the action of the soluble C3 convertase, iC3Bb, C3b fragments are bound to the microbial surface (see Figure 8.8). These bind factor B which is then cleaved by factor D to produce C3bBb, the surface-bound convertase of the alternative pathway. This enzyme cleaves C3 to produce further C3b fragments bound to the microbe and small soluble C3a fragments. The C3b fragments can be used either to make more C3 convertase, which amplifies the activation of C3, or to engage the CR1 receptors of a macrophage or neutrophil and promote phagocytosis of the microbe. The small, soluble C3a fragments attract phagocytes to sites of complement fixation.

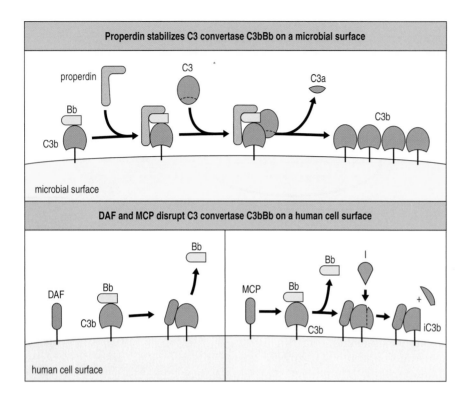

Figure 8.10 Control proteins stabilize the C3 convertase of the alternative pathway on microbial surfaces and destabilize it on human cell surfaces. The soluble protein properdin (factor P) binds to C3bBb and extends its lifetime on the microbial surface. When C3bBb is formed on a human cell surface it is rapidly disrupted by the action of one of two membrane proteins: decay-accelerating factor (DAF) or membrane cofactor protein (MCP). In combination, these regulatory proteins ensure that much complement is fixed to pathogen surfaces and little is fixed to human cell surfaces.

8-4 Several classes of plasma protein limit the spread of infection

In addition to complement, several other types of plasma protein impede the invasion and colonization of human tissues by microorganisms. Damage to blood vessels activates the coagulation system, a cascade of plasma enzymes that forms blood clots, which immobilize microorganisms and prevent them from entering the blood and lymph, as well as reducing the loss of blood and fluid. Platelets are a major component of blood clots and during clot formation they release a variety of highly active substances from their storage granules. These include prostaglandins, hydrolytic enzymes, growth factors, and other mediators that stimulate various cell types to contribute to antimicrobial defense, wound healing, and inflammation. Further mediators, including the vasoactive peptide bradykinin, are produced by the kinin system, a second enzymatic cascade of plasma proteins that is triggered by tissue damage. By causing vasodilation, bradykinin increases the supply of the soluble and cellular materials of innate immunity to the infected site.

Many pathogens use proteases to damage human tissues and human defenses, thereby facilitating invasion and dissemination. In some instances the proteins are made by the pathogen, in others the pathogen hijacks a human protease, for example plasmin, for its own purposes. To counter these mechanisms, human secretions and plasma contain protease inhibitors. Several of the microbicidal peptides of innate immunity are specific inhibitors of microbial proteases. These include the cryptdins secreted by Paneth cells in the gut and the serpins that inactivate serine proteases and regulate the clotting cascade (Figure 8.11, top panel). Most of these inhibitors act by binding irreversibly to the active site of the protease and are specific for a narrow spectrum of proteases.

About 10% of serum proteins are protease inhibitors. Among these are the α_2-macroglobulins, glycoproteins of 180 kDa that circulate as monomers, dimers, and trimers and are able to inhibit a broad range of proteases. α_2-Macroglobulins have structural similarities with complement component C3, including the presence of internal thioester bonds. The α_2-macroglobulin molecule lures a protease with a bait region that it is allowed to cleave. This activates the α_2-macroglobulin, producing two effects: first, the thioester is used to covalently bind the protease to the α_2-macroglobulin; second, the α_2-macroglobulin undergoes a conformational change by which it envelops the protease and prevents it attacking other substrates (Figure 8.11, bottom panel). The resulting complexes of protease and α_2-macroglobulin are rapidly cleared from the circulation by a receptor present on hepatocytes, fibroblasts, and macrophages.

8-5 Phagocytosis by macrophages provides a first line of cellular defense against invading microorganisms

When a pathogen invades a human tissue the first effector cells of the immune system it encounters are the resident macrophages. **Macrophages** are the mature forms of circulating monocytes that have left the blood and taken up residence in the tissues. They are prevalent in the connective tissues, the linings of the gastrointestinal and respiratory tracts, the alveoli of the lungs, and in the liver, where they are known as **Kupffer cells**. Macrophages are long-lived phagocytic cells that participate in both innate and adaptive immunity.

Although macrophages phagocytose bacteria and other microorganisms in a nonspecific fashion, the process is made more efficient by the participation of receptors on the macrophage surface that bind to specific ligands on the microbial surface. Deposition of C3b fragments on the surface of a pathogen, through activation of the alternative pathway of complement, provides ligands for the CR1 complement receptor of macrophages. When C3b is degraded by factor I to iC3b it then becomes a ligand for the macrophage's CR3 and CR4 receptors. The combination of opsonization by complement and phagocytosis by macrophages allows pathogens to be recognized and destroyed from the very beginning of an infection (see Section 7-21, p. 212).

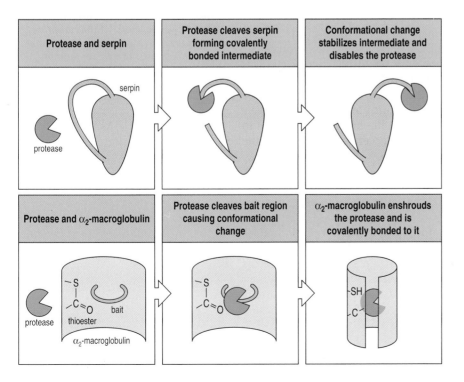

Figure 8.11 Serpins and α_2-macroglobulin inhibit potentially damaging proteases. Microbial invasion and colonization of human tissues is dependent on the actions of microbial proteases. In response, human plasma is loaded with protease inhibitors. One type of protease inhibitor, as exemplified by the serpins, permanently inactivates the protease by covalently binding into its active site (upper panels). The α_2-macroglobulins take a different approach: they first trap the protease with a 'bait' region and when the protease cleaves the bait the α_2-macroglobulin binds it covalently through activation of its thioester group and enshrouds the protease so that it cannot access other protein substrates, even though it is still active.

Phagocytosis of pathogens by macrophages is also aided by receptors of innate immunity that bind directly to components of the microbial surface that are characteristic of pathogens but absent from human cells. The CR3 and CR4 complement receptors recognize several microbial cell-surface molecules, including bacterial lipopolysaccharide (LPS), the lipophosphoglycan of *Leishmania*, the filamentous hemagglutinin of *Bordetella*, and cell-surface structures on yeasts such as *Candida* and *Histoplasma*. Several other phagocytic receptors are carbohydrate-binding proteins—lectins—that bind to particular combinations or configuration of sugars that are not a feature of human cells. The mannose receptor is a calcium-dependent (C-type) lectin that binds to certain bacteria and viruses, including HIV. The scavenger receptors are a heterogenous group of receptors with a preference for molecules rich in sialic acid (Figure 8.12).

8-6 Receptors that detect microbial products signal macrophage activation

In addition to receptors that facilitate the phagocytosis of microorganisms, macrophages also have receptors or sensors for pathogen components that signal the macrophage to make and secrete cytokines that recruit other cells to defend the infected tissue. In this category are the Toll-like receptors (TLR), which in humans comprise a family of 10 receptors with specificities for different microbial products (Figure 8.13). The best studied of these receptors is TLR-4, for which one ligand is LPS, the major cell-surface component of Gram-negative bacteria and the endotoxin that is most commonly responsible for **septic shock,** the most severe outcome of **sepsis,** or infection of the bloodstream. TLR-4 does not bind LPS directly. When LPS is released from bacterial surfaces, as will occur on phagocytosis, it is bound by the LPS receptor CD14 on the macrophage surface. Alternatively, LPS picked up in the plasma by the soluble LPS-binding protein is delivered to CD14 at the macrophage surface. The TLR-4 dimer and another protein called MD-2 then associate with the complex of CD14 and LPS to generate intracellular signals via the cytoplasmic signaling domain of TLR-4. Among the cytokines released by the macrophage in response to these signals is TNF-α, which is the cytokine directly responsible for causing septic shock.

Septic shock causes the death of more than 100,000 people in the United States each year, with Gram-negative bacteria being the most common cause. The role of TLR-4 in combating infections with Gram-negative bacteria is shown by a recently discovered association of a TLR-4 variant with septic shock. This variant has glycine at position 299 in the amino-acid sequence instead of the asparagine found in the common form of TLR-4, and is overrepresented in patients with septic shock. Indeed, the only person known to be homozygous for this TLR-4 allotype died in adolescence of infection of the kidneys caused by the commensal bacterium *E. coli*.

The TLR-4 receptor also responds to chemically or structurally related ligands present on other pathogens. Other members of the TLR family sense different microbial products from the TLR-4 ligands; in each case, they are constituents of common groups of pathogens and are not found in human cells. Double-stranded RNA, sensed by TLR-3, is an intermediate nucleic acid present in many viral infections, while the unmethylated CpG nucleotide motifs detected by TLR-9 are abundant in bacterial genomes but rare in the human genome. Some of the TLRs form homodimers exclusively, whereas others form heterodimers. Such combinations of different chains potentially increase the diversity of receptor specificities. The only Toll-like receptor that is known to bind directly to the product it detects is TLR-5, which binds to bacterial flagellin. For the other members of the family, including TLR-4, it is uncertain whether direct contact is made with a microbial product.

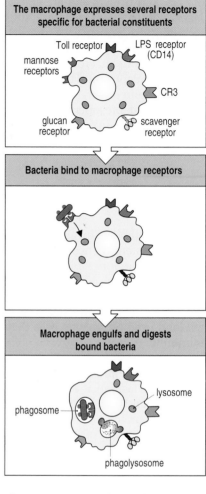

Figure 8.12 Macrophages have several types of surface receptor that bind to constituents of microbial surfaces and promote phagocytosis. Many of the receptors are lectins that bind carbohydrates that are characteristic of microorganisms.

Recognition of microbial products through Toll-like receptors			
Receptor	Ligands	Microorganisms recognized	Cell carrying receptor
TLR-4 homodimer	Lipopolysaccharide	Gram-negative bacteria	Macrophage
TLR-9 homodimer	DNA with unmethylated CpG motifs	Bacteria	
TLR-3 homodimer	Double-stranded DNA	Viral RNAs	
TLR-5 homodimer	Flagellin	Gram-negative bacteria	Intestinal epithelium
TLR-7 homodimer	Imidazoquinolines (antiviral drugs)		
TLR-2 (TLR-1,-6) heterodimers	Peptidoglycan, lipoarabinomannan, porins, bacterial lipoprotein, bacterial lipopeptide, yeast mannan, glycophosphatidyl-inositol anchors	Gram-positive bacteria Mycobacteria *Neisseria* Yeast Trypanosomes	Macrophage

Figure 8.13 Toll-like receptors recognize microbial constituents. Each of the known Toll-like receptors (TLRs) appears to recognize one or more microbial molecular patterns, but TLR-5 is the only case so far in which a direct interaction with a microbial product, the bacterial protein flagellin, has been demonstrated. TLR-9 is an intracellular receptor. Toll is the name of a receptor in the fruit fly *Drosophila melanogaster* with which the signaling domains of the human Toll-like receptors have structural similarity. Like many of the names given to genes and proteins by fruit fly geneticists Toll is a whimsical name and, in this case, derived from German slang.

The Toll-like receptors all trigger a common pathway of intracellular signaling, which is also shared by the IL-1 receptor and its relatives. It involves an adaptor protein called MyD88 and the interleukin-1 receptor associated kinase (IRAK) complex. Their actions lead to the translocation of the transcription factor nuclear factor κB (NFκB) from the cytoplasm to the nucleus, where it directs the transcription of genes for inflammatory cytokines (Figure 8.14). Patients deficient in IRAK function have impaired signaling through all the Toll-like receptors as well as through receptors for IL-1 and IL-18. These children suffer from recurrent infections with pyogenic bacteria and have generally poor inflammatory responses. However, they do not show any increased susceptibility to other types of pathogen, such as viruses, showing that signaling through Toll-like receptors is not necessary for providing passable protection against many common infections.

8-7 Activation of resident macrophages induces inflammation at sites of infection

On sensing the presence of pathogens through Toll-like and other receptors, macrophages are stimulated to secrete a battery of cytokines and other substances that recruit effector cells, prominently neutrophils, into the infected area. The infiltrating cells cause a state of **inflammation** to develop within the tissue. Inflammation describes the local accumulation of fluid accompanied

Figure 8.14 Toll receptors use a common pathway of signal transduction. This pathway is also used by IL-1 and IL-18. Common components of the pathway are the adaptor molecule MyD88, the interleukin-1 receptor-associated kinase 4 (IRAK4), the TNF-receptor-associated kinase 6 (TRAF6), the inhibitor of NFκB (IκB) and NFκB. MD2 is a soluble protein that associates with the extracellular domains of TLR-4, but not other TLR family members, and confers sensitivity to LPS. In the absence of a signal, NFκB is bound by IκB, which prevents it from entering the nucleus. In the presence of a signal, kinase activation causes IκB to be phosphorylated, which induces the release of NFκB and its passage to the nucleus where it activates genes encoding inflammatory cytokines.

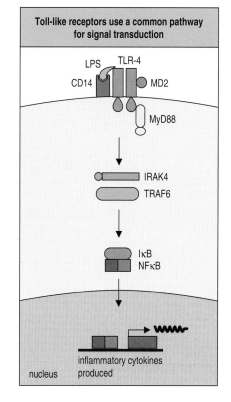

Toll-like receptors use a common pathway for signal transduction

LPS TLR-4
CD14   MD2
MyD88
IRAK4
TRAF6
IκB
NFκB

nucleus   inflammatory cytokines produced

by swelling, reddening, and pain. These effects stem from changes induced in the local blood capillaries that lead to an increase in their diameter (a process called dilation), reduction in the rate of blood flow, and increased permeability of the blood vessel wall. The increased supply of blood to the region causes the local redness and heat associated with inflammation. The increased permeability of blood vessels allows the movement of fluid, plasma proteins, and white blood cells from the blood capillaries into the adjoining connective tissues, causing the swelling and pain.

Translocation of NFκB to the macrophage nucleus (see Figure 8.14), initiates transcription of various cytokine genes. Prominent cytokines produced by activated macrophages are **IL-1**, **IL-6**, **CXCL8**, **IL-12**, and **tumor necrosis factor-α (TNF-α)**. These proinflammatory cytokines have powerful effects that can be localized in the infected tissue or be manifest systemically throughout the body; they also contribute to both the innate and the adaptive immune response (Figure 8.15).

CXCL8, also called IL-8, is one of a large family of around 40 chemoattractant cytokines, or **chemokines** (Figure 8.16). Some chemokines, including CXCL8, attract leukocytes into sites of tissue damage or infection. Others direct the traffic of leukocytes during their development or their recirculation through the tissues of the body. Chemokines are small, structurally similar proteins of

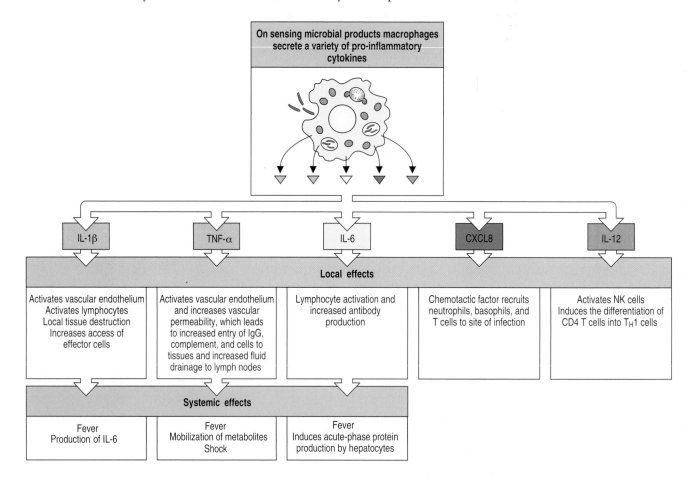

Figure 8.15 Important cytokines secreted by macrophages in response to bacterial products include IL-1, TNF-α, Il-6, CXCL8, and IL-12. TNF-α is an inducer of a local inflammatory response that helps to contain infections. It also has systemic effects, many of which are harmful. The chemokine CXCL8 is also involved in the local inflammatory response, helping to attract neutrophils to the site of infection. IL-1, IL-6, and TNF-α have a critical role in inducing the acute-phase response in the liver and induce fever, which favors effective host defense in various ways. IL-12 activates natural killer (NK) cells and favors the differentiation of CD4 T cells into the T$_H$1 subset in adaptive immunity.

around 60–140 amino acids. Two major subfamilies are defined on the basis of pairs of cysteine residues, which are either adjacent (CC) or separated by another amino acid (CXC). Cells are attracted from the blood into infected tissue by following a concentration gradient of chemokine produced by cells

Class	Chemokine	Produced by	Receptors	Cells attracted	Major effects
CXC	CXCL8 (IL-8)	Monocytes Macrophages Fibroblasts Keratinocytes Endothelial cells	CXCR1 CXCR2	Neutrophils Naive T cells	Mobilizes, activates and degranulates neutrophils Angiogenesis
	CXCL7 (PBP, β-TG NAP-2)	Platelets	CXCR2	Neutrophils	Activates neutrophils Clot resorption Angiogenesis
	CXCL1 (GROα) CXCL2 (GROβ) CXCL3 (GROγ)	Monocytes Fibroblasts Endothelium	CXCR2	Neutrophils Naive T cells Fibroblasts	Activates neutrophils Fibroplasia Angiogenesis
	CXCL10 (IP-10)	Keratinocytes Monocytes T cells Fibroblasts Endothelium	CXCR3	Resting T cells NK cells Monocytes	Immunostimulant Antiangiogenic Promotes T_H1 immunity
	CXCL12 (SDF-1)	Stromal cells	CXCR4	Naive T cells Progenitor (CD34$^+$) B cells	B-cell development Lymphocyte homing Competes with HIV-1
	CXCL13 (BLC)	Stromal cells	CXCR5	B cells	Lymphocyte homing
CC	CCL3 (MIP-1α)	Monocytes T cells Mast cells Fibroblasts	CCR1, 3, 5	Monocytes NK and T cells Basophils Dendritic cells	Competes with HIV-1 Antiviral defense Promotes T_H1 immunity
	CCL4 (MIP-1β)	Monocytes Macrophages Neutrophils Endothelium	CCR1, 3, 5	Monocytes NK and T cells Dendritic cells	Competes with HIV-1
	CCL2 (MCP-1)	Monocytes Macrophages Fibroblasts Keratinocytes	CCR2B	Monocytes NK and T cells Basophils Dendritic cells	Activates macrophages Basophil histamine release Promotes T_H2 immunity
	CCL5 (RANTES)	T cells Endothelium Platelets	CCR1, 3, 5	Monocytes NK and T cells Basophils Eosinophils Dendritic cells	Degranulates basophils Activates T cells Chronic inflammation
	CCL11 (Eotaxin)	Endothelium Monocytes Epithelium T cells	CCR3	Eosinophils Monocytes T cells	Role in allergy
	CCL18 (DC-CK)	Dendritic cells	?	Naive T cells	Role in activating naive T cells
C	XCL1 (Lymphotactin)	CD8>CD4 T cells	?	Thymocytes Dendritic cells NK cells	Lymphocyte trafficking and development
CXXXC (CX$_3$C)	CX3CL1 (Fractalkine)	Monocytes Endothelium Microglial cells	CX$_3$CR1	Monocytes T cells	Leukocyte–endothelial adhesion Brain inflammation

Figure 8.16 Properties of selected chemokines. Chemokines fall mainly into two related but distinct groups. The CC chemokines, which in humans are all encoded in one region of chromosome 4, have two adjacent cysteine residues, whereas the CXC chemokines, which are found in a cluster on chromosome 17, have an amino-acid residue between the equivalent two cysteines. A C chemokine with only one cysteine at this location, and fractalkine, a CX$_3$C chemokine, are encoded elsewhere in the genome. Each chemokine interacts with one or more receptors, and affects one or more types of cell.

within the infected site. Chemokines interact with their target cells by binding to specific cell-surface receptors, which in humans comprise a family of 16 seven-span transmembrane proteins that signal through associated GTP-binding proteins (Figure 8.17).

The principal function of CXCL8, a CXC chemokine (see Figure 8.16), is to recruit neutrophils from the blood into infected areas. CXCL8, for example, is bound by chemokine receptors CXCR1 and CXCR2. Interaction with a chemokine has two distinct effects on the targeted leukocyte: first, the cell's adhesive properties are altered so that it can leave the blood and enter tissue; second, its movement is guided towards the center of infection along a gradient of the chemokine, present both in solution and attached to the extracellular matrix and endothelial cell surfaces. Chemokines have structural and functional similarity to the defensins: some chemokines have antimicrobial activity while some defensins have chemoattractant properties and bind to chemokine receptors.

The cytokine IL-12 serves to activate natural killer (NK) cells, lymphocytes of the innate immune response that enter infected sites soon after infection. Cytokines IL-1 and TNF-α facilitate entry of neutrophils, NK cells, and other effectors into infected areas by inducing changes in the endothelial cells of the local blood vessels. Other effector molecules released by macrophages are plasminogen activator, phospholipase, and other enzymes, prostaglandins, oxygen radicals, peroxides, nitric oxide, leukotrienes, and platelet-activating factor (PAF), which all contribute to inflammation and tissue damage. In the course of complement activation, the soluble complement fragments C3a and C5a recruit neutrophils from the blood into infected tissues and stimulate mast cells to degranulate, releasing the inflammatory molecules histamine and TNF-α, among others. Molecules involved in the induction of inflammation are known generally as **inflammatory mediators**. The combined effect of all this activity is to produce a local state of inflammation with its characteristic symptoms.

In response to TNF-α, vascular endothelial cells make platelet-activating factor, which triggers blood clotting and blockage of the local blood vessels. This restricts the leakage of plasma from the blood and prevents pathogens from entering the blood and disseminating infection throughout the body. However, in those instances in which infection does spread to the blood, and endotoxins such as LPS provoke widespread production of TNF-α, the actions that contain local infection so effectively can become catastrophic (Figure 8.18). A systemic bacterial infection of the blood induces macrophages in the liver, spleen, and other sites to release TNF-α, which causes dilation of blood vessels and massive leakage of fluid into tissues throughout the body, leading to a profound state of shock (septic shock). One symptom is widespread blood clotting in capillaries—disseminated intravascular coagulation—which exhausts the supply of clotting proteins. More critically, septic shock frequently leads to failure of vital organs such as the kidneys, liver, heart, and lungs, which are soon compromised by the lack of a normal blood supply. Consequently, there is a high mortality rate for patients suffering septic shock.

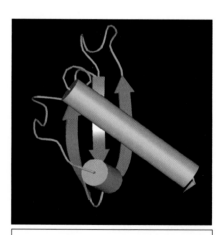

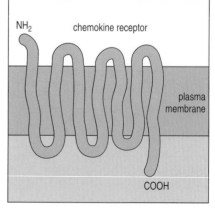

Figure 8.17 Chemokines bind to chemokine receptors that are G-protein-coupled receptors. The structure of the chemokine CXCL8 is shown in the upper panel. The receptors for the chemokines are members of the family of seven-span receptors, which also includes the photoreceptor protein rhodopsin and many other receptors. They have seven transmembrane helices, as depicted in the schematic in the lower panel, and all the family interact with G proteins.

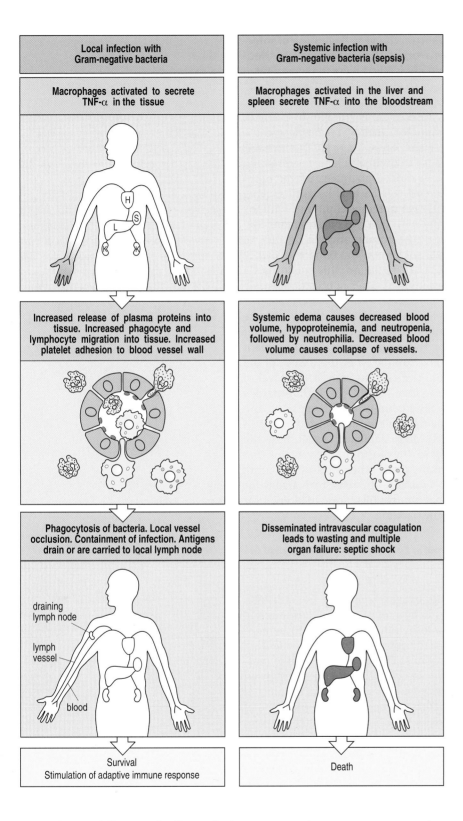

Figure 8.18 TNF-α released by macrophages induces protection at the local level but can lead to catastrophe when released systemically. The panels on the left describe the causes and consequences of the release of TNF-α within a local area of infection. In contrast, the panels on the right describe the causes and consequences of the release of TNF-α throughout the body. The initial effects of TNF-α are on the endothelium of blood vessels, especially venules. It causes increased blood flow, vascular permeability, and endothelial adhesiveness for white blood cells and platelets. These events cause the blood in the venules to clot, preventing the spread of infection and directing extracellular fluid to the lymphatics and lymph nodes, where the adaptive immune response is activated. When an infection develops in the blood, the systemic release of TNF-α and the effect it has on the venules in all tissues simultaneously induce a state of shock that can lead to organ failure and death. H, heart; K, kidney; L, liver; S, spleen.

8-8 Neutrophils are dedicated phagocytes that are summoned to sites of infection

By engulfing and killing microorganisms, phagocytic cells are the principal means by which the immune system destroys invading pathogens. The two kinds of phagocyte that serve this purpose—the macrophage and the neutrophil—have distinct and complementary properties. Macrophages are long-lived, they reside in the tissues, work from the very beginning of infection, raise the alarm, and have functions other than phagocytosis. **Neutrophils**, on the other hand, are short-lived dedicated killers that circulate in the blood awaiting a call from a macrophage to enter infected tissue.

Neutrophils are a type of granulocyte, having numerous granules in the cytoplasm, and are also known as **polymorphonuclear leukocytes** because of the variable and irregular shapes of their nuclei (see Figure 1.9, p. 10). Neutrophils were historically called microphages because of their smaller size compared with macrophages. What they lack in size they more than make up for in number: they are the most abundant white blood cells, a healthy adult having some 50 billion in circulation at any time. This abundance combined with the short life span of the circulating neutrophil—less than two days—means that around 60% of the hematopoietic activity of the bone marrow is devoted to neutrophil production. Mature neutrophils are kept in the bone marrow for about five days before being released into the circulation; this constitutes a large reserve of neutrophils that can be called on at times of infection.

Neutrophils are excluded from healthy tissue, but at infected sites the release of inflammatory mediators attracts neutrophils to leave the blood and enter the infected area in large numbers, where they soon become the dominant phagocytic cell. Every day, some 3×10^9 neutrophils enter the tissues of the mouth and throat, the most contaminated sites in the body. The arrival of neutrophils is the first of a series of reactions, called the **inflammatory response**, by which cells and molecules of innate immunity are recruited into sites of wounding or infection. Although neutrophils are specialized for working under the anaerobic conditions that prevail in damaged tissues, they still die within a few hours after entry. In doing so, they form the creamy **pus** that characteristically develops at infected wounds and other sites of infection. This is why extracellular bacteria such as *S. aureus*, which are responsible for the superficial infections and abscesses that neutrophils tackle in large numbers, are known as pus-forming or **pyogenic** bacteria.

8-9 The homing of neutrophils to infected tissues is induced by inflammatory mediators

During infection, inflammatory mediators cause changes in the adhesion molecules expressed by the vascular endothelium of blood capillaries in and around the infected site. The vascular endothelium is then said to be activated. Inflammatory mediators also induce the expression of complementary adhesion molecules on circulating neutrophils. These changes enable neutrophils in the blood to bind to activated vascular endothelium within an infected site and to squeeze between endothelial cells and enter the infected tissue. The release of inflammatory mediators by the increasing numbers of neutrophils steadily increases the inflammation within the infected tissue.

Neutrophils have receptors for inflammatory mediators on their surfaces: examples are receptors for C5a, cleaved from C5 during complement activation, and for the CXCL8 secreted by activated macrophages. In addition, neutrophils have a receptor for chemoattractants that are specifically produced by bacterial infections. This receptor is specific for peptides containing N-formylmethionine, a component of bacterial but not of human proteins. Binding of ligands to surface receptors induces the changes in neutrophil adhesion molecules.

The process by which neutrophils migrate out of blood capillaries and into tissues is called **extravasation** and it occurs in four steps. The first is the interaction between circulating leukocytes and blood vessel walls that slows down

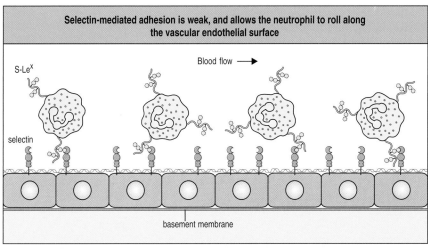

Selectin-mediated adhesion is weak, and allows the neutrophil to roll along the vascular endothelial surface

Blood flow →

S-Le^x

selectin

basement membrane

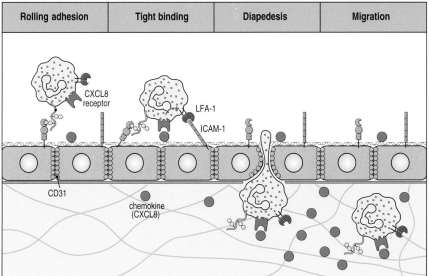

| Rolling adhesion | Tight binding | Diapedesis | Migration |

CXCL8 receptor

LFA-1

ICAM-1

CD31

chemokine (CXCL8)

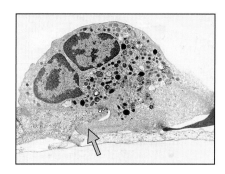

Figure 8.19 Neutrophils are directed to sites of infection through interactions between adhesion molecules. Inflammatory mediators and cytokines produced as the result of infection induce the expression of selectin on vascular endothelium, which enables it to bind leukocytes. The top panel shows the rolling interaction of a neutrophil with vascular endothelium due to transient interactions between selectin on the endothelium and sialyl-Lewis^x (s-Le^x) on the leukocyte. The bottom panel shows the conversion of rolling adhesion into tight binding and subsequent migration of the leukocyte into the infected tissue. The four stages of extravasation are shown. Rolling adhesion is converted into tight binding by interactions between integrins on the leukocyte (LFA-1 is shown here) and adhesion molecules on the endothelium (ICAM-1). Expression of these adhesion molecules is also induced by cytokines. A strong interaction is induced by the presence of chemoattractant cytokines (the chemokine CXCL8 is shown here) that have their source at the site of infection. They are held on proteoglycans of the extracellular matrix and cell surface to form a gradient along which the leukocyte can travel. Under the guidance of these chemokines, the neutrophil squeezes between the endothelial cells and penetrates the connective tissue (diapedesis). It then migrates to the center of infection along the CXCL8 gradient. The electron micrograph shows a neutrophil that has just started to migrate between adjacent endothelial cells but has yet to break through the basement membrane, which is at the bottom of the photograph. The blue arrow points to the pseudopod that the neutrophil is inserting between the endothelial cells. The dark mass in the bottom right-hand corner is an erythrocyte that has become trapped under the neutrophil. Photograph ($\times$ 5500) courtesy of I. Bird and J. Spragg.

the leukocytes. This interaction is mediated by selectins, adhesion molecules that bind to particular types of carbohydrate (see Figure 6.4, p. 149). In healthy tissue, vascular endothelial cells contain granules, known as **Weibel–Palade bodies**, which contain **P-selectin**. On exposure to inflammatory mediators, including leukotriene LTB_4, C5a, and histamine, the P-selectin in the Weibel–Palade bodies is transported to the cell surface. A second selectin, **E-selectin**, is also expressed on the endothelial cell surface a few hours after exposure to LPS or TNF-α. The two selectins bind to the sialyl-Lewis^x carbohydrates of cell-surface glycoproteins on the leukocyte. These reversible interactions allow the neutrophils to adhere to the blood vessel walls and to 'roll' slowly along them by forming new adhesive interactions at the front of the cell while breaking them at the back (Figure 8.19, top panel).

The second step in extravasation depends on interactions between the integrins LFA-1 and CR3 on the neutrophil and adhesion molecules on the endothelium, for example ICAM-1, whose expression is also induced by TNF-α. Under normal conditions, LFA-1 and CR3 interact only weakly with endothelial adhesion molecules, but exposure to the CXCL8 coming from cells in the inflamed tissues, induces conformational changes in the LFA-1 and CR3 on a rolling leukocyte that strengthen their adhesion. As a result, the neutrophil holds tightly to the endothelium and stops rolling (see Figure 8.19, bottom panels).

In the third step, the neutrophil crosses the blood vessel wall. LFA-1 and CR3 contribute to this movement, as does adhesion involving the immunoglobulin superfamily protein CD31, which is expressed by both neutrophils and endothelial cells at their junctions with one another. The leukocyte squeezes between neighboring endothelial cells, a maneuver known as **diapedesis**, and reaches the basement membrane, a part of the extracellular matrix. It then crosses the basement membrane by secreting proteases that break down the membrane. The fourth and final step in extravasation is movement of the neutrophil toward the center of infection in the tissue. This migration is accomplished on the gradient of CXCL8, which originates within the infected site (see Figure 8.19).

At various stages in their maturation, activation, and performance of effector function, all types of white blood cell leave the blood and home to particular tissues. All these migrations involve mechanisms analogous to those controlling the entry of neutrophils into infected tissue. Cytokines and chemokines induce changes in adhesion molecules on white blood cells and vascular endothelium that determine where and when extravasation occurs.

8-10 Neutrophils are potent killers of pathogens and are themselves programmed to die

Neutrophils phagocytose microorganisms by mechanisms similar to those used by macrophages. Neutrophils express both Fc receptors and complement receptors that facilitate the phagocytosis of pathogens opsonized with complement in the innate response, and with antibody and complement once specific antibody is available (Figure 8.20). The range of particulate material that neutrophils engulf is greater than that tackled by macrophages, as is the diversity of the microbicidal substances stored in their three types of granules. Because mature neutrophils are programmed to die young, they devote more of their resources to the storage and delivery of antimicrobial weaponry than the longer-living macrophage.

Almost immediately after a pathogen is engulfed by a neutrophil, a battery of degradative enzymes and other toxic substances is brought to bear upon it. Phagosomes containing recently captured microorganisms are fused with the preformed neutrophil granules (which are modified lysosomes). These contain hydrolytic degradative enzymes, NADPH-dependent oxidases, and antimicrobial peptides such as the α-defensins (Figure 8.21). In the resulting

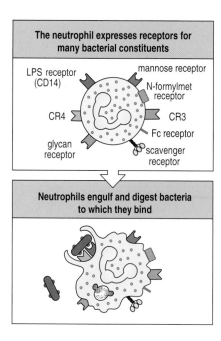

Figure 8.20 Bacteria binding to neutrophil receptors induce phagocytosis and microbial killing. The figure illustrates this type of mechanism for two such receptors, CD14 and CR4, which are specific for bacterial lipopolysaccharide (LPS). Bacteria binding to these receptors stimulate the phagocytosis and degradation of the bacterium.

Class of mechanism	Specific products
Acidification	pH=~3.5 – 4.0, bacteriostatic or bactericidal
Toxic oxygen-derived products	Superoxide (O_2^-), hydrogen peroxide (H_2O_2), singlet oxygen ($^1O_2^{\bullet}$), hydroxyl radical ($OH^{\bullet}$), hypohalite (OCl^-)
Toxic nitrogen oxides	Nitric oxide (NO)
Antimicrobial peptides	Defensins and cationic proteins
Enzymes	NAPDH-dependent oxidases: generate toxic oxygen derivatives Lysozyme: dissolves cell walls of some Gram-positive bacteria Acid hydrolases: further digest bacteria
Competitors	Lactoferrin (binds Fe) and vitamin B_{12}-binding protein

Figure 8.21 Bactericidal agents produced or released by phagocytic cells on the ingestion of microorganisms. Most of these agents are present in both macrophages and neutrophils. Some of them are toxic; others, such as lactoferrin, work by binding essential nutrients and preventing their uptake by bacteria.

phagolysosome, the NADPH-dependent oxidases generate toxic oxygen radicals and hydrogen peroxide, the latter being quickly converted into hypochlorous acid. Activated phagocytes also produce nitric oxide (NO), generated from arginine and oxygen by the inducible enzyme nitric oxide synthase (iNOS). NO reacts with oxygen radicals within the phagolysosome to produce highly toxic peroxynitrite.

This general oxidative attack on ingested microorganisms is accompanied by a transient increase in oxygen consumption called the **respiratory burst**. The combination of toxic oxygen metabolites, proteases, phospholipases, lysozyme, and antimicrobial peptides is able to kill both Gram-positive and Gram-negative bacteria, fungi, and even some of the enveloped viruses. The mature neutrophil cannot replenish its granule contents, so once they are used up the neutrophil dies by apoptosis and is ultimately phagocytosed by a macrophage. The dependence of the body's defenses upon neutrophils is seen from the miserable effects of their absence. Patients with neutrophil deficiencies suffer recurrent infections, often by the bacteria and fungi that form part of the normal flora of healthy people.

The toxic oxygen species produced by activated neutrophils can diffuse out of the cell and damage host cells. To limit the damage, the respiratory burst is also accompanied by the synthesis of enzymes that inactivate these potent small molecules: superoxide dismutase converts superoxide to hydrogen peroxide, and catalase degrades hydrogen peroxide to water and oxygen.

In some circumstances the contents of phagocyte lysosomes and granules are released to the outside of the cell. This enables neutrophils to attack antibody-coated pathogens that are too big for them to ingest, such as parasitic worms. The phagocyte adapts to this situation by secreting the contents of its granules onto the surface of the worm at the region of contact. This secretory reaction can also be provoked by contact with host cells, and phagocyte activation can thus cause extensive tissue damage during the course of an infection.

8-11 Inflammatory cytokines raise body temperature and activate hepatocytes to make the acute-phase response

A systemic effect of the inflammatory cytokines IL-1, IL-6, and TNF-α is to cause the rise in body temperature called **fever**. The cytokines act on temperature-control sites in the hypothalamus, and on muscle and fat cells, altering energy mobilization to generate heat (Figure 8.22). In this context,

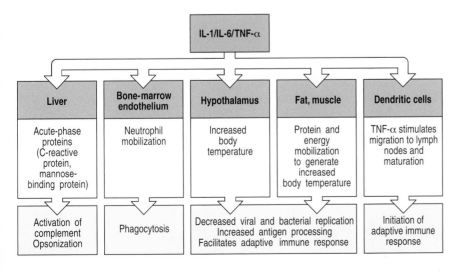

Figure 8.22 **The macrophage-produced cytokines TNF-α, IL-1, and IL-6 have a spectrum of biological activity that helps to coordinate the body's response to infection.**

such molecules are called 'endogenous pyrogens' to distinguish them from the 'exogenous' products of pathogens that also raise the body's temperature. On balance, a raised body temperature helps the immune system fight infection, because most bacterial and viral pathogens grow better at temperatures lower than that of the human body, and adaptive immunity becomes more potent at higher temperatures. At elevated temperatures, bacterial and viral replication is decreased, whereas processing of antigen is enhanced. In addition, human cells become more resistant to the deleterious effects of TNF-α when experiencing fever.

A further systemic effect of TNF-α, IL-1, and IL-6 is to change the spectrum of soluble plasma proteins secreted by hepatocytes in the liver, thus producing the **acute-phase response**. Those proteins whose synthesis and secretion is increased during the acute-phase response are called **acute-phase proteins** (Figure 8.23). Two of the acute-phase proteins—C-reactive protein and mannose-binding lectin—enhance the fixation of complement at pathogen surfaces. **C-reactive protein** is a pentamer of identical subunits and a member of the **pentraxin** family of proteins. C-reactive protein binds to the phosphorylcholine component of lipopolysaccharides in bacterial and fungal cell walls, but not to the phosphorylcholine present in the phospholipids of human cell membranes. In binding to bacteria, C-reactive protein acts as an opsonin, which can then bind C1q, initiating the classical pathway of complement fixation in the absence of specific antibody. The interaction of C-reactive protein with C1q involves the stalks of the C1q molecule, whereas the interaction of antibodies with C1q involves the globular heads (see Figure 7.31, p. 208). Despite these differences in the initial binding mechanism, the same sequence of complement reactions occurs when either C-reactive protein or antibody interacts with a pathogen.

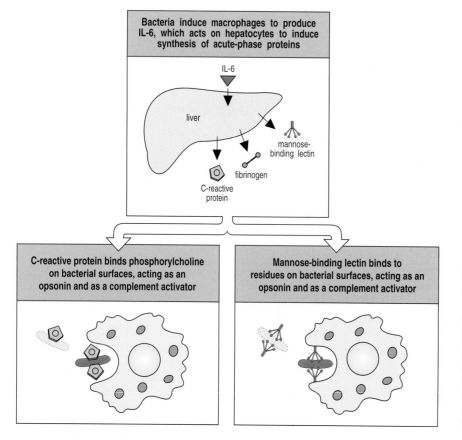

Figure 8.23 The acute-phase response increases the supply of the recognition molecules of innate immunity. Acute-phase proteins are produced by liver cells in response to the cytokines released by phagocytes in the presence of bacteria. In humans they include C-reactive protein, fibrinogen, and mannose-binding lectin. Both C-reactive protein and mannose-binding lectin bind to structural features of bacterial cell surfaces that are not found on human cells. On binding to bacteria, they act as opsonins and also activate complement, facilitating phagocytosis and also direct lysis (dashed lines) of the bacteria.

Step in pathway	Protein serving function in pathway			Relationship
	Alternative	Lectin	Classical	
Initiation	D	MBL-associated serine protease	C1s	Serine proteases
Covalent binding to cell surface	C3b	C4b		Homologous
C3/C5 convertase	Bb	C2a		Homologous
Control of activation	CR1 H	CR1 C4bp		Identical Homologous
Opsonization		C3b		Identical
Initiation of membrane-attack complex formation		C5b		Identical
Local inflammation		C5a, C3a		Identical

Figure 8.24 There are close structural relationships between components of the alternative, lectin-mediated, and classical pathways of complement activation. Although there are many different complement components they are built from a small number of structural modules. C3, C4, and C5 form one set of related components; C6–C9 form another. A third family comprises the serine proteases, which include factor D of the alternative pathway, C1r and C1s of the classical pathway, and the serine protease associated with the mannose-binding lectin (MBL). Factor B of the alternative pathway and C2 of the classical pathway are highly homologous. The major divergence between the three pathways is in their initiation. The antibodies that initiate the classical pathway, the MBL that initiates the lectin pathway, and the spontaneous hydrolysis of C3 that initiates the alternative pathway are distinct molecular mechanisms.

The second acute-phase protein that activates the complement system is **mannose-binding lectin** (**MBL**). It binds to mannose-containing carbohydrates of bacteria and yeast and is a calcium-dependent lectin. In its overall structure and domain organization, the MBL molecule is similar to C1q, although there are no obvious similarities in their amino-acid sequences. MBL resembles a bunch of flowers in which the carbohydrate-recognition domains form the flowers. Each flower contains three binding sites and each MBL molecule has five or six flowers, giving MBL either 15 or 18 potential sites for attachment to a pathogen's surface. Even relatively weak individual interactions with a carbohydrate structure can be developed into high-avidity interactions through the use of multipoint attachments. Although some carbohydrates on human cells contain mannose, they do not bind MBL because their geometry does not permit multipoint attachment to MBL.

Like C1q, MBL activates a proteolytic enzyme complex (MBL-associated serine protease or MASP) that cleaves C4 and C2 and thus initiates complement activation. This pathway is known as the lectin pathway of complement activation. MBL also serves as an opsonin that facilitates the uptake of bacteria by monocytes in the blood. These cells do not express the macrophage mannose receptor but have receptors that can bind to MBL molecules coating a bacterial surface. MBL is a member of a protein family called the **collectins**; the family also includes the pulmonary surfactant proteins A and D (SP-A and SP-D), which work in the lungs to opsonize pathogens such as *Pneumocystis carinii*.

In the absence of infection, C-reactive protein and MBL are present at low levels in plasma; their levels increase during the acute-phase response. Both proteins bind to structures that are commonly found on pathogens but not on human cells, and initiate complement activation and fixation by a pathway that is almost identical to the classical pathway used by antibodies. Such comparisons reveal how the antibodies of the adaptive immune response have a role analogous to that of C-reactive protein and MBL in innate immunity and serve to expand the range of pathogen-recognition molecules. Furthermore, nearly all of the complement components used in the classical and lectin pathways are structurally and functionally related to components of the alternative pathway, which was probably the first sequence of complement reactions to evolve (Figure 8.24).

Alleles encoding nonfunctional variants of MBL are present at frequencies greater than 10% in human populations. Consequently, deficiency of MBL is common and causes increased susceptibility to infection. Individuals who carry two variant alleles are more likely to develop severe meningitis caused by *Neisseria meningitidis*, a bacterium that is carried as a harmless commensal by about 1% of the population. Similar susceptibility is observed in people who are deficient for a terminal complement component, showing that complement-mediated killing of the bacteria is the mechanism by which healthy carriers keep their *N. meningitidis* in order.

8-12 Type I interferons inhibit viral replication and activate host defenses

When a human cell becomes infected with a virus it responds by making cytokines called **type I interferons** or simply **interferon**. These comprise a quite different family of proteins from the gamma interferon (IFN-γ) that is secreted by inflammatory effector lymphocytes (NK cells, CD8 T cells, and CD4 T_H1 cells; see Sections 8-13 and 8-19), and which is called type II interferon. The immediate effects of type I interferon are to interfere with viral replication by the infected cell, and to signal neighboring uninfected cells that they too should prepare for a viral infection. Further effects of type I interferon are to alert cells of the immune system that infection is about, and to make virus-infected cells more vulnerable to attack by killer lymphocytes. As almost all types of human cell are susceptible to viral infections, virtually all cells are equipped to make type I interferon and its cell-surface receptor. The receptor is always present on cell surfaces, ready to bind interferon newly made in response to infection. Although type I interferon is barely detectable in the blood of healthy people, upon infection it becomes abundant.

There are many different forms, or isotypes, of type I interferon. Humans have a single form of **interferon-β (IFN-β)**, multiple forms of **interferon-α (IFN-α)** and several additional isotypes: IFN-δ, -κ, -λ, -τ and -ω. The isotypes have a similar structure, bind to the same cell-surface receptor, and are specified by linked genes on human chromosome 9.

The induction of type I interferon synthesis is triggered by intracellular events that follow viral infection or ligand binding to a cell-surface receptor, for example double-stranded RNA binding to TLR-3. Double-stranded RNA, a type of nucleic acid not found in healthy human cells, is a component of some viral genomes, and an intermediary nucleic acid in viral life cycles. Infection or ligand binding is the trigger that leads to the phosphorylation of a cytoplasmic protein called interferon-response factor 3 (IRF3), which dimerizes and enters the nucleus to initiate transcription of the IFN-β gene, which is also dependent on the transcription factors NFκB and AP-1. Once IFN-β is secreted it acts in an autocrine fashion, binding to receptors on the same cell that made it, or in paracrine fashion, by binding to receptors on uninfected cells nearby (Figure 8.25).

When interferon binds to its receptor, the Jak1 and Tyk2 kinases associated with the receptor initiate intracellular reactions that change the expression of a variety of human genes, a process called the **interferon response** (Figure 8.26). Among the cellular proteins induced by interferon are some that interfere directly with viral genome replication. An example is the enzyme oligoadenylate synthetase, which polymerizes ATP by 2′–5′ linkages rather than the 3′–5′ linkages normally present in human nucleic acids. These unusual oligomers activate an endoribonuclease that degrades viral RNA. Also activated by IFN-α and IFN-β is a serine/threonine protein kinase (PI kinase) that phosphorylates the protein synthesis initiation factor eIF-2, thereby preventing viral protein synthesis and the production of new infectious virions.

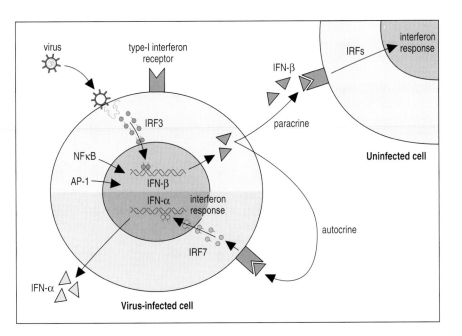

Figure 8.25 Virus-infected cells are stimulated to produce type I interferons. The cell on the left is infected with a virus that triggers signals that lead to the phosphorylation, dimerization and passage to the nucleus of the transcription factor interferon-response factor 3 (IRF3). Transcription factors NFκB and AP-1 are also mobilized and coordinate with IRF3 to turn on transcription of the interferon (IFN)-β gene. These events are depicted in the upper half of the cell. Secreted IFN-β binds to the interferon receptor on the infected cell surface, acting in autocrine fashion to mobilize other interferon-response factors and change patterns of gene expression to give the interferon response. These events are depicted in the lower half of the cell, being exemplified by IRF7 turning on transcription of the IFN-α gene, which it does without the need for AP-1 or NFκB. Secreted IFN-β will also bind to the interferon receptor expressed by nearby cells that are not infected by the virus, acting in paracrine fashion to induce the interferon response that helps these cells to resist infection.

Several of the interferon-induced proteins are transcription factors similar to IRF3, the only one of the group that is made constitutively. The other interferon-response factors are instrumental in turning on the transcription of many different genes, including those for interferons other than IFN-β. Interferon-response factor 7 (IRF7) initiates the transcription of IFN-α, which does not require the participation of NFκB and AP-1. In this manner, a positive feedback loop develops, in which a small amount of interferon serves to increase both the size and range of future production.

As well as interfering with viral replication, interferon also induces cellular changes that make the infected cell more likely to be attacked by killer lymphocytes. NK cells are lymphocytes of innate immunity that provide defense against viral infections by secreting cytokines and killing infected cells. When IFN-α or IFN-β bind to the interferon receptors on circulating NK cells these become activated and drawn into infected tissues, where they attack virus-infected cells. Interferon also causes changes in the cell-surface molecules of infected cells that make them more visible to cytotoxic CD8 T cells of the adaptive immune response (see Figure 8.26): it increases the transcription of MHC class I genes and the TAP peptide transporter proteins (see Section 3-9, p. 78), which together increase the overall abundance of MHC class I molecules at the infected cell's surface. Interferon also favors production of the LMP2 and LMP7 proteasome subunits, which by replacing other forms of

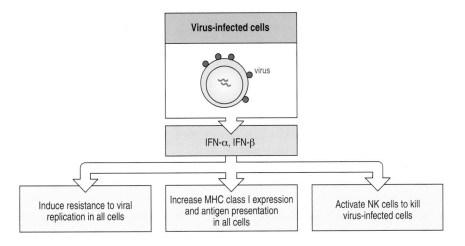

Figure 8.26 Major functions of the type I interferons. The interferons-α and -β (IFN-α and IFN-β) have three major functions. First, they induce resistance to viral replication by activating cellular genes that destroy viral mRNA and inhibit the translation of viral proteins. Second, they induce MHC class I expression in most cells of the body, thus increasing the level of antigen presentation to CD8 cytotoxic T cells by infected cells, and the resistance of uninfected cells to NK cells. Third, they activate NK cells to kill virus-infected cells.

proteasome subunit skew intracellular protein degradation towards the production of peptides that bind to MHC class I molecules (see Section 3-9, p. 78). All these interferon-regulated genes are located in the MHC.

Because of its power to boost the immune response, type I interferon has been explored as a treatment for human disease. It has been found to ameliorate several conditions: infections with hepatitis B or C virus; the degenerative autoimmune disease multiple sclerosis, which affects the central nervous system; and certain leukemias and lymphomas.

Although almost all human cells can secrete type I interferon, specialized cells called **interferon-producing cells** (**IPCs**) or **natural interferon-producing cells** (**NIPCs**) secrete up to 1000-fold more interferon than other cells. These lymphocyte-like cells are present in the blood, making up less than 1% of the total leukocytes, and are distinguished by cytoplasm resembling that of a plasma cell, another cell type engaged in the massive production of secreted protein (Figure 8.27). During an infection, interferon-producing cells congregate in the T-cell areas of draining lymph nodes, after having entered from the blood at high endothelial venules. The interferon-producing cell serves as a mobile and abundant source of interferon, making it available throughout the body. As well as responding to viral infections, interferon-producing cells have TLR-9 receptors for bacterial DNA, which enables them to secrete interferon in the presence of bacteria. The interferon response is, therefore, not limited to viral infections.

The interferon-producing cell resembles the dendritic cell in ways that seem important but are not yet fully understood. Because high concentrations of type I interferon can drive the maturation of dendritic cells, a function of the interferon-producing cells present in the T-cell areas of lymph nodes draining an infected site might be to complete the maturation of dendritic cells arriving at the node in the draining lymph. On the other hand, when interferon-producing cells are cultured with microbial products and inflammatory cytokines (but not type I interferon) they differentiate into cells resembling dendritic cells. The capacity, professional or otherwise, of these **plasmacytoid dendritic cells** (**PDC**) to present antigens to T cells remains to be determined, although they do have much of the necessary molecular equipment.

8-13 NK cells provide an early defense against intracellular infections

In Chapter 6 we saw how cytotoxic CD8 T cells kill virus-infected cells on recognition of virus-derived peptides presented by MHC class I molecules. The circulating lymphocyte population contains a second type of cytotoxic lymphocyte known as the **natural killer cell** (**NK cell**). This cell has many similarities in effector mechanisms to the cytotoxic T lymphocyte. What most clearly distinguishes NK cells from T cells is that the former do not rearrange or express their T-cell receptor genes; neither do NK cells rearrange or express immunoglobulin genes. NK cells represent a third type of lymphocyte that is distinct from B cells and T cells and active in innate immunity.

NK cells are large lymphocytes that circulate in the blood, but unlike circulating B cells and T cells they have a well developed cytoplasm containing cytotoxic granules. When first discovered, NK cells were called 'large granular lymphocytes;' they provide innate immunity against intracellular infections and migrate from the blood into infected tissues in response to inflammatory cytokines. Patients who lack NK cells suffer from persistent viral infections, particularly of herpes viruses, which they cannot clear without help from antiviral drugs, despite making a normal adaptive immune response. These

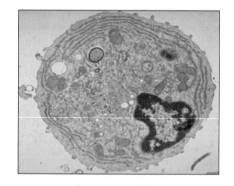

Figure 8.27 Type-I-interferon-producing cell from human peripheral blood. Note the extensive rough endoplasmic reticulum that is similar in appearance to that of a plasma cell and is due to the massive synthesis and secretion of interferon by these cells. Image courtesy of Dr Yong-Jun Liu.

rare individuals demonstrate the importance of NK cells in managing virus infections and show how the response of NK cells complements that of cyto-toxic T cells. NK cells have two types of effector function—cell killing and the secretion of cytokines—that are used in different ways depending on the pathogen.

When NK cells are freshly isolated from blood they can kill certain types of target cell. This base level of cytotoxicity is increased 20–100-fold on exposure to the IFN-α and IFN-β produced in response to viral infection. Type I inter-ferons also induce the proliferation of NK cells. NK cells are also activated by IL-12, which especially targets NK cells, and by TNF-α, both of which are pro-duced by macrophages early in many infections. The actions of these four cytokines produce a wave of activated NK cells during the early part of a virus infection that can either terminate the infection or contain it during the time required to develop cytotoxic CD8 T cells (Figure 8.28).

Stimulation of NK cells with IFN-α and IFN-β favors the development of cyto-toxic effector function, whereas stimulation with IL-12 favors the production of cytokines. The principal cytokine released by NK cells is IFN-γ, which acti-vates macrophages (see Section 6-14, p. 169). Macrophage secretion of IL-12 and NK-cell secretion of IFN-γ create a system of positive feedback that increases the activation of both types of cell within an infected tissue. Inter-actions between NK cells and dendritic cells can also lead to mutual activa-tion or to killing of dendritic cells, events that influence if and when dendritic cells will migrate to secondary lymphoid tissue and initiate the adaptive immune response. In the early stage of an infection, NK cells are the major producers of IFN-γ, which then activates macrophages to secrete cytokines that initiate the T-cell response. When effector T cells are formed and enter the infected site, they become the major source of IFN-γ and of cell-mediated cytotoxicity. With the arrival of effector T cells, NK-cell functions are turned off by IL-10, an inhibitory cytokine made by cytotoxic T cells.

8-14 NK-cell receptors differ in the ligands they bind and the signals they generate

NK cells circulate in the blood in a partially activated state, poised to enter infected tissue when macrophages sound the alarm. Keeping NK cells in this state of readiness for infection, and curbing their potential to attack healthy tissue, is an extensive array of cell-surface receptors, of which some deliver activating signals and others inhibitory signals. The NK-cell receptors fall into two broad structural types: in one the ligand-binding site is an immunoglob-ulin-like domain and the other has a lectin-like ligand-binding domain simi-lar to the carbohydrate-recognition domain of the mannose-binding lectin. Although it is convenient to call the latter group the **NK-cell lectin-like recep-tors**, many actually bind protein ligands rather than carbohydrates. The main families of NK-cell receptors are shown in Figure 8.29. Although some recep-tors are expressed by all NK cells, for example the lectin-like receptor NKG2D, most are expressed by subpopulations of NK cells. Consequently, individual NK cells express different combinations of receptors, imparting heterogene-ity to a person's NK-cell population and providing a repertoire of responses to pathogens. Among other cell types involved in the innate immune response, such as macrophages, dendritic cells, and neutrophils, individual cells also express different combinations of receptors.

Ligands for NK-cell receptors are usually cell-surface proteins whose expres-sion is altered in response to infection or other trauma. Many of them are MHC class I or MHC class I-like molecules. The activating lectin-like receptor NKG2D binds to MIC-A and MIC-B, which are class-I-like heavy chains that

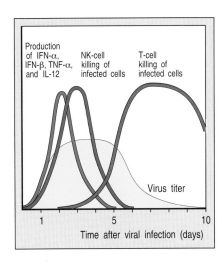

Figure 8.28 NK cells provide an early response to virus infection. The kinetics of the immune response to an experimental virus infection of mice are shown. As a result of infection, a burst of cytokines is secreted, including IFN-α, IFN-β, TNF-α, and IL-12 (green curve). These induce the proliferation and activation of NK cells (blue curve), which are seen as a wave emerging after cytokine production. NK cells control virus replication and the spread of infection while effector CD8 T cells (red curve) are developing. The level of virus (the virus titer) is given by the curve described by the yellow shading.

NK-cell receptors					
Inhibitory			**Activating**		
Receptor	Structural type	Ligand	Receptor	Structural type	Ligand
KIR2DL	Ig	HLA-C	KIR2DS	Ig	HLA-C
KIR3DL	Ig	HLA-B,C	KIR3DS	Ig	HLA-B?
LILRB1,2	Ig	HLA class I	LILRA3	Ig	?
CD94:NKG2A	Lectin	HLA-E	CD94:NKG2C/E	Lectin	HLA-E
LAIR-1	Ig	?	LAIR-2	Ig	?
			NKG2D	Lectin	MIC-A,B and others
			NKp30	Ig	?
			NKp44	Ig	?
			NKp46	Ig	?
			CD16	Ig	Fc

Figure 8.29 NK cells express diverse inhibitory and activating receptors. This diagram shows only some of the receptors and organizes them according to signaling potential, structural type (immunoglobulin-like (Ig) and lectin-like), and ligands, where known. In the nomenclature for the killer-cell immunoglobulin-like receptors (KIRs), 2DL and 3DL correspond to receptors having long cytoplasmic tails and two or three immunoglobulin-like domains respectively; 3DS corresponds to a receptor having a short cytoplasmic tail and three immunoglobulin-like domains. Whereas all KIRs and leukocyte-associated immunoglobulin-like receptors (LAIRs) are expressed by NK cells, only a minority of the leukocyte immunoglobulin-like receptors (LILRs) are expressed by NK cells. NKp30, NKp44, and NKp46 are collectively called the natural cytotoxicity receptors; they are important for the killing of tumor cells and are the only receptors uniquely expressed by NK cells.

do not associate with β_2-microglobulin and are encoded by genes in the HLA region. MIC-A and MIC-B are examples of stress-induced proteins. The only tissue in which they are made constitutively is intestinal epithelium, and there the amount is small. When any epithelial cell becomes infected, damaged, or cancerous, however, expression of MIC-A and MIC-B is induced, and for intestinal epithelium it increases. Once expressing MIC-A and MIC-B, epithelial cells become targets for NK-cell attack mediated through the NKG2D receptor. Through the agency of small proteins called adaptor molecules, which associate with the cytoplasmic tail of NKG2D, kinases are activated and lead to the release of cytotoxic granules and cytokines. Other ligands for the NKG2D receptor are a different type of class I-like molecule, encoded by a family of genes on human chromosome 6 away from the HLA region.

Other lectin-like NK-cell receptors are heterodimers formed from CD94 paired with either NKG2A or NKG2C. Both the CD94:NKG2A and CD94:NKG2C receptors recognize composite ligands composed of a peptide derived from the leader sequence of an HLA-A, -B, or -C allotype bound to the oligomorphic class I molecule HLA-E. As HLA-E is so finicky in the peptides it binds, the presence and abundance of HLA-E on the cell surface is completely dependent on the amounts of other HLA class I heavy chains being made by the cell. Although the CD94:NKG2A and CD94:NKG2C receptors have similar ligand-binding specificity, they differ radically in signaling function. Whereas NKG2C associates with an adaptor molecule and delivers activating signals, NKG2A has a cytoplasmic tail that recruits phosphatases to deliver inhibitory signals. These two receptors are rarely expressed by the same NK cell; CD94:NKG2A is more frequently expressed than CD94:NKG2C and serves to prevent NK cells from attacking healthy human cells expressing a normal complement of HLA class I molecules.

The **killer-cell immunoglobulin-like receptors** (**KIR**) are a family of activating and inhibitory NK-cell receptors that use either two or three immunoglobulin domains to bind to polymorphic determinants of HLA-A,

-B, or -C ligands. Of these ligand–receptor interactions, those between HLA-C and inhibitory KIR are best understood. Three-dimensional structures of the complex of HLA-C and KIR show that KIR interacts with the same face of the MHC class I molecule as the T-cell receptor, the face formed by the two α helices and the bound peptide (Figure 8.30). The difference is that KIR covers less of the face than the T-cell receptor, just that part containing the carboxy-terminal parts of the bound peptide and the $α_1$ helix, and the amino-terminal part of the $α_2$ helix. The presence of asparagine or lysine residues at position 80 in the HLA-C $α_1$ helix defines two groups of HLA-C allotypes that bind to different KIRs. Whereas all allotypes of HLA-C are KIR ligands, it is not so for HLA-A and -B, pointing to HLA-C being particularly specialized towards control of the NK-cell response.

Inhibitory KIRs with specificity for HLA-A, -B, and -C are used to protect healthy cells from NK-cell attack, in a manner analogous to CD94:NKG2A. Other members of the KIR family are activating receptors, some of which have a similar HLA-C specificity to the inhibitory receptors. Although likely, it has yet to be demonstrated that activating KIRs serve to activate NK cells against infection. Conversely, the **natural cytotoxicity receptors** (**NCR**), a distinct group of activating NK-cell receptors of the immunoglobulin superfamily, are known to mediate NK-cell killing of tumor cells, but their ligands remain poorly defined.

Although many NK-cell receptors have been identified, the ways in which they work together to recognize virus-infected cells are poorly understood, and are likely to depend on the type of virus and the type of cell infected. Many viruses interfere with MHC class I expression in the cells they infect, a tactic that prevents cytotoxic CD8 T cells from killing the infected cells. The reduced level of MHC class I can then make the infected cells vulnerable to killing by NK cells, which on interaction with the infected cells will not be inhibited through engagement of their inhibitory MHC class I receptors. For some viruses, such as human cytomegalovirus, the situation could be further complicated by viral proteins that mimic MHC class I molecules and interfere with NK-cell attack by ligation of the NK cells' inhibitory receptors.

8-15 Three genetic complexes contribute to NK-cell recognition of 'missing-self'

The families of genes encoding lectin-like NK-cell receptors cluster together in a region of chromosome 12 called the natural killer complex (NKC; Figure 8.31). Similarly, the KIR genes are found in a region of chromosome 14 called the leukocyte receptor complex (LRC), which also contains genes for other receptors of the immunoglobulin superfamily. Next to the KIR genes in the LRC is the family of leukocyte immunoglobulin-like receptor (LILR) genes, which include two that are present on NK cells and have a broad specificity for HLA class I molecules (see Figure 8.29). Receptors determined by NKC and LRC genes interact with class I and class I-like ligands that are the products of genes in the HLA complex.

As we saw in Chapter 3, some HLA class I genes in humans are highly polymorphic while others are present as only one or a very few alleles in the population. A similar division can be made for the NK-cell receptors. As a group, the human lectin-like receptors are less polymorphic than the immunoglobulin-like receptors. Most conserved is NKG2D, but it binds to diverse ligands, of which MIC-A is highly polymorphic.

KIRs are the most diverse of the NK-cell receptors—so diverse that unrelated people rarely (< 1%) have exactly the same set of KIR alleles. Three factors combine to create this diversity. First, individuals possess different numbers

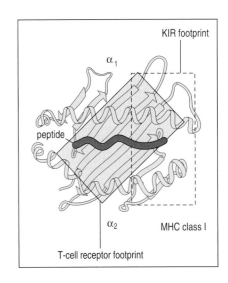

Figure 8.30 KIRs interact with the same face of the MHC class I molecule as the T-cell receptor. The ribbon diagram (yellow) is of the $α_1$ and $α_2$ domains of an MHC class I molecule and shows the bound peptide (red). The T-cell receptor interacts with that part of the face formed by the two α helices and the peptide, as outlined by the gray-shaded rectangle. A KIR interacts with the right-hand half of the same face as the T-cell receptor, as outlined by the dashed line.

Figure 8.31 Many NK-cell receptors are encoded by genes in one of two genetic complexes. The natural-killer complex (NKC) on human chromosome 12 encodes lectin-like NK-cell receptors, whereas the leukocyte receptor complex (LRC) on chromosome 19 encodes NK-cell receptors that are members of the immunoglobulin superfamily. The LRC comprises the killer-cell immunoglobulin-like receptor (KIR), the leukocyte immunoglobulin-like receptor (LILR), and the leukocyte-associated immunoglobulin-like receptor (LAIR) gene families. Flanking the LRC are genes encoding other immunoglobulin-like molecules of the immune system, including the NKp46 receptor of NK cells, and the DAP10 and DAP12 signaling adaptor molecules that transduce signals from activating NK-cell receptors. Figure based on data courtesy of John Trowsdale.

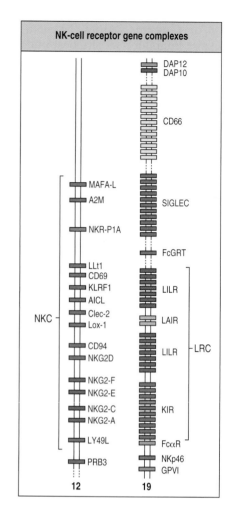

NK-cell receptor gene complexes

of KIR genes (ranging from 7 to 15); second, KIR haplotypes have different numbers of genes encoding activating and inhibitory receptors; third, the individual KIR genes are polymorphic. Some people have genotypes containing only genes specifying inhibitory KIRs, whereas others have an equivalent number of genes encoding activating and inhibitory KIRs. Differences in KIR type are beginning to be correlated with clinical parameters in infectious disease, autoimmune disease, and transplantation.

Only one of the KIR genes, KIR2DL4, is expressed by all human NK cells. The others are expressed on subpopulations of NK cells, with an almost random distribution in the number and combination of KIR genes expressed by any given cell. Once the pattern of KIR expression is set during NK-cell development it remains stable. Variation in KIR expression makes a prominent contribution to the NK-cell repertoire, and clones of NK cells can be defined on the basis of the combination of expressed KIR genes. The one requirement placed on NK-cell receptors is that they inhibit the NK cell from killing when confronted with healthy cells expressing self-HLA class I molecules (Figure 8.32). Because the ligands for KIRs are polymorphic MHC class I molecules, which KIRs that can serve as inhibitory receptors for a person's MHC class I molecules depends on that person's HLA class I type. However, this restriction

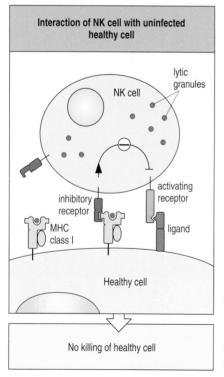

Interaction of NK cell with uninfected healthy cell

No killing of healthy cell

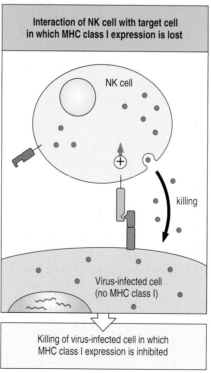

Interaction of NK cell with target cell in which MHC class I expression is lost

Killing of virus-infected cell in which MHC class I expression is inhibited

Figure 8.32 Possible mechanism by which NK cells distinguish uninfected cells from infected cells. NK cells have activating and inhibitory receptors that interact with ligands on the surfaces of human cells. The nature of the activating receptors and their ligands is uncertain. The inhibitory receptors interact with MHC class I molecules, which are present on virtually all healthy cells. On interaction with a healthy cell the inhibitory signals dominate and inhibit NK-cell cytotoxicity (left panel). Infection of a cell with an intracellular parasite might change the expression or conformation of MHC class I molecules such that the interaction of the cell with the receptors on the NK cell generates a stimulatory signal that causes the NK cell to kill the infected cell (right panel).

does not prevent NK cells expressing inhibitory KIRs for which there is no self-HLA class I ligand. In such cases, either another KIR or CD94:NKG2A or an LILR provides the necessary inhibitory receptor for self-HLA class I. By expressing at least one inhibitory receptor that binds to a self-HLA class I allotype, each NK cell is tolerant of the healthy cells of its own body (autologous cells) and does not attack them.

Although tolerant of healthy autologous cells, NK cells need not be tolerant of healthy allogeneic cells—cells from another individual. In culture, NK cells from one person will mount an attack on cells from another person that lack one, or more, of the HLA class I ligands that would engage their inhibitory receptors (Figure 8.33). In this situation there is a subpopulation of NK cells that cannot be inhibited by the HLA class I molecules on the target cell and it is these NK cells that kill the targets. Certain HLA-C-controlled alloreactions can be beneficial in eradicating residual leukemic cells and preventing relapse following allogeneic hematopoietic-cell transplantation to treat acute myelogenous leukemia. Whereas T-cell alloreactions are generated by the active recognition of a foreign MHC molecule, NK-cell alloreactions are caused by the absence of a self-MHC class I determinant. Alloreactive NK cells are, therefore, said to see 'missing-self'.

Figure 8.33 NK cells mediate alloreactions determined by HLA class I type. In humans the dominant NK-cell alloreactions are due to an amino-acid substitution at position 80 in HLA-C. This position is in the carboxy-terminal half of the helix of the α₁ domain (ribbon diagram at top left) and can be either asparagine (Asn) or lysine (Lys). HLA-C-mediated alloreactions arise from subpopulations of NK cells that are dependent on inhibitory HLA-C-specific KIRs for their tolerance to healthy autologous cells. The lower left panel shows such an NK cell, which is kept in check by interaction between its inhibitory KIR, which is specific for HLA-C Asn80, and an HLA-C Asn-80-containing allotype on the interacting cell. This NK cell will also be inhibited by healthy allogeneic cells that express HLA-C Asn80-containing allotypes (center panel), but will not be inhibited by allogeneic cells that only express HLA-C Lys80-containing allotypes. In the absence of the inhibitory signal from the HLA-C-specific KIR, the balance of signaling in the NK cell is tipped towards activation, which instructs the NK cell to kill the allogeneic cell (right panel). Because the killing is controlled by an inhibitory receptor, it appears as though the alloreactive NK cell is sensing the autologous HLA-C ligand that is missing from the allogeneic cell. NK-cell-mediated killing of autologous tumor cells that have lost expression of one or more HLA class I allotypes is similarly determined by recognition of missing-self.

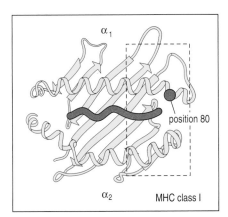

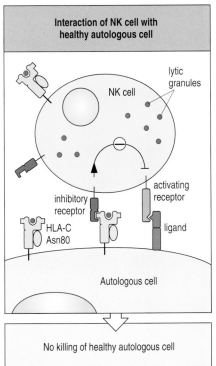
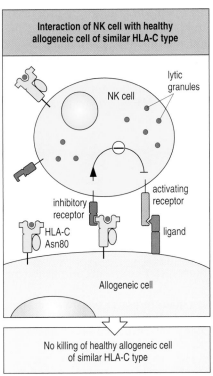
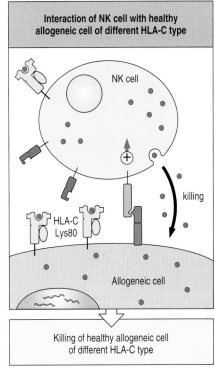

8-16 Minority subpopulations of B and T cells contribute to innate immunity

During B- and T-cell development, minority subpopulations are produced that can be distinguished on the basis of nonconventional cell-surface phenotype and a restricted repertoire of B- or T-cell receptors. Within these minority subpopulations, each lymphocyte clone is much larger than is usual in the majority subpopulation, so these minority subpopulations can respond to a primary infection without first undergoing a lengthy period of clonal expansion and differentiation. Two of these subpopulations have been introduced already: B-1 cells (see Section 4-6, p. 108) and γ:δ T cells (see Section 3-4, p. 71).

The B-1 lineage of B cells arises early in embryonic development and uses a limited set of V regions to produce surface IgM that is characterized by a rather general affinity for polysaccharide antigens (Figure 8.34). B-1 cells respond rapidly to antigen, producing detectable levels of antibody within 48 hours of infection. These responses are independent of cognate T-cell help, do not involve isotype switching or somatic hypermutation, and do not produce immunological memory. B-1 cells are characteristically found in the peritoneal cavity and do not depend on the bone marrow for renewal.

γ:δ T cells have less diverse T-cell receptors than α:β T cells and do not recognize peptide antigens presented by MHC molecules. About 4% of the T cells in human blood are γ:δ T cells with a restricted set of receptors encoded by $V_{\gamma2}V_{\delta2}$ gene segments (in an alternative nomenclature $V_{\gamma2}$ is $V_{\gamma9}$). These cells respond quickly to infections by releasing inflammatory cytokines and killing infected cells. In a variety of infectious diseases the abundance of $V_{\gamma2}V_{\delta2}$ T cells in the blood increases substantially, to become as much as 60% of all T cells. The MIC-A molecule appears to be a ligand for $V_{\gamma2}V_{\delta2}$ T-cell receptors, but the situation is complicated by the presence of the NKG2D receptor on these cells because it too binds MIC-A (Figure 8.35).

Other ligands for $V_{\gamma2}V_{\delta2}$ T-cell receptors are neither proteins nor peptides, but small organic phosphates and alkylamines. Such compounds are produced during infection as a consequence of microbial metabolism and breakdown, but can also be formed by human digestion and metabolism. In habitual tea drinkers, breakdown of the unusual amino acid theatin present in tea yields ethylamine, which stimulates γ:δ T cells. The absence of MHC restriction for γ:δ T cells, and the wide-ranging size and structure of the antigens they recognize, has led to the view that γ:δ T-cell receptors behave more like antibodies than like α:β T-cell receptors in the antigens they bind.

A minority subpopulation of α:β T cells also has a restricted T-cell receptor repertoire. These cells are called NK T cells because they express NKR-PIA, one of the lectin-like NK-cell receptors. The T-cell receptors of NK T cells comprise an α chain consisting of $V_{\alpha24}$ and $J_{\alpha15}$ segments associated with a $V_{\beta\,11}$-containing β chain. Unlike the majority of α:β T cells, NK T cells are not restricted to binding peptide antigens presented by polymorphic MHC class I molecules, but bind lipid antigens presented by CD1D, a β_2-microglobulin-associated class I molecule that is distinguished by its hydrophobic antigen-binding site. In the thymus, positive selection of NK T cells is accomplished by

Figure 8.34 B-1 cells provide a quick response to carbohydrate antigens such as bacterial polysaccharides. Almost all the antibody made by B-1 cells is of the IgM isotype and its principal effector function is the activation of complement. Although the B-1 cell response is stimulated by IL-5 made by T cells, it does not depend on cognate interactions with T cells.

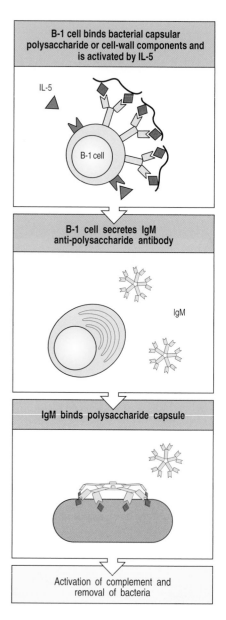

B-1 cell binds bacterial capsular polysaccharide or cell-wall components and is activated by IL-5

IL-5

B-1 cell

B-1 cell secretes IgM anti-polysaccharide antibody

IgM

IgM binds polysaccharide capsule

Activation of complement and removal of bacteria

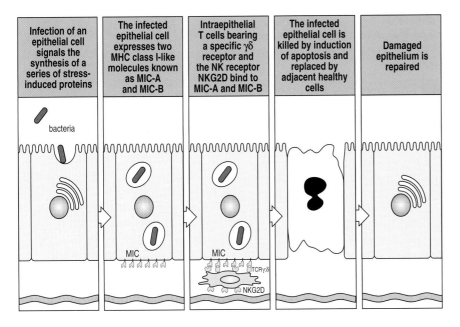

| Infection of an epithelial cell signals the synthesis of a series of stress-induced proteins | The infected epithelial cell expresses two MHC class I-like molecules known as MIC-A and MIC-B | Intraepithelial T cells bearing a specific γ:δ receptor and the NK receptor NKG2D bind to MIC-A and MIC-B | The infected epithelial cell is killed by induction of apoptosis and replaced by adjacent healthy cells | Damaged epithelium is repaired |

Figure 8.35 T cells of the mucosal immune system bearing γ:δ T-cell receptors and the activating NKG2D receptor recognize and kill injured enterocytes. Infection or other injury to enterocytes, the epithelial cells lining the lumen of the gut, stimulates a stress response, which causes expression on the cell surface of the MHC class-like molecules, MIC-A and MIC-B. Many intraepithelial T cells bear a γ:δ T-cell receptor that is specific for MIC-A and MIC-B; they also bear the MIC-A- and MIC-B-specific NK-cell receptor NKG2D. For NK cells, NKG2D is an activating receptor, but on these γ:δ T cells it acts as a co-receptor for the γ:δ receptor. When a stressed enterocyte express MIC-A and MIC-B the γ:δ T cells induce it to die by apoptosis. The dying enterocyte is removed from the epithelium, allowing the local tissue injury to be repaired.

complexes of self lipids and CD1D expressed by cortical thymocytes. Outside the thymus, CD1D is expressed by intestinal epithelium, where it presents lipids derived from pathogens infecting the gut. NK T cells have the machinery to kill infected cells and to release large quantities of IL-4 and IFN-γ within an hour or two of having their T-cell receptors engaged by complexes of CD1D and lipid antigens. The cytokines released by NK T cells contribute to the activation of NK cells, macrophages, and dendritic cells in the early phase of the immune response. Although their receptors recognize an MHC class I molecule, NK T cells do not express CD8: they express either CD4 or no co-receptor at all.

CD1D is one of five isoforms of CD1, the others being CD1A, B, C, and E. They specialize in the presentation of lipid antigens and are encoded by a family of genes on human chromosome 1. Microorganisms usually have lipids that are chemically and structurally distinct from those of human cells and they, like microbial cell-surface carbohydrates, offer good targets for antigen receptors of the human immune system. By binding to the hydrophobic alkyl and acyl chains of lipids, CD1 presents the hydrophilic head groups or sugars for interaction with T-cell receptors.

The CD1A, B, and C isoforms present lipid antigens to diverse, conventional CD8 and CD4 α:β T cells. These isoforms are not expressed in intestinal epithelium, but in dendritic cells and activated monocytes, where they cycle between the cell surface and intracellular endosomal and lysosomal vesicles. The three isoforms are distinguished by the types of lipid they bind and the endosomal compartments to which they traffic in search of lipid antigens. Unlike conventional MHC class I molecules, CD1 molecules can associate with the invariant chain in the course of their travels. An example of a lipid antigen presented by CD1 is the mycolic acid of mycobacteria (Figure 8.36). Although CD1E protein is made and is present inside cells its role in antigen presentation is not known.

Summary

Innate immunity provides a variety of defenses against infection that work immediately a pathogen is first confronted or very soon after. These fixed defenses are always available and do not improve with repeated exposure to the same pathogen. The skin and other epithelia form physical barriers that

Figure 8.36 Mycolic acid. This molecule is present in the cell walls of mycobacteria and is a ligand for the γ:δ T-cell receptor. Its physical and chemical properties are very different from the peptides recognized by α:β T cells.

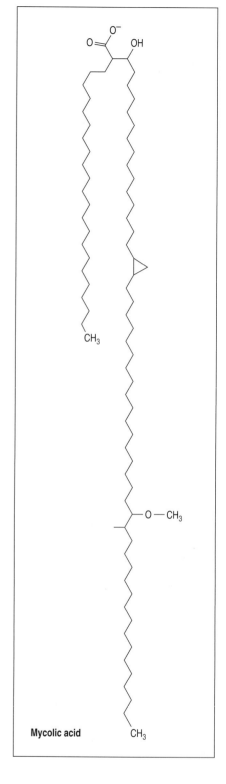

Mycolic acid

prevent entry to the body's tissue. Reactive ions, peptides, and other small molecules are chemical weapons that directly kill microorganisms. Inhibiting a pathogen's progress in colonizing tissues and spreading infection are the protease inhibitors, clotting cascade, and kinin reactions. A host of plasma proteins and cell-surface molecules provide systems for identifying microbiological invaders and distinguishing them from self. Complement provides a general means to tag almost any component at a microbial surface; more specific receptors bind common chemical aspects of microbial macromolecules that are not a part of the human body. As well as helping resident macrophages to phagocytose pathogens, these interactions induce the macrophages to pour out inflammatory cytokines that summon other effector cells of innate immunity—neutrophils, NK cells, γ:δ T cells, and NK T cells—to the site of infection. Interaction between these cells, and with resident macrophages and dendritic cells, produces mutual activation and cytokine secretion that heightens the state of inflammation in the infected tissue. Bacterial infections are frequently overcome by the phagocytic powers and potent poisons of abundant neutrophils. In viral infections, the production of type I interferons by infected cells and interferon-producing cells will often set the stage for NK cells to terminate infection. Only when the defenses of innate immunity are failing to contain an infection is the decision made to mobilize reinforcements from adaptive immunity.

Adaptive immune responses to infection

Through the activation of phagocytes, the production of cytokines, and the creation of a state of inflammation, innate immunity sets the scene for the development of the adaptive immune response. In this part of the chapter we shall consider the adaptive immune response to pathogens to which an individual has never been previously exposed. Such infections stimulate a **primary immune response**, involving the coordinated activation of naive T cells and B cells and the production of effector cells and molecules, including antibodies specific against the pathogen. These effectors help to clear the pathogen from the body and prevent reinfection in the short term, while the pathogen is still infecting others in the population. Usually, an adaptive immune response also leads to long-lasting immunity against the pathogen, which we shall consider in the last part of this chapter. Adaptive responses build on a foundation of effector mechanisms that are all part of innate immunity. Indeed, the inflammation produced by the innate immune response provides the essential environment for starting the adaptive immune response. The adaptive response refines and amplifies these effector mechanisms, providing the extra dimension of antigen specificity. In this part of the chapter we shall see how the component parts of adaptive immunity, which have been discussed in Chapters 2–7, work together in clearing infections.

8-17 Adaptive immune responses start with T-cell activation in secondary lymphoid tissues

The innate immune response develops locally at the site of infection through the action of plasma proteins, the stimulation of macrophages, the recruitment of effector cells, and the establishment of a state of inflammation. Movement of antigens, cytokines, and other inflammatory mediators away from the site of infection has an impact on more distant sites, particularly nearby lymphoid tissues, and can even have systemic effects. When the innate immune response fails to terminate infection in a timely fashion, molecular and cellular communications between the infected tissue and its most accessible secondary lymphoid tissue initiate an adaptive immune response. This involves an extensive process in which pathogen-specific lymphocyte clones are selected, expanded, and differentiated. All this takes place within the secondary lymphoid tissue over the course of several days and leads to the generation of several types of effector lymphocytes. The effector CD4 T_H1 and cytolytic CD8 T cells then migrate to the infected tissue to function, while CD4 T_H2 cells and plasma cells perform their functions in the lymphoid tissues.

Although all adaptive immune responses have much in common, there are also important differences that depend on the tissue infected and the secondary lymphoid tissue that it alerts. Infections of the skin and connective tissues drain through the lymphatics to lymph nodes, infections of the blood are pumped into the spleen, and infections of mucosal surfaces are dealt with by Peyer's patches and other lymphoid aggregations that lie immediately beneath the mucosal epithelium. In this part of the chapter we shall use the lymph node and Peyer's patches to exemplify the events that proceed in secondary lymphoid tissues.

The start of the adaptive immune response is marked by the migration in the lymph of antigen-laden or infected dendritic cells from the infected tissue to the draining lymph node. Dendritic cells have many properties of macrophages and throughout the innate immune response they have been stimulated or infected by the pathogen, taking up antigens and reacting to the cytokines made by other immune-system cells. Eventually, the sum of intracellular signals emanating from these stimuli cause a switch to be thrown and the dendritic cell migrates in the lymph and enters the lymph node in response to chemokines made by stromal cells in the node. With this journey, the dendritic cell matures, changing the focus of its activity from antigen uptake to antigen presentation and stimulation of naive T cells. The dendritic cell thus becomes a professional antigen-presenting cell (see Section 6-5, p. 151).

On entering lymph nodes in the lymph, dendritic cells are attracted by chemokines produced by the high endothelial venules in the T-cell areas, where the dendritic cells take up residence. As naive T cells enter the node from a high endothelial venule they are drawn toward the dendritic cells in the T-cell areas by the chemokine CCL18 (also called DC-CK) (see Figure 8.16), which is made by dendritic cells and specifically attracts naive T cells. The T cells scan the dendritic cell surface to find peptide:MHC complexes that bind to their antigen receptors. The small number of T cells that are specific for the antigens in the node are activated on engagement of their T-cell receptors; proliferation and differentiation begin and eventually lead to large clones of antigen-specific effector T cells.

Pathogens and their antigens are also carried passively in the lymph from infected tissue to the draining lymph node. There they are taken up by the macrophages resident in the node, processed, and presented to naive T cells. Lymph-node macrophages are ferociously efficient at removing pathogens from afferent lymph. In an experiment in which 3 billion streptococci were perfused through a single lymph node in a small volume of fluid (5 ml), over 99% of the organisms were removed. Lymph nodes are equally efficient in removing antigen-specific T cells from the blood. Within two days of an antigen being presented in a lymph node, all the antigen-specific T cells in the circulation were trapped there (Figure 8.37).

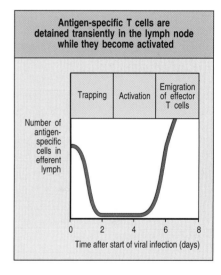

Antigen-specific T cells are detained transiently in the lymph node while they become activated

Trapping | Activation | Emigration of effector T cells

Number of antigen-specific cells in efferent lymph

Time after start of viral infection (days)

Figure 8.37 Antigen-specific T cells become trapped in lymph nodes. This graph charts the presence of virus-specific T cells in efferent lymph during the course of an experimental virus infection. Before infection, virus-specific T cells recirculate between blood, lymph nodes, and lymphatics. On infection, antigen-specific T cells are trapped, and collect in the lymph nodes draining the infected tissue. After 2 days they can no longer be detected in the efferent lymph. For 2–4 days from the start of the infection, the T cells remain in the lymph node, where they divide and differentiate into effector T cells. At that time, effector T cells leave in the efferent lymph to travel to the infected tissue. Because of clonal expansion, the number of antigen-specific T cells in the efferent lymph after day 4 becomes much larger than the numbers detected before infection.

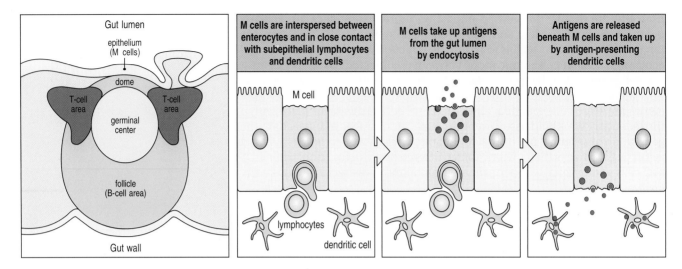

Figure 8.38 In the gut M cells deliver antigen to Peyer's patches. A typical region of gut-associated lymphoid tissue is shown in the schematic on the left. These areas are organized similarly to the lymph node and the white pulp of the spleen, with distinctive B- and T-cell zones, lymphoid follicles, and germinal centers. The other three panels show how the M cells of the gut epithelial wall deliver pathogens from the luminal side of the gut mucosa to the lymphoid tissue within the gut wall. The cell membrane at the base of these cells is folded around lymphocytes and dendritic cells within the Peyer's patches. M cells take up antigens from the lumen of the gut by endocytosis and the antigens are transported through the cells by transcytosis and delivered directly to antigen-presenting cells and lymphocytes of the mucosal immune system.

Dendritic cells that do not enter a node but are carried on by the lymph to the blood, take up residence in the T-cell areas of the spleen, where they initiate an adaptive immune response. Similarly, pathogens that fail to be removed from the lymph by lymph-node macrophages and pass into the blood will usually be taken out of circulation by macrophages in the spleen.

8-18 Microfold cells in the gut deliver antigens to Peyer's patches

Compared to the skin, the internal mucosal surfaces of the body are more vulnerable to infection, and most infections are caused by pathogens that have crossed a mucosal barrier. Mucosal surfaces are protected by lymphocytes and mucosa-associated lymphoid tissues (MALT) (see Chapter 1) that are dedicated to the purpose. The characteristics of mucosal defense are well illustrated by the small intestine, an organ devoted to the digestion and uptake of food, which is also home to some 14 million commensal microorganisms. Both food and commensal organisms represent an abundance of foreign antigens. They do not, however, stimulate an immune response except in situations where the gut epithelium has become inflamed due to damage, infection, or cancer.

Peyer's patches are the secondary lymphoid organs of the small intestine. They form domed follicles within the gut wall and are organized into B- and T-cell areas (Figure 8.38). Within the mucosal epithelium and above the dome of a Peyer's patch are specialized M cells that take up antigens and pathogens from the lumen of the gut and deliver them to the Peyer's patch by transcytosis. Unlike enterocytes, the predominant cell of the gut epithelium, these cells do not have microvilli but microfolds, hence their name of **microfold cells** or **M cells**. M cells also differ from enterocytes by not having the protection of a thick glycocalyx and a secreted mucus. M cells are, therefore, sites where pathogens find it easier to breach the intestinal wall, a lure that delivers them directly to a secondary lymphoid organ (see Figure 8.38).

Pathogens and antigens arriving at the basal side of the M cells encounter dendritic cells and macrophages that process and present their antigens. These will be scanned by circulating naive T cells that enter Peyer's patches from the blood and return to it via the lymphatics draining the intestines. Naive T cells that recognize specific antigens presented by the dendritic cells and macrophages in a Peyer's patch will be trapped there to proliferate and differentiate.

8-19 Primary CD4 T-cell responses are influenced by the cytokines made by cells of innate immunity

Effector CD4 T cells are of two types: T_H1, which promote a response dominated by inflammatory effector cells; and T_H2, which promote an antibody-mediated response. The decision as to which way a naive CD4 T cell will differentiate occurs when the T cell is first being stimulated by a professional antigen-presenting cell and is largely determined by the cytokine environment created by the ongoing innate immune response. When mutual interaction between dendritic cells, macrophages, and NK cells has favored production of IL-12 and IFN-γ, differentiation to T_H1 cells is favored. Thus, an innate response favoring inflammatory effector cells causes the adaptive immune response to continue with the strategy (Figure 8.39). By contrast, differentiation to T_H2 cells is favored when IL-4 dominates the cytokine environment produced by the innate response.

The direction of CD4 T-cell differentiation is also influenced by the abundance and density of the peptide:MHC complexes at the surface of professional antigen-presenting cells and their affinity for T-cell receptors. Peptides that are abundantly presented, or have strong affinity for T-cell receptors, tend to stimulate T_H1 responses, whereas peptides that are in low abundance, or have weak affinity for T-cell receptors, tend to stimulate T_H2 responses. Allergic responses, in which minute amounts of antigen preferentially stimulate T_H2 cells and IgE production, are of the latter type.

There is an inherent tendency for CD4 T cells to become polarized towards either a T_H1 or T_H2 response, because the cytokines produced by one type of effector CD4 T cell inhibit the differentiation and/or the function of the other type. The IL-10 produced by T_H2 cells prevents the development of a T_H1 response through a general inhibition of antigen processing and presentation by dendritic cells and macrophages. Thus, IL-12 is never produced, which in turn prevents activation of NK cells, differentiation of T_H1 cells, and IFN-γ production. The lack of IFN-γ further inhibits macrophages and their ability to act as professional antigen-presenting cells. Conversely, the IFN-γ produced by T_H1 cells prevents the activation of T_H2 cells (Figure 8.40). However, although highly polarized responses do occur, as in the two forms of leprosy (see Section 6-9, p. 159), the majority of immune responses involve contributions from both T_H1 and T_H2 CD4 cells.

CD8 T cells, which secrete cytokines as well as killing cells, can be influenced to make cytokines favoring either a T_H1 or T_H2 CD4 T-cell response. The unhelpful CD4 T_H2 response in lepromatous leprosy is thought to arise from the presence of CD8 T cells secreting IL-10 and transforming growth factor

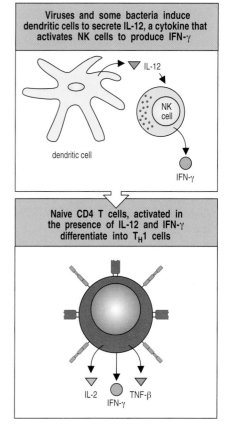

Figure 8.39 Cytokines elicited by the pathogen affect the differentiation of CD4 T cells. The innate immune response to different types of pathogen involves the secretion of cytokines that influence the course of the adaptive immune response. In response to viruses and intracellular bacteria, macrophages and dendritic cells produce IL-12. This cytokine stimulates NK cells to produce IFN-γ, which drives proliferating CD4 T cells to become of the T_H1 type.

Viruses and some bacteria induce dendritic cells to secrete IL-12, a cytokine that activates NK cells to produce IFN-γ

IL-12

NK cell

dendritic cell

IFN-γ

Naive CD4 T cells, activated in the presence of IL-12 and IFN-γ differentiate into T_H1 cells

IL-2   TNF-β
IFN-γ

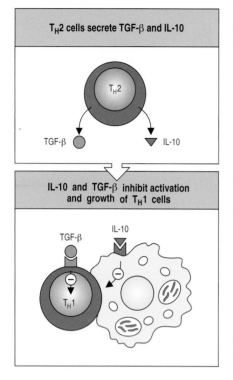

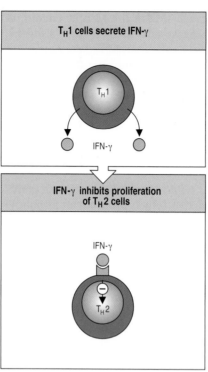

Figure 8.40 Each type of CD4 T cell produces cytokines that inhibit the other type of CD4 T cell. T$_H$2 cells make IL-10 and TGF-β (top left panel). IL-10 prevents macrophages from activating T$_H$1 cells, perhaps by preventing the synthesis of IL-12 by the macrophage. TGF-β acts directly on T$_H$1 cells to inhibit their growth (bottom left panel). T$_H$1 cells make IFN-γ (top right panel), which blocks the growth of T$_H$2 cells (bottom right panel).

(TGF)-β, which suppress the T$_H$1 response. In the less severe tuberculoid form of the disease, the cytotoxic T cells secrete IFN-γ, which helps polarize the CD4T cells to the more helpful T$_H$1 response that activates infected macrophages to kill the mycobacteria they contain.

8-20 Effector T cells are guided to sites of infection by newly expressed cell adhesion molecules

The activation of naive T cells and their differentiation into effector T cells takes 4–5 days. In the process, their complement of cell-surface adhesion molecules changes, enabling them to leave the secondary lymphoid tissue and travel to the site of infection. The expression of L-selectin, which enables naive T cells to enter lymph nodes, is lost, and the expression of other adhesion molecules is increased. A crucial change is the increased expression on effector T cells of the integrin VLA-4, which binds to VCAM-1 adhesion molecules induced on vascular endothelium by inflammatory cytokines (Figure 8.41).

The differential expression of adhesion molecules directs effector T cells to distinct anatomical sites. T cells activated in Peyer's patches by antigens from the gut leave in the lymph that drains the intestines, and after passing through mesenteric lymph nodes they ultimately reach the thoracic duct and

Figure 8.41 Changes in the adhesion molecules expressed by effector T cells enable them to leave the blood and enter infected tissues. As shown in the upper panel, naive T cells express L-selectin, which binds to the sulfated carbohydrates of vascular addressins (such as CD34) on high endothelial vessels in lymph nodes. As shown in the lower panel, effector T cells no longer express L-selectin but express VLA-4 and increased amounts of LFA-1. These adhesion molecules bind to VCAM-1 and ICAM-1 respectively, whose expression is induced on vascular endothelium at sites of inflammation. This change enables effector T cells to home to infected tissues.

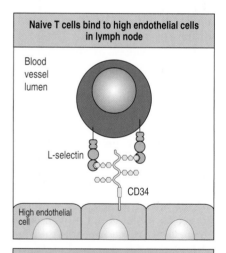

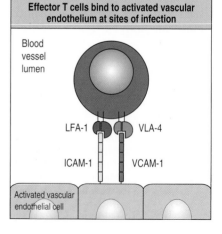

are drained into the blood. The effector T cells circulate in the blood and re-enter the mucosal tissue by recognizing the mucosal addressin **MAdCAM-1**, which is present on the endothelial cells of small blood vessels lining the walls of the gut and other mucosal surfaces. Within the lamina propria of the gut wall, effector T cells form small groups and aggregates as well as being scattered singly. With this pattern of lymphocyte traffic (Figure 8.42) an immune response initiated in one or a few Peyer's patches brings effector T cells to all mucosal tissues to prevent the spread of infection.

When effector T cells recognize specific antigen in the infected tissue they produce cytokines, such as TNF-α, which induce the expression of E-selectin, VCAM-1, ICAM-1, and the chemokine CCL5 (see Figure 8.16). These feed back onto the effector T cells to activate their adhesion molecules. The increased binding of endothelial VCAM-1 and ICAM-1 to T-cell VLA-4 and LFA-1 recruits more effector T cells into the inflamed tissue. The TNF-α and IFN-γ secreted by effector T cells work synergistically to enhance inflammation by changing the shape of vascular endothelial cells in ways that increase blood flow, vascular permeability, and supply of leukocytes and plasma to the infected tissue.

8-21 Antibody responses develop in lymphoid tissues under the direction of T$_H$2 cells

Whereas effector CD8 T cells and T$_H$1 cells work mainly at sites of infection, effector T$_H$2 cells activate B cells within the secondary lymphoid tissues, from whence secreted antibodies then travel in blood or lymph or across epithelia to the sites of infection. Early in an infection, pathogen-derived antigens are presented by the dendritic cells and macrophages in the draining lymph

Figure 8.42 Anatomy of mucosal immune responses. Naive lymphocytes enter the submucosal lymphoid tissues, such as a Peyer's patch, from the blood (left panel). Antigens (red triangles) derived from an infection of pathogenic organisms in the gut (red circles) are taken up into the Peyer's patch where they stimulate antigen-specific lymphocytes. The activated lymphocytes then leave the Peyer's patch in the draining lymph and return to the blood via the mesenteric lymph nodes and the thoracic duct (center panel). The activated lymphocytes circulate in the blood and enter all the submucosal lymphoid tissues, not just the ones in which the immune response originated (right panel). In this way all the mucosal tissues are defended against spread of the infection.

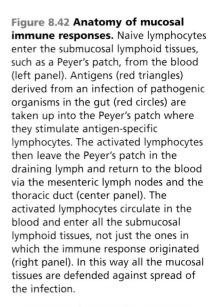

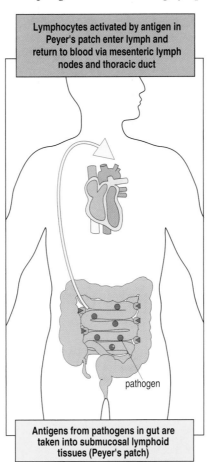

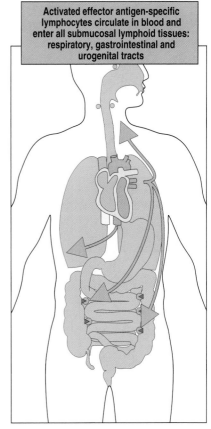

node. Clones of antigen-specific T cells are expanded within the T-cell areas of the node and some differentiate into effector T_H2 cells. Naive B cells also recirculate through the blood and lymph, entering lymph nodes by the same mechanisms used by naive T cells. B cells enter lymph nodes in the T-cell areas, where they bind and endocytose specific antigen by capturing it with their cell-surface immunoglobulin. The endocytosed antigen is then processed by the B cells and presented on MHC class II molecules to the effector T_H2 cells, which in response secrete the cytokines necessary for B-cell proliferation and differentiation.

After encounter with antigen, some 5 days are required for effector T_H2 cells to be produced from naive T cells. At about this time, antigen-specific B cells in contact with T_H2 cells begin to proliferate in the T-cell area to form a primary focus. Some of the B cells activated in these primary foci migrate to the medullary cords of the node, where they become short-lived plasma cells that produce a first wave of pathogen-specific antibody (Figure 8.43, top row). The corresponding B cells in the spleen move to those parts of the red pulp directly adjoining the T-cell zones of the white pulp. In these sites, B-cell numbers increase exponentially for 2–3 days; the cells undergo six or seven cell divisions before their progeny differentiate into antibody-producing plasma cells. Most of these plasma cells die by apoptosis after 2–3 days. Other

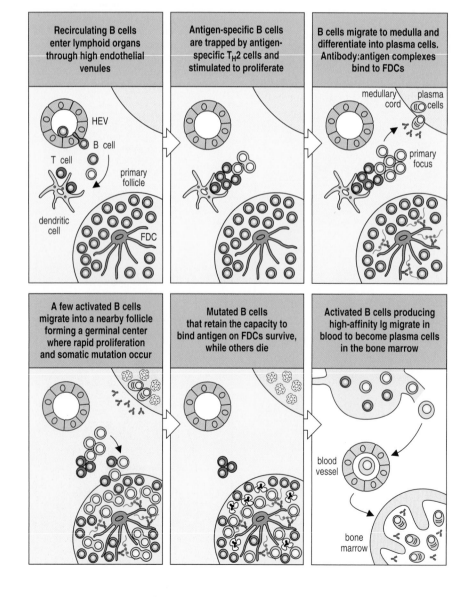

Figure 8.43 Secondary lymphoid tissues provide an environment in which antigen-specific B cells can interact with helper T_H2 cells specific for the same antigen. Top row: Naive B cells leave the blood at high endothelial venules (HEV) and meet effector T_H2 cells specific for the same antigen in the T-cell areas of the lymphoid tissues (a lymph node is illustrated here). B cells that engage with their cognate T_H2 cells become activated and proliferate. Some of the activated B cells move to the medullary chords of the lymph node, where they divide further and differentiate into plasma cells that over a few days secrete antibodies of the first wave. Bottom row: Other activated B cells migrate with their effector T_H2 cells to primary lymphoid follicles, where they are stimulated both by the cytokines released by the T_H2 cells and by the antigen:antibody complexes bound to the surface of follicular dendritic cells (FDC). The resultant B-cell proliferation forms a germinal center, the site of somatic hypermutation and selection of B cells expressing mutant immunoglobulins with higher affinity for antigen. The surviving B cells move to the medullary cords of the lymph node or migrate to the bone marrow, where they become plasma cells secreting a second wave of antibodies with higher overall affinity than those in the first wave.

activated B cells migrate to the follicles, where they proliferate further, forming a germinal center in which the B cells undergo somatic hypermutation (see Figure 8.43, bottom row).

The antibodies produced by the first wave of plasma cells form immune complexes with antigen and activate complement. Some of the immune complexes bind to follicular dendritic cells in the lymphoid follicles (see Figure 8.43). The variant antibodies produced by the mutated B cells are tested against the antigen in the immune complexes collected on the follicular dendritic cells. B cells having the highest-affinity antigen receptors compete most effectively for antigen and will be selected to survive, to be helped by CD4 T cells, and to become plasma cells (see Figure 7.12, p. 190). They secrete a second wave of antibody with a higher overall affinity for antigen than the antibody of the first wave.

8-22 Antibody secretion by plasma cells occurs at sites distinct from those at which B cells are activated by T_H2 cells

As we saw in the previous section, the first B cells to differentiate into plasma cells migrate to the medullary cords in lymph nodes or to the red pulp of the spleen. The other activated B cells enter the lymphoid follicles and form a germinal center. About 10% of the activated B cells in germinal centers leave as immature plasma cells called **plasmablasts**, and migrate to distant sites, where they differentiate into plasma cells that live, on average, for about a month. The remaining 90% die in the germinal center, having failed to capture antigen and cognate T-cell help (see Figure 7.12, p. 190).

Plasmablasts originating in the follicles of Peyer's patches and mesenteric lymph nodes migrate by the same route as the effector T cells that originate at those sites. They travel in draining lymph to the blood, in which they circulate, and then leave the blood through the walls of small blood vessels to enter the lamina propria of the gut and other epithelia. Plasmablasts originating in peripheral lymph nodes migrate to the medullary cords or the bone marrow, and those originating in the spleen migrate to the bone marrow (Figure 8.44). In immune responses that successfully terminate an acute infection, germinal centers are present for only 3–4 weeks after the supply of antigen is turned off. After this time, small amounts of antigen are retained on follicular dendritic cells and these continue to stimulate the proliferation of B cells in follicles. These B cells are likely to be the precursors of pathogen-specific plasma cells that can be found in the mucosa and bone marrow over periods of months to years. They are thus the source of the antibodies that provide protective immunity.

The primary adaptive immune response terminates most infections that elude the innate immune response. However, within the population suffering from an infectious disease, some individuals succumb to its effects before the

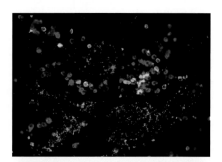

Figure 8.44 Plasma cells are dispersed in medullary cords of lymph nodes and in bone marrow. In these sites they secrete antibody directly into the blood for distribution to the rest of the body. In the upper micrograph, plasma cells in lymph node medullary cords are stained green (with fluorescein anti-IgA) if they are secreting IgA, and red (with rhodamine anti-IgG) if they are secreting IgG. The lymphatic sinuses are outlined by granular staining selective for IgA. In the lower micrograph, plasma cells in the bone marrow are revealed with light-chain specific antibodies (fluorescein anti-λ and rhodamine anti-κ stain). Plasma cells secreting immunoglobulins containing λ light chains are stained yellow-green; those secreting immunoglobulins containing κ light chains stain red. Photographs courtesy of P. Brandtzaeg.

	Cytotoxic CD8 T cell	T$_H$1 cell	T$_H$1/T$_H$2 cell
Typical pathogens	Vaccinia virus Influenza virus Rabies virus *Listeria*	*Mycobacterium tuberculosis* *Mycobacterium leprae* *Leishmania donovani* *Pneumocystis carinii*	*Clostridium tetani* *Staphylococcus aureus* *Streptococcus pneumoniae* Polio virus *Pneumocystis carinii*
Location	Cytosol	Macrophage vesicles	Extracellular fluid
Antigen recognition	Peptide:MHC class I on infected cell	Peptide:MHC class II on infected macrophage	Peptide:MHC class II on antigen-specific B cell
Effector action	Killing of infected cell	Activation of infected macrophages	Activation of specific B cell to make antibody

Figure 8.45 The role of the three kinds of effector T cell in the response to typical pathogens.

immune system can gain the upper hand. Those at greatest risk are the young, the old, the malnourished, and those already suffering from other diseases. Some infections can never be cleared, because certain pathogens successfully evade or subvert the immune response so that they can persist in the body for life. In this latter category are the protozoan parasites *Leishmania* and *Toxoplasma*, and the herpes viruses, including Epstein–Barr virus, herpes simplex virus, and cytomegalovirus. Depending on the way in which a pathogen exploits the human body, the type of immune response that eliminates or contains it is different (Figure 8.45).

Summary

Most infections are highly localized and can be cleared by the innate immune response. Only if infection spreads to secondary lymphoid tissues, or its antigens are carried there by dendritic cells, is an adaptive immune response made, and its initiation is dependent on the inflammatory cytokines produced by the cells of innate immunity. In the secondary lymphoid tissues the presentation of pathogen-derived antigens to naive circulating T cells causes pathogen-specific T cells to divide and differentiate into effector T cells within the lymphoid tissue. The relative production of T$_H$1 and T$_H$2 effector CD4 T cells and of CD8 T cells depends on the nature of the infection and the mix of inflammatory cytokines produced at the site of infection. Effector T$_H$1 cells and CD8 T cells leave the lymphoid tissues and travel in the lymph and blood to the infected tissues. Once there, T$_H$1 cells activate the destruction of extracellular pathogens by macrophages, while CD8 T cells kill infected human cells. Effector T$_H$2 cells remain mainly in the lymphoid tissues, where they help to activate pathogen-specific B cells and drive isotype switching and affinity maturation of their immunoglobulins. There are differences in the delivery of antigen and in lymphocyte traffic for infections arising at mucosal surfaces compared to skin infections.

Immunological memory and the secondary immune response

As well as clearance of the infection, a successful primary adaptive immune response also establishes a state of long-term **protective immunity**. A second or subsequent encounter with the same pathogen will provoke a faster, stronger, **secondary immune response**. This is produced by circulating antibody and by clones of long-lived B and T cells that have been formed during the primary response. In a person with protective immunity, the infection is usually cleared before it produces any symptoms. The purpose of vaccination is to produce a state of protective immunity against a particular pathogen before the pathogen itself is encountered.

8-23 Immunological memory after infection is long lived

In the period immediately after recovery from an infection, a person will usually encounter the pathogen again as it is passed around family, friends, colleagues, and other members of the community. Such reinfections are terminated quickly by the effector cells and antibodies built up during the first infection. Specific antibodies present in blood and extracellular fluids as a result of a primary immune response are one of the first immune defenses encountered by a pathogen on a subsequent infection. Their rapid action prevents the infection from becoming established. As a result of the reinfection, new supplies of antibody and effector cells will be made. In the absence of further infection the levels of antibody and effector T cells gradually decline, eventually reaching very low levels that are no longer effective. However, the B- and T-cell populations retain a 'memory' of the infection that enables a rapid and powerful secondary adaptive immune response to be made should reinfection occur, even many years after the primary infection. Memory is sustained by populations of pathogen-specific memory B and T cells that persist over long periods of time and can be reactivated when confronted with the pathogen. Memory B and T cells are activated more quickly and in greater numbers than naive lymphocytes were in the primary response. As a consequence, the infection is cleared quickly by the secondary response, with few or no symptoms of disease (Figure 8.46).

This phenomenon of **immunological memory** is classically illustrated by epidemiological studies of the inhabitants of one of the Faroe Islands in the North Atlantic Ocean, who were severely affected by a measles epidemic that infected the entire population. Many years later, when measles virus was again introduced to the island, only those members of the population born after the original epidemic developed disease. All those who had survived the first measles epidemic retained a protective immunity that prevented their second measles infection from becoming established and causing disease.

8-24 Pathogen-specific memory B cells are more abundant and make better antibodies than naive B cells

During a primary infection, proliferation of antigen-specific B cells produces plasma cells to confront the ongoing infection and **memory B cells** to deal with future infections. A few weeks after a primary infection has been cleared, the number of memory B cells reaches a maximum, which is sustained for

Figure 8.46 Protective immunity consists of preformed antibodies and existing effector T cells, and of immunological memory. The abundance of pathogen-specific antibodies and effector T cells is tracked here through the course of a person's successive exposures to a pathogen. During first infection the level rises, reaching a plateau as the immune system gains control. After clearance the levels decline, but for a time are sufficient to quickly terminate any reinfection. After an extended period without exposure to the pathogen, levels decline to the point where protective immunity is mediated by memory B and T cells, which respond rapidly to infection by producing new supplies of antibody and effector T cells.

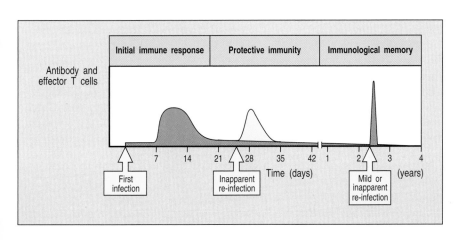

life. On secondary infection, 10–100 times more pathogen-specific B cells respond than did naive B cells in the primary response. Not only is the number of pathogen-specific B cells greater in the secondary response, the overall affinity of the antibodies they make is increased. Memory B cells are also distinguished from naive B cells by their ability to make antibodies of isotypes other than IgM. The population of memory B cells generated by the primary response includes cells that have switched to IgG, IgA, and IgE, so these isotypes emerge at the very beginning of a secondary response.

Both the amount of antibody and its average affinity continue to increase upon successive infections with, or vaccinations against, a pathogen (Figure 8.47). Increase in affinity is much greater for IgG than for IgM, because the IgG derives from isotype-switched B cells that have undergone somatic hypermutation and affinity maturation in germinal centers (see Section 7-4, p. 188). IgM, in contrast, is produced chiefly from B cells that differentiate early in the immune response and retain unchanged the antigen-binding sites of the naive B-cell receptors. The progressive increase in the affinity of antibody in the blood is therefore due to the replacement of IgM with IgG and IgA having mutated antigen-binding sites of higher affinity (see Section 7-4, p. 188).

The higher affinity of their antigen receptors makes memory B cells more efficient than naive B cells in binding and internalizing antigen for processing and presentation to helper T cells. Memory B cells also express high levels of MHC class II molecules on their surface compared with naive B cells, which also makes their presentation of antigen more efficient. These factors enable the secondary response to be more sensitive to the presence of antigen. Smaller pathogen populations are sufficient to trigger a B-cell response, which will therefore occur at a time earlier during an infection than was the case for the primary response.

By the generation and maintenance of populations of antigen-specific memory B cells, the immune system makes a heavy investment in protection against re-exposure to infections already experienced and survived. For these, a strong memory B-cell response can be recalled years after infection. There is an obvious advantage to this strategy. Infections are passed from person to person within human populations, often over extended periods, and the likelihood of repeated exposure to the same pathogen remains high. Antibodies produced in the primary response remain in the blood and lymph for some time after the infection has cleared. They are important contributors to a secondary response because they can bind to pathogens immediately the pathogen enters the body. The strength of B-cell memory is such that even with organisms that produce high mortality on first infection, for example the smallpox virus, the mortality on secondary exposure is very low.

8-25 T-cell memory is maintained by T cells that have different cell-surface markers from naive T cells

Memory T cells have been harder to study than memory B cells, because T cells do not undergo isotype switching and somatic hypermutation, processes that indelibly mark memory B cells as different from effector cells of the primary response. In the absence of cell-surface markers that absolutely distinguish memory T cells from effector T cells, the distinction is made using several cell-surface proteins that are expressed at distinguishable levels or in different forms on the surface of naive cells and memory cells (Figure 8.48). Of these changes, those affecting the tyrosine phosphatase CD45 are particularly important, both for T-cell function and for distinguishing naive and memory T cells. Isoforms of CD45 having differently sized extracellular domains are produced by alternative patterns of mRNA splicing. Naive T cells predominantly express the CD45RA isoform, which has a larger extracellular domain associated with

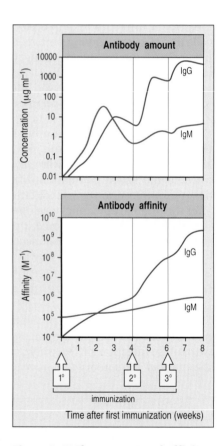

Figure 8.47 The amount and affinity of antibody increase after successive immunizations with the same antigen. The upper panel shows the increase in the amount of antibody present in blood serum over a course of three immunizations with the same antigen. The relative amounts of the IgM (green) and IgG (blue) isotypes are shown. The lower panel shows the changes in average antibody affinity that occur. Note that the vertical axis of each graph has a logarithmic scale.

Cell-surface molecule	Other names	Relative expression on cells of indicated subset		Comments
		Naive	Memory	
LFA-3	CD58	1	>8	Ligand for CD2, involved in adhesion and signaling
CD2	T11	1	3	Mediates T-cell adhesion and activation
LFA-1	CD11a/CD18	1	3	Mediates leukocyte adhesion and signaling
α_4 integrin	VLA4	1	4	Involved in T-cell homing to tissues
CD44	Ly24 Pgp-1	1	2	Lymphocyte homing to tissues
CD45RO		1	30	Lowest molecular weight isoform of CD45
CD45RA		10	1	High molecular weight isoform of CD45
L-selectin		High	Most high, some low	Lymph node homing receptor
CD3		1.0	1.0	Part of T-cell receptor complex

Figure 8.48 **Naive and memory T cells differ in their surface phenotype.** Many of the cell-surface molecules that distinguish memory T cells from naive T cells are also shared with effector T cells. The changes serve to increase the adhesion of the T cell to antigen-presenting cells and to endothelial cells and also to increase the sensitivity of the memory T cell to antigen stimulation.

weaker signals in response to specific antigen; memory T cells predominantly express the CD45RO isoform which has a smaller extracellular domain associated with stronger signals in response to antigen (Figure 8.49).

Direct functional evidence for a memory T cell with distinctive properties comes largely from the study of CD8 cytotoxic T cells. Memory CD8 T cells differ from effector cytotoxic CD8 T cells in that they need an induction period

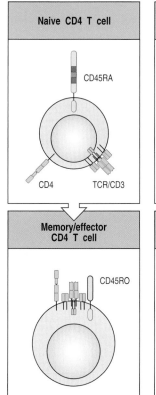

Naive CD4 T cell

CD45RA

CD4   TCR/CD3

Memory/effector CD4 T cell

CD45RO

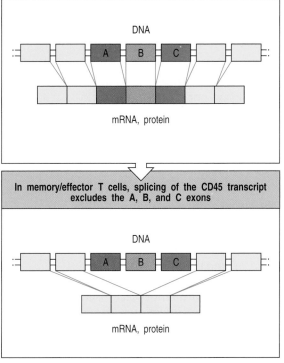

Splicing of the CD45 gene transcript in naive T cells includes the A, B, and C exons

DNA

A   B   C

mRNA, protein

In memory/effector T cells, splicing of the CD45 transcript excludes the A, B, and C exons

DNA

A   B   C

mRNA, protein

Figure 8.49 **Memory CD4 T cells express an altered CD45 isoform with modified interactions with the T-cell receptor and co-receptors.** CD45 is a transmembrane tyrosine phosphatase involved in T-cell activation. Through differential splicing of CD45 mRNA, the isoform of CD45 expressed by naive CD4 T cells, called CD45RA, is larger than the isoform expressed by memory CD4 T cells, called CD45RO. The absence of the sequences encoded by exons A, B, and C in CD45RO enables it to associate with both the T-cell receptor and the CD4 co-receptor and improve the efficiency of signal transduction.

Figure 8.50 Encounter with antigen generates effector T cells and long-lived memory T cells. Most of the effector T cells generated from antigen-stimulated naive T cells are relatively short lived, dying either from antigen overload or from lack of antigenic stimulus or necessary cytokines. Some, however, become long-lived memory T cells. In order to persist, the memory cells need periodic stimulation with cytokines; these are generated in the course of immune responses to antigens different from the antigens specific for the memory cells themselves.

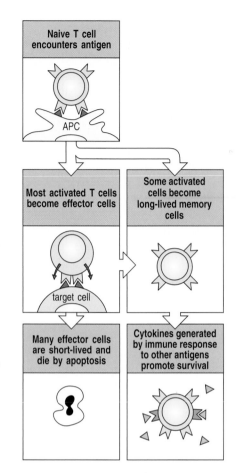

of 1–2 days before they have the capacity to lyse target cells. Effector CD8 T cells, on the other hand, can program a target cell for lysis in a matter of minutes. Memory CD8 T cells are also unlike naive CD8 T cells because their induction period is shorter than that for naive T cells and no cell division is involved. Although memory CD8 T cells can be stimulated without help from CD4 T cells, the quality and quantity of the resulting response is improved by the participation of CD4 T cells.

The evidence for memory CD4 T cells is less direct, principally because helper function (for example, the stimulation of antibody production by B cells) requires several days after antigen stimulation to become measurable, a period comparable to that needed for the presumed induction. By contrast, the cytolytic function of CD8 T cells can be determined within a few hours after completion of the induction period. What has been defined is a subset of CD4 T cells in immunized animals that have the characteristic cell-surface marker molecules of effector cells but different activation requirements. Unlike effector CD4 T cells, these presumed memory T cells can be stimulated to secrete effector cytokines only by antigen presented by a professional antigen-presenting cell.

A healthy adult human has 10^{12} peripheral $\alpha{:}\beta$ T cells, of which half are naive T cells and half are memory T cells. From sequence analysis of T-cell receptors it is estimated that the naive T cells have 2.5×10^7 antigen specificities and the memory T cells have only 1.5×10^5 antigen specificities. The acquisition of T-cell memory for a pathogen means that, on average, 100-fold more T cells will respond to a secondary infection with a pathogen than responded to the primary infection.

8-26 Maintenance of immunological memory does not require stimulation with antigen

A general property of lymphocytes is their requirement for regular stimulation in order to survive. If such survival signals are not received, then lymphocytes die through apoptosis. For naive circulating T cells, survival signals are generated through interactions of the T-cell receptor with complexes of MHC and self-peptides on the surface of healthy cells. These interactions are similar to those that caused positive selection of the lymphocytes during development in the thymus.

Memory CD8 T cells do not receive their survival signals from receptor-mediated interactions with peptide:MHC complexes but through antigen-independent stimulation involving the cytokine IL-15, which is induced as part of the interferon response. The cellular receptor for IL-15 has two of its polypeptides in common with the IL-2 receptor, and the two cytokines have complementary properties: IL-15 causes memory cells to proliferate, whereas IL-2 inhibits their proliferation. It seems that memory CD8 T-cell populations are maintained in the absence of secondary infection with specific pathogen, and this is achieved by continuing cell division and turnover, which is stimulated by cytokines produced as a general result of any immune activity (Figure 8.50). By this process, which is also implicated in sustaining the memory B-cell population, the maintenance of immunological memory does not require the presence of specific antigen or its persistence in the body following resolution of infection.

Although the overall size of the memory CD8 T-cell population remains stable, the number of memory cells that represent memory to a particular pathogen changes with time and history of infection. To accommodate the memory T cells acquired upon infection with a current virus, some of the memory T cells corresponding to previous infections with other viruses will be lost. Such losses can be recouped on subsequent reinfections. Although recent or recurrent infections will be better represented in the memory-cell population, this does not unduly compromise the memory response to other infections: elderly humans can make effective secondary responses to pathogens which were only encountered in childhood.

8-27 The second and subsequent responses to a pathogen are mediated solely by memory lymphocytes and not by naive lymphocytes

The secondary adaptive immune response to a pathogen involves the reactivation of clones of B and T cells that were first activated and expanded during the primary response. Activation of naive lymphocytes with specificity for the pathogen does not occur in a secondary immune response. This is because the cells and molecules made during the primary response prevent the activation of naive lymphocytes. This phenomenon is put to practical use in prevention of the **hemolytic anemia of the newborn** caused by antibodies specific for the rhesus (Rh) antigen on erythrocytes. If a Rh$^-$ mother carrying a Rh$^+$ child is infused with anti-Rh IgG antibody before she begins to make an immune response to her child's red cells, the response will be inhibited. Fetal red cells that enter the maternal circulation bind the anti-Rh IgG antibodies. When these red cells subsequently bind to the surface immunoglobulin of naive maternal B cells that are Rh specific, a cross-linking of B-cell receptors and Fc receptors occurs. This cross-linking delivers an inhibitory signal to the naive B cell. In contrast, when the same antigen:antibody complexes interact with the antigen receptors and the Fc receptors on the surface of memory B cells, an activating signal is delivered (Figure 8.51).

Such a mechanism, whereby the primary response to a pathogen or antigen prevents the further activation of naive cells during a secondary response, is believed to underlie the phenomenon of **original antigenic sin**. This implicitly puritanical term was coined on the basis of observations that the first influenza strain encountered by a person constrained the types of antibody

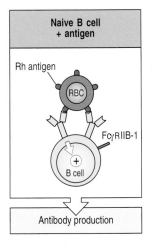
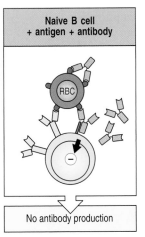
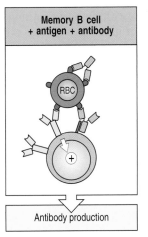

Figure 8.51 IgG antibody suppresses the activation of naive B cells by cross-linking the B-cell receptor to the receptor FcγRIIB-1 on the B-cell surface. Antigen binding to the B-cell antigen receptor delivers an activating signal (left panel) via the B-cell receptor; simultaneous signaling via the antigen receptor and FcγRIIB-1 delivers a negative signal to naive B cells (middle panel). Such cross-linking does not affect memory B cells (right panel). In this example the antigen is the Rh$^+$ alloantigen of erythrocytes. In pregnancy a fetus that is Rh$^+$ can stimulate an Rh$^-$ mother to make Rh$^-$-specific antibody. This response can be inhibited by prior infusion of some anti-Rh IgG antibody.

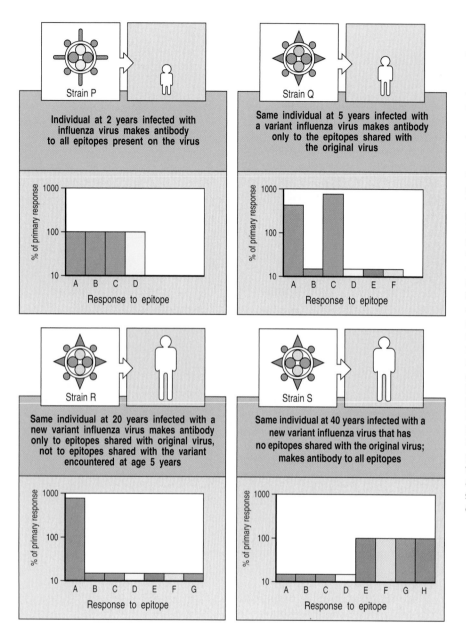

Strain P

Individual at 2 years infected with influenza virus makes antibody to all epitopes present on the virus

Strain Q

Same individual at 5 years infected with a variant influenza virus makes antibody only to the epitopes shared with the original virus

Strain R

Same individual at 20 years infected with a new variant influenza virus makes antibody only to epitopes shared with original virus, not to epitopes shared with the variant encountered at age 5 years

Strain S

Same individual at 40 years infected with a new variant influenza virus that has no epitopes shared with the original virus; makes antibody to all epitopes

Figure 8.52 Highly mutable viruses such as influenza can escape from the protective immunity of the human host without eliciting new immunity to the changes in the virus. A child 2 years old is infected with influenza virus strain P and responds by making antibodies against all the viral epitopes: A, B, C, and D (top left panels). At 5 years old the child is infected with the Q strain of the virus, which shares epitopes A and C with strain P but has new epitopes E, and F. The child's immune system uses only memory B cells resulting from the earlier infection with strain P in defense against strain Q and so the response is restricted to antibodies against epitopes A and C (top right panels). At 20 years old, the same individual is infected with a third strain of influenza, R, which has epitope A in common with strain P, epitopes E and F in common with strain Q, and a new epitope, G. Even 18 years after the first influenza infection, the antibody response concentrates solely on epitope A, which is shared with strain P (bottom left panels). Only when the person is infected with influenza strain S, which lacks epitopes A, B, C, and D, can a primary response be made to epitopes E, F, G, and H of strain S (bottom right panels). Strain S is seen as a novel virus to the person's immune system and as a consequence the person will likely suffer a full blown influenza as a result of this infection.

made in response to future infections with different strains. During a second infection only those epitopes that are common to the infecting strain and the original strain stimulate antibody production. This dependence on the memory cells that developed during the first infection is seen even when subsequent infections involve influenza strains with highly immunogenic epitopes that are different from those on the original strain. The strategy of the immune system is to use only B cells that are rapidly and easily mobilized. This allows the highly mutable influenza virus to increasingly escape from the protective immunity of a host while the host's immune system is paralyzed and prevented from responding to the changes in the virus. As a result, protective immunity to influenza is not retained for life, and the virus successfully infects millions of human beings every year. The imprint made by original antigenic sin is broken only on infection with a strain of influenza that lacks all the B-cell epitopes of the strain to which a person was first exposed (Figure 8.52).

Summary

In the course of the primary immune response, effector T cells and antibodies accumulate. Once an infection has been cleared, these cells and molecules will prevent reinfection by the same pathogen in the short term. In the long term, persistent clones of memory B and T cells expanded during the primary response afford protective immunity if the same antigen is encountered. When that happens they mount a secondary response that is both stronger and more rapid than the primary response. Antibodies produced in the secondary response are of higher affinity and of isotypes other than IgM. Activation of naive lymphocytes is suppressed and only memory lymphocytes are reactivated. In a secondary immune response, the immune system, therefore, devotes all its resources to producing high-affinity antibodies and specific T cells that rapidly clear the invading pathogen before the infection can become established. The maintenance of memory cells does not appear to require the persistence of the original antigen; instead, survival signals for memory lymphocytes are provided by cytokines such as IL-15.

Summary to Chapter 8

The human body has three types of defense, all of which must be overcome if a pathogen is to establish an infection and then exploit its human host for the remainder of that person's life. The first defense is made up of the protective epithelial surfaces of the body, which successfully prevent most pathogens from ever gaining entry to the rich resources of the body's interior. Any pathogen that succeeds in penetrating an epithelial surface is immediately faced by the recognition molecules and effector cells of innate immunity. Some of these elements of immunity have been refined over hundreds of millions of years to respond to the distinguishing features of bacteria and other groups of pathogens. Viruses pose a different challenge, because they are made by human cells and, unlike bacteria, do not have chemically distinctive surface macromolecules that nonspecifically distinguish them from human cells. However, NK cells have receptor systems that can sense subtle changes in the surface of infected cells and, by killing the cells, eliminate or contain viral infections.

Most infections are efficiently cleared by the innate immune response and lead to neither disease nor incapacitation. In the minority of infections that escape innate immunity and spread from their point of entry, the pathogen then faces the combined forces of innate and adaptive immunity. The week or so required for the primary adaptive immune response to develop is the time during which the body is most vulnerable and during which infections can progress to the point of causing disease, damage, and even death. This delay, during which the pathogens have the upper hand and can be transmitted to other people, is largely responsible for the devastation caused by epidemic infections. Most infections, however, never reach that stage, but are successfully terminated by the adaptable and specific recognition systems of adaptive immunity. A successful recovery from infectious disease and the consequent development of immunological memory provide the body with enhanced defenses that will repel future attack from the causal agent (Figure 8.53). Once a state of immunity has been established, a subsequent encounter with the pathogen provokes a stronger and more rapid attack that clears the pathogen from the body before the infection can take hold.

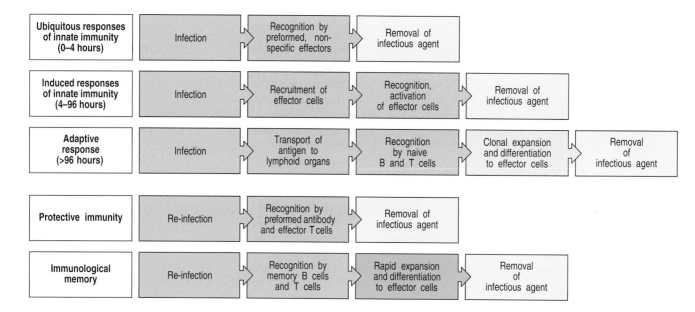

Figure 8.53 The immune response to an initial infection. The innate immune response can be divided into mechanisms that start immediately on infection, for example complement activation and macrophage phagocytosis, and those induced within a few hours by inflammatory cytokines. The effector functions of adaptive immunity begin to work only after 4 or more days. In the period after the termination of an infection, the effector cells and molecules made in the primary response continue to provide protective immunity but gradually decline. In the long term, immunological memory enables the strength of the secondary response to far surpass the primary response.

Questions

Question 8–1

Describe (a) three direct and (b) three indirect mechanisms through which pathogens cause damage to their hosts. (c) Give an example of an infectious agent and the disease it causes through each mechanism. (Refer to Figure 8.3.)

Question 8–2

When an infection establishes itself at a particular anatomical site the microenvironment influences which immune response will be induced and used to eradicate the pathogen.

    A. Name the specific extracellular and intracellular locations where infections can be established, indicating which are extracellular and which are intracellular.

    B. Viruses can be found both outside and inside cells. Describe the immune mechanisms used to eradicate them in each case.

    C. Which intracellular site is colonized by trypanosomes? What immune mechanism is used against them when in this location?

    D. Which class of antibodies would be most likely to act against pathogens that colonize epithelial surfaces? Give an example of such a pathogen. (Refer to Figure 8.4.)

Question 8–3

Surface epithelia lining the skin and respiratory, urogenital and gastrointestinal tracts provide protective barriers for the body against the external environment where pathogens are encountered.

    A. Describe the three main ways in which epithelia act as barriers to infection, giving details of the various mechanisms employed.

    B. How can antibiotics upset the barrier function of intestinal epithelia? Give a specific example. (Refer to Figures 8.6 and 8.7)

Question 8–4

    A. What are the main (i) similarities and (ii) differences in the general properties and roles of macrophages and neutrophils?

    B. How do they both destroy extracellular pathogens? Give details of the process. (Refer to Figures 8.20 and 8.21.)

Question 8–5

Describe in chronological order the four steps involved in the extravasation of neutrophils to infected tissue sites during an innate immune response. Use the following terms in your description: rolling adhesion, tight binding, diapedesis, migration, inflammatory mediators, integrins, adhesion molecules, chemokines, Weibel–Palade bodies, P-selectin, E-selectin, sialyl-Lewis^x and basement membrane protease. (Refer to Figure 8.19.)

Question 8–6

A. Using the table below match the local and systemic effects in Column A with the appropriate cytokine in Column B. Note that more than one answer in Column B may be used.

B. Which of these cytokines are produced by macrophages? Which cells produce the other(s)? (Refer to Figures 8.15, 8.22, 8.25, and 8.27.)

Column A	Column B
a. Activation of blood vessel endothelium	1. IL-1
b. Lymphocyte activation	2. IL-6
c. Fever	3. CXCL8
d. Induction of IL-6 synthesis	4. IL-12
e. Increase in vascular permeability	5. TNF-α
f. Localized tissue destruction	6. Type I interferons
g. Production of acute-phase proteins by hepatocytes	
h. Induction of resistance to viral replication	
i. Increase in levels of MHC class I molecules on cell surfaces	
j. Activation of NK cells	
k. Biased differentiation of naive T cells into T$_H$1 cells	
l. Enhanced antibody production	
m. Leukocyte chemotaxis	
n. Activation of binding by β_2 integrins (LFA-1, CR3)	
o. Septic shock	
p. Mobilization of metabolites	

Question 8–7

A. What induces type I interferon production?

B. Do normal cells produce this inducer? Why, or why not?

C. Identify three effects of type I interferons during viral infection. (Refer to Figures 8.25 and 8.26.)

Question 8–8

Natural killer cells (NK cells) carry so-called activating and inhibitory receptors on their surface.

A. What property of NK cells do these receptors activate and inhibit, respectively? Explain your answer.

B. How are NK cells thought to use these receptors to recognize and eliminate virus-infected cells?

C. Why are the actions of NK cells considered nonspecific and categorized as innate immunity?

D. Why do the NK cells of the recipient of an organ transplant sometimes attack the transplanted tissue? (Refer to Figures 8.29, 8.32, and 8.33.)

Question 8–9

A. Explain briefly how immunological memory operates in (i) the short term and (ii) the long term.

B. Why is a secondary immune response, which may occur many years after the first exposure to the pathogen, initiated much more rapidly than a primary immune response?

Question 8–10

List some of the differences in cell-surface molecules between naive T cells and memory T cells, and explain how these differences help memory T cells to mount a strong secondary immune response. (Refer to Figures 8.48 and 8.49.)

Question 8–11

During inflammation, host tissue may be damaged owing to the release of toxic oxygen derivatives produced by activated macrophages and neutrophils. Explain how host mechanisms are induced to limit these damaging bystander effects.

Question 8–12

Explain specifically how the systemic release of TNF-α by macrophages causes (a) septic shock, (b) disseminated intravascular coagulation, (c) organ failure, and (d) hemorrhaging during infection with blood-borne pathogens. (Refer to Figure 8.18.)

Question 8–13

A. In the context of antibody production, explain how the phenomenon of 'original antigenic sin' is exploited clinically to prevent hemolytic anemia of the newborn.

B. Which receptors must be cross-linked for this to occur? (Refer to Figure 8.51.)

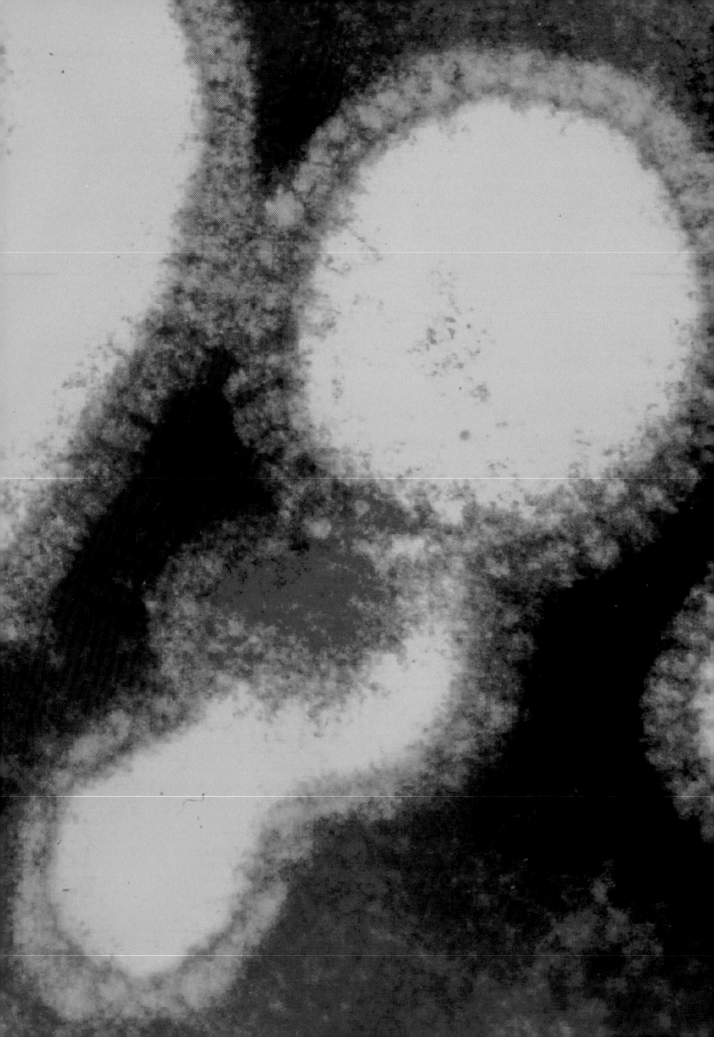

Chapter 9
Failures of the Body's Defenses

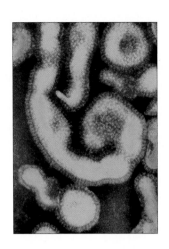

In Chapter 8 we saw how most pathogens that threaten the human body are prevented from establishing infection, and that those infections that do occur are usually terminated by the actions of innate and adaptive immunity. In this situation there is strong pressure on pathogens to evolve ways of escaping or subverting the immune response. Microorganisms with such advantages can compete more successfully with other potential pathogens to exploit the resources of the human body. The first part of this chapter describes examples of the different types of mechanism they use.

The body's defenses against infection can also fail because of inherited deficiencies of the immune system. Some of these are described in the second part of the chapter. Within the human population there are mutant alleles for many of the genes encoding components of the immune system. These mutant genes cause immunodeficiency diseases, which vary in severity depending on which gene is defective. Correlation of the molecular defects in immunodeficiency diseases with the types of infection to which patients become vulnerable reveals the effectiveness of the various arms of the immune response against different kinds of pathogen.

In the third part of the chapter we explore one particular host–pathogen relationship that combines themes from the first two parts of the chapter. This concerns the human immunodeficiency virus (HIV), which is extraordinarily effective at both escaping and subverting the immune response. During the course of an infection, which can last for decades, HIV gradually but inexorably wears down the immune system to the point where it no longer works. The long-term consequence of HIV infection is that patients become severely immunodeficient and develop the fatal disease known as acquired immune deficiency syndrome (AIDS).

Evasion and subversion of the immune system by pathogens

The immune response to any pathogen involves complex molecular and cellular interactions between the pathogen and its host, and any stage in this interaction could be targeted by a pathogen and used for its own benefit. The systematic study of pathogen genomes reveals that most, if not all, pathogens have means of escaping or subverting immune defenses, and that some of them have many genes devoted to this purpose.

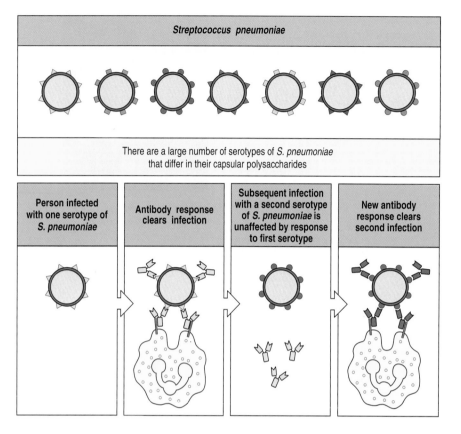

Figure 9.1 **Protective immunity towards *Streptococcus pneumoniae* is serotype-specific.** Different strains, or serotypes, of *S. pneumoniae* have antigenically different capsular polysaccharides, as shown in the upper panel. Antibodies against capsular polysaccharides opsonize the pathogen and enable it to be phagocytosed. A person infected with one serotype of *S. pneumoniae* clears the infection with type-specific antibody, as shown in the lower panels. These antibodies, however, have no protective effect when the same person is infected with a different type of *S. pneumoniae*. The second infection can be cleared only by a new primary response.

9-1 Genetic variation within some species of pathogen prevents effective long-term immunity

Antibodies directed against macromolecules on the surfaces of pathogens are the most important source of long-term protective immunity to many infectious diseases. Some species of pathogen evade such protection by existing as numerous different strains, which differ in the antigenic macromolecules on their outer surfaces. One such is the bacterium *Streptococcus pneumoniae*, which causes pneumonia. Different genetic strains of *S. pneumoniae* differ in the structure of the capsular polysaccharides and compete with each other to infect humans. These strains, of which at least 90 are known, are called **serotypes** because antibody-based serological assays are used to define the differences between them. After resolution of infection with a particular serotype, or type, of *S. pneumoniae*, a person will have made antibodies that prevent reinfection with that type but which will not prevent primary infection with another type (Figure 9.1). *S. pneumoniae* is a common cause of bacterial pneumonia because its genetic variation prevents individuals from developing an effective immunological memory against all strains. Genetic variation in *S. pneumoniae* has evolved as a result of selection by the immune response of its human hosts.

9-2 Mutation and recombination allow influenza virus to escape from immunity

Some viruses also display genetic variation, influenza virus being a well-studied example. This virus infects epithelia of the respiratory tract and passes easily from one person to another in the aerosols generated by coughs and sneezes. Protective immunity to influenza is provided principally by antibodies that bind to the hemagglutinin and neuraminidase glycoproteins of the viral envelope. These antibodies are made during the primary immune response to the virus. The course of a primary infection is short (1–2 weeks) and the virus is cleared from the system by a combination of cell-mediated immunity and antibody. The pattern of infection of influenza virus characteristically causes **epidemics**, in which the virus spreads rapidly through the population and then quickly subsides. Long-term survival of the influenza virus is ensured by the generation of new viral strains that evade the protective immunity generated during past epidemics.

Influenza is an RNA virus with a genome consisting of eight RNA molecules. RNA replication is relatively error-prone and generates many point mutations on which selection can act. New viral strains that lack the hemagglutinin or neuraminidase epitopes that induced protective immunity in the previous epidemic regularly emerge and cause an influenza epidemic every other winter or so. An individual's protective immunity against influenza is determined by the strain of virus to which they were first exposed—the phenomenon of 'original antigenic sin' (see Section 8-27, p. 273). The history of exposure to particular strains of the virus differs within the population, largely according to age, and so there are subpopulations of people with differing degrees of immunity to the current strain of influenza. The people to suffer most at any particular time will be those whose protective immunity has been lost because of the new mutations present in the current strain. This type of evolution of influenza, which causes relatively mild and limited disease epidemics, is called **antigenic drift** (Figure 9.2, left panels).

In contrast, every 10–50 years an influenza virus emerges that is structurally quite different from its predecessors and is able to infect almost everyone. Besides spreading more widely to cause a **pandemic** (a worldwide epidemic), such viruses cause more severe disease and a greater number of deaths than do the viruses that result from antigenic drift. The influenza strains that cause pandemics are recombinant viruses that have some of their RNA genome derived from an avian influenza virus and the remainder from a human influenza virus. In these recombinant strains, the hemagglutinin and/or the neuraminidase are encoded by RNA molecules of avian origin and are antigenically very different from those against which people have protective immunity. New pandemic strains often arise in parts of south-east Asia where farmers live in close proximity to their livestock such as pigs, chickens, and ducks. One theory is that the recombinant viruses arise in pigs that have become simultaneously infected with both avian and human viruses. If such a recombinant jumps back into the human population it has a tremendous competitive advantage, and in sweeping through the human population will rapidly replace other viral strains. Recombinant influenza viruses can similarly sweep through bird populations and are greatly feared by poultry farmers. This mode of evolution is called **antigenic shift** (see Figure 9.2, right panels).

9-3 Trypanosomes use gene rearrangement to change their surface antigens

Mutation and recombination are not the only means by which pathogens can change the face they present to the immune system. Certain protozoans regularly change their surface antigens by a process of gene rearrangement, the most striking example being the African trypanosomes (for example, *Trypanosoma brucei*), which cause sleeping sickness. The trypanosome life cycle involves both mammalian and insect hosts. Insect bites transmit trypanosomes to humans, in whom the parasites replicate in the extracellular spaces. The trypanosome's surface is formed of a glycoprotein, of which there are numerous variants, each encoded by a different gene. The trypanosome

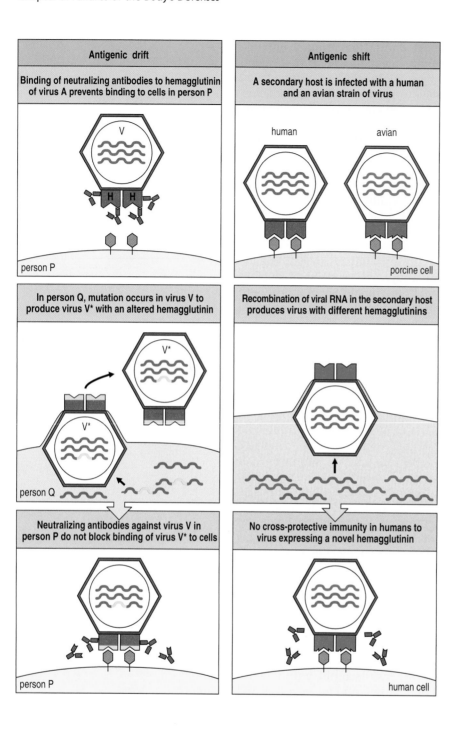

Figure 9.2 Variation in influenza virus due to mutation and recombination allows repeated infections. The left panels show the process of antigenic drift in the influenza virus. Only the viral hemagglutinin (H) is shown for simplicity. A person P infected with influenza strain V produces neutralizing antibodies against the viral hemagglutinin. When person P is further challenged with strain A, these antibodies bind to the hemagglutinin and prevent the virus from infecting cells (top panel). Person P is thus immune to strain V. Influenza virus mutates frequently, as depicted in the center panel where point mutation (yellow) in the hemagglutinin gene of strain V produces strain V* in which the hemagglutinin epitopes recognized by the preexisting protective antibodies are eliminated. This shift allows strain V* of influenza virus to infect cells of person P without obstruction from the antibodies previously made against strain V (bottom panel). For P to clear infection with strain V* requires a primary immune response to this strain with development of specific neutralizing antibodies. The right panels show the process of antigenic shift. A human influenza virus (red) and an avian influenza virus (blue) simultaneously infect a cell within the secondary host, a pig (top panel), and their RNA segments become reassorted to produce a recombinant virus (center panel). The recombinant virus expresses hemagglutinin of avian origin that is antigenically very different from that in the original human virus. The recombinant virus readily infects humans because antibodies protective against the original hemagglutinin cannot bind to the new hemagglutinin (bottom panel). For simplicity, only three of the eight RNA molecules are drawn.

genome contains more than 1000 genes encoding these **variable surface glycoproteins** (**VSG**s). At any time, an individual trypanosome produces only one form of VSG. This is because rearrangement of a VSG gene into a unique site in the genome—the expression site—is required for its expression. Rearrangement occurs by a process of gene conversion (Figure 9.3). The vast majority of the rapidly replicating trypanosomes that emerge after initial infection will express the same dominant form of VSG. A very small minority will, however, have changed the expressed VSG gene, and now express other forms. The host makes an antibody response to the dominant form of VSG, but not to the minority forms. Antibody-mediated clearance of trypanosomes expressing the dominant VSG facilitates the growth of those expressing the minority forms, of which one will come to dominate the trypanosome population. In time, the numbers of trypanosomes expressing the new dominant form are sufficient to stimulate antibody production, whereupon they too are cleared and yet other variants selected.

This mechanism of immune evasion causes trypanosome infections to produce a dramatic cycling in the number of parasites within an infected person (see Figure 9.3, bottom panel). The chronic cycle of antibody production and antigen clearance leads to a heavy deposition of immune complexes and inflammation. Neurological damage occurs and eventually leads to coma, the so-called sleeping sickness. Trypanosome infections are a major health problem for humans and cattle in large parts of Africa. Indeed it is largely because of trypanosomes that wild populations of big game animals still survive in Africa and have not been replaced by domestic cattle. Malaria, another disease caused by a protozoan parasite that escapes immunity by varying its surface antigens, is also a major cause of human mortality in equatorial Africa.

Trypanosomes change their coats by gene rearrangements similar to the gene conversions used by chicken B lymphocytes to produce their primary repertoire of immunoglobulins (see Section 4-1, p. 100). Similar strategies of antigenic variation are also used by several species of bacteria whose ability to escape from the human immune response makes them successful pathogens and major public-health problems. *Salmonella typhimurium*, a common cause of food poisoning, can alternate expression of two antigenically distinct flagellins, proteins of the bacterial flagella. This occurs by reversible inversion of part of the promoter of one of the flagellin genes, which inactivates that gene and allows the expression of the second gene. *Neisseria gonorrhoeae*, the cause of the widespread sexually transmitted disease gonorrhea, has several variable antigens, the most impressive being the pilin protein, a component of the adhesive pili on the bacterial surface. Like the VSGs of African trypanosomes, pilin is encoded by a family of variant genes, only one of which is expressed at a time. Different versions of the pilin gene introduced into the expression site provide a minority population of variant bacteria. When the host's immune response places pressure on the dominant type, another is ready to take its place.

9-4 Herpes viruses persist in human hosts by hiding from the immune response

To terminate an established viral infection, infected cells must be killed by cytotoxic CD8 T cells. For this to occur, some of the peptides presented by MHC class I molecules at the surface of infected cells must be of viral origin, a condition easily fulfilled by rapidly replicating viruses such as influenza. Consequently, influenza infections are efficiently cleared by the immune system by a combination of cytotoxic T cells and antibodies, the latter neutralizing extracellular virus particles. In contrast, some other viruses are difficult to clear because they enter a quiescent state within human cells, one in which they neither replicate nor generate enough virus-derived peptides to signal their presence to cytotoxic T cells. Development of this dormant state, which is called **latency** and does not cause disease, is a favored strategy of the herpes viruses. Later on, when the initial immune response has subsided, the virus will reactivate, causing an episode of disease.

Herpes simplex virus, the cause of cold sores, first infects epithelial cells and then spreads to sensory neurons serving the area of infection. The immune response clears virus from the epithelium, but the virus persists in a latent state in the sensory neurons. Various stresses can reactivate the virus, including sunlight, bacterial infection, or hormonal changes. After reactivation, the virus travels down the axons of the sensory neurons and reinfects

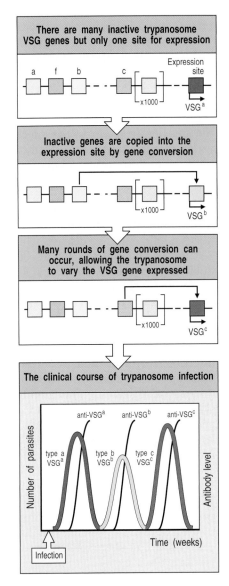

Figure 9.3 Antigenic variation by African trypanosomes allows them to escape from adaptive immunity. In the top panel, the VSG^a gene (shown in red) is in the expression site and the VSG^b gene (yellow) and VSG^c gene (blue) are inactive. In the second panel, gene conversion has replaced VSG^a with VSG^b at the expression site; in the third panel, VSG^c has replaced VSG^b at the expression site. The bottom panel shows how the patient's antibody response (thin black lines) to the VSG proteins selects for low-frequency variants and causes a cycle in which parasite populations (red, yellow, and blue lines) alternate between boom and bust over a number of weeks.

the epithelial tissue (Figure 9.4). Viral replication in epithelial cells and production of viral peptides restimulates CD8 T cells, which kill the infected cells, creating a new sore. This cycle can be repeated many times throughout life. Neurons are a favored site for latent viruses to lurk because they express very small numbers of MHC class I molecules, further reducing the potential for presentation of viral peptides to CD8 T cells.

The herpes virus varicella-zoster (also called herpes zoster) remains latent in one or a few ganglia, chiefly dorsal root ganglia, after the acute infection of epithelium—chickenpox—is over. Stress or immunosuppression can reactivate the virus, which moves down the nerve and infects the skin. Reinfection causes the reappearance of the classic varicella rash of blisters, which cover the area of skin served by the infected ganglia. The disease caused by reactivation of varicella-zoster is commonly known as shingles. In contrast to herpes simplex virus, reactivation of varicella-zoster usually occurs only once in a lifetime.

A third herpes virus that causes a persistent infection is the Epstein–Barr virus (EBV), to which most humans are exposed. First exposure in childhood produces a mild cold-like disease, whereas adolescents or adults encountering EBV for the first time develop infectious mononucleosis (also known as glandular fever), an acute infection of B lymphocytes. EBV infects B cells by binding to the CR2 component of the B-cell co-receptor complex (see Figure 7.3, p. 184). Most of the infected B cells proliferate and produce virus, leading in turn to the stimulation and proliferation of EBV-specific T cells. The result is an unusually large number of mononuclear white blood cells (lymphocytes, mostly T cells), which gives the disease its name. After some time, the acute infection is brought under control by CD8 cytotoxic T cells, which kill the virus-infected B cells. The virus persists in the body, however, because a minority of B cells become latently infected. This involves shutting off the synthesis of most viral proteins except EBNA-1, which maintains the viral genome in these cells. Latently infected cells do not present a target for attack by CD8 cytotoxic cells because the proteasome is unable to degrade EBNA-1 into peptides that can be bound and presented by MHC class I molecules.

After recovery from the initial exposure to EBV it is unusual for reactivation of the virus to lead to disease. It seems likely that CD8 T cells quickly control episodes of viral reactivation. In immunosuppressed patients, however, reactivation of the virus can cause a disseminated EBV infection, and infected B cells can also undergo malignant transformation, causing B-cell lymphoproliferative disease.

9-5 Certain pathogens sabotage or subvert immune defense mechanisms

Pathogens also exploit the immune-system cells that are ranged against them. For example, *Mycobacterium tuberculosis* commandeers the macrophage's pathway of phagocytosis for its own purposes. On being phagocytosed, *M. tuberculosis* prevents fusion of the phagosome with the lysosome, thus protecting itself from the bactericidal actions of the lysosomal contents. It then survives and flourishes within the cell's vesicular system. *Listeria monocytogenes*, in contrast, escapes from the phagosome into the macrophage's cytosol, where it grows and replicates. However, the intracytosolic way of life elicits cytotoxic CD8 T-cell responses against *L. monocytogenes*, which eventually terminate the infection.

The protozoan parasite *Toxoplasma gondii*, the cause of **toxoplasmosis**, creates its own specialized environment within the cells that it infects. It

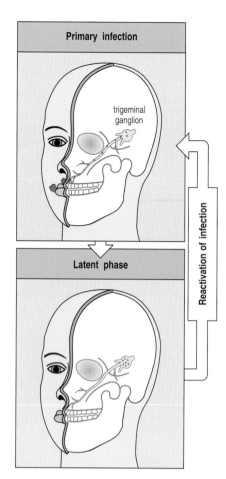

Figure 9.4 Persistence and reactivation of herpes simplex virus infection. The initial infection around the lips is cleared by the immune response and the resulting tissue damage is manifest as cold sores (upper panel). The virus (small red dots) has meanwhile entered sensory neurons, for example those in the trigeminal ganglion whose axons innervate the lips, where it persists in a latent state (lower panel). Various types of stress can cause the virus to leave the neurons and reinfect the epithelium, once again reactivating the immune response and causing cold sores. People infected with herpes simplex viruses periodically get cold sores as a result of this process. During its active phase the virus can be passed from one person to another.

surrounds itself with a membrane-bounded vesicle that does not fuse with other cellular vesicles or cell membranes. Such isolation prevents the binding of *T. gondii*-derived peptides to MHC molecules and their presentation to T cells. The spirochete *Treponema pallidum*, the cause of syphilis, evades specific antibody by coating itself with human proteins. This is also a strategy pursued by the schistosome, a parasitic helminth.

Of the four groups of pathogens (see Figure 1.2, p. 3), viruses have evolved the greatest variety of mechanisms for subverting or escaping immune defenses. This is because their replication and life cycle depend completely on the metabolic and biosynthetic processes of human cells. Viral self-defense strategies include: capture of cellular genes encoding cytokines or cytokine receptors, which when expressed by the virus can divert the immune response; synthesis of proteins that inhibit complement fixation; and synthesis of proteins that inhibit antigen processing and presentation by MHC class I molecules. Examples of defensive mechanisms used by herpes viruses and pox viruses are shown in Figure 9.5.

Viral strategy	Specific mechanism	Result	Virus examples
Inhibition of humoral immunity	Virally encoded Fc receptor	Blocks effector functions of antibodies bound to infected cells	Herpes simplex Cytomegalovirus
	Virally encoded complement receptor	Blocks complement-mediated effector pathways	Herpes simplex
	Virally encoded complement control protein	Inhibits complement activation of infected cell	Vaccinia
Inhibition of inflammatory response	Virally encoded cytokine homolog, e.g., β-chemokine receptor	Sensitizes infected cells to effects of β-chemokine; advantage to virus unknown	Cytomegalovirus
	Virally encoded soluble cytokine receptor, e.g., IL-1 receptor homolog, TNF receptor homolog, interferon-γ receptor homolog	Blocks effects of cytokines by inhibiting their interaction with host receptors	Vaccinia Rabbit myxoma virus
	Viral inhibition of adhesion molecule expression, e.g., LFA-3 ICAM-1	Blocks adhesion of lymphocytes to infected cells	Epstein–Barr virus
	Protection from NFκB activation by short sequences that mimic TLRs	Blocks inflammatory responses elicited by IL-1 or bacterial pathogens	Vaccinia
Blocking of antigen processing and presentation	Inhibition of MHC class I expression	Impairs recognition of infected cells by cytotoxic T cells	Herpes simplex Cytomegalovirus
	Inhibition of peptide transport by TAP	Blocks peptide association with MHC class I	Herpes simplex
Immunosuppression of host	Virally encoded cytokine homolog of IL-10	Inhibits T_H1 lymphocytes Reduces interferon-γ production	Epstein–Barr virus

Figure 9.5 Mechanisms used by herpes and pox viruses to subvert the immune response.

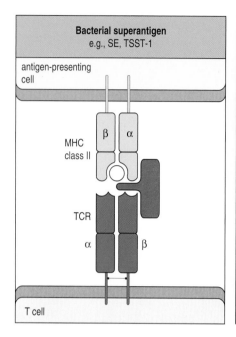

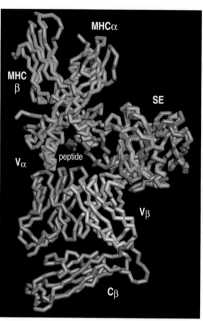

Figure 9.6 **Bacterial superantigens bridge α:β T-cell receptors and MHC class II molecules in the absence of a specific peptide.** In the diagram of the interaction (left panel) the superantigen is red. The molecular model (right panel) shows the interactions of the superantigen staphyloccal enterotoxin (SE) with an MHC class II molecule and an α:β T-cell receptor. Although the MHC class II molecule has a bound peptide it is usually not one for which the T-cell receptor is specific.

9-6 Bacterial superantigens stimulate a massive but ineffective T-cell response

Some pathogens induce a general suppression of a person's immune response. For example, staphylococci produce toxins such as the **staphylococcal enterotoxins** and **toxic shock syndrome toxin-1**, which can bind simultaneously to MHC class II molecules and T-cell receptors in the absence of a specific peptide antigen (Figure 9.6). By forming a bridge between a CD4 T cell's receptors and the MHC class II molecules of an antigen-presenting cell these toxins mimic specific antigen and cause the T cell to divide and differentiate into effector T cells. Because the toxins bind to sites shared by many different T-cell receptors they stimulate an excessive polyclonal response that can involve 2–20% of the total number of circulating CD4 T cells. Because of this property these toxins are called **superantigens**. The consequence of superantigen stimulation is a massive production and release of cytokines, particularly IL-1, IL-2, and TNF-α, which causes systemic shock (see Figure 8.18, p. 243). At the same time, a useful adaptive immune response is suppressed. After proliferation, T cells that have bound superantigens undergo apoptosis in the absence of any further specific stimulation, removing many antigen-specific T-cell clones from the peripheral circulation.

9-7 Immune responses can contribute to disease

Since the immune response is a powerful and destructive force, some of the disease symptoms, or **pathology**, in most infections are due to the immune response. For some infectious diseases all pathology is due to the immune response, an example being the wheezy bronchiolitis caused by T$_H$2 cells responding to infection with **respiratory syncytial virus** (**RSV**). This virus accounts for a large proportion of hospital admissions for infants in the developed countries—some 90,000 admissions and 4500 deaths each year in the United States alone. A failed attempt at developing a vaccine against RSV revealed that infants vaccinated with a killed virus preparation suffered worse disease on subsequent infection with RSV than did unvaccinated infants. This unfortunate outcome was because the vaccine failed to induce neutralizing antibodies but successfully activated virus-specific T$_H$2 cells. On subsequent infection with RSV, the secondary response of the T$_H$2 cells produced IL-3, IL-4, and IL-5 in quantities that exacerbated the disease-causing aspects of the immune response—the induction of bronchospasm, increased secretion of mucus and recruitment of tissue-damaging eosinophils into the tissues of the respiratory tract.

Tissue damage and disease symptoms can also result from the immune response to parasites, as illustrated by *Schistosoma mansoni*. These blood flukes lay their eggs in the hepatic portal vein. Some eggs reach the intestine and are shed in the feces, thereby enabling the infection to spread to other people. Other eggs lodge in the portal circulation of the liver, where they elicit a powerful immune response that leads to chronic inflammation, hepatic fibrosis, and eventual liver failure. Underlying this progression is an excessive activation of T_H2 cells.

Summary

From the human perspective the ideal immune system would be one that terminates infection before the pathogen damages tissues or saps the body's resources. In contrast, an ideal situation for a pathogen is one in which the immune system does not interfere with growth and replication, while other parts of the body provide food and shelter. To further their cause, pathogens have evolved ways of reducing the effectiveness of the human immune response. Antigenic variation in the pathogen prevents the maturation of the adaptive response and the development of useful immunological memory. Latency, a means of avoiding the immune response, allows viruses to lie low within cells until immunity has waned. More active strategies are for pathogens to interfere with key elements of the immune response, either to inhibit normal immune function or to recruit the response to the pathogen's advantage. Immune responses to pathogens can themselves be a significant cause of pathology.

Inherited immunodeficiency diseases

Inherited defects in genes for components of the immune system cause **immunodeficiency diseases**, which reveal themselves by enhanced susceptibility to infection. Before the advent of antibiotic therapy during the 1940s, most individuals with inherited immune defects died from infection during infancy or early childhood. Because many normal infants also succumbed to infection in that earlier era, death from immunodeficiency did not stand out until the 1950s when the first such disease was described. Since then, many inherited immunodeficiency diseases have been identified and correlated with susceptibility to particular classes of pathogen. Each disease is due to a defect in a particular protein or glycoprotein, and the precise symptoms depend on the role of that component in the immune response (Figure 9.7).

9-8 Most inherited immunodeficiency diseases are caused by recessive gene defects

Before the 1950s, any dominant trait causing a severe immunodeficiency would probably have been eliminated from the population with the death of the child in whom it first occurred. Thus, most of the inherited immunodeficiency syndromes that have been identified are due to recessive mutations in single genes. One of the few dominant immune defects known is a defect in the receptor for interferon-γ (IFN-γ), in which the defective receptor still binds IFN-γ but cannot produce an intracellular signal. Infants with this defect are susceptible to disseminated infection with mycobacteria, including the normally innocuous strain used for vaccination against tuberculosis.

For recessive defects in autosomal genes, that is genes on chromosomes other than the sex chromosomes, only children who inherit a defective allele from both parents are immunodeficient. The parents are heterozygous for the defect and are usually healthy. Such people are referred to as **carriers**. Recessive defects in genes on the X chromosome are a special case, because women

Name of deficiency syndrome	Specific abnormality	Immune defect	Susceptibility
Severe combined immune deficiency	ADA deficiency	No T or B cells	General
	PNP deficiency	No T or B cells	General
	X-linked *scid*, γ_c chain deficiency	No T cells	General
	Autosomal *scid* DNA repair defect	No T or B cells	General
DiGeorge's syndrome	Thymic aplasia	Variable numbers of T and B cells	General
MHC class I deficiency	TAP mutations	No CD8 T cells	Chronic lung and skin inflammation
MHC class II deficiency	Lack of expression of MHC class II	No CD4 T cells	General
Wiskott–Aldrich syndrome	X-linked; defective WASP gene	Defective anti-polysaccharide antibody and impaired T cell activation responses	Encapsulated extracellular bacteria
X-linked agamma-globulinemia	Loss of Btk tyrosine kinase	No B cells	Extracellular bacteria, viruses
X-linked hyper-IgM syndrome	Defective CD40 ligand	No isotype switching	Extracellular bacteria *Pneumocystis carinii* *Cryptosporidium parvum*
Common variable immunodeficiency	Unknown; MHC-linked	Defective IgA and IgG production	Extracellular bacteria
Selective IgA	Unknown; MHC-linked	No IgA synthesis	Respiratory infections
Phagocyte deficiencies	Many different	Loss of phagocyte function	Extracellular bacteria and fungi
Complement deficiencies	Many different	Loss of specific complement components	Extracellular bacteria especially *Neisseria* spp.
Natural killer (NK) cell defect	Unknown	Loss of NK function	Herpes viruses
X-linked lympho-proliferative syndrome	SH2D1A mutant	Inability to control B cell growth	EBV-driven B cell tumors
Ataxia telangiectasia	Gene with PI 3-kinase homology	T cells reduced	Respiratory infections
Bloom's syndrome	Defective DNA helicase	T cells reduced Reduced antibody levels	Respiratory infections

Figure 9.7 Inherited immunodeficiency syndromes of humans, their gene defects, and their pathogen susceptibilities. ADA, adenosine deaminase; PNP, purine nucleotide phosphorylase; TAP, transporter associated with antigen processing; WASP, Wiskott–Aldrich syndrome protein; EBV, Epstein–Barr virus; NK cell, natural killer cell; PI-3 kinase, phosphoinositol-3 kinase.

have two copies of the X chromosome whereas men have only one. As there is no other copy of the gene to compensate, inheritance of one defective copy of an X-linked gene by a male child is sufficient to cause disease. A woman who inherits a single defective gene will also have the normal gene and usually will be healthy. Because of this mode of inheritance, X-linked immunodeficiencies are far more common in men than in women and are also more common than immunodeficiencies linked to autosomes.

A wide range of defects in immune system genes exists in the human population. Whereas those listed in Figure 9.7 were discovered in patients with severe disease, other immune system gene defects were found first in seemingly healthy blood donors. Examples of the latter are the lack of certain complement components (see Section 7-19, p. 210) or of one of the MHC class I molecules. Those immune system genes that can be lost with little noticeable effect are usually members of multigene families in which another family member compensates for the defective gene. Modern genetic methods and knowledge of the full set of human genes have facilitated both the identification of genetic defects and their correlation with immunodeficiency and altered immune response.

9-9 Antibody deficiency leads to an inability to clear extracellular bacteria

The first immunodeficiency disease to be described is characterized by antibody deficiency and X-linked inheritance and is named **X-linked agammaglobulinemia** (**XLA**). The defect in XLA is in a protein tyrosine kinase, which is called Bruton's tyrosine kinase (Btk) to honor the discoverer of the syndrome. Btk contributes to intracellular signaling from the B-cell receptor and is necessary for the growth and differentiation of pre-B cells (see Section 4-4, p. 106). Btk is normally also expressed in monocytes and T cells, but these cells are not functionally compromised by its absence from XLA patients. Other immunodeficiency diseases affecting antibody production have been described, and almost all are caused by failure of either the development or the activation of B lymphocytes.

Women with one functional and one nonfunctional copy of the *Btk* gene are themselves healthy, but are carriers who pass XLA on to half their male children. In females, one X chromosome is randomly inactivated in each cell. In carriers, those B-cell precursors in which the inactivated X has the defective *Btk* gene develop normally, whereas B-cell precursors in which the inactivated X has the functional *Btk* gene fail to develop. Thus, all circulating B cells in heterozygous females have as their active X chromosome the one having the functional *Btk* gene; consequently, carrier females make entirely normal B-cell responses (Figure 9.8).

Patients who lack antibodies are vulnerable to infection with extracellular pyogenic bacteria that have polysaccharide capsules resistant to phagocytosis. These include *Haemophilus influenzae*, *Streptococcus pneumoniae*, *Streptococcus pyogenes*, and *Staphylococcus aureus*. In people who make normal antibody responses, such organisms are cleared by phagocytosis after opsonization by antibody and complement. Antibody-deficient patients are also more susceptible to viral infections, particularly those caused by enteroviruses, which enter the body through the gut and which in normal individuals are neutralized by antibodies.

Genetic defects can also cause deficiency in a particular isotype or class of antibody. The most common kind of inherited deficiency of this type is IgA deficiency, which is experienced by about 1 person in 800. No obvious

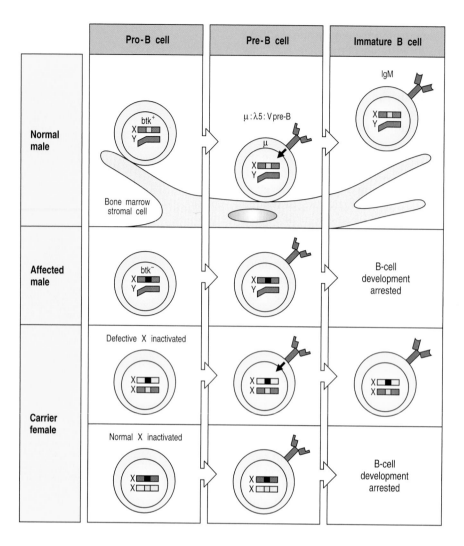

Figure 9.8 In patients with X-linked agammaglobulinemia B cells do not develop beyond the pre-B cell stage. In XLA, Bruton's tyrosine kinase is defective. In patients with this disease B cells become arrested at the pre-B cell stage because intracellular signals cannot be generated by the pre-B cell receptor. Most patients with this disease are males because males have only one copy of the X chromosome. Heterozygous females are carriers of the disease trait although healthy themselves. During development, female cells randomly inactivate one of their X chromosomes. Consequently, half of the developing B cells in a female carrier become arrested at the pre-B cell stage because they inactivate the X chromosome that carries the good copy of the btk gene; the other half develop to become functional B cells because they inactivate the X chromosome that carries the bad copy of the btk gene and use the X chromosome having the good copy.

increased susceptibility to infectious disease is associated with this condition, which might reflect the fact that modern humans tend to eat cooked and processed foods, which greatly reduces the threat posed by pathogens entering the gastrointestinal tract. In contrast, chronic lung disease is more common in people with IgA deficiency, suggesting that the trend towards poorer air quality in the cities, where most people live, makes the role of IgA in the respiratory tract of increasing importance.

Patients who have immunodeficiencies confined to B-cell functions are able to resist many pathogens successfully, and those to which they are susceptible can be treated with antibiotics. However, although pyogenic infections can be cured in this way, the successive rounds of infection and treatment sometimes lead to permanent tissue damage caused by the excessive release of proteases from both the infecting bacteria and the defending phagocytes. These effects are particularly pronounced in the airways, where the bronchi lose their elasticity and become sites of chronic inflammation. This condition, called **bronchiectasis**, can lead to chronic lung disease and eventual death. To prevent such developments, patients with XLA are given monthly injections of gamma globulin, an antibody-containing preparation made from the plasma of healthy blood donors. Such preparations contain antibodies against the range of common pathogens and provide what is called **passive immunity** against those pathogens.

9-10 Diminished antibody production also results from inherited defects in T-cell help

Diminished antibody production is also a symptom of defective genes for the membrane-associated cytokine CD40 ligand. As discussed in Chapter 7, interaction of CD40 ligand on activated T cells with B-cell CD40 is a crucial part of the T-cell help given to B cells. This stimulates B-cell activation, the development of germinal centers, and isotype switching. CD40 ligand is encoded on the X chromosome and so most patients with a hereditary deficiency in CD40 ligand are males. In the absence of CD40 ligand, virtually no specific antibody is made against T-cell-dependent antigens. IgG, IgA, and IgE levels are extremely low, and IgM levels are abnormally high. Recognition of this latter characteristic led to the condition being named **X-linked hyper-IgM syndrome**. Patients with this immunodeficiency are inherently susceptible to infection with pyogenic bacteria, but these infections can usually be prevented by regular injections of gamma globulin and cleared by antibiotics when they do occur. A further consequence of the disease is the absence of germinal centers in the lymph nodes and other secondary lymphoid tissues (see Figure 7.15, p. 192).

Macrophage activation by T cells also depends on the interaction of CD40 ligand on the T cell with CD40 on the macrophage. The lack of this interaction in patients with X-linked hyper-IgM syndrome impairs the inflammatory response and the mobilization of leukocytes by inflammatory cytokines. Whereas infection normally induces an increase in the number of white cells in the blood (**leukocytosis**), this cannot occur in patients lacking CD40 ligand. On the contrary, their blood can become profoundly deficient in neutrophils. This state, called **neutropenia**, leads to severe sores and blisters in the mouth and throat. These anatomical sites are always infested with bacteria and their health depends on continual surveillance by phagocytes. These symptoms of neutropenia can be cured by the intravenous administration of granulocyte–macrophage colony-stimulating factor (GM-CSF), a cytokine that stimulates the production and release of phagocytes by the bone marrow. In normal individuals, GM-CSF is secreted by macrophages in response to activation by T cells through interactions with CD40 and CD40 ligand.

9-11 Defects in complement components impair antibody responses and cause the accumulation of immune complexes

The effector functions recruited by antibodies to clear pathogens and antigens are all facilitated by complement activation. Consequently, the spectrum of infections associated with complement deficiencies overlaps substantially with that associated with defective antibody production. Defects in the activation of C3, and in C3 itself, are associated with susceptibility to a wide range of pyogenic infections, emphasizing the important role of C3 as an opsonin that promotes the phagocytosis of bacteria by macrophages and neutrophils. In contrast, defects in C5–C9, the terminal complement components that form the membrane-attack complex, have more limited effects, of which susceptibility to *Neisseria* is the best example. The most effective defense against *Neisseria* is complement-mediated lysis of extracellular bacteria, and this requires all the components of the complement pathway (Figure 9.9).

The early components of the classical pathway are necessary for the elimination of immune complexes. As discussed in Section 7-22, p. 214, attachment of complement components to soluble immune complexes allows them to be transported, or ingested and degraded, by cells bearing complement receptors. Immune complexes are mainly transported by erythrocytes,

Complement protein	Effects of deficiency
C1, C2, C4	Immune-complex disease
C3	Susceptibility to capsulated bacteria
C5–C9	Only effect is susceptibility to *Neisseria*
Factor D, properdin (factor P)	Susceptibility to capsulated bacteria and *Neisseria* but no immune-complex disease
Factor I	Similar effects to deficiency of C3
DAF, CD59	Autoimmune-like conditions including paroxysmal nocturnal hemoglobinuria

Figure 9.9 Diseases caused by deficiencies in the pathways of complement activation.

which capture the complexes with the CR1 complement receptor that binds to C4b or C3b. Deficiencies in complement components C1–C4 impair the formation of C4b and C3b and lead to the accumulation of immune complexes in the blood, lymph, and extracellular fluid and their deposition within tissues. In addition to directly damaging the tissues in which they deposit, immune complexes activate phagocytes, causing inflammation and further tissue damage.

Deficiencies in the proteins that control complement activation can also cause immunodeficiency. People who lack the plasma protein factor I in effect lack C3. Because factor I is absent, the conversion of C3 to C3b is unchecked, and supplies of C3 are rapidly depleted (see Section 7-25, p. 217). Patients who lack properdin (factor P), a plasma protein that enhances the activity of the alternative pathway, have a heightened susceptibility to *Neisseria*, because reduced deposition of C3 prevents formation of the membrane-attack complex and bacterial lysis. In contrast, a deficiency in decay-accelerating factor (DAF) or CD59 causes an autoimmune-like condition. The cells of patients lacking DAF or CD59 are not protected from activating the alternative pathway of complement activation. The resultant complement-mediated lysis of erythrocytes causes paroxysmal nocturnal hemoglobinuria (see Section 7-26, p. 220).

Another disease caused by a deficiency of a complement regulator is **hereditary angioneurotic edema**. In this condition the absence of the C1 inhibitor C1INH leads to uncontrolled activation of the classical pathway (see Section 7-25, p. 217). One consequence is overproduction of the vasoactive C2a fragment, causing the accumulation of fluid in the tissues and epiglottal swelling, which can lead to death by suffocation.

9-12 Defects in phagocytes result in enhanced susceptibility to bacterial infection

Phagocytosis by macrophages and neutrophils is the principal method by which the immune system destroys bacteria and other microorganisms. Any defect that compromises phagocyte activity will, therefore, have a profound effect on the capacity to clear infections (Figure 9.10). One kind of deficiency arises from mutations in the gene encoding CD18, the common β subunit of the leukocyte integrins CR2, CR4, and LFA-1 (see Figure 6.4, p. 149). These cell adhesion molecules are needed for phagocytes to leave the blood and enter sites of infection (see Figure 8.19, p. 245). Phagocytes lacking functional integrins are, therefore, unable to migrate to where they are needed. This defect is known as **leukocyte adhesion deficiency**.

Leukocyte adhesion deficiency is associated with persistent infection with extracellular bacteria, which cannot be cleared because of the defective phagocyte function. Children with this defect have recurrent pyogenic infections and problems with wound healing; if they survive long enough, they develop severe inflammation of the gums. Their neutrophils and macrophages cannot migrate into tissues and also, because CR2 and CR4 are complement receptors as well as adhesion molecules, they cannot take up and destroy bacteria opsonized with complement (see Section 7-21, p. 212). Patients deficient in CD18 suffer from infections that respond poorly to antibiotic treatment and which persist despite the generation of normal B-cell and T-cell responses. Neutropenia caused by chemotherapy, malignancy, or aplastic anemia produces a similar susceptibility to severe pyogenic bacterial infections.

Other kinds of defect in phagocytes affect their ability to kill ingested bacteria. In **chronic granulomatous disease**, the antibacterial activity of phagocytes is compromised by their inability to produce the superoxide radical O_2^- (see Section 8-10, p. 246). Mutations affecting any of the four proteins of the NADPH oxidase system can produce this phenotype. Patients with this disease suffer from chronic bacterial infections, often leading to granuloma formation. Deficiencies in the enzymes glucose-6-phosphate dehydrogenase and myeloperoxidase also impair intracellular bacterial killing, leading to a similar but less severe phenotype. A different phenotype characterizes **Chédiak–Higashi syndrome**, in which phagocytosed materials are not delivered to lysosomes because of a defect in the vesicle fusion mechanism. The lack of phagocyte function has effects in many different organs as well as leading to persistent and recurrent bacterial infections. The gene defects that cause this disease are still unknown.

Syndrome	Cellular abnormality	Immune defect	Associated infections and other diseases
Leukocyte adhesion deficiency	Defective CD18 (cell adhesion molecule)	Defective migration of phagocytes into infected tissues	Widespread infections with capsulated bacteria
Chronic granulomatous disease (CGD)	Defective NADPH oxidase. Phagocytes cannot produce O_2^-	Impaired killing of phagocytosed bacteria	Chronic bacterial and fungal infections. Granulomas
Glucose-6-phosphate dehydrogenase (G6PD) deficiency	Deficiency of glucose-6-phosphate dehydrogenase. Defective respiratory burst	Impaired killing of phagocytosed bacteria	Chronic bacterial and fungal infections. Anemia is induced by certain agents
Myeloperoxidase deficiency	Deficiency of myeloperoxidase in neutrophil granules and macrophage lysosomes and impaired production of toxic oxygen species	Impaired killing of phagocytosed bacteria	Chronic bacterial and fungal infections
Chédiak–Higashi syndrome	Defect in vesicle fusion	Impaired phagocytosis due to inability of endosomes to fuse with lysosomes	Recurrent and persistent bacterial infections. Granulomas. Effects on many organs

Figure 9.10 Defects in phagocytic cells cause persistent bacterial infections.

9-13 Defects in T-cell function result in severe combined immune deficiencies

Whereas B cells contribute only to the antibody response, T cells function in all aspects of adaptive immunity. This means that inherited defects in the mechanisms of T-cell development and T-cell function have a general depressive effect on the immune system's capacity to respond to infection. Patients with T-cell deficiencies tend to be susceptible to persistent or recurrent infections with a broader range of pathogens than patients with B-cell deficiencies. Those patients who make neither T-cell-dependent antibody responses nor cell-mediated immune responses are said to have **severe combined immune deficiency** (**SCID**).

T-cell development and function depend on the action of many proteins and so the SCID phenotype can arise from defects in any one of a number of genes. Because of the inheritance pattern of the X chromosome, X-linked diseases are more easily discovered, and at least two forms of SCID are of this type. The first form of X-linked SCID is due to mutation in a gene on the X chromosome that encodes the common γ chain of several cytokine receptors, including the receptors for IL-2, IL-4, IL-7, IL-9, and IL-15. In each of these receptors the γ chain transduces ligand binding into signals that alter gene expression. This is accomplished through interaction of the γ chain with the Jak3 kinase (see Section 6-12, p. 164). Indeed, as would be predicted, patients defective in the Jak3 kinase have an autosomally inherited immunodeficiency similar in phenotype to that of patients with X-linked SCID. The phenotype of SCID is so severe that affected infants survive only if they are isolated in a pathogen-free environment, or if their immune system is replaced by bone marrow transplantation and the passive administration of antibodies.

Another X-linked deficiency of T-cell function is **Wiskott–Aldrich syndrome** (**WAS**), which involves the impairment of platelets as well as lymphocytes. The relevant gene on the X chromosome encodes a protein called Wiskott–Aldrich syndrome protein (WASP). This protein is involved in the cytoskeletal reorganization needed before T cells can deliver cytokines and signals to the B cells, macrophages, and other target cells with which they routinely interact during the immune response.

SCID due to an absence of T-cell function is also caused by defects in **adenosine deaminase** (**ADA**) or **purine nucleotide phosphorylase** (**PNP**), which are enzymes involved in purine degradation. Although the absence of these enzymes causes an accumulation of nucleotide metabolites in all cells, the effects of this are particularly toxic to developing T cells and, to a lesser extent, to developing B cells. Infants with these immunodeficiencies have an underdeveloped thymus that contains few lymphocytes. ADA and PNP deficiencies are autosomally inherited (Figure 9.11).

Another kind of SCID is caused by the lack of expression of HLA class II molecules. The disease was named **bare lymphocyte syndrome** because the defect was first discovered on B lymphocytes, the major population of peripheral blood cells that expresses HLA class II. In these patients, CD4 T cells fail to develop (see Section 5-7, p. 133), which compromises all aspects of adaptive immunity. Bare lymphocyte syndrome arises from defects in transcriptional regulators essential for the expression of all HLA class II loci. A homozygous defect in any one of four proteins produces the condition. One protein is the class II transactivator (CIITA), the other three are components of RFX, a transcriptional complex that binds to a conserved sequence in the promoter of HLA class II genes called the X box.

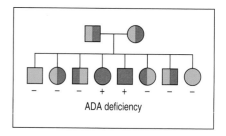

ADA deficiency

Figure 9.11 Adenosine deaminase (ADA) deficiency is inherited in autosomal fashion. The inheritance of ADA deficiency in a family. Both parents are healthy carriers who have one functional copy of the ADA gene (red) and one defective copy (green). Two of the eight children inherited defective copies of the ADA gene from both parents and have ADA deficiency (+). Males are indicated by squares and females by circles.

A defect in either of the two genes encoding the TAP peptide transporter impedes the binding of peptides by HLA class I molecules, leading to an unusually low abundance of HLA class I molecules on cell surfaces. This form of immunodeficiency, called **bare lymphocyte syndrome** (**MHC class I**), is less severe than the SCID caused by absence of HLA class II, its principal effect being selective loss of CD8 T cells (see Section 5-7, p. 133) and of cytotoxic T-cell responses to intracellular infections.

Defects in various proteins and enzymes that contribute to the rearrangement of immunoglobulin and T-cell receptor genes also cause autosomally inherited forms of SCID. These include the RAG-1 and RAG-2 proteins and also the DNA-dependent protein kinase (DNA-PK) implicated in the resolution of the hairpins produced during DNA rearrangement.

9-14 Some inherited immunodeficiencies lead to specific disease susceptibilities

Patients who lack either the receptor for IFN-γ or that for IL-12 can suffer persistent and ultimately fatal infections of common intracellular bacteria, such as the ubiquitous, nontuberculous strain of mycobacteria, *Mycobacterium avium*. The IFN-γ made by NK cells during the innate immune response and by CD4 and CD8 T cells during the adaptive immune response is the key cytokine that activates macrophages to kill the bacteria that infect them. The IL-12 secreted by infected macrophages is essential for activating the secretion of IFN-γ by NK cells, CD4 T cells, and CD8 T cells. A consequence of this type of immunodeficiency is that vaccination against tuberculosis with the live vaccine strain of *Mycobacterium bovis*, Calmette-Guèrin, can actually cause disseminated infection and disease.

As we saw in Section 9-4, many healthy people maintain a persistent EBV infection of B cells, which is held in check by EBV-specific T cells. For patients, mostly boys, with a defect in the SH2D1A gene on the X chromosome, a balance is never achieved and childhood EBV infections can become overwhelming and even progress to lymphoma. This immunodeficiency is called **X-linked lymphoproliferative syndrome**. Although the SH2D1A protein is believed to be a regulator of lymphocyte-activating signals, its precise functions and contribution to the control of EBV infection are uncertain.

9-15 Hematopoietic stem cell transplantation is used to correct genetic defects of the immune system

Many immunodeficiencies are due to gene defects that principally affect hematopoietic cells. These conditions can, in principle, be corrected by transplantation of hematopoietic stem cells obtained from the bone marrow or peripheral blood of a healthy donor. Such therapy is not undertaken lightly and the potential benefits from correcting the immunodeficiency must be weighed against the risks associated with the immunosuppressive treatments required for successful transplantation. The benefits of hematopoietic stem cell transplantation are directly correlated with the degree of HLA matching between patient and donor. Matching serves two purposes: first, it reduces the graft-versus-host disease (GVHD) caused by alloreactive T cells in the transplant that attack nonhematopoietic cells of the recipient; second, it ensures effective reconstitution of the adaptive immune system (see Figure 5.11, p. 134). After transplantation, all bone marrow derived cells in the recipient are of donor HLA type, whereas all other cells are of the recipient's HLA type. Positive selection of T cells is exclusively on thymic epithelium of the recipient's HLA type. Thus, the extent to which the mature T cells respond to antigens presented by professional antigen-presenting cells of donor HLA type is directly correlated with the number of shared HLA class I and II allotypes (see Section 5-6, p. 132).

Now that the genes responsible for immunodeficiency diseases are being identified, another type of therapeutic strategy is being explored. In **somatic gene therapy**, a functional copy of the defective gene is introduced into stem cells that have been isolated from the patient's bone marrow. The stem cells in which the defect has been corrected are then reinfused into the patient, where they provide a self-renewing source of immunocompetent lymphocytes and other hematopoietic cell types. Although highly attractive in principle, the practical development of gene therapy is at an early and experimental stage.

Summary

The best characterized gene defects affecting the immune system are those that show up in early childhood and confer exceptional vulnerability to common infections. The characterization of immunodeficiency diseases and the gene defects that cause them is almost the only way in humans of determining the relative importance of different cells and molecules in immune defenses, and of testing current models of how the human immune system works. The most severe immunodeficiencies are due to gene defects that cause an absence of all T-cell function and thus, directly or indirectly, impair B-cell function as well. Such deficiencies are known as severe combined immune deficiencies (SCID). The absence of antibodies due to genetic defects in B-cell development or function leads to particular susceptibility to pyogenic bacteria. Deficiencies in the early components of complement pathways cause a failure to opsonize. This results in increased susceptibility to bacterial infection, as do defects in phagocytes.

Acquired immune deficiency syndrome

Acquired immune deficiency syndrome (**AIDS**) was first described by physicians early in the 1980s. The disease is characterized by a massive reduction in the number of CD4 T cells, accompanied by severe infections of pathogens that rarely trouble healthy people, or by aggressive forms of Kaposi's sarcoma or B-cell lymphoma. All patients diagnosed as having AIDS eventually die from the effects of the disease. In 1983 the virus now known to cause AIDS, the **human immunodeficiency virus** (**HIV**), was first isolated. Two types of HIV are now distinguished—HIV-1 and HIV-2. In most countries HIV-1 is the principal cause of AIDS. HIV-2 is less virulent, causing a slower progression to AIDS. It is endemic in West Africa and has spread widely through Asia.

AIDS is a disease new to the medical profession and also to the human species. The earliest evidence for HIV comes from samples from African patients obtained in the late 1950s. It is believed that the viruses first infected humans in Africa by jumping from other primate species—HIV-1 coming from the chimpanzee, HIV-2 from the sooty mangabey. In neither of these species does the endogenous HIV-related virus cause disease.

As commonly occurs when a naive host population is hit with a new infectious agent, the effects of HIV on the human population have been immense and AIDS is now a disease of pandemic proportions. The World Health Organization currently estimates that 42 million people are infected with HIV. Although advances continue to be made in understanding the nature of the disease and its origins, the number of people infected with HIV continues to grow—5 million new infections in 2002—and tens of millions of people will die from AIDS in the years to come (Figure 9.12).

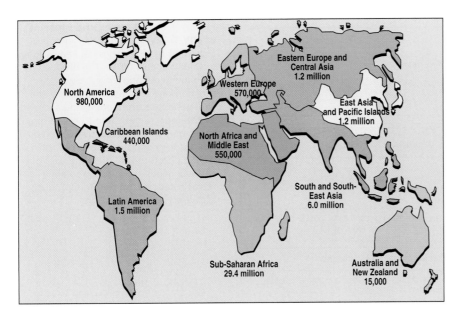

Figure 9.12 Incidence of HIV infection. Numbers of adults and children living with HIV/AIDS at the end of 2002, as estimated by the World Health Organization.

9-16 HIV is a retrovirus that causes slowly progressing disease

HIV is an RNA virus with an RNA nucleoprotein core (the nucleocapsid) surrounded by a lipid envelope derived from the host-cell membrane and containing virally encoded envelope proteins (Figure 9.13). HIV is an example of a retrovirus, so called because these viruses use an RNA genome to direct the synthesis of a DNA intermediate, a situation backwards or 'retro' from that used by most biological entities. One nucleocapsid protein is a protease used to cleave the gp41 and gp120 envelope glycoproteins from a larger precursor polyprotein; other nucleocapsid proteins are the enzymes reverse transcriptase and integrase, which are required for viral replication.

When HIV infects a cell, the RNA genome is first copied into a complementary DNA (cDNA) by reverse transcriptase. The viral integrase then integrates the cDNA into the genome of the host cell to form a **provirus**, a process facilitated by repetitive DNA sequences called long terminal repeats (LTRs) that flank all retroviral genomes. Proviruses use the transcriptional and translational machinery of the host cell to make viral proteins and RNA genomes, which assemble into new infectious virions. The genes and proteins of HIV are illustrated in Figure 9.14. HIV belongs to a group of retroviruses that cause slowly progressing diseases. They are collectively called the **lentiviruses**, a name derived from the latin word *lentus*, meaning slow.

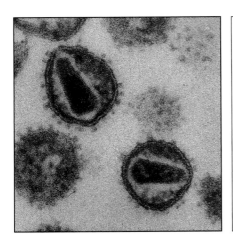

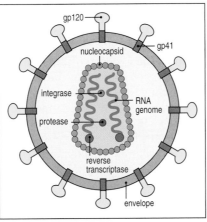

Figure 9.13 The virion of human immunodeficiency virus (HIV). The left panel is an electron micrograph showing three virions. The right panel is a diagram of a single virion. gp120 and gp41 are virally encoded envelope glycoproteins of molecular mass 120 kDa and 41 kDa, respectively. Photograph courtesy of Hans Gelderblom.

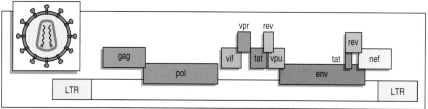

Gene		Gene product/function
gag	Group-specific antigen	Core proteins and matrix proteins
pol	Polymerase	Reverse transcriptase, protease, and integrase enzymes
env	Envelope	Transmembrane glycoproteins. gp120 binds CD4 and CCR5; gp41 is required for virus fusion and internalization
tat	Transactivator	Positive regulator of transcription
rev	Regulator of viral expression	Allows export of unspliced and partially spliced transcripts from nucleus
vif	Viral infectivity	Affects particle infectivity
vpr	Viral protein R	Transport of DNA to nucleus. Augments virion production. Cell cycle arrest
vpu	Viral protein U	Promotes intracellular degradation of CD4 and enhances release of virus from cell membrane
nef	Negative-regulation factor	Augments viral replication *in vivo* and *in vitro*. Downregulates CD4 and MHC class II

Figure 9.14 The genes and proteins of HIV-1. HIV-1 has an RNA genome consisting of nine genes flanked by long terminal repeats (LTRs). The products of the nine genes and their known functions are tabulated. Several of the viral genes are overlapping and are read in different frames. Others encode large polyproteins which after translation are cleaved to produce several proteins having different activities. The gag, pol, and env genes are common to all retroviruses and their protein products are all present in the virion.

In almost all people, HIV produces an infection that cannot be successfully terminated by the immune system and continues for many years. Although the initial acute infection is controlled to the point at which disease is not apparent, the virus persists and replicates in a manner that gradually exhausts the immune system, leading to immunodeficiency and death.

9-17 HIV infects CD4 T cells, macrophages, and dendritic cells

Macrophages, dendritic cells, and CD4 T cells are vulnerable to HIV infection because they express CD4, which the virus uses as a receptor. Chimpanzees, our closest living relative, are resistant to HIV infection because of a small difference in the structure of their CD4 glycoprotein in comparison to ours. The gp120 envelope glycoprotein of HIV binds tightly to human CD4, enabling virions to attach to CD4-expressing human cells. Before entry of the virus into the cell, gp120 must also bind to a co-receptor in the host-cell membrane. Once the co-receptor is bound, the gp41 envelope glycoprotein mediates fusion of the viral envelope with the plasma membrane of the host cell, allowing the viral genome and associated proteins to enter the cytoplasm.

The viral co-receptors are normal human chemokine receptors that are subverted by HIV to further its own propagation at the expense of the human host. There are different variants of HIV, and the cell types that they infect largely depend upon which co-receptor they bind. The HIV variants that spread infection from one person to another bind to the CCR5 co-receptor present on macrophages, dendritic cells, and CD4 T cells. Although they infect several types of human cell these HIV variants are called 'macrophage-tropic' for want of a better term. The HIV variants that infect activated CD4 T cells bind to the CXCR4 co-receptor and are called 'lymphocyte-tropic'. Whereas infection by macrophage-tropic HIV variants requires only modest levels of cell-surface CD4, infection by the lymphocyte-tropic viruses requires the higher levels present on activated CD4 T cells. Macrophages and dendritic cells at the site of virus entry are the first cells to be infected. Subsequently, the virus produced by the macrophages starts to infect the CD4 T-cell population. In about 50% of cases, the viral phenotype switches to the lymphocyte-tropic type late in infection. This is followed by a rapid decline in CD4 T-cell count and progression to AIDS. In their mutual interaction with the human population the two types of HIV variant have complementary roles: the lymphocyte-tropic viruses cause the disease, while the macrophage-tropic variants make it a pandemic.

The production of infectious virions from HIV provirus requires the activation of an infected CD4 T cell. Activation induces the synthesis of the transcription factor NFκB, which binds to promoters in the provirus. This directs the infected cell's RNA polymerase to transcribe viral RNAs. At least two of the proteins encoded by the virus serve to promote replication of the viral genome. Among other activities, the **Tat** protein binds to a sequence in the LTR of the viral mRNA, known as the transcriptional activation region (TAR), where it prevents transcription from shutting off and thus increases the transcription of viral RNA. The **Rev** protein controls the supply of viral RNA to the cytoplasm and the extent to which that RNA is spliced. At early times in infection Rev delivers RNA that encodes the proteins necessary for making virions. Later, complete viral genomes are supplied, which assemble with the viral proteins to form complexes that bud through the plasma membrane to give infectious virions (Figure 9.15).

9-18 Most people who become infected with HIV progress in time to develop AIDS

Infection with HIV usually occurs after the transfer of bodily fluids from an infected person to an uninfected recipient. Provirus can be carried in infected CD4 T cells, dendritic cells, and macrophages, whereas virions can be transmitted via blood, semen, vaginal fluid, or mother's milk. Infection is commonly spread by sexual intercourse, intravenous administration of drugs with contaminated needles, breast feeding, or transfusion of human blood or blood components from HIV-infected donors.

Immediately after HIV infection a person can either be asymptomatic or experience a transient 'flu-like' illness. In either case, virus becomes abundant in the peripheral blood, while the number of circulating CD4 T cells markedly declines (Figure 9.16). This acute viremia is almost always accompanied by activation of an HIV-specific immune response, in which anti-HIV antibodies are produced and cytotoxic T cells become activated to kill virus-infected cells. This response reduces the load of virus carried by the infected person and causes a corresponding increase in the number of circulating CD4 T cells. When an infected person first exhibits detectable levels of anti-HIV antibodies in their blood serum, they are said to have undergone **seroconversion**. The amount of virus persisting in the blood after the symptoms of acute viremia have passed is directly correlated with the subsequent course of disease.

The initial phase of infection is followed by an asymptomatic period, also called 'clinical latency'. During this phase, which can last for 2–15 years, there is persistent infection and replication of HIV in CD4 T cells, causing a gradual decrease in T-cell numbers. Eventually, the number of CD4 T cells drops below that required to mount effective immune responses against other infectious agents. That transition marks the end of clinical latency, the beginning of the period of increasing immunodeficiency, and the onset of AIDS. Patients with AIDS become susceptible to a range of opportunistic infections and some cancers, and it is from the effects of these that they die.

At the beginning of the AIDS epidemic in North America and Europe, viral transmission through infected blood products caused hemophiliacs and other patients dependent on blood products to become infected with HIV. Because hemophiliacs are so dependent on the medical profession, whether they be HIV-infected or not, it was possible to study the progress of their HIV infections in a systematic and rigorous manner. The results, which were

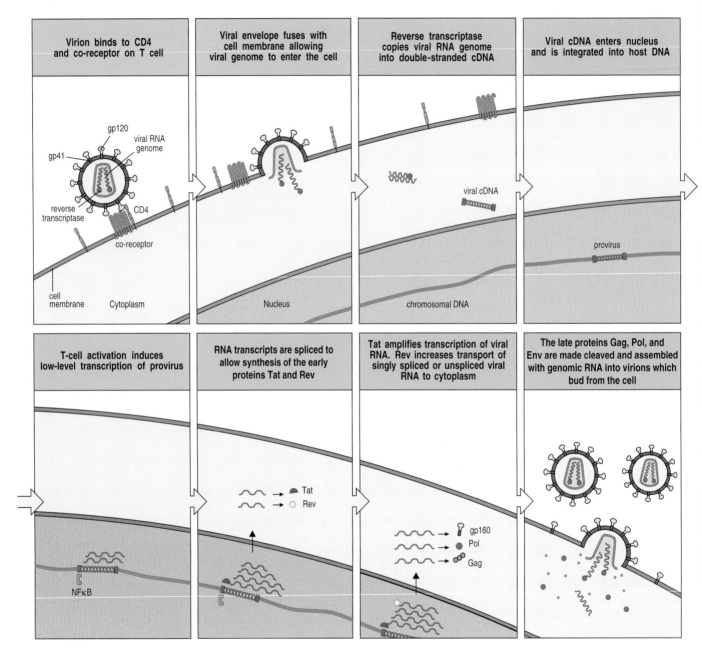

Figure 9.15 The life cycle of HIV in human cells. The virus binds to CD4 on a cell surface by using the envelope protein gp120, which is altered by CD4 binding so that it now also binds a specific chemokine co-receptor on the cell. This binding releases gp41, which then causes fusion of the viral envelope with the plasma membrane and release of the viral core into the cytoplasm. The RNA genome is released and is reverse transcribed into double-stranded cDNA. This migrates to the nucleus in association with the viral integrase, and is integrated into the cell genome, becoming a provirus. Activation of a T cell causes low-level transcription of the provirus that directs the synthesis of the early proteins Tat and Rev, which then expand and change the pattern of provirus transcription to produce mRNA encoding the protein constituents of the virion and RNA molecules corresponding to the HIV genome. Envelope proteins travel to the plasma membrane, whereas other viral proteins and viral genomic RNA assemble into nucleocapsids. New virus particles bud from the cell, acquiring their lipid envelope and envelope glycoproteins in the process.

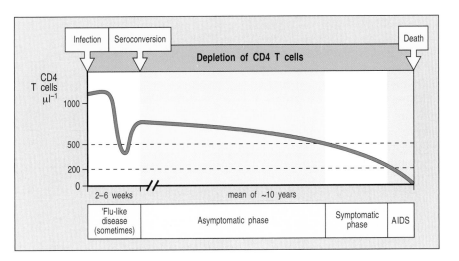

Figure 9.16 After infection with HIV there is a gradual extinction of CD4 T cells. The number of CD4 T cells (green line) refers to those present in peripheral blood. Opportunistic infections and other symptoms become more frequent as the CD4 T-cell count falls, starting at around 500 cells μl^{-1}. The disease then enters the symptomatic phase. When CD4 T-cell counts fall below 200 cells μl^{-1} the patient is said to have AIDS.

obtained during the time before effective treatments were available, demonstrate that most HIV-infected individuals are destined to progress to AIDS in the absence of effective medical intervention (Figure 9.17). Today, HIV infection through contaminated blood products has largely been eliminated in the richer countries by routine screening of individual units of blood for the presence of HIV.

Although the vast majority of infected people gradually progress to AIDS, a small minority do not; they are called 'long-term nonprogressors'. A small percentage seroconvert but their CD4 T-cell counts and other measures of immune competence are maintained. They have exceptionally low levels of circulating virus, and are being studied intensively to determine how they are able to control the infection. Another small group of people remain seronegative and disease-free despite extensive exposure to the virus. Some of this group have specific cytotoxic lymphocytes and T_H1 cells directed against infected cells, which suggests that at some time they have either been infected with the virus or have been exposed to noninfectious HIV antigens. The mechanisms that allow these people to resist AIDS are just beginning to be understood. As described in the next section, some people have a well-defined inherent resistance to AIDS: they are deficient in the viral co-receptor CCR5.

Figure 9.17 Once an HIV infection is established it usually leads to AIDS. Hemophilia is an inherited disease in which components of the blood clotting pathway do not function. Any wound can result in excessive bleeding for these patients and is potentially fatal. Effective treatment for hemophilia is provided by regular blood infusions of clotting factors purified from the blood of healthy donors. In the early 1980s, when the AIDS epidemic was well underway but the cause of the disease was unknown, some seemingly healthy blood donors were actually infected with HIV. The virus from such donations ended up contaminating some batches of clotting factor, with the result that many hemophiliacs became infected with HIV. The graph shows the health of HIV-infected and uninfected hemophiliacs over a period of some 15 years following infection. The development of AIDS is dependent upon HIV infection and is an almost inevitable consequence of infection, as seen from the fact that none of the plots tapers off. The time taken for AIDS to become manifest decreases with the patients' age.

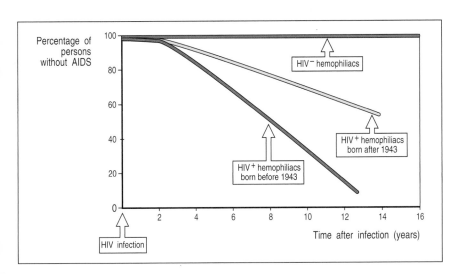

9-19 Genetic deficiency of the CCR5 co-receptor for HIV confers resistance to infection

Some people who are heavily exposed to HIV never become infected. Likewise, their isolated macrophages and lymphocytes cannot be deliberately infected with macrophage-tropic variants of HIV, the viral strains responsible for the spread of infection. These people are resistant to HIV infection because none of their cells has CCR5, the co-receptor for macrophage-tropic HIV variants, on the cell surface. The genetic basis for this defect is a mutated allele of the CCR5 gene in which a 32-nucleotide deletion from the coding region leads to an altered reading frame, premature termination of translation, and a nonfunctional protein. So far, the mutant allele has been found only in caucasoid populations, in which approximately 10% of the population are heterozygous for the mutant allele and 1% are homozygous. Only people who are homozygous for the mutant allele are resistant to HIV infection. A few individuals who are homozygous for the nonfunctional CCR5 variant have become infected with HIV, apparently as a result of primary infection by lymphocyte-tropic strains of the virus that use the CXCR4 co-receptor.

CCR5 is a normal component of the human immune system, in which it is a receptor for the chemokines CCL3 (MIP-1α), CCL4 (MIP-1β) and CCL5 (RANTES). The fact that almost everyone has a functional CCR5 gene strongly argues that this receptor has made useful contributions to human immunity, survival, and reproduction. On the other hand, for individuals facing exposure to HIV the survival advantage of not having CCR5 clearly outweighs that of having it. In the absence of the HIV pandemic, homozygosity for the CCR5 deletion would be considered a form of immunodeficiency, like the milder immunodeficiencies described in the previous part of this chapter, but in today's world it becomes an important asset of disease resistance. This type of evolutionary process, in which an immune system component that was useful in fighting past wars against pathogens becomes detrimental during a current conflict, has happened throughout human history. A pathogen subverts an immune system component and consequently selects for the survival of those humans who have genetic variants that resist that subversion. What we experience is a never-ending arms race, forever selecting for polymorphism and change in the human immune system.

That the CCR5 deletion mutant was already at high frequency in the caucasoid population before HIV began its colonization of the human species is a strong indication that this mutation enhanced human survival during a previous disease epidemic. One idea is that the mutant was selected and driven to high frequency by the Black Death, the pandemic of plague that ravaged Europe during the Middle Ages, causing up to 30% mortality in some communities. In the current age HIV will probably have its strongest selective effect in sub-Saharan Africa, where some 30 million people are infected with HIV and constitute more than 30% of the population of several countries. As treatment is rarely available for African patients, mortality rates will be high and survival will be greatly enhanced by genetic variants in any immune system component that prevents infection or reduces disease severity.

9-20 HIV escapes the immune response and develops resistance to antiviral drugs by rapid mutation

People infected with HIV make adaptive immune responses that can prevent the overt symptoms of disease for many years. Included in these responses are T_H1 and T_H2 cells, B cells that make neutralizing antibodies, and CD8 cytotoxic T cells that kill virus-infected cells (Figure 9.18). However, the virus is rarely eliminated. One reason the virus keeps ahead of the human immune response is its high rate of mutation during the course of an infection.

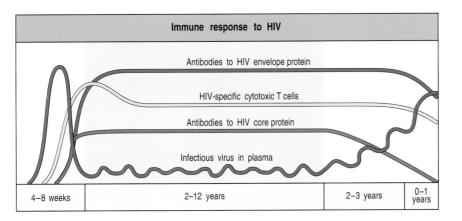

Figure 9.18 The adaptive immune response to HIV limits the effects of infection. During the course of infection, the levels of HIV virions and of components of adaptive immunity in the blood plasma change. Early in infection, while the adaptive immune response is being activated, the virus reaches high levels (red line). With the production of HIV-specific antibodies (blue lines) and HIV-specific cytotoxic T cells (yellow line), the virus is kept at a low level but is not eliminated. When the destruction of CD4 T cells outstrips their rate of renewal, adaptive immunity gradually collapses and virus levels increase again.

HIV and other retroviruses have high mutation rates because their reverse transcriptases lack proofreading mechanisms of the type possessed by cellular DNA polymerases. Consequently, reverse transcriptases are prone to making errors and these nucleotide substitutions soon accumulate to give new variant viral genomes. Even though the infection of a person might start from a single viral species, mutation throughout the infection produces many viral variants, called quasi-species, that coexist within the infected person.

The presence of variant viruses increases the difficulty of terminating the infection by immune mechanisms. Immune attack by neutralizing antibody will select for the survival of viral variants that have lost the epitope recognized by the antibody. Similarly, pressure from virus-specific cytotoxic T cells selects for viruses in which the peptide epitope recognized by the cytotoxic T cell has changed. In some instances the homologous peptide derived from the variant virus interferes with the presentation of the antigenic peptide from the original virus, thereby allowing both viral species to escape the cytotoxic T cell.

The high mutation rate of HIV greatly complicates the task of developing a vaccine against it. It also limits the effectiveness of antiviral drugs. Potential targets for drugs are the viral reverse transcriptase that is essential for the synthesis of provirus, and the viral protease that cleaves viral polyproteins to give viral enzymes and proteins. Inhibitors of reverse transcriptase and protease have been found, and these drugs prevent further infection of healthy cells. Unfortunately, mutation inevitably produces HIV variants with proteins resistant to the action of the drugs. For protease inhibitors, resistance appears within days (Figure 9.19) whereas resistance to the reverse transcriptase inhibitor zidovudine (AZT) takes months to develop. The difference is because zidovudine resistance requires the accumulated effects of three or four independent mutations in the reverse transcriptase, whereas the protease can acquire resistance with just one mutation. One situation in which a relatively short course of drug treatment has long-term benefit is pregnancy. Treating HIV-infected pregnant women with zidovudine can prevent HIV transmission to their children *in utero* and at birth.

Because HIV can so easily escape from the effects of any single drug, several antiviral drugs are now used together in what is called **combination therapy**.

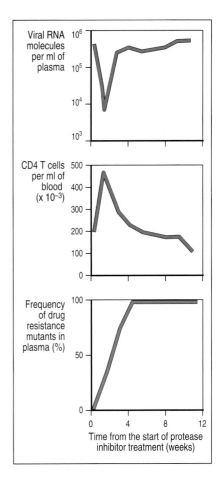

Figure 9.19 HIV can rapidly acquire resistance to protease inhibitor drugs. When a patient is treated with a single protease inhibitor drug the decrease in viral load (top panel) and increase in CD4 T-cell numbers (center panel) is only transient. The drug's benefit is short-lived because it selects for existing minority populations of drug-resistant HIV variants that rapidly expand and then continue the progress to AIDS (bottom panel).

The ideal is to destroy the entire population of viruses before any one of them has accumulated enough mutations to resist all the drugs. Such combination therapy, also called **highly active anti-retroviral therapy (HAART)** is effective at reducing the abundance of virus (the viral load) (Figure 9.20) and retarding disease progression. The drugs do not stop virus production by cells that are already infected, but prevent new infections from forming a provirus and becoming productive. Two weeks after starting combination therapy the amount of virus in the blood has decreased to about 5% of the level before treatment. The rapidity of this decline is because both activated CD4 T cells and free virions have short lifetimes. At this point no CD4 T cells are making virus and the level now present in the blood is due to production by longer-lived HIV-infected cells: macrophages, dendritic cells, and memory CD4 T cells. Continuing treatment further decreases the abundance of virus, but at a slower rate than before. Although the virus eventually becomes undetectable this does not reflect its eradication, as the virus re-emerges in patients who stop taking their medicine.

9-21 Clinical latency is a period of active infection and renewal of CD4 T cells

The administration of antiviral drugs to HIV-infected individuals has revealed the active nature of the infection during the period of clinical latency. Within two days of starting a course of drugs the amount of virus in the blood decreases dramatically (see Figure 9.20). At the same time, the number of CD4 T cells increases substantially (Figure 9.21). This shows, first, that virus is being produced and cleared continuously within infected individuals, and second, that in the face of HIV infection the body continues to produce new CD4 T cells, which quickly become infected with HIV. Thus the period of clinical latency, when the overall numbers of CD4 T cells in the blood are gradually declining, is actually a time of immense immune activity. Vast numbers of T cells are produced and die and virions are neutralized, the latter most probably through antibody-mediated opsonization and phagocytosis.

Although measurement of lymphocyte and virion numbers in the blood is used clinically to monitor the progress of HIV infection, the secondary lymphoid

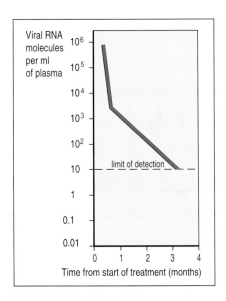

Figure 9.20 Combination drug therapy reduces HIV in the blood to below detectable levels. The abundance of HIV in patients' blood at different times after starting a course of combination drug therapy is shown.

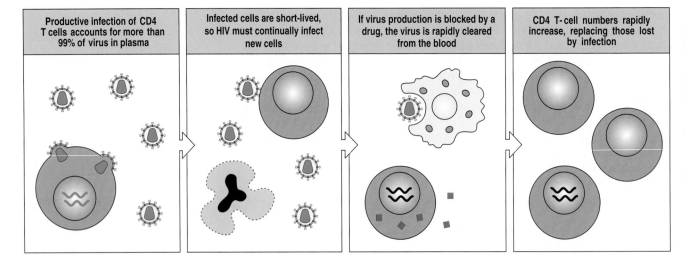

Figure 9.21 Antiviral drugs rapidly clear virus from the blood and increase the number of circulating CD4 T cells. The first and second panels show that maintenance of HIV levels in the blood depends on the continual infection of newly produced CD4 T cells. This is because cells live for only a few days once infected. The third and fourth panels show the effects of administering a drug (red squares) that blocks the viral life cycle. The existing virions in the blood are rapidly cleared by the actions of neutralizing antibody, complement, and phagocytes. Newly produced CD4 T cells are not infected, whereupon they live longer and accumulate in the circulation.

tissues are the sites where most lymphocytes are found and where CD4 T cells are activated and produce virus. In HIV-infected people these tissues are loaded with virions, many of which are trapped on the surface of follicular dendritic cells.

9-22 HIV infection leads to immunodeficiency and death from opportunistic infections

Within a few days of being infected with HIV, a CD4 T cell dies. Three kinds of mechanism are thought to contribute to the death toll. One is direct killing as a result of the viral infection or virions binding to cell-surface receptors, the second is increased susceptibility of infected cells to apoptosis, and the third is killing by cytotoxic CD8 T cells specific for viral peptides presented by HLA class I molecules on the infected CD4 T cell.

Throughout clinical latency the daily loss of CD4 T cells is for the most part compensated for by the supply of new CD4 T cells. However, with time a steady decline in the CD4 T-cell number is evident, showing how the virus gradually wins this war of attrition. Eventually, the numbers of CD4 T cells become so low that immune responses to all foreign antigens are compromised and HIV-infected people become exceedingly susceptible to other infections. Because CD4 T cells are required for all aspects of adaptive immunity, the vulnerability of patients with AIDS due to late-stage HIV infection resembles that of children with inherited severe combined immune deficiency.

The infections that most frequently affect AIDS patients are due to infectious agents that are already present in or on the body and that are actively kept under control in healthy people. Such agents are called opportunistic pathogens and the infections they cause are known as opportunistic infections (Figure 9.22). There is a rough hierarchy in the times at which particular opportunistic infections occur in AIDS, which is correlated with the order in which the different types of immunity collapse. Cellular immunity due to CD4 T_H1 cells tends to be lost before either antibody responses or cytotoxic CD8 T cells.

The oral and respiratory tracts are soft tissues loaded with organisms and in many AIDS patients they are the site of the first opportunistic infections. For example, *Candida* causes oral thrush and *Mycobacterium tuberculosis* causes tuberculosis. Later, patients can suffer from diseases caused by the reactivation of latent herpes viruses that are no longer under control by CD8 T cells. Such diseases include shingles due to varicella-zoster, B-cell lymphoma due to EBV, and an endothelial tumor called Kaposi's sarcoma, which is caused by the herpes virus HHV8. *Pneumocystis carinii*, a common environmental fungus that is hardly ever a problem to healthy people, frequently causes pneumonia and death in AIDS patients. In the later stages of AIDS, reactivation of cytomegalovirus, a herpes virus with similar effects to those of EBV, can cause B-cell lymphoproliferative disease. Infection with the opportunistic pathogen *Mycobacterium avium* also becomes prominent. The opportunistic infections suffered by individual AIDS patients vary greatly. With the collapse of the immune system, only drugs and other interventions can be used to treat the opportunistic infections. These are rarely completely effective or without

Infections	
Parasites	*Toxoplasma* species *Cryptosporidium* species *Leishmania* species *Microsporidium* species
Bacteria	*Mycobacterium tuberculosis* *Mycobacterium avium* *intracellulare* *Salmonella* species
Fungi	*Pneumocystis carinii* *Cryptococcus neoformans* *Candida* species *Histoplasma capsulatum* *Coccidioides immitis*
Viruses	Herpes simplex Cytomegalovirus Varicella–zoster

Malignancies

Kaposi's sarcoma (associated with
  herpes virus HHV8)
Non-Hodgkin's lymphoma, including
  EBV-positive Burkitt's lymphoma
Primary lymphoma of the brain

Figure 9.22 A variety of opportunistic infections kill patients with AIDS. The most common of the opportunistic infections that kill AIDS patients in developed countries are listed. The malignancies are listed separately but they are also the result of impaired responses to infectious agents.

their own deleterious effects. Eventually, the accumulated tissue damage resulting from the direct effects of HIV infection, opportunistic infections, and medical intervention causes death.

Summary

A major cause of acquired immunodeficiency in the human population today is the human immunodeficiency virus (HIV). This is a slow-acting retrovirus that remains latent for years after the initial infection. HIV infects CD4 T cells, macrophages, and dendritic cells, which all carry the CD4 glycoprotein on their surface, which is the receptor for the virus. The effects of HIV are due chiefly to a gradual and sustained destruction of CD4 T cells, which leads eventually to a profound T-cell immunodeficiency known as acquired immune deficiency syndrome (AIDS). CD4 T cells proliferate as part of their normal function and are continually being renewed; these properties allow HIV to maintain a long-lasting infection in which individuals remain relatively healthy for years. Patients with AIDS succumb to a range of opportunistic infections and to some otherwise rare, virus-associated cancers. HIV has only begun to infect humans in the past 50 years and the lack of any accommodation between host and pathogen is reflected in the strong correlation between infection and death. Genetic polymorphisms in the human population can also influence susceptibility to HIV infection. Because infection with HIV is usually dependent on the presence of the chemokine receptor CCR5, a viral co-receptor, people who lack this cell-surface protein have considerable resistance to HIV infection.

A feature of HIV infection that encourages its dissemination in the population is that infected individuals can live normal lives for many years without knowing they are infected. This poses problems for social approaches to reducing the frequency of sexual transmission. A second approach is to develop vaccines that provide protective immunity, terminating the initial acute HIV infection. Determination of the most effective aspects of the initial immune response will be crucial to this approach, which is also bedeviled by the mutability of HIV. The third approach is that of designing drugs that interfere with the growth and replication of HIV while not affecting normal functions of human cells. This pharmacological approach is the one that has seen most investment and the results have been mixed. The mutability of the virus, which enables it to escape from individual drugs, is currently perceived as the major problem and its solution is the use of combination drug therapy.

Summary to Chapter 9

In the course of their long relationship with humans, successful pathogens have developed mechanisms that allow them to exploit the human body to the full. Indeed, a pathogen is, by definition, an organism that is habitually able to overcome the body's immune defenses to such an extent that it causes disease. One class of adaptations is those whereby the pathogen changes itself or its behavior to evade the ongoing immune response. This prevents the immune system from adapting to the pathogen and improving the response. In a second type of adaptation, the pathogen is able to impair or prevent the immune response. Pathogens can have more than one such adaptation, and for some pathogens a considerable fraction of their genome is devoted to foiling the immune system. Highly successful pathogens are not necessarily the most virulent. Nonlethal ubiquitous host–pathogen relationships, such as that of humans and the Epstein–Barr virus, have generally evolved over a long period of association. However, even these pathogens can cause life-threatening disease when the immune system is compromised.

Inherited immunodeficiencies are caused by a defect in one of the genes necessary for the development or function of the immune system. Depending on the gene involved, immunodeficiencies range from manageable susceptibilities to particular pathogens to a general vulnerability created by the complete absence of adaptive immunity. The most severe inherited immunodeficiencies are very rare—evidence of the importance of the immune system to human survival.

Immunodeficiency can also be acquired as the result of infection. The human immunodeficiency virus (HIV) infects macrophages, dendritic cells, and CD4 T cells and eventually reduces the number of CD4 T cells to a level at which severe immunodeficiency results—the acquired immune deficiency syndrome (AIDS). This retroviral pathogen has only just begun to exploit the human species, but it already has a variety of effective adaptations acquired during its history in other primate hosts. Most people infected with HIV have no overt symptoms of disease for years, which facilitates the spread of HIV by sexual transmission, and HIV infection has now reached epidemic proportions within the human population. The discovery of natural genetic traits that endow individual humans with resistance to HIV indicates possibilities for accommodation in the longer term.

Questions

Question 9–1
A. Explain what is meant by serotype-specific immunity.
B. How does *Streptococcus pneumoniae* exploit serotype-specific immunity to evade detection? (Refer to Figure 9.1.)

Question 9–2
A. Which antigens are most important in the immune response to the influenza virus?
B. Explain the difference between antigenic drift and antigenic shift in the influenza virus.
C. Which is most likely to lead to a major worldwide pandemic?
D. What is the role of the phenomenon of 'original antigenic sin' in immunity to this virus? (Refer to Figures 9.2 and 8.52, and Section 8-27.)

Question 9–3
During the course of a trypanosome infection, a periodic rise and fall in the number of parasites is seen. Say what feature of trypanosomes is causing this, and explain these observations. (Refer to Figure 9.3.)

Question 9–4
Using Table Q9–4 match the mechanism of evasion and subversion of the immune system in Column 1 with the pathogen in Column 2.

Question 9–5
A. Explain the mechanism by which human immunodeficiency virus (HIV) enters a host cell.
B. Explain the cellular tropism of HIV, discussing the difference between macrophage-tropic and lymphocyte-tropic HIV.

Table Q9–4

a. Variant pilin protein expression	1. *Staphylococcus aureus*
b. Induction of quiescent (latent) state in neurons	2. *Toxoplasma gondii*
c. Reactivation of infected ganglia after stress or immunosuppression	3. *Salmonella typhimurium*
d. Alternate expression of two antigenic forms of flagellin	4. Influenza virus
e. Recombination of RNA genomes of avian and human origins	5. *Mycobacterium tuberculosis*
f. Escape from phagosome and growth and replication in cytosol	6. Varicella-zoster
g. Survival in a membrane-bounded vesicle resistant to fusion with other cellular vesicles	7. *Neisseria gonorrhoeae*
h. Coating its surface with human proteins	8. *Treponema pallidum*
i. Inhibiting fusion of phagosome with lysosome and survival in the host cell's vesicular system	9. *Listeria monocytogenes*
j. Immunosuppression caused by nonspecific proliferation and apoptosis of T cells	10. Herpes simplex virus

C. Some people seem to be resistant to HIV infection because a primary infection cannot be established in macrophages. What is the reason for this? (Refer to Figure 9.15.)

Question 9–6

A. What does the term seroconversion mean in relation to an HIV infection?

B. What relationship does seroconversion have to the time course of an HIV infection? (Refer to Figures 9.16 and 9.18.)

Question 9–7

A. Deficiencies in antibody production can be due to a variety of underlying genetic defects. Name two immunodeficiency diseases, other than the severe combined immunodeficiences, in which a defect in antibody production is the cause of the disease, and for which the underlying genetic defect is known. For each disease, say how (i) antibody production is affected, and (ii) what the underlying defect is and why it has this effect.

B. What is the main clinical manifestation of immunodeficiency diseases in which antibody production is defective but cell-mediated immune responses are intact? (Refer to Sections 9-9, 9-10, and Figure 9.7.)

Question 9–8

A. Explain why an inherited deficiency in complement components C3 or C4 can result in less efficient clearance of immune complexes from the blood and lymph than normal.

B. What clinical symptoms can this lead to?

C. What is the only known effect of deficiencies in complement components C5–C9? Explain this effect. (Refer to Section 9-11, and Figure 9.9.)

Question 9–9

A. Name three immunodeficiency diseases caused by defects in phagocytes.

B. Which immunodeficiency disease is caused by a defect in the phagocyte NADPH oxidase system, and what is the cellular effect of this defect?

C. What are the main clinical effects of defects in phagocyte function? (Refer to Section 9-12, and Figure 9.10.)

Question 9–10

A. What type of immune deficiency would you see in a child lacking the common γ chain of the receptor for cytokines IL-2, IL-4, and IL-7, among others? Explain your answer.

B. Why would you see the same type of immunodeficiency in a child lacking Jak3 kinase function?

C. What treatment might be possible to remedy this immunodeficiency? (Refer to Section 9-13.)

Question 9–11

Which property of HIV renders the virus difficult to eradicate by the body's immune defenses, and also limits the efficacy of drug therapies?

Question 9–12

Explain why a staphylococcal infection might produce a medical emergency.

Question 9–13

What would you predict might happen to the course of the HIV infection in a person who developed toxic-shock syndrome while in the latent phase of HIV? Explain your answer. (Refer to Figure 9.15.)

Chapter 10

Over-reactions of the Immune System

The protective functions of the immune system depend on recognition events that distinguish molecular components of infectious agents from those of the human body. In this context the immune system is said to recognize a pathogen's molecules as foreign. Besides infectious agents, humans come into daily contact with numerous other molecules that are equally foreign but do not threaten health. Many of these molecules derive from the plants and animals that we eat or are present in the environments where we live, work, and play. For most people for most of the time, contact with these molecules stimulates neither inflammation nor adaptive immunity.

In some circumstances, however, certain kinds of innocuous molecule stimulate an adaptive immune response and the development of immunological memory in predisposed members of the population. On subsequent exposures to the antigen the immune memory produces inflammation and tissue damage that is at best an irritation, and at worst a threat to life. The person feels ill, as though fighting off an infection, when no infection exists. The over-reactions of the immune system to harmless environmental antigens are called either **hypersensitivity reactions** or **allergic reactions**. The environmental antigens that cause these reactions are termed **allergens** and they induce a state of hypersensitivity or **allergy**, the latter word being of Greek derivation and meaning 'altered reactivity.'

Unfortunately, the antigens that provoke these over-reactions are often common in the human environment: in developed countries some 10–40% of the inhabitants are allergic to one or more environmental antigens. Some of the antigens responsible for common hypersensitivities are listed in Figure 10.1.

10-1 Four types of hypersensitivity reaction are caused by different effector mechanisms of adaptive immunity

Hypersensitivity reactions are conventionally grouped into four types according to the effector mechanisms that produce the reaction (Figure 10.2). Some antigens (penicillin, for example) cause different types of hypersensitivity reaction, depending on the circumstances in which they are encountered. **Type I hypersensitivity reactions** result from the binding of antigen to antigen-specific IgE bound to its Fc receptor, principally on mast cells. This interaction causes the degranulation of mast cells and the release of inflammatory mediators. Type I reactions are commonly caused by inhaled particulate antigens, of which plant pollens are good examples. Type I reactions have effects of varying severity, ranging from a runny nose to breathing difficulties and

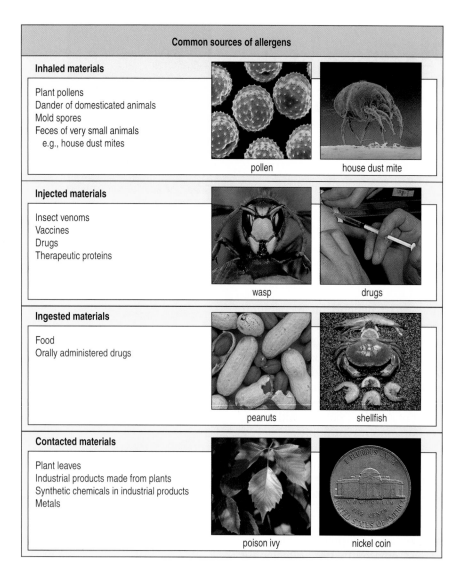

Common sources of allergens

Inhaled materials

Plant pollens
Dander of domesticated animals
Mold spores
Feces of very small animals
  e.g., house dust mites

pollen   house dust mite

Injected materials

Insect venoms
Vaccines
Drugs
Therapeutic proteins

wasp   drugs

Ingested materials

Food
Orally administered drugs

peanuts   shellfish

Contacted materials

Plant leaves
Industrial products made from plants
Synthetic chemicals in industrial products
Metals

poison ivy   nickel coin

Figure 10.1 Some substances that are a common cause of hypersensitivity reactions.

even death by asphyxiation. Because of their prevalence and diversity, a large part of this chapter will be devoted to the IgE-mediated hypersensitivity reactions.

Type II hypersensitivity reactions are caused by small molecules that bond covalently to cell-surface components of human cells, producing modified structures that are perceived as foreign by the immune system. The B-cell response to these new epitopes produces IgG, which on binding to the modified cells causes their destruction through complement activation and phagocytosis. The antibiotic penicillin is an example of a small reactive molecule that can induce a type II hypersensitivity reaction.

Type III hypersensitivity reactions are due to small soluble immune complexes formed by soluble protein antigens binding to the IgG made against them. Some of these immune complexes become deposited in the walls of small blood vessels or the alveoli of the lungs. The immune complexes activate complement and initiate an inflammatory response that damages the tissue, impairing its physiological function. When antibodies or other proteins derived from nonhuman animal species are given therapeutically to patients, type III hypersensitivity reactions are a potential side-effect.

The effector molecules initiating type I, II, and III hypersensitivity reactions are all antibodies. In contrast, **type IV hypersensitivity reactions** are caused by the products of antigen-specific effector T cells. Most reactions are caused by CD4 T_H1 cells. For example, the inflammatory reaction around the site of an insect bite or sting is caused by CD4 T_H1 cells that respond to peptide epitopes derived from venom and other insect proteins introduced by the bite. A minority of type IV hypersensitivity reactions are due to cytotoxic CD8 T cells. They arise when small, reactive, lipid-soluble molecules pass through cell membranes and bond covalently to intracellular human proteins. Degradation of these chemically modified proteins yields abnormal peptides that bind to HLA class I molecules and stimulate a cytotoxic T-cell response. For example, the allergic response to poison ivy involves cytotoxic T cells. These recognize peptides derived from intracellular proteins that are modified by chemical reaction with pentadecacatechol, a chemical acquired by touching the plant's leaves.

Type I hypersensitivity reactions

A prerequisite for a type I hypersensitivity reaction is that the person makes IgE antibody when he or she first encounters the antigen. This is how a person becomes **sensitized** to the antigen. IgE differs from other antibody isotypes in being located predominantly in tissues, where it is bound to mast cells by high-affinity surface receptors known as FcεRI. We first look at the effector cells and molecules that produce a type I response, and then consider the characteristic properties of allergens and how sensitization occurs.

10-2 IgE binds irreversibly to Fc receptors on mast cells, basophils, and activated eosinophils

IgE antibody causes allergic reactions because of its exceptionally high affinity for its Fc receptor and the cellular distribution of the receptor. Binding of the IgE constant region to its high-affinity receptor—**FcεRI**—is the tightest of the antibody–Fc receptor interactions ($K_d \approx 10^{10}$ M^{-1}, see Figure 7.24, p. 200) and can be considered for all practical purposes as irreversible. Also, unlike other isotypes, IgE binds to its receptor in the absence of antigen. FcεRI is expressed constitutively by mast cells and basophils, and by eosinophils after

	Type I	Type II		Type III	Type IV		
Immune reactant	IgE	IgG		IgG	T_H1 cells	T_H2 cells	CTL
Antigen	Soluble antigen	Cell- or matrix-associated antigen	Cell-surface receptor	Soluble antigen	Soluble antigen	Soluble antigen	Cell-associated antigen
Effector mechanism	Mast-cell activation	Complement, FcR^+ cells (phagocytes, NK cells)	Antibody alters signaling	Complement Phagocytes	Macrophage activation	Eosinophil activation	Cytotoxicity
Example of hypersensitivity reaction	Allergic rhinitis, asthma, systemic anaphylaxis	Some drug allergies (eg penicillin)	Chronic urticaria (antibody to FαRIα)	Serum sickness, Arthus reaction	Contact dermatitis, tuberculin reaction	Chronic asthma, chronic allergic rhinitis	Contact dermatitis

Figure 10.2 Hypersensitivity reactions fall into four classes on the basis of their mechanism. The types of preexisting immunity causing the reaction are listed along with the types of antigen that provoke the response, the underlying effector mechanism, and the nature of the ailment produced.

Figure 10.3 Cross-linking of FcεRI on the surface of mast cells by antigen and IgE causes mast-cell activation and degranulation. The high-affinity IgE receptor (FcεRI) on mast cells, basophils, and activated eosinophils is formed from one α chain, one β chain, and two γ chains. The binding site for IgE is formed by the two extracellular immunoglobulin-like domains of the α chain. The β chain and the two disulfide-bonded γ chains are largely intracellular and contribute to signaling. When receptors are cross-linked by antigen binding to IgE on the surface of a mast cell, a signal is transmitted that leads to the release of preformed granules containing histamine and other inflammatory mediators.

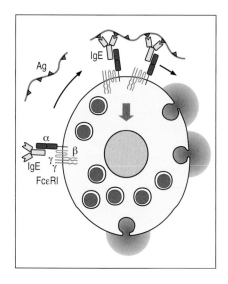

they have been activated by cytokines. After a primary IgE response has subsided and the antigen has been cleared by the usual means, all antigen-specific IgE molecules that have not encountered antigen will be bound by their Fc regions to FcεRI on these cells. The stable complexes of IgE and FcεRI effectively provide mast cells, basophils, and eosinophils with antigen-specific receptors. All these cells have granules containing preformed inflammatory mediators. When the antigen is next encountered it will bind to the receptors and activate the cells to release their inflammatory mediators (Figure 10.3). Initial degranulation is followed by induced synthesis of a wider range of mediators.

Two important features distinguish the 'antigen receptors' on mast cells, basophils, and eosinophils from those of B cells and T cells. The first is that the cell's effector function becomes operational immediately after antigen binds to the 'receptor' and does not require a phase of cell proliferation and differentiation. The second is that each cell is not restricted to carrying receptors of a single antigen specificity but can carry a range of IgE representing that present in the circulation. These features combine to quicken and strengthen the response to any antigen to which a person has been sensitized. When any one of those antigens enters a tissue, all the mast cells in the vicinity bearing a sufficient number of molecules of IgE specific for that antigen will be triggered to degranulate immediately.

IgE-mediated activation of mast cells, basophils, and activated eosinophils is thought to provide protection against parasites (see Section 7-13, p. 202), which are prevalent in tropical regions. This type of environment is where the human species originated and where it has lived for most of its existence. In the developing countries of the tropics, parasites are abundant and a one-third of the world's population is infected by one or more parasites.

In contrast, in the developed countries where parasite infections are rare, IgE responses tend to be stimulated by contact with nonthreatening substances in the environment. In North America and Europe, the impact of allergic disease is increasingly being felt in all aspects of human life. Some physicians have called it an 'epidemic of allergy'. Because allergy has directed medical research in this area we know far more about the allergic effects of IgE than we do about its benefits in controlling infection.

10-3 Tissue mast cells orchestrate IgE-mediated allergic reactions through the release of inflammatory mediators

Mast cells are resident in mucosal and epithelial tissues lining the body surfaces and serve to alert the immune system to local trauma and infection. Present in all vascularized tissues except the central nervous system and the retina, their initial response to activation is to release inflammatory molecules that are stored in 50–200 large granules that fill the cytoplasm (Figure 10.4). The name **mast cell** derives from the German word *Mastzellen*, meaning

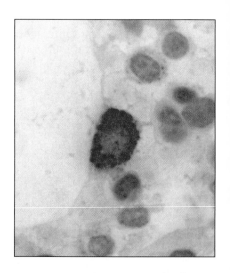

Figure 10.4 Light micrograph of mast cell stained for the granule protease chymase to show the numerous granules that fill its cytoplasm. The mast cell is in the center of the picture: its nucleus is stained pink and the cytoplasmic granules are stained red. Magnification × 1000. Photograph courtesy of D. Friend.

'fattened' or 'well-fed cells', which is how they appear under the microscope. Mast-cell granules contain principally histamine, heparin, tumor necrosis factor-α (TNF-α), chondroitin sulfate, neutral proteases, and other degradative enzymes and inflammatory mediators (Figure 10.5). Heparin, an acidic proteoglycan, is largely responsible for the characteristic staining of the mast-cell granules with basic dyes.

On leaving the bone marrow, immature mast cells travel in the blood as agranular cells that enter tissues and settle near small blood vessels. There they mature and form their characteristic granules. This differentiation is mediated by stem-cell factor (SCF), which interacts with CD117 (also called Kit) on the mast-cell surface. The importance of mast cells for IgE-mediated inflammatory reactions is well illustrated by the phenotype of mice lacking functional CD117. These mice have no differentiated mast cells and none of the inflammatory responses normally caused by IgE.

Two kinds of human mast cell are distinguished: a mucosal mast cell, which produces the protease tryptase, and a connective tissue mast cell, which produces chymotryptase. In patients with T-cell immunodeficiencies, only connective tissue mast cells are present, suggesting that the development of mucosal mast cells is dependent on T cells. Mast-cell activation occurs in the presence of any antigen that can cross-link the IgE molecules bound to FcεRI at the cell surface. This can be accomplished by antigens with repetitive epitopes that cross-link IgE molecules of the same specificity, or by antigens possessing two or more different epitopes that cross-link IgE molecules of different specificities. Once a mast cell's receptors have been cross-linked, degranulation occurs within a few seconds, releasing the stored mediators into the immediate extracellular environment (see Figure 10.5).

Prominent among these chemicals is histamine, an amine derivative of the amino acid histidine (Figure 10.6). Histamine exerts a variety of physiological effects through three kinds of histamine receptor—H1, H2, and H3—which

Class of product	Product	Biological effects
Enzyme	Tryptase, chymase, cathepsin G, carboxypeptidase	Remodeling of connective tissue matrix
Toxic mediator	Histamine, heparin	Toxic to parasites Increase vascular permeability Cause smooth muscle contraction
Cytokine	TNF-α (some stored preformed in granules)	Promotes inflammation, stimulates cytokine production by many cell types, activates endothelium
	IL-4, IL-13	Stimulate and amplify T_H2-cell response
	IL-3, IL-5, GM-CSF	Promote eosinophil production and activation
Chemokine	CCL3	Chemotactic for monocytes, macrophages, and neutrophils
Lipid mediator	Leukotrienes C_4, D_4 and E_4	Cause smooth muscle contraction Increase vascular permeability Cause mucus secretion
	Platelet-activating factor	Chemotactic for leukocytes Amplifies production of lipid mediators Activates neutrophils, eosinophils, and platelets

Figure 10.5 **Molecules released by mast cells on stimulation by antigen binding to IgE.** Proteases, histamine, heparin, and TNF-α (shown in red) are prepackaged in granules and are released immediately the mast cell is stimulated by antigen binding. TNF-α is also synthesized after mast-cell activation. The other molecules are synthesized and released only as a result of mast-cell activation.

have been defined on different cell types. Acute allergic reactions involve histamine binding to the H1 receptor on nearby smooth muscle cells and on endothelial cells of blood vessels. Stimulation of the H1 receptor on endothelial cells induces vessel permeability and the entry of other cells and molecules into the allergen-containing tissue, causing inflammation. Smooth muscle cells are induced to contract on binding histamine, which constricts airways, for example, and histamine also acts on the epithelial linings of mucosa to induce the increased secretion of mucus. All these actions produce different effects depending upon the tissue exposed to allergen. Sneezing, coughing, wheezing, vomiting, and diarrhea can all be induced in the course of an allergic reaction.

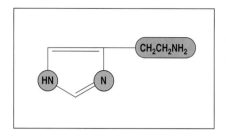

Figure 10.6 The chemical structure of histamine.

Other molecules released from mast-cell granules include mast-cell chymotryptase, tryptase, and other neutral proteases that activate metalloproteases in the extracellular matrix. Collectively, these enzymes break down extracellular matrix proteins. The action of histamine is complemented by that of TNF-α, which is also released from mast-cell granules. TNF-α activates endothelial cells, causing an increased expression of adhesion molecules, thus promoting leukocyte traffic from the blood into the increasingly inflamed tissue (see Figure 8.19, p. 245). Mast cells are unique in their capacity to store TNF-α and release it on demand. At the start of an inflammatory response, mast cells are usually the main source of this cytokine.

Besides the preformed inflammatory mediators in granules, mast cells synthesize and secrete other mediators in response to activation (see Figure 10.5). These include chemokines, cytokines—IL-4 and more TNF-α—prostaglandins, and leukotrienes. The latter are synthesized from fatty acids (Figure 10.7, top panel). All these mediators act locally within the area of tissue surrounding the

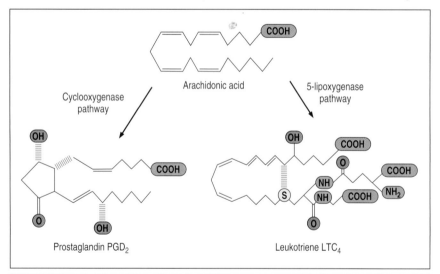

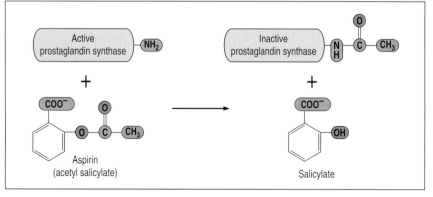

Figure 10.7 Mast cells synthesize prostaglandins and leukotrienes from arachidonic acid by different enzyme pathways. Arachidonic acid, an unsaturated fatty acid, is used as the substrate for the cyclooxygenase pathway to make prostaglandins, and as the substrate for the 5-lipoxygenase pathway to make leukotrienes, as shown in the top panel. Arachidonate is itself made from linolenic acid, one of the essential unsaturated fatty acids that must be provided in the diet because it cannot be made by human cells. Aspirin (acetyl salicylate) prevents the synthesis of prostaglandins by the cyclooxygenase pathway by irreversibly inhibiting the enzyme prostaglandin synthase, as shown in the bottom panel.

activated mast cells. The leukotrienes have activities similar to those of histamine, but are more than a hundred times more potent on a molecule-for-molecule basis. The two kinds of mediator are therefore complementary. Histamine provides a rapid response while the more potent leukotrienes are being made. In the later stages of allergic reactions, leukotrienes are principally responsible for inflammation, smooth muscle contraction, constriction of airways, and mucus secretion from mucosal epithelium. Mast cells also secrete prostaglandin D_2 (PGD_2), which promotes dilation and increased permeability of blood vessels and also acts as a chemoattractant for neutrophils. Aspirin reduces inflammation by inactivating prostaglandin synthase, the first enzyme in the cyclooxygenase pathway. The inactivation is irreversible because aspirin binds covalently to the active site of the enzyme (see Figure 10.7, bottom panel).

The combined effect of the chemical mediators released by mast cells is to attract circulating leukocytes to the site of mast-cell activation, where they amplify the reaction initiated by antigen and IgE on the mast-cell surface. These effector leukocytes include eosinophils, basophils, neutrophils, and T_H2 lymphocytes. In parasite infections these cells work together either to promote explosive reactions that kill the parasite or expel it from the body (see Section 7-13, p. 202). In allergic reactions the same reactions give only detrimental side-effects—damage to the body's tissues and impairment of their function.

10-4 Eosinophils and basophils are specialized granulocytes that release toxic mediators in IgE-mediated responses

Eosinophils are granulocytes whose granules contain arginine-rich basic proteins. In histological sections these proteins stain heavily with eosin, hence the name eosinophil, meaning 'liking eosin' (Figure 10.8). Normally, only a minority of eosinophils are circulating in the blood. Most are resident in tissues, especially in the connective tissue immediately underlying epithelia of the respiratory, gastrointestinal, and urogenital tracts. As with mast cells, the activation of eosinophils by external stimuli leads to the staged release of toxic molecules and inflammatory mediators. Preformed chemical mediators and proteins present in the granules are released first (Figure 10.9, top panel). The normal function of these highly toxic molecules is to kill invading microorganisms and parasites directly. This is followed by a slower induced synthesis and secretion of prostaglandins, leukotrienes, and cytokines, which amplify the inflammatory response by activation of epithelial cells and leukocytes, including more eosinophils (see Figure 10.9, bottom panel).

The eosinophil response is highly toxic and potentially damaging to the host as well as to parasites. Limiting the damage are various control mechanisms.

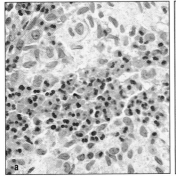 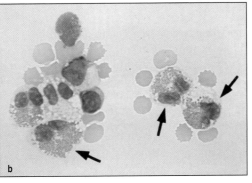

Figure 10.8 Eosinophils have a characteristic staining pattern in histological sections. Panel a shows a light micrograph of a section through a Langerhans' cell histiocytosis in skin that is heavily infiltrated with eosinophils. The eosinophils have bilobed nuclei and stain pink with the stain eosin. Panel b shows a higher-power light micrograph of a blood smear in which partly degranulated eosinophils (arrows) are surrounded by erythrocytes. Panel a courtesy of T. Krausz; panel b courtesy of F. Rosen and R. Geha.

Class of product	Product	Biological effects
Enzyme	Eosinophil peroxidase	Toxic to targets by catalyzing halogenation Triggers histamine release from mast cells
	Eosinophil collagenase	Remodeling of connective tissue matrix
Toxic protein	Major basic protein	Toxic to parasites and mammalian cells Triggers histamine release from mast cells
	Eosinophil cationic protein	Toxic to parasites Neurotoxin
	Eosinophil-derived neurotoxin	Neurotoxin
Cytokine	IL-3, IL-5, GM-CSF	Amplify eosinophil production by bone marrow Cause eosinophil activation
Chemokine	CXCL8	Promotes influx of leukocytes
Lipid mediator	Leukotrienes C_4, D_4 and E_4	Cause smooth muscle contraction Increase vascular permeability Cause mucus secretion
	Platelet-activating factor	Chemotactic to leukocytes Amplifies production of lipid mediators Activates neutrophils, eosinophils, and platelets

Figure 10.9 Activated eosinophils secrete toxic proteins contained in their granules and also produce cytokines and inflammatory mediators. The enzymes and toxic proteins are contained preformed in the granules and are released immediately on eosinophil activation (shown in red). The cytokines, chemokines, and lipid mediators are synthesized only after activation.

When the body is healthy, the number of eosinophils is kept low by limiting their production in the bone marrow. When infection or antigenic stimulation activates T_H2 cells, the IL-5 and other cytokines released by the T_H2 cells stimulate the bone marrow to increase both the production of eosinophils and their release into the circulation. The migration of eosinophils into tissues is controlled by a group of chemokines (CCL5, CCL7, CCL11, and CCL13) that bind to the receptor CCR3, expressed by eosinophils. Of these chemokines, CCL11, also called eotaxin, is particularly important in the migration of eosinophils and is produced by activated endothelial cells, T cells, and monocytes.

The activity of eosinophils is further controlled by modulation of their sensitivity to external stimuli. This is achieved through the regulated expression of FcεRI. In the resting state, eosinophils do not express FcεRI, do not bind IgE, and cannot therefore be induced to degranulate by antigen. Once an inflammatory response is set in motion, cytokines and chemokines in the inflammatory site induce eosinophils to express FcεRI. The expression of Fcγ receptors and complement receptors on the eosinophil surface also increases, facilitating the binding of eosinophils to pathogen surfaces coated with IgG and complement.

The potential of eosinophils to cause tissue damage is clearly seen in patients who have abnormally high numbers of eosinophils. Certain T-cell lymphomas, for example, constitutively secrete IL-5, which continually expands the eosinophil population. This expansion is detected by a greatly increased number of eosinophils in the blood (**hypereosinophilia**). With so many circulating eosinophils, only a small proportion need to be activated for their effects to be seriously disruptive. Patients with hypereosinophilia can suffer damage to the endocardium of the heart (Figure 10.10) and to nerves, leading to heart failure and neuropathy. Both of these effects are thought to be due to cytotoxic and neurotoxic proteins released from eosinophil granules.

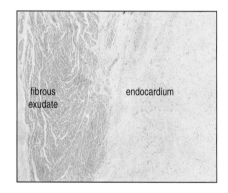

fibrous exudate   endocardium

Figure 10.10 The presence of an abnormally large number of circulating eosinophils causes damage to the heart. The photograph shows a section of endocardium from a patient with hypereosinophilic syndrome. The damaged tissue is characterized by an organized fibrous exudate and an endocardium thickened by fibrous tissue. Although the patient has large numbers of circulating eosinophils, these cells are not seen in the injured endocardium, which is thought to be damaged by the contents of granules released by circulating eosinophils. Photograph courtesy of D. Swirsky and T. Krausz.

In localized allergic reactions, mast-cell degranulation and the activation of T_H2 cells cause the accumulation of activated eosinophils at the site. Their presence is characteristic of chronic allergic inflammation; for example, eosinophils are considered the principal cause of the airway damage that occurs in chronic asthma.

Basophils are granulocytes whose granules stain with basic dyes such as hematoxylin (Figure 10.11). The granules of basophils package a similar, but not identical, set of mediators to those of mast cells. Although basophils resemble mast cells in some ways, they are developmentally more closely related to eosinophils. They share a common stem-cell precursor, as well as requiring similar growth factors, including IL-3, IL-5, and GM-CSF. The production of eosinophils and basophils seems to be reciprocally regulated, so that a combination of TGF-β and IL-3 promotes the maturation of basophils while suppressing that of eosinophils. Basophils are normally present in very low numbers in the circulation and are therefore more difficult to study than other leukocytes. Their functions in defense against parasites seem similar to those of eosinophils. Like eosinophils, basophils are recruited into sites of allergic reactions, where they are activated to degranulate by antigen cross-linking the IgE bound to the FcεRI on the basophil cell surface.

Mast cells, eosinophils, and basophils often act in concert. Mast-cell degranulation initiates the inflammatory response, which then recruits eosinophils and basophils. Eosinophil degranulation releases **major basic protein**, which in turn causes the degranulation of mast cells and basophils. This latter effect is augmented by any of the cytokines—IL-3, IL-5, and GM-CSF—that affect the growth, differentiation, or activation of eosinophils and basophils.

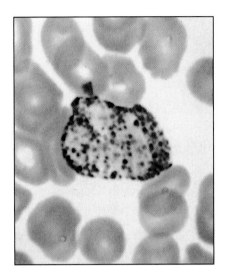

Figure 10.11 Light micrograph of basophil stained with Wright's Giemsa. The basophil in the center is surrounded by erythrocytes in this blood smear. Magnification × 1000. Photograph courtesy D. Friend.

10-5 Mast cells, basophils, and eosinophils can amplify an IgE response started by T_H2 cells

IgE is responsible for most allergic responses (Figure 10.12). The production of IgE is favored when the immune system is challenged with small quantities

IgE-mediated allergic reactions			
Syndrome	Common allergens	Route of entry	Response
Systemic anaphylaxis	Drugs Serum Venoms Peanuts	Intravenous (either directly or following rapid absorption)	Edema Increased vascular permeability Tracheal occlusion Circulatory collapse Death
Wheal and flare	Insect bites Allergy testing	Subcutaneous	Local increase in blood flow and vascular permeability
Allergic rhinitis (hay fever)	Pollens (ragweed, timothy, birch) Dust-mite feces	Inhaled	Edema of nasal mucosa Irritation of nasal mucosa
Bronchial asthma	Pollens Dust-mite feces	Inhaled	Bronchial constriction Increased mucus production Airway inflammation
Food allergy	Shellfish Milk Eggs Fish Wheat	Oral	Vomiting Diarrhea Pruritis (itching) Urticaria (hives) Anaphylaxis

Figure 10.12 Allergic reactions mediated by IgE.

of antigen and when the cytokine IL-4 is present at the time that naive CD4 T cells are presented with antigen. Under these circumstances CD4 T cells tend to make a T$_H$2 response (see Section 8-19, p. 263), which then produces more IL-4 and additional cytokines that stimulate B cells to switch their immunoglobulin isotype to IgE (see Section 7-5, p. 192). Initial sensitization to an allergen is thus favored by circumstances that promote the production of antigen-specific T$_H$2 cells and the production of IgE, and disfavored by conditions that produce T$_H$1 cells. The principal function of IL-4 seems to be to facilitate the IgE response; in mice unable to make this cytokine the main defect is diminished IgE synthesis.

Once started, an IgE response can be amplified through the participation of mast cells, basophils, and activated eosinophils. These cells bind IgE to their FcεRI, which can then be cross-linked by antigen, causing cellular activation. As part of this activation the cells secrete IL-4 and express CD40 ligand. Like T$_H$2 cells, the activated mast cells, basophils, and eosinophils can drive class switching and IgE production by B cells (Figure 10.13). These cellular interactions can occur at the site of allergic reactions, where B cells are seen to form germinal centers.

10-6 Common allergens are small proteins inhaled in particulate form that stimulate an IgE response

In the previous sections we have considered the effector cells responsible for the symptoms of allergies. Here we turn to the types of antigen to which people become sensitized. Because T cells are stimulated by antigen-derived peptides, the allergens that provoke type I responses are invariably proteins. Much human allergy is caused by a limited number of airborne proteins that are inhaled. Small amounts of the protein enter the body by crossing the mucosa of the respiratory tract and then stimulate T$_H$2 responses in the local lymphoid tissues. Allergens are only a small proportion of all the proteins that humans inhale, and a central goal of research on allergy is to define what distinguishes proteins that become allergens from those that do not. A definitive answer has yet to be found. Figure 10.14 shows some of the properties that characterize inhaled allergens.

Most allergens are small, soluble proteins that are present in dried-up particles of material derived from plants and animals. Examples are pollen grains, the mixture of dried cat skin and saliva that forms dander, and the dried feces of the house dust mite *Dermatophagoides pteronyssimus*. The light dry particles become airborne and are inhaled by humans. Once inhaled, the particles become caught in the mucus bathing the epithelia of the airways and lungs. They then rehydrate, releasing the antigenic proteins. These antigens are carried to professional antigen-presenting cells within the mucosa. The antigens

Figure 10.13 Antigen binding to IgE on mast cells leads to amplification of IgE production. IgE secreted by plasma cells binds to the high-affinity IgE receptor on mast cells (illustrated here) and basophils. When the surface-bound IgE is crosslinked by antigen, these cells express CD40 ligand (CD40L) and secrete IL-4, which in turn binds to IL-4 receptors (IL-4R) on the activated B cell, stimulating isotype switching by B cells and the production of more IgE. These interactions can occur *in vivo* at the site of allergen-triggered inflammation, for example in bronchial-associated lymphoid tissue.

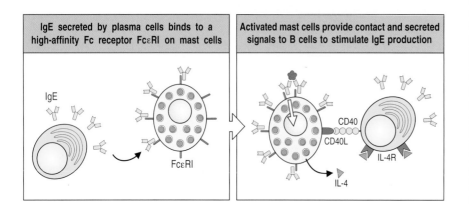

are then processed and presented by these cells to CD4 T cells, stimulating a T_H2 response that leads to the production of IgE and its binding to mast cells (Figure 10.15). Small soluble protein antigens are more efficiently leached out of particles and penetrate the mucosa. A substantial proportion of allergens are proteases, and it is likely that their enzymatic activities facilitate the breakdown of the particle, the release of allergen, and the generation of peptides that stimulate T_H2 cells.

The major allergen responsible for more than 20% of the allergies in the human population of North America is a cysteine protease derived from *D. pteronyssimus*. Advances in the heating and cooling of homes, offices, and other buildings are believed to be responsible for the prevalence of this allergy because they provide an environment that encourages both the growth of *D. pteronyssimus* and the desiccation of its feces. The air currents created by forced-air heating, air conditioners, and vacuum cleaners all help to move the particles into the air, where they will be breathed in by the buildings' human inhabitants.

The cysteine protease of *D. pteronyssimus* is related to the protease papain, which comes from the papaya fruit and is used in cooking as a meat tenderizer. Workers involved in the commercial production of papain become allergic to the enzyme, an example of an occupational immunological disease. Similarly, the protease subtilisin, the 'biological' component of some laundry detergents, causes allergy in laundry workers. Chymopapain, a protease related to papain, is used in medicine to degrade intervertebral disks in patients with sciatica. A rare complication of this procedure affects patients who are sensitized to chymopapain; they experience an acute systemic allergic response to the enzyme, an example of systemic anaphylaxis.

10-7 Predisposition to allergy has a genetic basis

In the caucasoid populations of Europe and North America, up to 40% of people are more likely than the rest of the population to make IgE responses to common environmental antigens. Allergists call this predisposed state **atopy**. As a group, atopic people have higher levels of soluble IgE and circulating eosinophils than nonatopic people. Family studies indicate a genetic basis for atopy, with an involvement of genes on chromosomes 5 and 11.

The candidate gene on chromosome 11 encodes the β subunit of FcεRI, and the region implicated on chromosome 5 contains a cluster of genes including

Features of inhaled allergens that may promote the priming of T_H2 cells that drive IgE responses	
Molecular type	Proteins, because only they induce T-cell responses
Function	Allergens are often proteases
Low dose	Favors activation of IL-4-producing CD4 T cells
Low molecular weight	Allergen can diffuse out of particle into mucus
High solubility	Allergen is readily eluted from particle
High stability	Allergen can survive in desiccated particle
Contains peptides that bind host MHC class II	Required for T-cell priming

Figure 10.14 Properties of inhaled allergens.

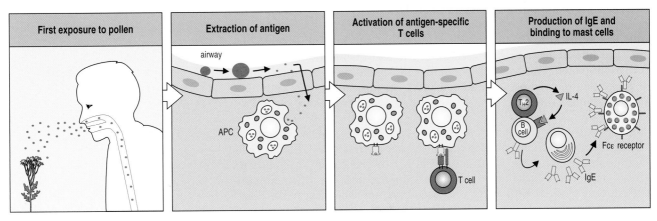

Figure 10.15 Sensitization to an inhaled allergen. Antigens that leach from inhaled pollen are taken up by antigen-presenting cells (APC) in the mucosa of the airways. These activate naive T cells to become T_H2 effector cells, which secrete IL-4. IL-4 binds to the B cell's IL-4 receptor, causing the B cell to switch its immunoglobulin isotype and secrete IgE. The IgE binds to FcεRI on mast cells.

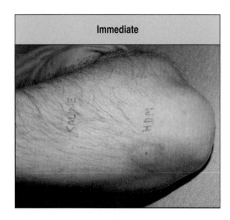
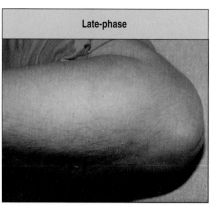

Figure 10.16 Allergic reactions consist of an immediate reaction followed later by a late-phase reaction. The photographs show the immediate reaction (left panel) and the late-phase reaction (right panel) that develop from the injection of house dust mite (HDM) antigen into the epidermis of an allergic individual. In the left panel the site of allergen injection is labeled HDM and the site of the control injection of saline is labeled 'saline'. HDM produces a wheal-and-flare reaction. The wheal is the raised area of skin with the injection site at the center, the flare is the redness (erythema) spreading out from the wheal. In the right panel, the swelling from the HDM injection has spread to involve surrounding tissue. Photographs courtesy of A.B. Kay.

those for IL-3, IL-4, IL-5, IL-9, IL-12, IL-13, and GM-CSF. These cytokines are directly involved in isotype switching, eosinophil survival, and mast-cell proliferation. For example, an inherited sequence difference in the promoter region of the IL-4 gene is correlated with the raised IgE levels seen in atopic people. HLA class II polymorphism also affects the IgE response to certain allergens. An IgE response to several pollen antigens from ragweed is correlated with the expression of the HLA class II allotype DRB1*1501. Such associations imply that certain HLA class II:peptide combinations predispose to stimulation of a T_H2 response.

10-8 IgE-mediated allergic reactions consist of an immediate response followed by a late-phase response

One way in which clinical allergists detect a person's sensitivity to allergens is to observe the reaction when small quantities of common allergens are injected into the skin. Substances to which a person is sensitive produce a characteristic inflammatory reaction called a **wheal and flare** at the site of injection within a few minutes (Figure 10.16, left panel). Substances to which the person is not allergic produce no such reaction. Because of its rapid appearance, the wheal and flare is called an **immediate reaction**. Such reactions are the direct consequence of IgE-mediated mast-cell degranulation in the skin. Released histamine and other mediators cause increased permeability of local blood vessels, whereupon fluid leaves the blood and produces local swelling (edema). The swelling produces the wheal at the injection site, and the increased blood flow into the surrounding area produces the redness that is the flare. Immediate reactions can last for up to 30 minutes and the relative intensity of the wheal and flare varies.

Some 6–8 hours after the immediate reaction has subsided, a second reaction—the **late-phase reaction**—occurs at the site of injection (see Figure 10.16, right panel). This consists of a more widespread swelling and is due to the leukotrienes, chemokines, and cytokines synthesized by mast cells after IgE-mediated activation.

Although skin tests are a good way of determining whether a patient has developed an allergen-specific IgE response, they cannot assess, for example, whether the response to a particular allergen is the cause of a patient's asthma—an allergic reaction in which the airways become inflamed, constricted, and blocked with mucus. To ascertain this, allergists make direct measurements of a person's breathing capacity in the absence and presence of inhaled allergen (Figure 10.17). On inhalation of an allergen to which a patient is sensitized, mucosal mast cells in the respiratory tract degranulate. The released mediators cause immediate constriction of the bronchial smooth muscle, which results in the expulsion of material from the lungs by

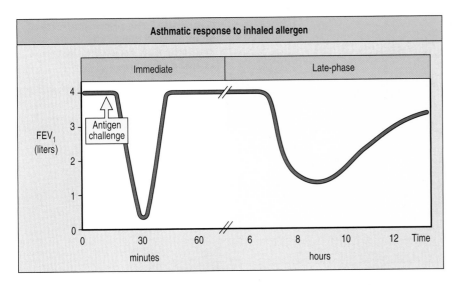

Figure 10.17 The course of an asthmatic response to inhaled allergen. The response is measured in terms of breathing capacity: the forced expiratory volume of air in 1 second (FEV_1). This is the maximum volume of air that a person can forcibly breathe out of the lungs in one second after taking a deep breath. The immediate response in the lungs finishes within an hour, to be followed some 6 hours later by the late-phase response.

coughing, and difficulty in breathing. As in the skin test, the immediate response in the lungs finishes within an hour, to be followed some 6 hours later by the late-phase response, which is due to leukotrienes and other mediators. In allergies to inhaled antigens, such as chronic asthma, the late-phase reaction is the more damaging. It induces the recruitment of leukocytes, particularly eosinophils and T_H2 lymphocytes, into the site and, if antigen persists, the late-phase response can easily develop into a chronic inflammatory response in which allergen-specific T_H2 cells promote eosinophilia and IgE production.

10-9 The effects of IgE-mediated allergic reactions vary with the site of mast-cell activation

When sensitized people are reexposed to an allergen, the effect of the IgE-mediated reaction varies depending on the allergen and the tissues with which it comes in contact. Only those mast cells at the site of exposure degranulate and, once released, the preformed mediators are short-lived. Their effects on blood vessels and smooth muscles are therefore confined to the immediate vicinity of the activated mast cells. The more sustained effects of the late-phase response are also restricted to the site of allergen exposure, because the leukotrienes and other induced mediators are also short-lived. The anatomy of the site of contact also determines how quickly the inflammatory reaction subsides. The tissues most commonly exposed to allergens are the mucosa of the respiratory and gastrointestinal tracts, the blood, and connective tissues. Airborne allergens irritate the respiratory tract and food-borne antigens the gastrointestinal tract; the blood and connective tissues receive allergens through insect bites and other wounds and also from absorption via the gut and respiratory mucosa (Figure 10.18).

For parasite infections, interactions between antigen, IgE, and mast cells trigger violent muscular contractions that can expel worms from the gastrointestinal tract and increase fluid flow to wash them out. In the lungs, the muscular spasms and increased mucus secretion that result from mast-cell activation can be seen as a way to expel organisms attached to respiratory epithelium, such as lung flukes. In allergy, this crude defense is mistakenly ranged against nonthreatening particles and proteins that are often smaller than any microorganism. The violence and power of the IgE-mediated response means that the clinical manifestations of an allergic reaction can vary markedly with the amount of IgE that a sensitized person has made, the amount of allergen triggering the reaction, and the route by which it entered

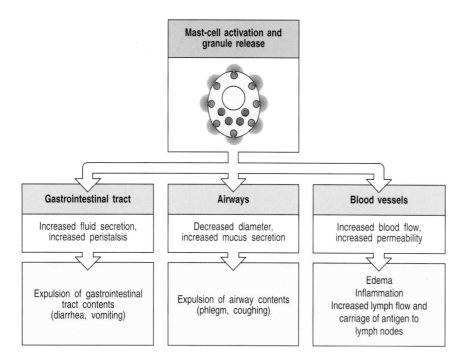

Figure 10.18 The physical effects of IgE-mediated mast-cell degranulation vary with the tissue exposed to allergen.

the body. In the next four sections we examine how allergic reactions vary when they occur in different tissues.

10-10 Systemic anaphylaxis is caused by allergens in the blood

When an allergen enters the bloodstream it can cause widespread activation of the connective tissue mast cells associated with blood vessels. This causes a dangerous hypersensitivity reaction called **systemic anaphylaxis**. During systemic anaphylaxis, disseminated mast-cell activation causes both an increase in vascular permeability and a widespread constriction of smooth muscle. Fluid leaving the blood causes the blood pressure to drop drastically, a condition called **anaphylactic shock**, and the connective tissues to swell. Damage is sustained by many organ systems and their function is impaired. Death is usually caused by asphyxiation due to constriction of the airways and swelling of the epiglottis (Figure 10.19). In the USA, more than 160 deaths a year are the result of anaphylaxis, which is the most extreme over-reaction of the body's defenses. Because this form of immunity is fatal rather than protective it was called '*ana*phylaxis', meaning anti-protection, to contrast with the '*pro*phylaxis' afforded by protective immunity.

Potential allergens are introduced directly into the blood by stings from wasps, bees, and other venomous insects; these account for about a quarter of the fatalities from anaphylaxis in the USA. Drug injections can also lead to anaphylaxis. Systemic anaphylaxis can also be provoked by food or drugs taken orally if the allergens that they contain are absorbed rapidly from the gut into the blood. Foods that can cause anaphylaxis include peanuts and brazil nuts. Anaphylactic reactions can be fatal, but treatment with an injection of epinephrine (adrenaline) will usually bring them under control. Epinephrine stimulates the reformation of tight junctions between endothelial cells. This reduces their permeability and prevents fluid loss from the blood, diminishing tissue swelling and raising blood pressure. Epinephrine also relaxes constricted bronchial smooth muscle and stimulates the heart (Figure 10.20). The danger of anaphylaxis is such that patients with known anaphylactic sensitivity to insect venoms or food are advised to carry a syringe full of epinephrine at all times.

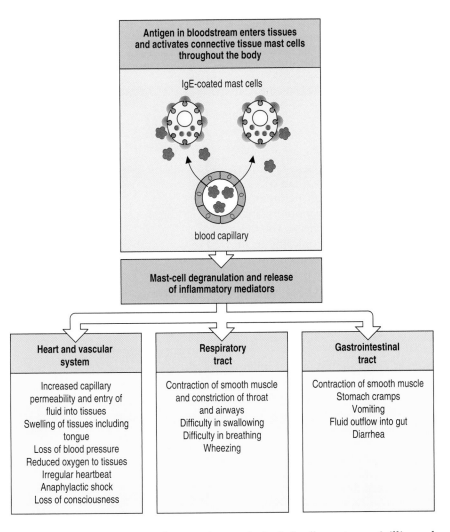

Figure 10.19 Systemic anaphylaxis is caused by allergens that reach the bloodstream and activate mast cells throughout the body.

The most common cause of systemic anaphylaxis is allergy to penicillin and related antibiotics, which accounts for about 100 fatalities a year in the USA. Penicillin is a small organic molecule with a reactive β-lactam ring. On ingestion or injection of the drug, the β-lactam ring can be opened up to produce covalent conjugates with proteins of the body, creating new 'foreign' epitopes. These modified proteins can then stimulate a variety of hypersensitivity reactions, as described in later sections. In some individuals, the response is dominated by T_H2 cells and thus by B cells producing IgE specific for the new epitopes. When penicillin is given to a person who has been sensitized in this manner, it causes anaphylaxis and even death. For this reason, clinicians take care to avoid prescribing any drug to patients who have a history of allergy to that drug. Unfortunately, penicillin cannot be modified to remove its allergic potential because the reactive β-lactam ring is essential for its antibiotic activity.

Reactions resembling anaphylaxis can occur in the absence of specific interaction between an allergen and IgE. Such **anaphylactoid reactions** are caused by other stimuli that induce mast-cell degranulation and can be occasioned by exercise or by certain drugs and chemicals. Anaphylactoid reactions are also treated with epinephrine.

10-11 Rhinitis and asthma are caused by inhaled allergens

Allergens most commonly enter the body by inhalation. Mild allergies to inhaled antigens are common, being manifested as violent bursts of sneeezing

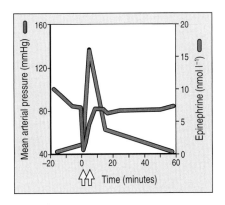

Figure 10.20 The change in blood pressure during systemic anaphylaxis and its treatment with epinephrine. Time 0 indicates the time at which the anaphylactic reaction was reported by the patient. The arrows indicate the times at which injections of epinephrine were given.

Figure 10.21 Allergic rhinitis is caused by allergens entering the respiratory tract. Histamine and other mediators released by activated mast cells increase the permeability of local capillaries and activate nasal epithelium to produce mucus. Eosinophils attracted into the tissues from the blood become activated and release their inflammatory mediators. Activated eosinophils are shed into the nasal passages.

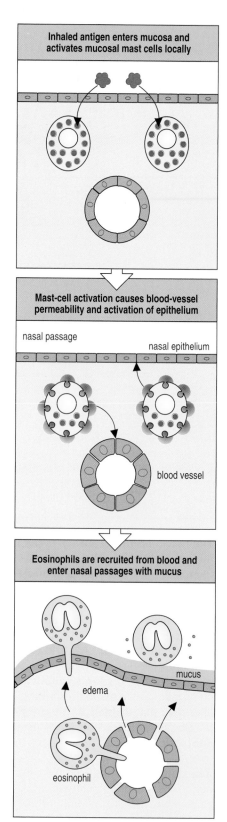

and a runny nose, a condition called **allergic rhinitis** or hay fever. It is caused by allergens that diffuse across the mucous membrane of the nasal passages and activate mucosal mast cells beneath the nasal epithelium (Figure 10.21). Allergic rhinitis is characterized by local edema leading to obstruction of the nasal airways and a nasal discharge of mucus that is rich in eosinophils. There is also a generalized irritation of the nose due to histamine release. The reaction can extend to the ear and throat, and the accumulation of fluid in the blocked sinuses and eustachian tubes is conducive to bacterial infection. The same exposure to allergen that produces rhinitis can affect the conjunctiva of the eyes, where the reaction is called **allergic conjunctivitis**. It produces itchiness, tears, and inflammation. Although these reactions are uncomfortable and distressing, they are generally of short duration and cause no long-lasting tissue damage.

Much more serious is **allergic asthma**, a condition suffered by 130 million people worldwide, in which allergic reactions cause chronic difficulties in breathing, such as shortness of breath and wheezing. Asthma is triggered by allergens activating submucosal mast cells in the lower airways of the respiratory tract. Within seconds of mast-cell degranulation there is an increase in the fluid and mucus being secreted into the respiratory tract, and bronchial constriction due to contraction of the smooth muscle surrounding the airway. Chronic inflammation of the airways is a characteristic feature of asthma, involving a persistent infiltration of leukocytes, including T_H2 lymphocytes, eosinophils, and neutrophils (Figure 10.22). The overall effect of the asthmatic attack is to trap air in the lungs, making breathing more difficult. Patients with allergic asthma often need treatment, and asthmatic attacks can prove fatal.

Although allergic asthma is initially driven by a response to a specific allergen, the chronic inflammation that subsequently develops seems to be perpetuated in the absence of further exposure to the allergen. In **chronic asthma** the airways can become almost totally occluded by mucus plugs (Figure 10.23). A generalized hypersensitivity in the airways also develops, and environmental factors other than reexposure to specific allergen can trigger asthmatic attacks. Typically, the airways of chronic asthmatics are hyper-responsive to chemical irritants commonly present in air, such as cigarette smoke and sulfur dioxide. Disease can be exacerbated by immune responses to bacterial or viral infections of the respiratory tract, especially when they are dominated by T_H2 cells. For this reason, chronic asthma is classified as a type IV hypersensitivity reaction caused by T cells.

10-12 Urticaria, angioedema, and eczema are allergic reactions in the skin

Allergens that activate mast cells in the skin to release histamine cause raised itchy swellings called **urticaria** or **hives**. Urticaria means 'nettle-rash' and the word derives from *Urtica*, the Latin name for stinging nettles; the origin of the word hives is unknown. This reaction is essentially the same as the immediate wheal-and-flare reaction caused by the deliberate introduction of allergens into the skin in tests to determine allergy (see Section 10-8), which can also be produced by the injection of histamine alone (Figure 10.24). Activation of mast

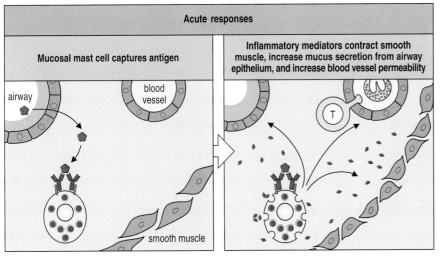
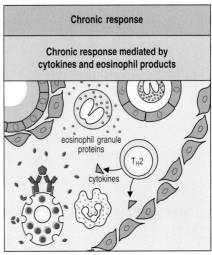

Acute responses

Chronic response

Mucosal mast cell captures antigen

Inflammatory mediators contract smooth muscle, increase mucus secretion from airway epithelium, and increase blood vessel permeability

Chronic response mediated by cytokines and eosinophil products

airway

blood vessel

smooth muscle

eosinophil granule proteins

T$_H$2

cytokines

Figure 10.22 The acute response in allergic asthma leads to T$_H$2-mediated chronic inflammation of the airways. In sensitized individuals, mast cells carrying IgE specific for the allergen are present in the mucosa of the airways, as shown in the first panel. As shown in the second panel, cross-linking of specific IgE on the surface of mast cells by inhaled allergen triggers them to secrete inflammatory mediators, causing bronchial smooth muscle contraction and increased mucus secretion from mucosal epithelium, which together lead to airway obstruction. Increased blood vessel permeability also caused by inflammatory mediators leads to edema and an influx of inflammatory cells, including eosinophils and T$_H$2 lymphocytes. Activated mast cells and T$_H$2 cells secrete cytokines that also augment eosinophil activation and degranulation, which causes further tissue injury and influx of inflammatory cells, as shown in the third panel. The end result is chronic inflammation, which can then cause irreversible damage to the airways.

cells in deeper subcutaneous tissue leads to a similar but more diffuse swelling called **angioedema** (or angioneurotic edema). Urticaria and angioedema can arise as a result of any allergy to a food or drug if the allergen gets carried to the skin by the bloodstream, and they are among the many reactions that occur during systemic anaphylaxis. Insect bites are a common cause of urticaria, and such local reactions usually occur without inducing a more general anaphylaxis.

A more prolonged allergic response in the skin is observed in some atopic children. This condition is called **atopic dermatitis** or **eczema**. The word eczema is of Greek derivation and means to 'break out' or 'boil over'. The condition is characterized by an inflammatory response that causes a chronic and itching skin rash with associated skin eruptions and fluid discharge. This response has similarities to that occurring in the bronchial walls of asthmatics. Eczema frequently presents in families with a history of asthma and allergic rhinitis, and is often associated with high IgE levels. However, the severity of the dermatitis is not readily correlated with exposure to particular allergens or to the levels of allergen-specific IgE. Thus the etiology of eczema remains poorly understood. Neither is it understood why eczema usually clears in adolescence, whereas rhinitis and asthma more often persist throughout life.

Figure 10.23 Inflammation of the airways in chronic asthma restricts breathing. Panel a shows a light micrograph of a section through the bronchus of a patient who died of asthma; there is almost total occlusion of the airway by a mucus plug (MP). The small white circle is all that is left of the lumen of the bronchus. In panel b, a light micrograph at higher magnification gives a closer view of the bronchial wall. It shows injury to the epithelium lining the bronchus, accompanied by a dense inflammatory infiltrate that includes eosinophils, neutrophils, and lymphocytes. L, lumen of the bronchus. Photographs courtesy of T. Krausz.

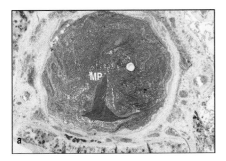
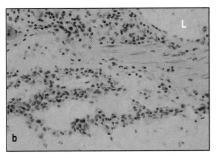

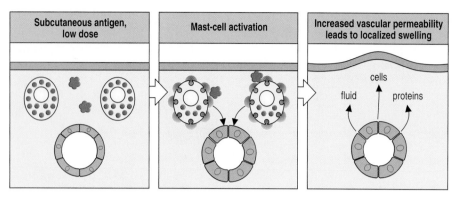

| Subcutaneous antigen, low dose | Mast-cell activation | Increased vascular permeability leads to localized swelling |

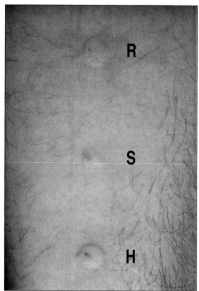

Figure 10.24 Allergen-induced release of histamine by mast cells in skin causes localized swelling. As shown in the panels on the left, allergen introduced into the skin of a sensitized individual causes mast cells in the connective tissue to degranulate. The histamine released dilates local blood vessels, causing rapid swelling due to leakage of fluid and proteins into tissues. The photograph shows the raised swellings (wheals) that appear 20 minutes after intradermal injections of ragweed pollen antigen (R) or histamine (H) into a person who is allergic to ragweed. The small wheal at the site of saline injection (S) is due to the volume of fluid injected into the dermis. Photograph courtesy of R. Geha.

10-13 Food allergies cause systemic effects as well as gut reactions

Humans probably eat a greater variety of foodstuffs than any other living animal. All human food is derived from plants and animals and contains a large number of different proteins, many of which are potentially immunogenic. As food passes down the gastrointestinal tract, the proteins are degraded by proteases into peptides of ever-decreasing size. These are a potential source of peptides for presentation to T_H2 cells. Despite the quantity and variety of food that humans eat, IgE is made against only an extremely small proportion of the proteins ingested. However, people sensitized to a particular protein will be allergic to any food containing that protein. Foods that commonly cause allergies include grains, nuts, fruits, legumes, fish, shellfish, eggs, and milk.

Once sensitized to a food allergen, any subsequent intake causes a marked and immediate reaction that can drive the person from the dining room. The allergen passes across the epithelial wall of the gut and binds to IgE on the mucosal mast cells associated with the gastrointestinal tract. The mast cells degranulate, releasing their mediators, principally histamine. The local blood vessels become permeable, and fluid leaves the blood and passes across the gut epithelium into the lumen of the gut. Meanwhile, contraction of smooth muscles of the stomach wall produces cramps and vomiting, while the same reaction in the intestine produces diarrhea (Figure 10.25). These reactions probably evolved originally to expel gut parasites; in the allergic reaction they serve to expel the allergen-containing food from the human body. This goal is indeed accomplished, but at the expense of dehydration, weakness, and waste of food.

In addition to allergic reactions localized in the gut, food allergens also produce reactions in other tissues, notably the skin. Depending on the timing and course of the gut reaction and of the uptake of allergen from the gut, allergen can enter the circulation and be transported elsewhere in the body. Mast cells in the connective tissue in the deeper layers of the skin tend to be activated by such blood-borne allergens, and their degranulation produces urticaria and angioedema. In this context, orally administered drugs behave similarly to food; they too can produce intestinal reactions, urticaria, and angioedema in sensitized individuals.

10-14 People with parasite infections and high levels of IgE rarely develop allergic disease

In tropical countries, parasitic helminth infections are endemic; more than 1 billion people worldwide are heavily and persistently infected. A universal feature of helminth infection is the stimulation of CD4 T_H2-like responses that produce raised levels of IgE, and increased numbers of eosinophils in the

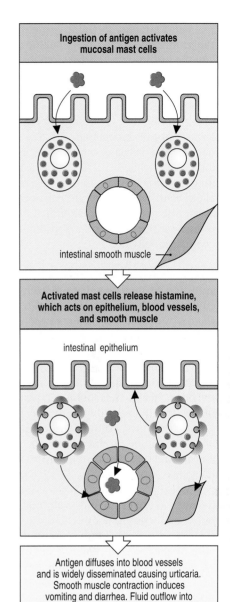

Figure 10.25 Ingested allergen can cause vomiting, diarrhea, and urticaria. Localized reactions are caused by histamine acting on intestinal epithelium, blood vessels and smooth muscles. Urticaria is caused by antigen that enters blood vessels and is carried to the skin.

blood and mast cells in tissues. Only a small fraction of the IgE is parasite-specific; the remainder is highly heterogeneous and represents the product of nonspecific, polyclonal B- and T-cell activation by the parasite. Despite the abundance and variety of IgE in their system, people with helminth infections are rarely afflicted by allergic disease.

Several facts may contribute to the resistance of helminth-infected individuals to allergy. One is that nonspecific IgE competes with parasite-specific IgE, or any allergen-specific IgE, for binding to the FcεRI receptors on mast cells, basophils, and activated eosinophils. This limits the extent to which parasite antigen binding to specific IgE triggers IgE-mediated effector mechanisms, allowing the parasite to escape killing by activated eosinophils or expulsion from the body by the action of activated mast cells. Another contributing factor is that T-cell responses are generally suppressed in chronic parasitic infections, probably owing to the participation of regulatory T cells and the inhibitory effects of IL-10, TGF-β, and nitric oxide.

For the populations of western Europe and North America, helminth and many other parasite infections have largely been eradicated. In these populations the prevalence of IgE-mediated allergy and asthma has been steadily increasing. The 'hygiene hypothesis' proposes that this increase has been caused by better hygiene, vaccination to prevent infection, and the increased use of antibiotics and other drugs to stop infections. Children are exposed to fewer and less heavy infections than were their parents, with the result that their immune systems are insufficiently used and become less successfully regulated. In other words, a lack of practice in dealing with real infections can reveal a propensity to perceive danger where it does not exist. Supporting the hygiene hypothesis are family studies showing that exposure of children to more infections, and at a younger age, reduces the likelihood that they will develop atopic allergic reactions.

10-15 Allergic reactions are prevented and treated by three complementary approaches

Three distinct strategies are used to reduce the effects of allergic disease. The first strategy is one of prevention—to modify a patient's behavior and environment so that contact with the allergen is avoided. Allergen-containing foods are avoided, houses are refurnished in ways that discourage mites, pets are kept outside, and desert vacations or sea cruises are taken during the pollen season.

The second strategy is pharmacological—to use drugs that reduce the impact of any contact with allergen. Such drugs block the effector pathways of the allergic response and limit the inflammation after IgE-induced activation of mast cells, eosinophils, and basophils. Antihistamines reduce rhinitis and urticaria by preventing histamine from binding to H1 histamine receptors on vascular endothelium and thus increasing vascular permeability. Corticosteroids, which suppress leukocyte function generally, are often administered topically or systemically to suppress the chronic inflammation of asthma, rhinitis, or eczema. Cromolyn sodium prevents the degranulation of activated mast cells and granulocytes and is inhaled by asthmatics as a prophylactic to prevent attacks. Epinephrine is used to treat anaphylactic reactions.

The third strategy in the treatment of allergy is immunological—to prevent the production of allergen-specific IgE. One way of achieving this is to modulate the antibody response so that it shifts from one dominated by IgE to one dominated by IgG. A procedure called **desensitization**, which was first described in 1911 and is used in a similar way today, can achieve this for some patients and some allergies. Patients are given a series of allergen injections in which the dose is initially very small and is gradually increased. Successful treatments are associated with the production of allergen-specific antibodies of the IgG4 isotype, which, like IgE, is a product of a T_H2 response, and increased levels of IL-10. An occasional consequence of the injections used for desensitization is anaphylaxis, because the patient is being exposed to the allergen to which they are sensitized. For this reason 'allergy shots' should always be given under carefully controlled conditions in which patients are monitored for the early symptoms of systemic anaphylaxis and, if need be, given epinephrine.

A more recent approach to desensitization is to vaccinate patients with allergen-derived peptides that are known to be presented by HLA class II molecules to CD4 T_H2 cells. The aim is to induce anergy of allergen-specific T cells *in vivo* by decreasing the expression of the CD3:T-cell receptor complex at the T_H2 cell surface. In principle, the advantage of this method over the older approach is that anaphylaxis should never be triggered by the injections, as only the native allergen protein and not the peptide vaccine can interact with allergen-specific IgE. Factors complicating this approach are the HLA class II polymorphisms determining which allergen-derived peptides can be presented by any individual. Any vaccine with general applicability to a human population must contain sufficient different peptides so that an anergizing response can be produced irrespective of HLA class II type. Alternatively, vaccines could be custom-made from peptides selected on the basis of a patient's HLA class II type.

Because allergies are a prevalent and increasing problem for humans in the richer countries in the world, there is much interest within the biotechnology and pharmaceutical industries in developing new approaches to the relief or cure of allergy. One potential set of targets for drugs are the signaling pathways that cells of the immune system use to enhance the IgE response. For example, inhibitors of the cytokines IL-4, IL-5, or IL-13 could block such pathways. Conversely, the administration of cytokines that promote T_H1 responses could shift the antibody response away from IgE and towards IgG. In experiments on mice, IFN-γ and IFN-α have both been shown to reduce IL-4-stimulated IgE synthesis. The high-affinity IgE receptor is also a potential target for drugs that bind the receptor and prevent the arming of mast cells with allergen-specific IgE.

Summary

Type I hypersensitivity reactions are caused by protein allergens binding to IgE molecules and activating mast cells. The activated mast cells release a variety of chemical mediators that orchestrate a local state of inflammation. Smooth muscles constrict, blood vessels dilate, and eosinophils and basophils enter the affected area. The effects of IgE-mediated reactions vary with the route of allergen entry to the body and the tissue affected. Inhaled allergens, for example plant pollens and animal dander, activate mast cells in the respiratory tract. Rhinitis is caused by reactions in the upper airways; asthma by reactions in the lower airways. Insect stings deliver allergens into the skin, where mast-cell mediators cause hives and urticaria. Allergens in certain foods, such as peanuts or shrimp, trigger mast cells of the gastrointestinal tract, resulting in vomiting and diarrhea. When food allergens are absorbed into the blood they become disseminated throughout the body, causing systemic mast-cell activation and resulting in widespread urticaria or even systemic anaphylaxis. The violent reactions activated by IgE are believed to have evolved as a defense against parasites. In allergic reactions this defensive mechanism is misguidedly aimed at environmental proteins that pose no threat.

Type II, III, and IV hypersensitivity reactions

IgG antibodies and specific T cells can cause both acute and chronic adverse hypersensitivity reactions. A variety of different effector mechanisms are involved, which cause inflammatory reactions and tissue destruction.

10-16 Type II hypersensitivity reactions are caused by antibodies specific for altered components of human cells

Occasional side-effects seen after the administration of certain drugs are hemolytic anemia caused by the destruction of red blood cells, or thrombocytopenia caused by the destruction of platelets. These are examples of type II hypersensitivity reactions and have been associated with the antibiotic penicillin, quinidine (a drug used to treat cardiac arrhythmia), and methyldopa, which is used to reduce high blood pressure. In each case, chemically reactive drug molecules bind to surface components of red blood cells or platelets and create new epitopes to which the immune system is not tolerant (Figure 10.26).

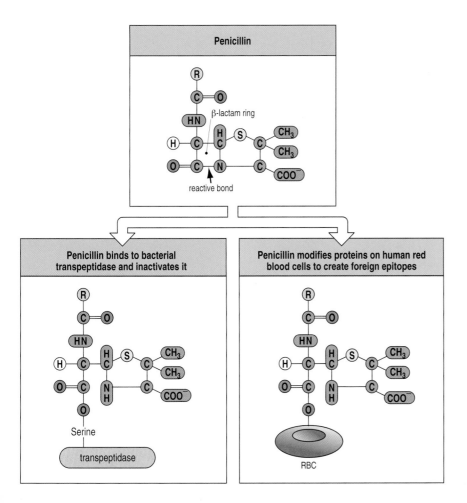

Figure 10.26 **Penicillin and other small-molecule drugs can modify human cells so that they display foreign epitopes**. Penicillin exerts its antibacterial action by mimicking the substrate for bacterial transpeptidase, which is required for bacterial cell wall synthesis. On binding to transpeptidase, a reactive bond in the β-lactam ring of penicillin opens and forms a covalent bond with an amino-acid residue in the active site of the transpeptidase, thereby inactivating the enzyme permanently (lower left panel). The same mechanism occasionally results in molecules of penicillin becoming covalently bonded to surface proteins of human cells (lower right panel). This modification of human proteins creates new epitopes that can act like foreign antigens. Red blood cells (RBC) are the cells most commonly modified in this way.

Figure 10.27 Penicillin–protein conjugates stimulate the production of anti-penicillin antibodies. Red cells that have been covalently bonded to penicillin (P) are phagocytosed by macrophages, which process the penicillin-modified proteins and present peptide antigens to specific CD4 T cells. These are activated to become effector T_H2 cells, which stimulate antigen-specific B cells to produce antibodies against the penicillin-modified epitope.

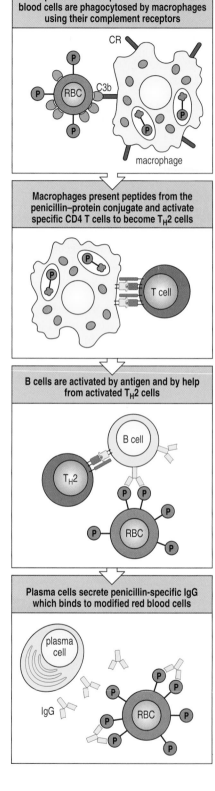

These epitopes stimulate the formation of IgM and IgG antibodies that are specific for the conjugate of drug and cell-surface component.

The penicillin-modified red cells acquire a coating of the complement component C3b as a side-effect of complement activation by the bacterial infection for which the drug has been given. This facilitates their phagocytosis by macrophages via complement receptors. These cells process the penicillin-modified protein and present peptides from it to specific CD4 T cells, which are activated to become effector T_H2 cells. These then stimulate antigen-specific B cells to produce antibodies against the penicillin-modified epitope (Figure 10.27). Binding of antibodies to the drug-conjugated cells activates complement by the classical pathway, resulting in either cell lysis by the terminal complement components or receptor-mediated phagocytosis by macrophages in the spleen (Figure 10.28).

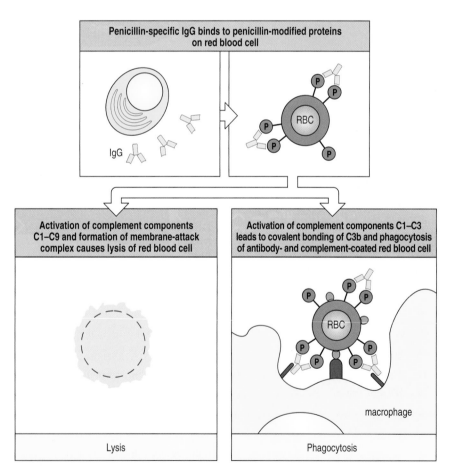

Figure 10.28 Binding of antibodies to penicillin-modified red cells makes them susceptible to complement- **mediated lysis or to phagocytosis via Fc receptors and complement receptors.**

10-17 Type III hypersensitivity reactions are caused by immune complexes formed from IgG and soluble antigens

Complexes of soluble protein antigens and their high-affinity IgG antibodies are generated in almost all immune responses and in most situations they are cleared without causing tissue damage. Immune complexes vary greatly in size, from a simple complex of one antigen and one antibody molecule through to large aggregates containing millions of antigen and antibody molecules. The larger aggregates fix complement efficiently and are readily taken up by phagocytes and removed from the circulation. Smaller immune complexes are less efficient at fixing complement; they tend to circulate in the blood and become deposited in blood vessel walls. When these complexes accumulate at such sites, they become capable of fixing complement and initiating tissue-damaging inflammatory reactions through their interactions with the Fc receptors and complement receptors on circulating leukocytes. This type of hypersensitivity is called type III hypersensitivity. Complement activation produces C3a and C5a. The former stimulates mast cells to release histamine, causing urticaria; the latter recruits inflammatory cells into the tissue. Platelets accumulate around the site of immune-complex deposition, and the clots that they form cause the blood vessels to burst, producing hemorrhage in the skin.

The size of the immune complexes formed at a particular time and place is strongly influenced by the relative concentrations of soluble antigen and antibody (Figure 10.29). Whether large immune complexes can be formed also depends on the size and complexity of the antigen; most antigens that will be encountered in normal circumstances contain multiple epitopes and thus can, in principle, form extensive immune complexes by cross-linking antibodies. When antigen is in excess, as occurs early in the immune response, each antigen-binding site on an antibody binds an antigen molecule, producing small immune complexes that often contain a single antibody molecule and two antigen molecules. Late in the immune response, antibodies are in excess; in these circumstances each antigen molecule binds to several antibody molecules. At intermediate times, when the amounts of antigen and antibody are more evenly balanced, larger immune complexes are formed in

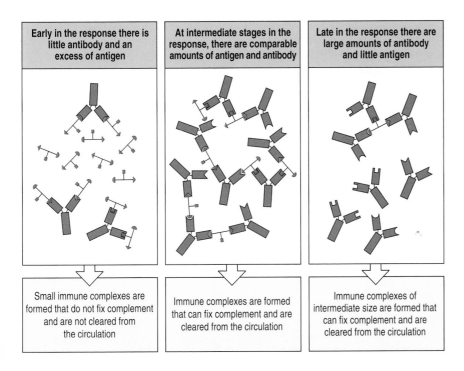

Early in the response there is little antibody and an excess of antigen	At intermediate stages in the response, there are comparable amounts of antigen and antibody	Late in the response there are large amounts of antibody and little antigen
Small immune complexes are formed that do not fix complement and are not cleared from the circulation	Immune complexes are formed that can fix complement and are cleared from the circulation	Immune complexes of intermediate size are formed that can fix complement and are cleared from the circulation

Figure 10.29 Immune complexes of different sizes and stoichiometries are formed during the course of an immune response.

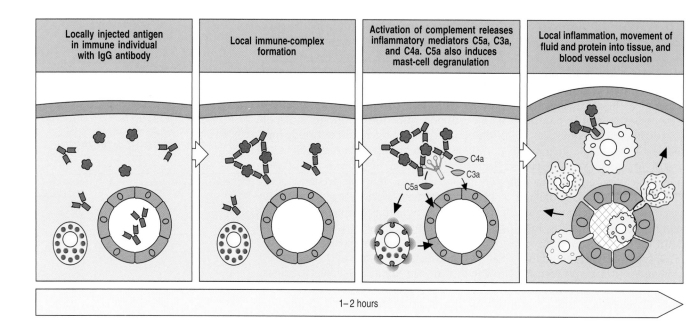

| Locally injected antigen in immune individual with IgG antibody | Local immune-complex formation | Activation of complement releases inflammatory mediators C5a, C3a, and C4a. C5a also induces mast-cell degranulation | Local inflammation, movement of fluid and protein into tissue, and blood vessel occlusion |

1–2 hours

which several antibody and antigen molecules are cross-linked. A minimum of two IgG molecules per complex is needed to fix complement, so it is at the beginning of the immune response, when soluble immune complexes fix complement poorly, that they are most likely to circulate in the blood and become deposited in blood vessel walls.

In people who have made IgG against a soluble protein, a type III hypersensitivity reaction can be experimentally induced in the skin by subcutaneous injection of the antigen. Specific IgG diffuses from the blood into the connective tissue at the site of injection and combines with antigen to form immune complexes. The complexes activate complement, creating an inflammatory reaction that draws leukocytes and antibodies into the site of injection. Here the Fc and complement receptors of leukocytes engage the immune complexes, activating the cells and further propagating the local inflammatory reaction. This type of reaction was first described by Nicholas-Maurice Arthus and is called an **Arthus reaction** (Figure 10.30). In humans, Arthus reactions usually appear as localized areas of erythema and hard swelling (induration) that subside within a day. Such reactions can often be seen at the site of injections used to desensitize IgE-mediated allergies. The dependence of the Arthus reaction on immune-complex interactions with Fc receptors is demonstrated by the failure to produce Arthus reactions in mice lacking the γ chain common to all Fc receptors.

10-18 Systemic disease caused by immune complexes can follow the administration of large quantities of soluble antigens

During the late nineteenth century and the first half of the twentieth century, diphtheria, scarlet fever, tetanus, and other life-threatening bacterial infections were treated by injecting patients with serum taken from horses that had been immunized with these bacteria or their toxins. The horse antibodies helped human patients to control and clear the infection but could also produce a systemic type III hypersensitivity reaction that became known as **serum sickness**. This condition occurred some 7–10 days after the administration of horse serum and was characterized by chills, fevers, rash, arthritis, vasculitis, and sometimes glomerulonephritis. The cause of serum sickness is the formation of antibodies against the foreign horse proteins and the

Figure 10.30 Localized deposition of immune complexes within a tissue causes a type III hypersensitivity reaction. In sensitized individuals, the introduction of allergen into a tissue leads to the formation of immune complexes with IgG in the extracellular fluid. The immune complexes activate complement and recruit inflammatory cells to the site, causing a hard swelling. Platelets accumulate in the capillary, leading to occlusion and rupture of the vessel, causing erythema.

Route	Resulting disease	Site of immune-complex deposition
Intravenous (high dose)	Vasculitis	Blood vessel walls
	Nephritis	Renal glomeruli
	Arthritis	Joint spaces
Subcutaneous	Arthus reaction	Perivascular area
Inhaled	Farmer's lung	Alveolar/capillary interface

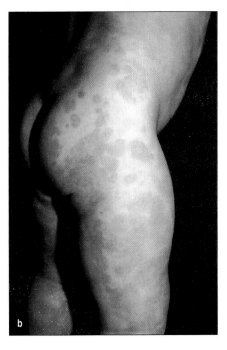

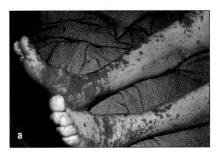

Figure 10.31 The pathology of type III hypersensitivity reactions is determined by the sites of immune-complex deposition. The table shows the types of reaction that result from different routes of antigen entry to the body. Serum sickness follows the intravenous administration of large amounts of foreign antigen. Photographs show hemorrhage in the skin (panel a) and urticarial rash (panel b) resulting from serum sickness. Photographs courtesy of R. Geha.

deposition of small immune complexes in tissues; the symptoms depend on which tissues are affected (Figure 10.31).

Therapeutic administration of serum from immunized horses is rarely used today. One remaining application is the use of horses to prepare anti-venom antisera for neutralizing the effects of snake-bite. However, the symptoms of serum sickness are now seen in other circumstances in which patients have received an infusion of large amounts of a foreign protein. Transplant patients given mouse monoclonal antibodies specific for human T cells to prevent rejection can get serum sickness. It also occurs occasionally in patients who have had a myocardial infarction (heart attack) and are treated with the bacterial enzyme streptokinase to degrade their blood clots. Serum sickness can also result from the intravenous administration of large amounts of a drug such as penicillin, which binds to host proteins on, for example, erythrocytes (see Section 10-16) and provokes an IgG response. This type of reaction can occur in people with no history of allergy to penicillin. Drug-induced serum sickness is now the most common example of this condition.

The onset of serum sickness coincides with the synthesis of antibodies, which form immune complexes with the antigenic proteins. Because the serum is loaded with antigen, large quantities of immune complexes are formed and dispersed throughout the body. The complexes fix complement and activate leukocytes bearing Fc receptors or complement receptors. These activated cells create an inflammatory response that causes widespread damage (Figure 10.32).

The formation of immune complexes induces the clearance of the antigenic proteins by the normal phagocytic pathways; consequently, serum sickness is of limited duration unless additional injections of the foreign antigen are given. If a second dose of antigen is given after the effects of the first dose have subsided, a secondary response will follow, with disease symptoms being manifested within a day or two of the second injection.

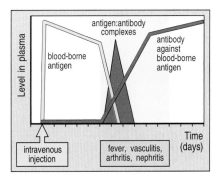

Figure 10.32 Serum sickness is a classic example of a transient immune-complex mediated syndrome. An injection of a large amount of a foreign antigen into the circulation leads to an antibody response. These antibodies form immune complexes with the circulating foreign antigens. The complexes are deposited in small blood vessels and activate complement and phagocytes, inducing fever and the symptoms of vasculitis, nephritis, and arthritis. All these effects are transient and resolve once the foreign antigen has been cleared.

A disease similar to serum sickness can be seen in certain infections in which the immune system fails to clear the pathogen, and both the infection and the immune response persist. For example, in subacute bacterial endocarditis, or in chronic viral hepatitis, the multiplying pathogens continue to produce antigens, and plasma cells continue to make antibodies. Immune complexes are continually being generated, deposited, and cleared, processes that can cause injury to small blood vessels and nerves in many organs, including the skin and kidneys.

Some common inhaled antigens tend to provoke an IgG rather than an IgE response and cause type III hypersensitivity reactions. Continued exposure to the antigen leads to the formation of immune complexes and their deposition in the walls of the alveoli in the lungs. Such deposits stimulate an inflammatory response. The resulting accumulation of fluid, antigen, and cells impedes the lungs' normal function of gas exchange, and the patient experiences difficulty in breathing. Occupations in which workers are exposed daily to quantities of the same airborne antigens can lead to this condition. The immune systems of farm workers exposed to hay dust and mold spores are often provoked in this manner, giving rise to the occupational disease called **farmer's lung**. Without a change in work habits the continuing deposition of immune complexes in the alveolar membranes leads to irreversible lung damage.

10-19 Type IV hypersensitivity reactions are mediated by antigen-specific effector T cells

Hypersensitivity reactions caused by effector T cells specific for the sensitizing antigen are known as type IV hypersensitivity or **delayed-type hypersensitivity reactions** (**DTH**), because they occur 1–3 days after contact with antigen. This time course contrasts with those of antibody-mediated hypersensitivity, which are generally apparent within a few minutes. The amount of antigen required to elicit a type IV hypersensitivity reaction is 100–1000 times greater than that required to produce antibody-mediated hypersensitivity reactions. This difference reflects the intrinsic inefficiency in generating peptide epitopes from protein antigens for presentation by HLA molecules. Antigens that cause common type IV hypersensitivity reactions are shown in Figure 10.33.

The best studied example of a type IV hypersensitivity reaction is the **tuberculin test**, the clinical test used to determine whether a person has been infected with *Mycobacterium tuberculosis*. In the tuberculin test a small amount of protein antigen extracted from *M. tuberculosis* is injected intradermally or intracutaneously. People with immunity to *M. tuberculosis*—those

Type IV hypersensitivity reactions are mediated by antigen-specific effector T cells		
Syndrome	Antigen	Consequence
Delayed-type hypersensitivity	Proteins: Insect venom Mycobacterial proteins (tuberculin, lepromin)	Local skin swelling: Erythema Induration Cellular infiltrate Dermatitis
Contact hypersensitivity	Haptens: Pentadecacatechol (poison ivy) Small metal ions: Nickel Chromate	Local epidermal reaction: Erythema Cellular infiltrate Contact dermatitis

Figure 10.33 Examples of common type IV hypersensitivity reactions. Hapten is the name given to any small molecule that when covalently bonded to a protein stimulates an immune response.

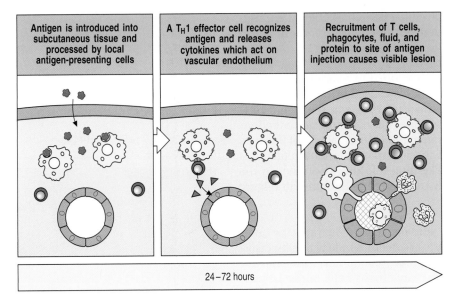

| Antigen is introduced into subcutaneous tissue and processed by local antigen-presenting cells | A T$_H$1 effector cell recognizes antigen and releases cytokines which act on vascular endothelium | Recruitment of T cells, phagocytes, fluid, and protein to site of antigen injection causes visible lesion |

24–72 hours

Figure 10.34 The stages and time course of a type IV hypersensitivity reaction. The first phase involves uptake, processing, and presentation of the antigen by local antigen-presenting cells. In the second phase, antigen-specific memory T cells produced during previous exposure to the antigen migrate into the site of injection and become activated. Because these antigen-specific cells are rare, and there is no inflammation to attract them into the site, it can take several hours for a T cell of the correct specificity to arrive. Activated T$_H$1 cells release mediators that activate local endothelial cells, recruiting an inflammatory cell infiltrate dominated by macrophages and causing the accumulation of fluid and protein. At this point, the lesion becomes apparent.

who have existing tuberculosis, or who have resolved an infection, or or who have been vaccinated with the BCG strain—develop an inflammatory reaction around the site of injection 24–72 hours later. The response is mediated by T$_H$1 cells that recognize peptides derived from the *M. tuberculosis* protein that are presented by HLA class II molecules. The peptides are presented by macrophages and dendritic cells in the vicinity of the injection and initially stimulate tuberculin-specific memory T cells that have left the blood and entered the tissue; these generate effector T$_H$1 cells locally. After activation, the T$_H$1 cells initiate further inflammatory reactions that recruit fluid, proteins, and other leukocytes to the site (Figure 10.34). Each of these phases takes several hours, accounting for the time taken before the response is seen or felt. The activated T$_H$1 cells produce cytokines that mediate these effects (Figure 10.35). In the United States the diagnostic value of the tuberculin skin test for existing infection currently outweighs the benefits of vaccination against tuberculosis, and hence vaccination is not routine. In the UK and

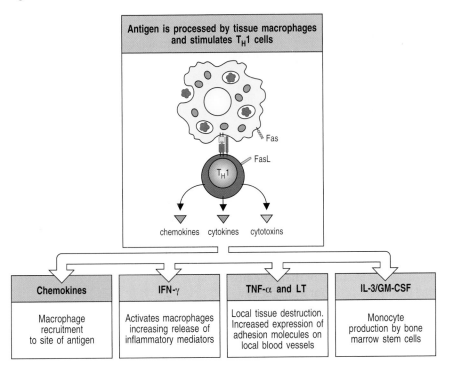

Antigen is processed by tissue macrophages and stimulates T$_H$1 cells

Fas
FasL
T$_H$1

chemokines   cytokines   cytotoxins

Chemokines	IFN-γ	TNF-α and LT	IL-3/GM-CSF
Macrophage recruitment to site of antigen	Activates macrophages increasing release of inflammatory mediators	Local tissue destruction. Increased expression of adhesion molecules on local blood vessels	Monocyte production by bone marrow stem cells

Figure 10.35 Most type IV hypersensitivity reactions are orchestrated by the cytokines released by T$_H$1 CD4 cells in response to antigen. Macrophages or tissue dendritic cells recruited to the site of inflammation by chemokines present antigen and amplify the response. The release of TNF-α and the cytotoxin lymphotoxin (LT, formerly known as TNF-β) affect local blood vessels through their effects on endothelial cells. IL-3 and GM-CSF stimulate the production of macrophages. IFN-γ and TNF-α activate macrophages. Macrophages are killed by LT and by the interaction of Fas on the macrophage with Fas ligand (FasL) on the T cell.

some other European countries, BCG is given to children as one of their routine vaccinations. It is preceded by a tuberculin test, which serves to detect any preexisting infection or immunity.

Type IV hypersensitivity responses can be developed to a variety of antigens in the environment. For example, the dermatitis caused by contact with the North American plants poison ivy (*Toxicodendron radicans*; see Figure 10.1) and poison oak (*T. diversilobium*, whose leaves are oval and untoothed, or shaped rather like oak leaves on young plants) is due to a reaction of this type and involves both CD4 and CD8 T cells. The reaction is caused by pentadecacatechol, a small, highly reactive lipid-like molecule that is present in the leaves and roots of the plant and is easily transferred to human skin (Figure 10.36).

When a person contacts poison ivy, pentadecacatechol penetrates the outer layers of the skin and indiscriminately forms covalent bonds with extracellular proteins and skin cell-surface proteins. On degradation of the chemically modified proteins by skin macrophages and Langerhans' cells, antigenic peptides that carry the pentadecacatechol adduct are generated and presented by HLA class II molecules to T_H1 cells. The cytokines secreted by the T_H1 cells activate macrophages and produce inflammation (see Figure 10.35). When pentadecacatechol penetrates the skin it also crosses the plasma membranes of cells and chemically modifies intracellular proteins. The processing of such modified proteins results in the presentation of chemically modified peptides by HLA class I molecules to CD8 T cells. On activation, the CD8 T cells have the potential to kill any cells that contacted the chemical and present modified peptides on their surfaces.

On first contact with poison ivy, a person might experience only a minor or undetectable reaction. During this primary or sensitizing response, Langerhans' cells or dendritic cells carry pentadecacatechol-modified proteins to the draining lymph node, where the T-cell response is initiated. Once immunological memory has been developed, subsequent contacts with the plant yield unpleasant reactions of the type shown in Figure 10.36. The red, raised, weeping skin lesions are due to heavy infiltration of the contact sites with blood cells, combined with the localized destruction of skin cells and the extracellular matrix that holds the skin together. Because of the delayed nature of the reaction there is plenty of time for a person to transfer pentadecacatechol from the initial site of contact to other parts of the body, a process that often exacerbates the extent of the reaction.

The reaction to poison ivy is an example of **contact sensitivity**, so called because contact with the skin is necessary to initiate the allergic response. Contact sensitivity can also develop to coins, jewelry, and other metallic objects containing nickel. In this case the bivalent nickel ions are chelated by histidine in human proteins; processing of these proteins forms T-cell epitopes to which the immune system responds.

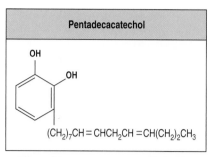

Pentadecacatechol

$(CH_2)_7CH = CHCH_2CH = CH(CH_2)_2CH_3$

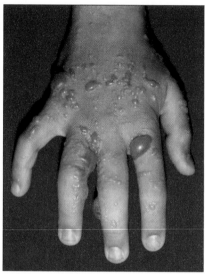

Figure 10.36 Physical contact with poison ivy transfers pentadecacatechol, which causes dermatitis. The chemical structure of pentadecacatechol is shown in the top panel. The photograph shows characteristic blistering skin lesions on the hand of a patient with dermatitis caused by contact with poison ivy (see Figure 10.1 for a photograph of the plant). Photograph courtesy of R. Geha.

Summary

Type II hypersensitivity reactions are mediated by IgG antibodies directed against cell-surface or matrix antigens and are commonly caused by drugs that are being given as treatment for other diseases. Because of their chemical reactivity, drugs bind to surface proteins of human cells, for example erythrocytes or platelets, creating new epitopes that stimulate an antibody response. On binding to the drug-modified cells, the antibodies activate complement, which leads to cell destruction. Type III hypersensitivity reactions are caused by soluble immune complexes formed by IgG binding to the soluble antigens against which they were made. They can be seen when non-human proteins, such as mouse monoclonal antibodies, are given therapeutically. Antibodies specific for the nonhuman proteins are made, and early in the response small immune complexes are formed that are not efficiently cleared from the circulation. These tend to deposit in blood vessels, where they activate complement and inflammation. Depending on the sites of activation, these reactions can cause vasculitis, nephritis, arthritis, or lung disease. Type IV hypersensitivity reactions are caused by effector T cells, often responding to reactive chemicals transferred into the skin by physical contact to cause a contact dermatitis. In some conditions, tissue damage is caused by the activation of macrophages by T_H1 cells and the actions of cytotoxic T cells; in others, such as chronic asthma, cytokines produced by T_H2 cells specific for an inhaled allergen activate eosinophils and other inflammatory cells.

Summary to Chapter 10

The immune system provides the body with powerful defenses against infection. These include adaptive immune responses that have the potential to respond to any structure that is not a normal part of the body. Inevitably, all humans develop adaptive immunity to some foreign substances, such as animal and plant proteins in foods, that are not associated with infection. In general, such immune responses are harmless, but this is not always so, and various allergies or hypersensitivities are caused by the immune system's overreactions to nonthreatening environmental antigens. A first exposure to an allergen is rarely noticeable, but hypersensitivity reactions are brought on by subsequent exposures in which the allergen interacts with previously formed antibodies or stimulates memory lymphocytes. Four types of hypersensitivity reaction are conventionally defined on the basis of the effector mechanisms that cause them. Hypersensitivity reactions of types I, II, and III are triggered by antibodies, whereas type IV hypersensitivity reactions are triggered by effector T cells. In all four types of reaction, recognition of the allergen triggers an unwanted inflammatory response of varying severity and duration.

Questions

Question 10–1
Identify four different ways that an individual may come into contact with an allergen and provide two examples of allergens for each type of contact. (Refer to Figure 10.1.)

Question 10–2
A. Describe how the components of the immune system involved differ between the following subsets of hypersensitivity reactions. (Refer to Figure 10.2.) (i) Type I versus type II/type III; (ii) type I/type II/type III versus type IV.

B. Describe how the antigens differ between the following subsets of hypersensitivity reactions. (i) Type I/type III versus type II; (ii) T_H1- or T_H2-dependent type IV versus type IV involving cytotoxic T cells.

Question 10–3
A. Describe in detail the effector mechanism responsible for mast-cell activation during a type I hypersensitivity reaction.

B. What are the products of mast-cell activation? (Refer to Figures 10.3 and 10.5.)

Question 10–4
A. What type of hypersensitivity reaction typically occurs when penicillin is administered to an allergic individual?

B. Describe in chronological order the events leading to this reaction.

C. What type of hypersensitivity reaction can penicillin cause in a nonallergic individual? (Refer to Figures 10.26, 10.27, 10.28, and 10.31.)

Question 10–5
Explain what causes farmer's lung and its pathology. State what type of hypersensitivity reaction is involved. (Refer to Figures 10.29, 10.30, and 10.31.)

Question 10–6
Describe four ways in which a type III hypersensitivity reaction differs from a type I reaction.

Question 10–7
A. What is the tuberculin test used for and how does it work?

B. What type of hypersensitivity reaction is involved?

C. Why does vaccination render this test useless? (Refer to Figures 10.33, 10.34, and 10.35.)

Question 10–8

Most extracellular antigens are taken up and processed by antigen-presenting cells via phagolysosomes and presented with MHC class II molecules to T_H1 and T_H2 cells.

A. Explain how pentadecacatechol, the antigen responsible for the hypersensitivity reaction against poison ivy, is an exception to this route of antigen processing and gains entry into the cytosolic route of antigen processing.

B. Which T cells are activated and what is the immunological consequence? (Refer to Figures 10.33 and 10.36.)

Question 10–9

A. Explain the effects of histamine binding to the H1 receptor on smooth muscle, mucosal epithelia, and endothelial cells of blood vessels.

B Why are antihistamines used to treat chronic asthma and allergic rhinitis?

Question 10–10

An immunological strategy in the treatment of allergy is called desensitization. Explain (a) how desensitization is carried out and (b) the potential risk associated with this treatment.

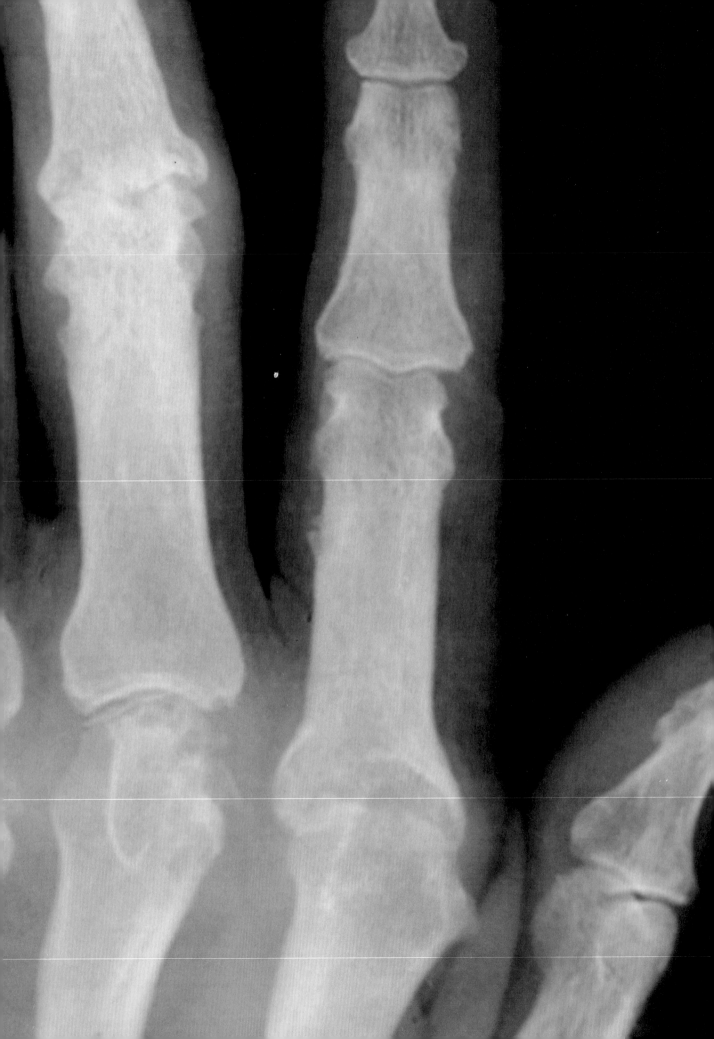

Chapter 11

Disruption of Healthy Tissue by the Immune Response

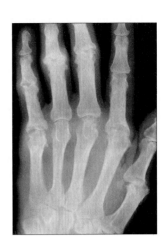

Chapter 10 described how hypersensitivity to harmless environmental antigens leads to acute or chronic disease, depending on the type of antigen and the frequency of exposure to it. In this chapter another set of chronic immunological diseases is considered, diseases caused by adaptive immunity that becomes misdirected at healthy cells and tissues of the body. Such diseases are known as **autoimmune diseases**, and many different types have been described. They are not uncommon; around 5% of the population of developed countries has one or more of these conditions, and their incidence is increasing. Autoimmune diseases can be caused by antibodies that perturb a normal physiological function or by inflammatory T cells that damage healthy cells or tissue at a rate that is beyond the capacity of the body to repair. When the targeted tissue is involved in essential day-to-day functions of the body, the autoimmune disease can become a threat to life.

Autoimmune diseases are caused by unwanted adaptive immune responses; they represent failures of the mechanisms that maintain **self-tolerance**—the prevention of attack on the body's own cells and tissues—in the populations of circulating B and T cells. Although much is known of the effects of autoimmune disease, less is understood about the events that break **tolerance** and cause an autoimmune response. The first part of the chapter describes some of the more common autoimmune diseases; the second part discusses some of the predisposing factors for autoimmune disease and the mechanisms by which tolerance can be broken and lead to disease.

Autoimmune diseases

There are many chronic diseases in which the immune system is active, often causing the affected tissues to be inflamed and abnormally infiltrated by lymphocytes and other leukocytes, but in which there seems to be no associated active infection. These diseases are caused by the immune system itself, which attacks cells and tissues of the body as though they were infected, causing chronic impairment of tissue and organ function. Chronic diseases of this kind are collectively known as autoimmune diseases because they are caused by immune responses directed towards autologous (self) components of the body. An immune response that causes an autoimmune disease is called an **autoimmune response** and it produces a state of **autoimmunity**.

Autoimmune diseases vary widely in the tissues they attack and the symptoms they cause. Some focus on a particular organ or cell type, others act

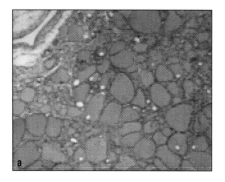 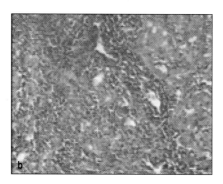

Figure 11.6 Hashimoto's thyroiditis. In a healthy thyroid gland the epithelial cells form spherical follicles containing thyroglobulin (panel a). In patients with Hashimoto's thyroiditis the thyroid gland becomes infiltrated with lymphocytes which destroy the normal architecture of the thyroid gland and can become organized into structures resembling secondary lymphoid tissue (panel b). Courtesy of Ian Lauder.

The proteins thyroglobulin, thyroid peroxidase, TSH receptor and the thyroid iodide transporter are uniquely expressed in thyroid cells. Immune responses to all these autoantigens have been detected in autoimmune thyroid disease. However, the disease symptoms caused by different types of autoimmunity to thyroid antigens are very different. Some conditions cause a loss of thyroid hormones; others increase their production.

In **Graves' disease**, the autoimmune response is focused on antibody production and the symptoms are caused by antibodies that bind to the TSH receptor. By mimicking the natural ligand, the bound antibodies cause a chronic overproduction of thyroid hormones that is independent of regulation by TSH and insensitive to the metabolic needs of the body (see Figure 11.5). This **hyperthyroid** condition causes heat intolerance, nervousness, irritability, warm moist skin, weight loss, and enlargement of the thyroid. Other aspects of Graves' disease are outwardly bulging eyes and a characteristic stare, symptoms caused by the binding of autoantibodies that react with the muscles of the eye. The autoimmune response in Graves' disease is biased towards a CD4 T_H2 response.

In **chronic thyroiditis**, also called **Hashimoto's disease**, the thyroid loses the capacity to make thyroid hormones. Hashimoto's disease seems to involve a CD4 T_H1 cell response, and both antibodies and effector T cells specific for thyroid antigens are produced. Lymphocytes infiltrate the thyroid, causing a progressive destruction of the normal thyroid tissue (Figure 11.6). The formation of organized lymphoid tissue within the thyroid, including germinal centers, gives the impression that the thyroid is being converted into a secondary lymphoid tissue. In Graves' disease, in comparison, there are fewer infiltrating lymphocytes and relatively little tissue destruction. Patients with Hashimoto's disease become **hypothyroid**, and eventually are unable to make thyroid hormone.

Treatment for Hashimoto's disease is replacement therapy with synthetic thyroid hormones taken orally on a daily basis. For Graves' disease the short-term treatment is drugs that inhibit thyroid function. The long-term treatment is removal of the thyroid by surgery, or its destruction by uptake of the radioisotope ^{131}I, followed by daily doses of thyroid hormones.

11-4 The cause of autoimmune disease can be revealed by the transfer of disease with immune effectors

A central goal in the study of autoimmune disease is the identification of autoantigens and effector mechanisms that cause the disease. This is no trivial task. One problem is that various types of autoimmunity are demonstrable in healthy people; a second is that once cell and tissue destruction has begun, it will often initiate further autoimmune responses that are consequences, not causes, of the disease.

The causes of autoimmune diseases involving antibodies are more easily identified than those due to effector T cells. For the former, the transfer of autoantibody from the patient to another human or to an animal will induce disease. Pregnant women who have an antibody-mediated autoimmune disease transport IgG molecules, but not lymphocytes, across the placenta to the fetal circulation. When the mother has an antibody-mediated disease such as Graves' disease, the baby is born with the symptoms of disease. As the baby grows and maternal IgG degrades, disease symptoms gradually go away. For diseases that could affect the baby's growth, treatment involves the removal of antibody from the circulation by total exchange of blood plasma (Figure 11.7).

When healthy rats are injected with serum from patients with Graves' disease they start to overproduce thyroid hormones and show symptoms of Graves' disease. In the past, such laboratory assays were used for the immunological diagnosis of Graves' disease. This whole-animal approach has now been superseded by assays that use a cultured line of rat thyroid cells. IgG fractions prepared from patients' plasma are added to thyroid-cell cultures and their capacity to cause cell activation and proliferation is assessed by measurement of the production of cyclic AMP and the synthesis of DNA, respectively. A complementary assay determines whether antibodies in a patient's plasma compete with TSH for binding to the TSH receptor of thyroid membranes isolated from pigs.

Because lymphocytes cannot pass from the maternal to the fetal circulation, babies born to mothers who have T cell-mediated autoimmune diseases do not show disease symptoms. Observations on human pregnancy therefore provide no positive information on the role of effector T cells in human autoimmune disease; neither do experiments in which T cells from patients are transferred into experimental animals. Such experiments fail to work because human T cells cannot recognize antigens presented by the MHC molecules of most other species. Only in models of autoimmune disease in inbred strains of rodents has it been possible to transfer disease with T cells and show that they cause the disease.

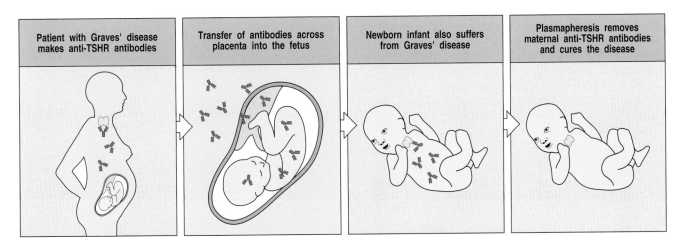

| Patient with Graves' disease makes anti-TSHR antibodies | Transfer of antibodies across placenta into the fetus | Newborn infant also suffers from Graves' disease | Plasmapheresis removes maternal anti-TSHR antibodies and cures the disease |

Figure 11.7 Temporary symptoms of antibody-mediated autoimmune diseases can be passed from affected mothers to their newborn babies. TSHR, thyroid-stimulating hormone receptor.

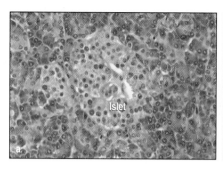
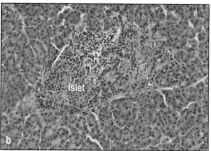

Figure 11.8 Comparison of histological sections of a pancreas from a healthy person and a patient with insulin-dependent diabetes mellitus (IDDM). Panel a shows a micrograph of a tissue section through a healthy human pancreas, showing a single islet. The islet is the discrete light-staining area in the center of the photograph. It is composed of hormone-producing cells, including the β cells that produce insulin. Panel b shows a micrograph of an islet from the pancreas of an IDDM patient with acute onset of disease. The islet shows insulitis characterized by a mainly lymphocytic infiltration at the islet periphery. Both tissue sections are stained with hematoxylin and eosin; magnification × 250. Photographs courtesy of G. Klöppel.

11-5 Insulin-dependent diabetes mellitus is caused by the selective destruction of insulin-producing cells in the pancreas

Insulin is secreted from the pancreas in response to the increased blood glucose level arising after a meal. By binding to surface receptors, insulin stimulates the body's cells to take up glucose and incorporate it into carbohydrates and fats. **Insulin-dependent diabetes mellitus (IDDM)**, also called type I diabetes or juvenile-onset diabetes, is an autoimmune disease caused by the selective destruction of the insulin-producing cells of the pancreas. Because insulin is a major regulator of cellular metabolism it is essential for children's normal growth and development. Symptoms of disease are usually manifested in childhood or adolescence and they rapidly progress to coma and death in the absence of treatment. IDDM principally affects populations of European origin, in which one person in 300 is a sufferer. This distribution, and the impact of the disease on young children, has made IDDM a major target for research in the countries of western Europe, North America, and Australia.

Scattered within the exocrine tissue of the pancreas are the **islets of Langerhans**, small clumps of endocrine cells that make the hormones insulin, glucagon, and somatostatin. The pancreas contains about half a million islets, each consisting of a few hundred cells. Each islet cell is programmed to make a single hormone: α cells make glucagon, β cells make insulin, and δ cells make somatostatin.

In patients with IDDM, antibody and T-cell responses are made against insulin, glutamic acid decarboxylase, and other specialized proteins of the pancreatic β cell. Which of these responses cause disease remains unclear. CD8 T cells specific for some peptide antigens unique to β cells are believed to mediate β-cell destruction, gradually decreasing the number of insulin-secreting cells. Individual islets become successively infiltrated with lymphocytes, a process called **insulitis**. β cells comprise about two-thirds of the islet cells; as they die, the architecture of the islet degenerates. A healthy person has about 10^8 β cells, providing an insulin-making capacity much greater than that needed by the body. This excess, and the slow rate of β-cell destruction, means that disease symptoms do not become manifest until years after the start of the autoimmune response. Disease commences when there are insufficient β cells to provide the insulin necessary to control the level of glucose in the blood (Figure 11.8).

The usual treatment for patients with IDDM is daily injection with insulin purified from the pancreas of pigs or cattle. Because of amino-acid sequence differences between human insulin and animal insulins, some patients develop an immune response to animal insulin. The antibodies they make against insulin can have two effects: they reduce the activity of the insulin, and they form soluble immune complexes that can lead to further tissue

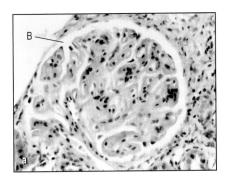
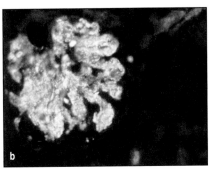
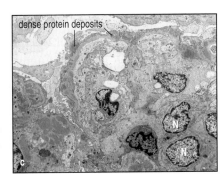

Figure 11.9 Deposition of immune complexes in the kidney glomeruli in systemic lupus erythematosus (SLE). Panel a shows a section through a glomerulus of a patient with SLE. Deposition of immune complexes causes thickening of the basement membrane (B). In panel b a similar kidney section is stained with fluorescent anti-immunoglobulin antibodies, revealing the presence of immunoglobulin in the basement membrane deposits. Panel c is an electron micrograph of part of a glomerulus. Dense protein deposits are seen between the glomerular basement membrane and the renal epithelial cells. Neutrophils (N) are also present, attracted by the deposited immune complexes. Photographs courtesy of M. Kashgarian.

damage. Recombinant human insulin produced in the laboratory from the cloned insulin gene is prescribed to patients who make antibodies against animal insulins.

A good animal model for human IDDM is the **non-obese diabetic (NOD)** strain of inbred mice. In these animals diabetes arises spontaneously and has clinical and immunological characteristics that parallel those seen in the human disease. There is also a similar bias towards females in the incidence of disease.

11-6 Autoantibodies against common components of human cells can cause systemic autoimmune disease

So far, this chapter has concentrated on diseases in which a single type of cell is targeted for autoimmune attack. At the other end of the spectrum is **systemic lupus erythematosus (SLE)**, a disease in which the autoimmune response is directed at autoantigens present in almost every cell of the body. SLE is an example of a **systemic autoimmune disease**.

Characteristic of SLE are circulating IgG antibodies specific for constituents of cell surfaces, cytoplasm, and nucleus, including nucleic acids and nucleoprotein particles. The binding of autoantibodies against cell-surface components initiates inflammatory reactions that cause cell and tissue destruction. In turn, these processes release soluble cellular antigens that form soluble immune complexes. On being deposited in blood vessels, kidneys, joints, and other tissues, such complexes can initiate further inflammatory reactions (Figure 11.9). All this provides a disrupted environment in which the immune system is increasingly stimulated to respond to common self components. SLE is thus a chronic inflammatory disease that can affect all tissues of the body. The disease commonly follows a course in which outbreaks of intense inflammation alternate with periods of relative calm. For individual patients the course of the disease is highly variable, both in its severity and in the organs and tissues involved. Many patients with SLE eventually die of the disease because of failure of vital organs such as the kidneys or brain.

Systemic lupus erythematosus was first described as 'lupus erythematosus' on the basis of the butterfly-shaped skin rash (erythema) that can occur on the face and gives the face an appearance like a wolf's head (*lupus* is Latin for wolf) (Figure 11.10). The rash is caused by the deposition of immune

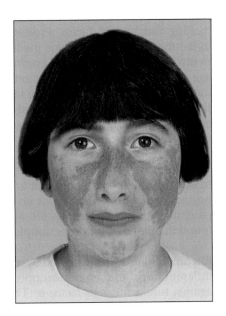

Figure 11.10 The characteristic facial rash of systemic lupus erythematosus. Although this butterfly-shaped rash was first used to recognize the disease, it is only seen in a proportion of patients who have the disease when defined immunologically. Photograph courtesy of M. Walport.

Figure 11.14 Autoantibodies against the acetylcholine receptor cause myasthenia gravis. In a healthy neuromuscular junction, signals generated in nerves cause the release of acetylcholine, which binds to the acetylcholine receptors of the muscle cells, causing an inflow of sodium ions that indirectly causes muscle contraction (upper panel). In patients with myasthenia gravis, autoantibodies specific for the acetylcholine receptor reduce the number of receptors on the muscle-cell surface by binding to the receptors and causing their endocytosis and degradation (lower panel). Consequently, the efficiency of the neuromuscular junction is reduced, which is manifested as muscle weakening.

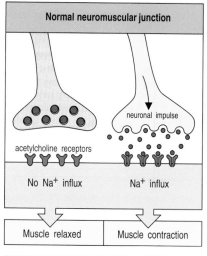

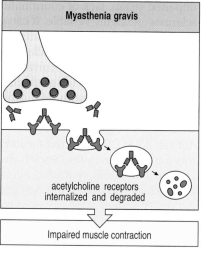

Autoantibodies against cell-surface receptors can act either to stimulate or to inhibit the receptor (Figure 11.15). The autoantibodies in myasthenia gravis prevent the function of the acetylcholine receptor and are called receptor **antagonists**. In contrast, the autoantibodies in Graves' disease facilitate receptor function (see Section 11-3) and are called receptor **agonists**. Autoantibodies of both types can be made against the insulin receptor; they lead to different symptoms. The cells of patients with antagonistic autoantibodies are unable to take up glucose, which accumulates in the blood, causing hyperglycemia and a form of diabetes mellitus resistant to treatment with insulin. In contrast, in patients with agonistic antibodies, cells deplete blood glucose to abnormally low levels. This state of hypoglycemia deprives the brain of glucose, causing light-headedness.

Summary

In autoimmunity the mechanisms of antigen recognition and effector function are identical to those used in the response to pathogens. In contrast, the symptoms of autoimmune disease are highly variable, depending on the triggering autoantigen and the target tissue. Autoimmunity diseases can be classified into three main types corresponding to the type II, type III, and type IV categories of hypersensitivity reactions, with which they share common effector mechanisms. Some autoimmune diseases, such as insulin-dependent diabetes or Graves' disease of the thyroid, are directed against antigens of one particular organ or tissue, and are known as organ- or tissue-specific autoimmune diseases. In contrast, systemic lupus erythematosus (SLE) is a systemic autoimmune disease, which is directed against components common to all cells. Autoimmune diseases caused by antibodies have been more readily studied than those caused by effector T cells, because antibodies can exert an effect when transferred from human patients to experimental animals. Understanding how effector T cells function in human autoimmune disease remains largely dependent on extrapolation from mouse models of autoimmune disease. Many autoimmune diseases tend to

Diseases mediated by antibodies against cell-surface receptors			
Syndrome	**Antigen**	**Antibody**	**Consequence**
Graves' disease	Thyroid-stimulating hormone receptor	Agonist	Hyperthyroidism
Myasthenia gravis	Acetylcholine receptor	Antagonist	Progressive muscle weakness
Insulin-resistant diabetes	Insulin receptor	Antagonist	Hyperglycemia, ketoacidosis
Hypoglycemia	Insulin receptor	Agonist	Hypoglycemia

Figure 11.15 Diseases mediated by antibodies against cell-surface receptors. Antibodies act as agonists when they stimulate a receptor on binding it, and as antagonists when they block a receptor's function on binding it.

Figure 11.18 **E**
centers. During
in the top pane
Ligation of thes
of the autoreac
receptor in the

Although defective alleles of
they are sufficiently commo
ian Jews—for substantial nu
defective alleles. For these i
teins is not made in the thyn
the T-cell repertoire. Startin
responses develop against
endocrine glands. Patients
patients suffer disorders of r
called inherited **autoimmun**
autoimmune polyendocri
(**APECED**). The ectodermal
hair, and fingernails (Figure

Despite the range and severit
stantial lifespans, the most l
cell carcinoma and fulminai
tolerance to major compone
catastrophic disease, but on
ties to the more common a
course of APECED is highly
other genetic and environme
for example, the autoimmun
tion with the fungus *Candid*
body response to *Candida* a
symptom of disease that is c
ure 11.19).

11-11 Insufficient contro
autoimmunity

Even when negative selectio
escape deletion and ente
encounter healthy tissue cel
the surface they will usuall
express the B7 co-stimulato
cells (see Section 6-4, p. 15C
can lead to tolerance by way
one mechanism by which t
that have reached the periph

During T-cell activation the
check by CTLA-4 molecules
Section 6-4, p. 150). Altern
duces soluble and membra
functional in dampening T-

Figure 11.19 **I**
regulator pro
diseases. This
candidiasis–ect
polyglandular (

be chronic conditions in which episodes of acute disease are interspersed
with periods of recovery. Once a tissue becomes inflamed and damaged by an
autoimmune response, the resulting increase in the processing and presentation of self-antigens frequently leads to expansion and diversification of the
autoimmune response. In addition, infections can exacerbate autoimmunity
in a nonspecific manner by inducing the production of inflammatory
cytokines within the infected tissues.

Genetic and environmental factors that predispose to autoimmune disease

At various stages in the development of the immune system and in the development and execution of an immune response, mechanisms are brought into
play that prevent attack on healthy cells and tissues. Together, these result in
the self-tolerance of the immune system. The tolerance-inducing mechanisms involved in adaptive immunity are more elaborate than those for
innate immunity. This is because potentially self-reactive lymphocytes are
continually being generated and a person's population of B cells and T cells
changes according to their experience of infection, vaccination and other
antigenic challenge. Various mechanisms contribute to self-tolerance. During
development, many clones of self-reactive B and T cells are deleted from the
repertoire and die. Of the self-reactive cells that do enter the peripheral circulation, some become anergic, some remain physically separated from the
self antigens to which they could respond, and others are suppressed by regulatory T cells. Naive T-cell activation requires co-stimulation, which itself
depends on infection, and this limits the circumstances under which autoreactive T cells can be activated. Also, in the absence of infection, T cells have
very limited access to tissues other than the blood and lymphoid tissues, and
thus may never encounter tissues expressing potential autoantigens. All
autoimmune diseases involve a breach of one of these mechanisms of self-
tolerance. Both genetic and environmental factors contribute to loss of self-
tolerance and the development of disease-causing autoimmunity. That a
large majority of the human population never suffers autoimmune disease is
testament to the protection provided by the immunological mechanisms of
self-tolerance.

11-9 All autoimmune diseases involve breaking T-cell tolerance

Autoimmune diseases that are caused by autoreactive inflammatory T cells
clearly involve a breach of T-cell tolerance. This is also true for autoimmune
diseases caused by antibodies. The B cells involved have switched isotype and
undergone affinity maturation by somatic hypermutation, and this means
they will have received help from antigen-specific T cells.

During B-cell maturation in the bone marrow, clonal deletion and inactivation of self-reactive B cells prevent the emergence of cells with antigen receptors that bind common molecules of human cell surfaces or plasma. This
process does not prevent the emergence of B cells with specificity for numerous other self antigens that are not present in the bone marrow or plasma.
The activation of these autoreactive B cells is prevented by additional mechanisms, the most important of which is T-cell tolerance, which deprives them
of T-cell help. When an autoreactive B cell is stimulated by its autoantigen it
migrates to the T-cell area of a secondary lymphoid tissue. Because antigen-
activated helper T cells are not available, the antigen-stimulated B cell fails to
enter a primary lymphoid follicle and becomes trapped in the T-cell zone,
where it dies by apoptosis (Figure 11.16).

encode proteins of identical amino-acid sequence, but differ in the quantity of soluble CTLA-4 they make. The allele producing less of the soluble CTLA-4 is associated with susceptibility to Graves' disease, Hashimoto's disease, and insulin-dependent diabetes, whereas the allele producing more of the soluble CTLA-4 is associated with resistance to these autoimmune diseases. Unlike APECED, in which homozygosity for defective AIRE alleles guarantees that the disease will develop, neither CTLA-4 allele can be considered 'defective' and the impact they have on disease is a subtle one. Thus the 'susceptible' allele accounts for 63.4% of the alleles in patients with Graves' disease, compared with 53.2% in healthy controls. Nonetheless, the difference is statistically significant and points to the importance of the balance struck in T-cell activation through the actions of CD28 and CTLA-4.

11-12 Regulatory T cells protect cells and tissues from autoimmunity

Potentially autoreactive circulating CD4 T cells against common autoantigens are present even in healthy people. These cells respond to autoantigens in culture but are kept in check in the body. Among these potentially autoreactive T cells are a class of CD4 T cells that express CD25, the α chain of the low-affinity IL-2 receptor, and are distinct from naive T cells. On contacting self-antigens presented by MHC class II molecules these cells themselves do not proliferate but can suppress the proliferation of naive T cells responding to autoantigens presented on the same antigen-presenting cell (Figure 11.21). They are therefore called **regulatory CD4 T cells (T$_R$)** and they constitute 1–3% of the CD4 T cell population. Their suppressive effects require contact between the two T cells and also involve the secretion of non-inflammatory cytokines such as IL-4, IL-10, and transforming growth factor (TGF)-β. Although once controversial, this active form of tolerance mediated by regulatory CD4 T cells is now considered a major mechanism for protecting the integrity of the body's tissues and organs. The suppressive function of regulatory T cells is dependent on CTLA-4 but not on CD28, which is consistent with a mechanism in which co-stimulation of regulatory T cells involves the binding of B7 on the antigen-presenting cell to CTLA-4 on the regulatory T cell.

Figure 11.20 Dystrophic fingernails in a patient with APECED. Photograph courtesy of Mark S. Anderson.

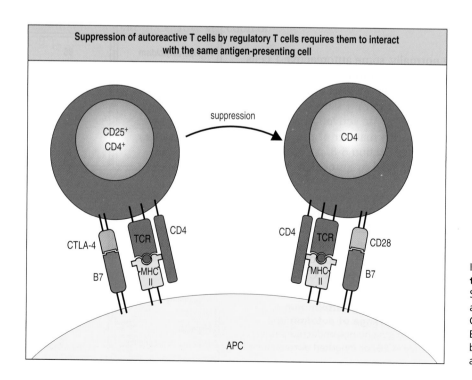

Suppression of autoreactive T cells by regulatory T cells requires them to interact with the same antigen-presenting cell

CD25⁺ CD4⁺ → suppression → CD4

CTLA-4 · TCR · CD4 · B7 · MHC II

CD4 · TCR · CD28 · MHC II · B7

APC

Figure 11.21 CTLA-4 is involved in the action of regulatory T cells. Suppression of an autoreactive T cell by a regulatory T cell is dependent on CTLA-4 on the regulatory T cell binding B7 on the antigen-binding cell, and on both T cells interacting with the same antigen-presenting cell.

Other mechanisms con
encounter their soluble
decrease in IgM expres
pathways of signal trans
tive as a consequence of
in germinal centers are

11-10 Incomplete de
causes autoimmune

Thymic selection of t
immunological self-tole
removes T cells that res
cules of thymic cells. In
peptides normally prese

Emphasizing the impor
vention of autoimmunit
autoimmune regulator
a transcription factor th
are principally expresse
scribed by a subpopulat
The presence of small a
means that peptides der
I and II molecules to for
the T-cell repertoire (see

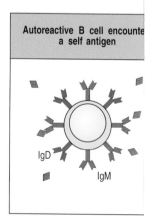

Autoreactive B cell encounte a self antigen

IgD

IgM

Figure 11.17 Peripheral B
encounters its soluble autoa
which leads to B-cell anergy
by a reduced amount of sur
its ligation. A mechanism fo

Figure 11
secondar
entering t
venules (h
in yellow;
autoantig
Because tl
the autore
foreign an
primary fo
receive su
zone (bot

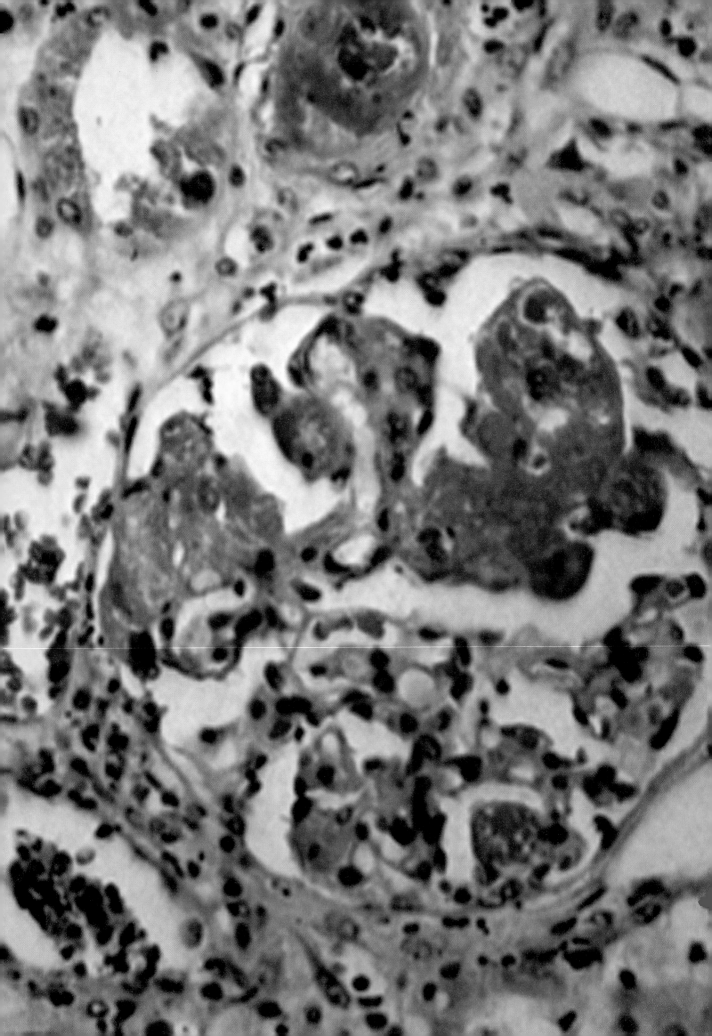

Chapter 12
Manipulation of the Immune Response

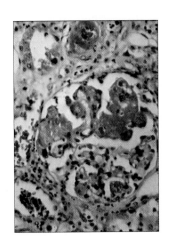

The preceding four chapters of this book describe the various ways in which the immune system responds helpfully or otherwise to environmental antigens. Given the spectrum of disease that results from inadequate or inappropriate immunity, clinical immunologists are driven to search for ways of manipulating the immune response to the benefit of their patients. This chapter examines three areas of medicine where this quest is being pursued—vaccination, organ transplantation, and cancer. In vaccination, the adaptive immune system is manipulated in an antigen-specific manner to stimulate protective immunity, and vaccination has protected billions of people against certain infectious diseases. In transplantation, nonspecific immunosuppression allows a diseased organ to be replaced by a healthy one from another individual, procedures that have extended the lives of many thousands of patients. In contrast, manipulation of the immune response in cancer patients is still at the stage of research and development. The immune response to cancer provides too little too late; the challenge in this case is to enhance a patient's response in ways that prevent the proliferation and spread of malignant disease.

Prevention of infectious disease by vaccination

The modern era of vaccination began in the 1780s with Edward Jenner's use of cowpox as a **vaccine** against smallpox. Although Jenner's procedure was widely embraced, vaccines for other diseases did not emerge until well into the nineteenth century. That was the period when microorganisms were first isolated, grown in culture, and shown to cause disease. Largely by a process of trial and error, methods of inactivating microbes or impairing their ability to cause disease were developed and vaccines were produced. Vaccines were eventually developed against most of the epidemic diseases that had plagued the populations of western Europe and North America. Indeed, by the middle of the twentieth century it began to seem that the combination of vaccines and antibiotics would solve the problem of infection once and for all. Such optimism was soon tempered by the difficulty of developing effective vaccines for some diseases, and by the emergence of new diseases and antibiotic-resistant forms of old ones. This part of the chapter first surveys the vaccines in current use and then turns to the challenge of pathogens that have yet to be tamed by vaccination.

12-1 Viral vaccines are made from whole viruses or viral components

The first medically prescribed vaccine was against smallpox, a disease characterized by a rash of spots that develop into scarring pustules. The very first smallpox vaccines were made from dried pustules taken from people who seemed to have less severe symptoms of the disease. This early vaccine thus contained the smallpox virus itself. Small amounts of this material were given to healthy people, either intranasally or intradermally through a scratch on the arm—a procedure known as **variolation**, from the word **variola**, the Latin name given to both the pustule and to the disease itself. Although successful in many cases, the drawback of variolation was the frequency with which it produced full-blown smallpox, resulting in the death of around one in a hundred of those vaccinated. However, despite the risk, variolation was widely used in the eighteenth century because the threat from smallpox was so much greater. At that time smallpox killed one in four of those infected and there were regular epidemics. In London, for example, more than a tenth of all deaths were due to smallpox.

Jenner's innovation towards the end of the eighteenth century was to use the related cowpox virus as a vaccine for smallpox. The cowpox virus, called **vaccinia**, causes only very mild infections in humans, but the immunity produced gives effective protection against smallpox as well as cowpox because the two viruses have some antigens in common (Figure 12.1). In the nineteenth century, Jenner's vaccine replaced variolation and was eventually responsible for the eradication of smallpox in the twentieth century. The words vaccinia and **vaccination** derive from *vaccus*, the Latin word for cow. Although the term vaccination was once used specifically in the context of smallpox, it now refers to any deliberate immunization that induces protective immunity against a disease.

Jenner's strategy for vaccination against smallpox is not possible for most pathogenic viruses because very few have a natural 'safe' counterpart. Most vaccines in use today are composed of preparations of the disease-causing virus for which the ability to cause disease has been destroyed or weakened. One type of vaccine consists of virus particles that have been chemically treated with formalin or physically treated with heat or irradiation so that they are no longer able to replicate. These are called **killed** or **inactivated virus vaccines**. The vaccines for influenza and rabies, and the Salk polio vaccine, are of this type. Only viruses whose nucleic acid can be reliably inactivated make suitable killed virus vaccines. Such vaccines also have the drawback that large amounts of pathogenic virus must be produced during their manufacture.

A second type of vaccine consists of live virus that has mutated so that it has a reduced ability to grow in human cells and is no longer pathogenic to

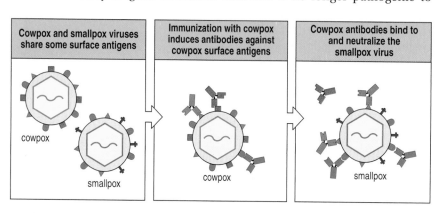

| Cowpox and smallpox viruses share some surface antigens | Immunization with cowpox induces antibodies against cowpox surface antigens | Cowpox antibodies bind to and neutralize the smallpox virus |

cowpox

smallpox

cowpox

smallpox

Figure 12.1 Vaccination with cowpox virus elicits neutralizing antibodies that react with antigenic determinants shared with smallpox virus. Shared antigenic determinants of cowpox also elicit protective T-cell immunity against smallpox (not shown here).

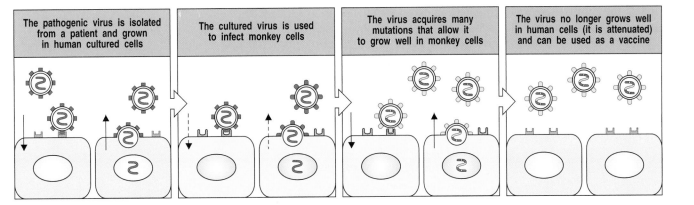

| The pathogenic virus is isolated from a patient and grown in human cultured cells | The cultured virus is used to infect monkey cells | The virus acquires many mutations that allow it to grow well in monkey cells | The virus no longer grows well in human cells (it is attenuated) and can be used as a vaccine |

humans. These vaccines, called **live-attenuated virus vaccines**, are usually more potent at eliciting protective immunity than killed virus vaccines because the attenuated virus can usually replicate to a limited extent and thus mimics a real infection. Most of the viral vaccines currently used to protect humans are live-attenuated vaccines. Attenuated virus is produced by growing the virus in cells of nonhuman animal species. Such conditions select for variant viruses that grow well in the nonhuman host and are consequently less fit for growth in humans (Figure 12.2). The measles, mumps, polio (Sabin vaccine), and yellow fever vaccines consist of live-attenuated viruses. Attenuated viral strains can also arise naturally. As a viral infection passes though the human population the virus generally diversifies by mutation, sometimes producing a strain with reduced pathogenicity. These strains represent natural forms of live-attenuated virus and also provide candidates for vaccines. One of the three live-attenuated polio virus strains that make up the oral polio vaccine is of this type (the Sabin strain 2).

In the human immune response to viruses, protective neutralizing antibodies tend to be directed toward particular surface components of the virus. Vaccines against some viruses have been made by using just these antigenic viral components and they are known as **subunit vaccines**. For example, the vaccine for hepatitis B virus (HBV) consists of the hepatitis B surface antigen (HBsAg), with which most anti-HBV antibodies react. The hepatocytes of patients infected with HBV secrete the surface antigen into the blood as minute particles. The first anti-hepatitis B vaccine was made from HBsAg purified from the blood of infected people, a procedure requiring the careful removal of infective virions. The vaccine now used is made from HBsAg produced by recombinant DNA technology, which avoids this problem. This vaccine provides about 85% of those vaccinated with protective immunity against HBV infection. People who do not respond may lack an effective CD4 T-cell response because their HLA class II allotypes fail to present HBsAg-derived peptides.

There remain, however, viruses for which no vaccine has yet been developed, despite considerable effort. Some are those, such as HIV, that naturally persist in the body for long periods, evading the efforts of the immune system to dislodge them. Others, like the viruses that cause the common cold, naturally provoke rather weak immune responses and also exist in such a large number of different strains that it is virtually impossible to develop a vaccine that will protect against all of them.

Figure 12.2 Attenuated viruses are selected by growing human viruses in non-human cells. To produce an attenuated virus, the virus must first be isolated by growing it in cultured human cells. This in itself can cause some attenuation; the rubella vaccine, for example, was made in this way. In general, however, the virus is then grown in cells of a different species, such as a monkey, until it has become fully adapted to those cells and grows only poorly in human cells. The adaptation is a result of mutation, usually a combination of several point mutations. It is usually hard to tell which of the mutations in the genome of an attenuated viral stock are critical to attenuation. An attenuated virus grows poorly in the human host, where it produces immunity but not disease.

12-2 Bacterial vaccines are made from whole bacteria, their secreted toxins, or capsular polysaccharides

Vaccines against bacterial diseases have been developed by using the same approaches as outlined above for viruses. Despite much research and development, the number of live-attenuated bacterial vaccines remains small. The oldest and most commonly used is the Bacille Calmette–Guérin (BCG) vaccine, which was derived from a bovine strain of *Mycobacterium tuberculosis* and provides protection against tuberculosis. The efficacy of this vaccine varies in different populations and although routinely given to children in some European countries, BCG is not used in the USA. More recently, live-attenuated vaccines against species of *Salmonella* have been introduced for both medical and veterinary use. An attenuated strain of *Salmonella typhi*, the bacterium that causes typhoid fever, was made by mutagenesis and selection for loss of a lipopolysaccharide necessary for pathogenesis. In this strain, which is now used as a vaccine, an enzyme necessary for lipopolysaccharide synthesis is defective.

Many bacterial diseases result from the effects of toxic proteins secreted by the bacteria. Examples are diphtheria, caused by *Corynebacterium diphtheriae*, and tetanus, caused by *Clostridium tetani*. Vaccines against these pathogens are made by purifying the respective toxin—**diphtheria toxin** or **tetanus toxin**—and treating it with formalin to destroy its toxic activity. The inactivated proteins, called **toxoids**, retain sufficient antigenic activity to provide protection against disease. Diphtheria and tetanus vaccines are thus comparable to the viral subunit vaccines. In medical practice, tetanus and diphtheria toxoids are combined in a single vaccine with a killed preparation of the bacterium *Bordetella pertussis*, the agent that causes whooping cough. This **combination vaccine** is called DTP, signifiying *D*iphtheria, *T*etanus, *P*ertussis. The immune response to the diphtheria and tetanus toxoids is enhanced by the adjuvant effect of the whole pertussis bacteria, which produces a strong inflammatory reaction at the site of injection. In some countries, Japan being the first, the DTP vaccine has been superseded by DTaP vaccines in which pertussis bacteria are replaced by an acellular pertussis component (aP). This acellular component contains a modified inactivated form of pertussis toxin—pertussis toxoid—and one or more of the other pertussis antigens, such as filamentous hemagglutinin, pertactin, and fimbrial antigen.

Many pathogenic bacteria have an outer capsule composed of polysaccharides that define species-specific and strain-specific antigens. For these **encapsulated bacteria**, which include the pneumococcus (*Streptococcus pneumoniae*), salmonellae, the meningococcus (*Neisseria meningitidis*), *Haemophilus influenzae*, *Escherichia coli*, *Klebsiella pneumoniae*, and *Bacteroides fragilis*, the capsule determines both the pathogenicity and the antigenicity of the organism. In particular, the capsule prevents fixation of complement by the alternative pathway. Only when antibodies bind to the capsule does complement fixation lead to bacterial clearance. Consequently, the aim of vaccination against such bacteria is to produce complement-fixing antibodies that bind to the capsule.

Subunit vaccines composed of purified capsular polysaccharides have been developed for some of these bacteria and they are effective in adults, in whom they provoke a T-independent antibody response (see Section 7-2, p. 183). In contrast, these vaccines are ineffective in children under 18 months old, probably because children do not develop good T-independent responses to polysaccharide antigens until a few years after birth (see Section 7-8, p. 197), and in the elderly. Both these groups are particularly susceptible to infection by encapsulated bacteria. The solution has been to convert the bacterial polysaccharide into a T-dependent antigen. This is done by covalently coupling the polysaccharide to a carrier protein, for example tetanus or diphtheria toxoid, which provides antigenic peptides for stimulating a CD4 T-cell response. On vaccination, T cells responding to the protein carrier can then stimulate polysaccharide-specific B cells to make antibodies. Vaccines of this type are

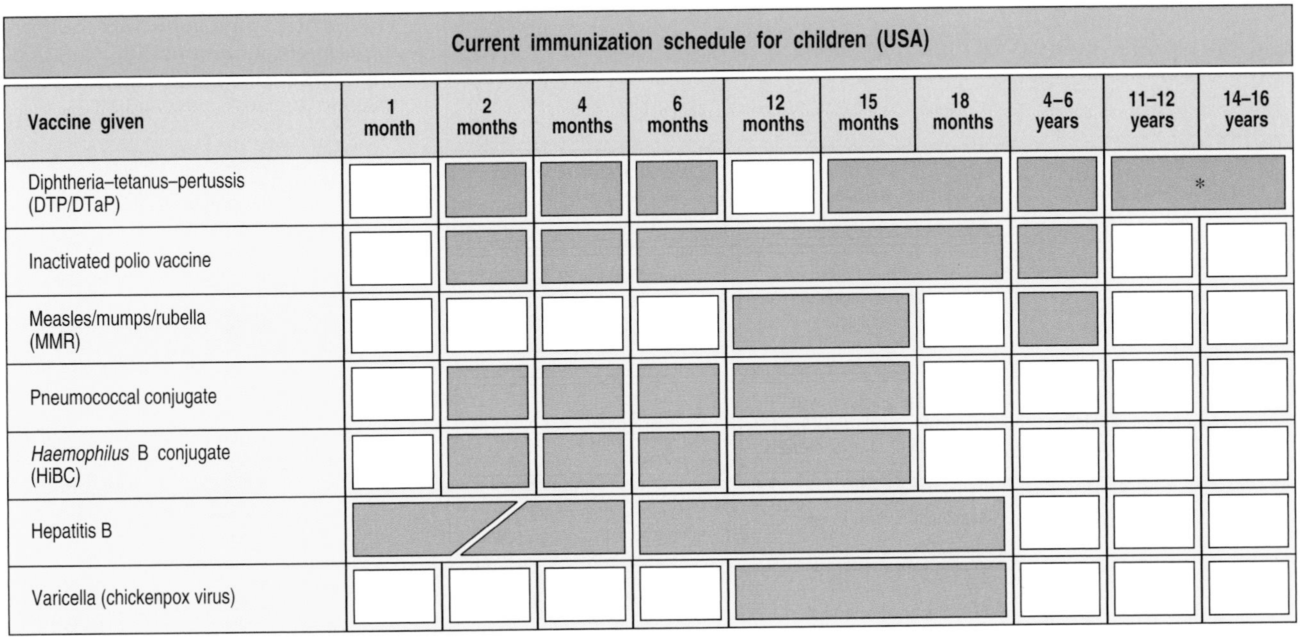

Current immunization schedule for children (USA)										
Vaccine given	1 month	2 months	4 months	6 months	12 months	15 months	18 months	4–6 years	11–12 years	14–16 years
Diphtheria–tetanus–pertussis (DTP/DTaP)		▨	▨	▨		▨		▨	*	
Inactivated polio vaccine		▨	▨	▨				▨		
Measles/mumps/rubella (MMR)					▨			▨		
Pneumococcal conjugate		▨	▨	▨	▨					
Haemophilus B conjugate (HiBC)		▨	▨	▨	▨					
Hepatitis B	▨	▨		▨						
Varicella (chickenpox virus)					▨					

called **conjugate vaccines**. A conjugate vaccine against *H. influenzae* (see Figure 6.37, p. 174) has reduced the incidence and severity of childhood meningitis caused by this bacterium.

A current schedule for recommended childhood immunizations in the USA is given in Figure 12.3. Multiple immunizations with most vaccines are recommended because they are needed to develop the memory B cells and T cells necessary for long-term protection.

Figure 12.3 Recommended childhood vaccination schedules in the United States. Each red bar denotes a time at which a vaccine dose should be given. Bars spanning multiple months indicate a range of times during which the vaccine may be given. DTaP, diphtheria, tetanus, and an acellular pertussis vaccine. *Tetanus and diphtheria toxoids only.

12-3 Adjuvants nonspecifically enhance the immune response

A prerequisite for a good immune response is a state of inflammation. During an infection this is initiated by microbial products that activate macrophages and recruit inflammatory cells (see Chapter 8). To work effectively, vaccination must also create a state of inflammation at the site in the body where the antigens are injected. In general, immunization with purified proteins leads to a poor immune response. The response can be enhanced by substances that induce inflammation by antigen-independent mechanisms. Such enhancing substances are called adjuvants, a word meaning helpers (Figure 12.4). In experimental immunology the most effective adjuvant is **Freund's complete adjuvant**, an emulsion of killed mycobacteria and mineral oil into which antigens are vigorously mixed. The active components contributed by the mycobacteria are a dipeptide, *N*-acetylmuramyl-L-alanine-D-isoglutamine (MDP), and elements of the bacterial cell-wall skeleton. In addition to stimulating inflammation, adjuvants cause soluble protein antigens to aggregate and precipitate to form particles, which facilitates their efficient uptake by antigen-presenting cells. The particulate nature of the antigen also reduces the rate at which antigen is cleared from the system.

Although many substances and preparations are known to be adjuvants, the only adjuvants approved for use in human vaccines are alum—a form of aluminum hydroxide—and an emulsion of squalene, oil, and water called MF59. Because vaccines are given to large numbers of healthy people, the safety standards for vaccines are high and they do not permit the side-effects of the most potent adjuvants. Although alum is a safe adjuvant, it is not particularly powerful, especially in stimulating immunity mediated by CD4 T_H1 cells and

Adjuvants that enhance immune responses		
Adjuvant name	Composition	Mechanism of action
Freund's incomplete adjuvant	Oil-in-water emulsion	Delayed release of antigen; enhanced uptake by macrophages
Freund's complete adjuvant	Oil-in-water emulsion with dead mycobacteria	Delayed release of antigen; enhanced uptake by macrophages; induction of co-stimulators in macrophages
Freund's adjuvant with MDP	Oil-in-water emulsion with muramyldipeptide (MDP), a constituent of mycobacteria	Similar to Freund's complete adjuvant
Alum (aluminum hydroxide)	Aluminum hydroxide gel	Delayed release of antigen; enhanced macrophage uptake
Alum plus *Bordetella pertussis*	Aluminum hydroxide gel with killed *B. pertussis*	Delayed release of antigen; enhanced uptake by macrophages; induction of co-stimulators
Immune stimulatory complexes (ISCOMs)	Matrix of lipid micelles containing viral proteins	Delivers antigen to cytosol; allows induction of cytotoxic T cells
MF 59	Squalene–oil–water emulsion	Delayed release of antigen

Figure 12.4 Some adjuvants used in experimental immunology. Only two adjuvants are licensed for use in vaccines given to humans—alum and MF59.

CD8 T cells. Adjuvant function can also be provided by the bacterial components of vaccines, such as the *B. pertussis* component of the DTP vaccine. **ISCOMs (immune stimulatory complexes)** are lipid carriers that act as adjuvants but have minimal toxicity. They seem to load peptides and proteins into the cell cytoplasm, allowing class I-restricted T-cell responses to peptides (Figure 12.5). These carriers are being considered for use in human immunization.

Another important practical aspect of vaccination is the route by which a vaccine is introduced into the human body. Today, most vaccines are given by injection or scarification. These procedures are disliked because of the pain that they cause, and for most pathogens they do not mimic the normal route of infection, which is via mucosal surfaces. Vaccination by the oral or nasal routes would be less traumatic and potentially more effective in stimulating protective immunity. It would also be cheaper, because less skill and time is required of the person administering the vaccine. That oral vaccines can be effective is shown by the live-attenuated polio vaccine. Moreover, just as the disease-causing polio virus can be transmitted by the orofecal route, so can the vaccine, for example by fecal contamination of swimming pools.

12-4 Vaccination can inadvertently cause disease

Live-attenuated viruses have made the best vaccines because they challenge the immune system in ways most like the natural pathogen. Consequently, the immune system is best prepared for a real infection after vaccination with a live virus vaccine. However, because of their similarity to the pathogen, attenuated viruses can revert to becoming pathogenic. For example, the

Sabin polio vaccine, which markedly reduced the incidence of polio in North America and Europe, induces polio and paralysis in three people per million vaccinated. The Sabin vaccine (trivalent oral polio vaccine, TVOP) consists of three different live-attenuated polio strains, all of which are needed to give protection against the natural variants of polio virus. Of these, strain 3 is the one responsible for causing disease after vaccination.

Strain 3 differs from a natural strain of polio by 10 nucleotide substitutions. Unfortunately, mutation back to the original nucleotide at just one of the substituted positions in strain 3 is sufficient to cause reversion to pathogenicity, and this happens at a low frequency either during preparation of the vaccine (which is carefully monitored) or after vaccination. In contrast, strain 1 differs from natural strains by 57 or more nucleotide substitutions, and reversion to pathogenicity is much less likely because it requires more than one back-mutation. As the incidence of natural polio infection decreases in a population, fear of the side-effects of vaccination can become greater than fear of the disease itself. For this reason there is considerable social pressure to improve the polio vaccine and some parents refuse to have their children vaccinated. A modified protocol for vaccination helps to overcome the problem; a killed polio vaccine (inactivated polio vaccine; IPV) is first given to induce some immunity; this is then followed by the live vaccine. In the United States, TVOP is no longer recommended for routine vaccination, and IPV is the vaccine of choice.

12-5 The need for a vaccine and the demands placed on it change with the prevalence of the disease

In eighteenth-century Europe, the high probability of death or permanent facial scarring from smallpox made the risk of variolation acceptable to those who could afford it. Later, the rare side-effects of the cowpox vaccine were tolerated while smallpox still posed a threat. In the late twentieth century, smallpox was eradicated and vaccination was discontinued. Smallpox vaccination seemed to be an example of a preventive measure that was so successful it put itself out of business. In the early twenty-first century, however, a new fear that smallpox might be used as a weapon has led to a program of renewed vaccine production and vaccination.

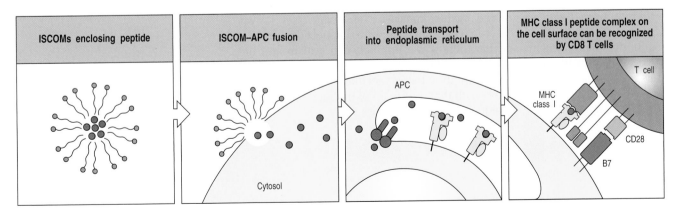

Figure 12.5 ISCOMs (immune stimulatory complexes) can be used to deliver peptides to the MHC class I processing pathway. ISCOMs are lipid micelles that will fuse with cell membranes. Peptides trapped in ISCOMs can be delivered to the cytosol of an antigen-presenting cell (APC), allowing the peptide to be transported into the endoplasmic reticulum, where it can be bound by newly synthesized MHC class I molecules and hence transported to the cell surface as peptide:MHC class I complexes. This is a possible method of delivering vaccine peptides to activate CD8 cytotoxic T cells. ISCOMs can also be used to deliver proteins to the cytosol of other types of cell, where they can be processed and presented as though they were a protein produced by the cell.

Concern for the safety of a vaccine can sometimes lead to a resurgence of disease, as seen for whooping cough in the 1970s. At the beginning of the twentieth century 1 in 20 children in the USA died from whooping cough. The DTP vaccine containing whole killed *B. pertussis* bacteria was introduced in the 1940s and was routinely given to infants at 3 months of age. The vaccination program produced a hundredfold decline in the annual incidence of whooping cough, from 2000 cases per million to 20 cases per million.

But as people's fear of the disease abated, their concern for the side-effects of the vaccine increased. All children vaccinated with pertussis vaccine develop inflammation at the injection site, and some develop a fever that induces persistent crying. Very rarely, the vaccinated child suffers fits and either a short-lived sleepiness or a transient floppy unresponsive state, all of which cause parental anxiety. In the 1970s, awareness of these established neurological side-effects was heightened by anecdotal reports of vaccination causing encephalitis and permanent brain damage, a connection that has never been conclusively proved. Distrust of pertussis vaccination grew, most notably in Japan.

In Japan, DTP vaccination was introduced in 1947. By 1974 the incidence of pertussis had been reduced by more than 99% and in that year no deaths were ascribed to the disease. In the following year, two children died soon after vaccination, raising fears that the vaccine was the direct cause of the deaths. During the next five years, the number of Japanese children being vaccinated fell from 85% to 15% and, as a consequence, the incidence of whooping cough increased by about twentyfold, as did the number of deaths from the disease. Japanese companies then developed vaccines containing antigenic components of pertussis instead of whole bacteria; acellular pertussis vaccines replaced the whole-cell vaccines in 1981 in Japan. By 1989 the incidence of pertussis was again at the very low levels of 1974. Acellular vaccines are now being increasingly used in other countries. These vaccines have a reduced incidence of the common side-effects such as inflammation, pain, and fever.

The incidence of measles infections in the UK population is currently increasing owing to public distrust of the combined vaccine against the measles, mumps, and rubella viruses (MMR vaccine). Measles is a highly infectious and potentially dangerous disease, which caused around 100 deaths a year in the UK before mass vaccination against measles was started in 1968. The combined MMR vaccine was introduced in 1988 and in the early 1990s 91% of children were being vaccinated against measles. Ten years later it was claimed, on the basis of 12 cases in which children were diagnosed as autistic soon after being vaccinated, that there was a link between autism and the MMR vaccine. Although the link has not been substantiated, distrust of MMR has meant that the proportion of children being vaccinated against measles has steadily decreased, with the result that outbreaks of measles infections in the UK began to increase in frequency and size (Figure 12.6).

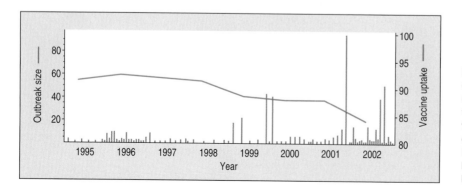

Figure 12.6 Increase in measles outbreaks in the UK since 1998 compared with uptake of the MMR vaccine. The vaccine uptake (red line) is the percentage of children that had completed a primary course of the MMR vaccine at their second birthday. Data courtesy of V.A.A. Jansen and M.E. Ramsey.

Available vaccines for infectious diseases in humans			
Bacterial diseases	**Types of vaccine**	**Viral diseases**	**Types of vaccine**
Diphtheria	Toxoid	Yellow fever	Attenuated virus
Tetanus	Toxoid	Measles	Attenuated virus
Pertussis (*Bordetella pertussis*)	Killed bacteria. Subunit vaccine composed of pertussis toxoid and other bacterial antigens	Mumps	Attenuated virus
Paratyphoid fever (*Salmonella paratyphi*)	Killed bacteria	Rubella	Attenuated virus
Typhus fever (*Rickettsia prowazekii*)	Killed bacteria	Polio	Attenuated virus (Sabin) or killed virus (Salk)
Cholera (*Vibrio cholerae*)	Killed bacteria or cell extract	Varicella (chickenpox)	Attenuated virus
Plague (*Yersinia pestis*)	Killed bacteria or cell extract	Influenza	Inactivated virus
Tuberculosis	Attenuated strain of bovine *Mycobacterium tuberculosis* (BCG)	Rabies	Inactivated virus (human). Attenuated virus (dogs and other animals). Recombinant live vaccinia-rabies (animals)
Typhoid fever (*Salmonella typhi*)	Vi polysaccharide subunit vaccines. Live-attenuated oral vaccine	Hepatitis A	Subunit vaccine (recombinant hepatitis antigen)
Meningitis (*Neisseria meningitidis*)	Purified capsular polysaccharide	Hepatitis B	Subunit vaccine (recombinant hepatitis antigen)
Bacterial pneumonia (*Streptococcus pneumoniae*)	Purified capsular polysaccharide		
Meningitis (*Haemophilus influenzae*)	*H. influenzae* polysaccharide conjugated to protein		

Figure 12.7 Diseases for which vaccines are available. Note that not all of these vaccines are equally effective, and not all are in routine use.

In 1998 the vast majority of the UK population had protective immunity against measles virus as a result of previous vaccination or infection. In this situation, a population has what is called **herd immunity**, which also indirectly protects the minority of people who have not been vaccinated. The pathogen cannot create a epidemic because of the low probability of finding susceptible individuals and creating a chain of infection. If the proportion of children being immunized against measles continues to decrease, however, the proportion of the UK population with no immunity to measles will become sufficiently large for herd immunity to be lost. An outbreak of measles could then more easily become an epidemic.

12-6 Vaccines have yet to be found for many chronic pathogens

The diseases for which we already have vaccines are ones in which the infection is acute and resolves in a matter of weeks, either by elimination of the pathogen or by the death of the patient. In the absence of vaccination, many of those infected would clear the infection, showing that the human immune system can defeat the invader. Even at its worst, smallpox killed only a third of those it infected. For such diseases, a vaccine that mimics the pathogen and provokes an immune response similar to that raised against the pathogen itself is likely to provide protective immunity. The diseases for which vaccines are available are listed in Figure 12.7.

Some diseases for which effective vaccines are not yet available		
Disease	Estimated annual mortality	Estimated annual incidence
Malaria	1.1 million	300–500 million
Schistosomiasis	15,000	No numbers available
Worm infestation	22,000	No numbers available
Tuberculosis	1.64 million	8 million
Diarrheal disease	4–6 million	4–5 billion
Respiratory disease	3.9 million	~360 million
HIV/AIDS	3.1 million	5 million
Measles*	745,000	30–40 million
Hepatitis C	46,000	~170 million †

Figure 12.8 Diseases for which better vaccines are needed. *The measles vaccines currently used are effective but they are heat sensitive and require carefully controlled refrigeration, as well as reconstitution, before use. In some tropical countries this reduces their usefulness. †This figure is the number of people worldwide estimated by the World Health Organization to be chronically infected with the hepatitis C virus. Annual mortality is difficult to estimate because hepatitis C is associated with chronic conditions such as cirrhosis of the liver and liver cancer, from which death eventually results. Estimated mortality data for 2001 are from *World Health Report 2002* (World Health Organization).

In contrast, most of the diseases for which vaccines have been difficult to find are due to chronic infections (Figure 12.8). Such pathogens are adept at evading and subverting the immune system, and so can live for years in a human host (see Chapter 9). Despite considerable research and investment, the approaches to vaccine design that have worked so well for acute bacterial and viral infections have failed to find vaccines for the chronic infections that plague humanity, especially infections with parasites. In most chronic infections there is little evidence that the immune system can clear the infection unaided, and, as a group, pathogens that cause chronic infections divert the immune system into making responses that do not clear the infection. Consequently, successful vaccines against these diseases must stimulate different immune responses from those resulting from most natural exposures to the pathogen.

For many chronic infections, a minority of people exposed to the pathogen become resistant to long-term infection. A study of their resistance should reveal examples of successful immune responses that might be emulated by vaccination. For example, of people infected with the hepatitis C virus, less than 30% clear the infection quickly, whereas the majority develop a chronic infection in which the liver goes through cycles of destruction and regeneration (Figure 12.9). Vaccine design for hepatitis C virus could be helped by a comparison of the unsuccessful immunity in patients with chronic hepatitis with the successful immunity in people who clear hepatitis C infection. This, however, is more easily said than done. Studying people who make unsuccessful responses is relatively easy, because they become sick and seek medical help. In contrast, people who successfully repel the virus are healthy, do not show up in a doctor's office, and are consequently difficult to identify. As a result, knowledge of the immune response against pathogens causing chronic infections encompasses mainly the failures of the immune system and not its successes.

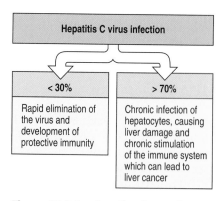

Figure 12.9 A minority of people exposed to hepatitis C virus resist the virus, whereas the majority develop chronic infection.

12-7 Genome sequences of human pathogens open up new avenues of vaccine design

Complete genome DNA sequences for an increasing number and variety of human pathogens have been determined. In the past 10 years, 140 bacterial genomes and 1600 viral genomes have been sequenced, many of them specifying human pathogens. Knowledge of the genetic blueprint for a pathogen provides a foundation for defining the biology of its pathogenesis and its interactions with the human immune system. With this knowledge, direct manipulation of a pathogen's genes can be designed to produce attenuated strains having the properties required of a vaccine (Figure 12.10). Such an approach should enable strain 3 of the Sabin polio vaccine to be mutated in ways that reduce its reversion frequency.

Cloned genes from pathogens can be safely expressed in bacterial or other cell cultures to produce subunit vaccines, thus avoiding the handling of large amounts of infectious agents or material purified from infected blood, as was used for the first hepatitis B vaccine. The ease with which genes can be moved from one organism to another opens up the possibility of using current bacterial and viral vaccine strains as vehicles for antigens that could stimulate protective immunity against other pathogens. Natural pathogens have acquired many mechanisms for evading and escaping the immune system; the removal of these functions could also generate improved strains for vaccination. In a complementary fashion, microbial genes encoding adjuvants could be incorporated into vaccine strains of other organisms.

The ideal starting point for vaccine design would be a comprehensive knowledge of how the human immune system responds to the particular infection and which mechanisms enable the pathogen to be quickly cleared from the body. With this understanding, candidate vaccines could be designed to stimulate the clones of B and T cells that make up a successful immune response. Most of the traditional vaccines have seemed to work by the induction of protective antibodies and, until recently, vaccines have been evaluated purely in those terms. An understanding of the importance of T-cell responses in protective immunity has stimulated the study of vaccines containing peptide epitopes that bind MHC molecules and thus can stimulate T cells specific for the pathogen. Although this approach has been shown to work, its application is complicated by the diversity of human HLA types and the peptides they are able to present.

The success of vaccines against chronic infections will depend critically on their ability to steer the CD4 T-cell response in ways that are helpful. In natural infections the bias towards either a CD4 T_H1 or T_H2 response can determine whether an infection resolves promptly or degenerates into chronic disease (see Section 8-19, p. 263). The inclusion of cytokines in vaccines might help to drive immunity in the desired direction.

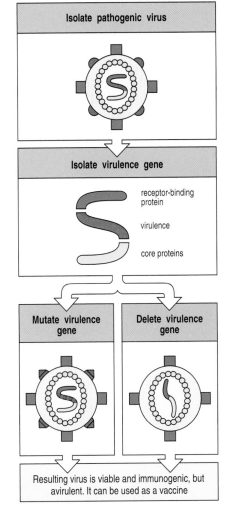

Figure 12.10 Production of live-attenuated viral strains by recombinant DNA techniques. If a viral gene that is necessary for virulence but not for growth or immunogenicity can be identified, this gene can be either mutated (left lower panel) or deleted (right lower panel) by using recombinant DNA techniques. This procedure creates an avirulent (nonpathogenic) virus that can be used as a vaccine. The mutations made in the virulence gene are usually many, so that the virus cannot easily revert to the wild type.

12-8 A useful vaccine against HIV has yet to be found

Since HIV was discovered in 1983 there have been intensive efforts in several countries to make a vaccine against it. At present, some 20 different HIV vaccines are at various stages of development. So far, only one vaccine has been thoroughly tested for its capacity to protect humans from HIV infection. This vaccine, which was targeted at gp120, the viral protein that binds to CD4 (see Section 9-17, p. 298), was tested in 5000 volunteers whose daily life puts them at risk of exposure to and infection by HIV. The result of this trial, which was announced in 2003, was that vaccination gave no protection. This modern example shows how unpredictable and difficult vaccine development can be.

In the development of vaccines against polio, measles, and other viruses, it was known from the beginning that infected humans had the capacity to recover from acute infection with the acquisition of protective immunity. For HIV there is no documented case of a person who has recovered from acute infection and for whom the protective immunity can be analyzed and set as the goal to be attained through vaccination. Thus, the types of immune effectors that can terminate HIV infection are unknown.

From the study of antibodies in the serum of HIV-infected people, several monoclonal antibodies have been made that can neutralize a range of HIV strains in experimental situations. The challenge is to find an immunogen that will stimulate the production of such antibodies when humans are vaccinated. A further complication is that although neutralizing antibodies are made in HIV-infected people, the virus readily mutates and escapes the antibodies.

Because HIV-infected individuals make more effective T-cell responses than antibody responses, much vaccination effort is directed at the induction of T-cell responses. This is based on the assumption that effective CD8 and CD4 T cells will terminate HIV infection at an early stage or keep it under control and at a low level. Evidence to support this idea comes from the approximately 5% of people who remain uninfected despite being frequently exposed to HIV. These people make a T-cell response to HIV but not an antibody response.

Summary

Vaccination involves the deliberate immunization of healthy people with some form of a pathogen or its component antigens. It induces a protective immunity that prevents subsequent infection with the same pathogen from causing disease. Vaccines can consist of killed whole pathogens, live-attenuated strains, nonpathogenic species related to the pathogen, or secreted and surface macromolecules of the pathogen. Vaccination has saved millions of lives and reduced the incidence of many common infectious diseases, particularly in the industrialized countries. By reducing the incidence of disease, successful vaccination programs inevitably lead to decreasing public awareness of the effects of disease and increasing concern with the safety of vaccination. Vaccine development has largely been a process of trial and error, one in which knowledge of immunological mechanisms played little part and the guiding principle was for the vaccine to resemble the natural pathogen as closely as possible. Although this approach has worked for pathogens causing acute infections, it has failed to produce vaccines against pathogens that establish chronic infections and cause chronic disease. These pathogens are adept at fooling the immune system; most people infected make immune responses that fail to clear the infection. Successful vaccines against these pathogens might need to push the immune system to respond in ways that are different from those seen in most natural infections with the pathogen.

Transplantation of tissues and organs

The replacement of diseased, damaged or worn-out tissue was for a long time a dream of the medical profession. Achieving the reality required the solution of three basic problems. First, transplants must be introduced in ways that allow them to perform their normal functions. Second, the health of both the recipient and the transplant must be maintained during surgery and the other procedures used in transplantation. Third, the immune system of the patient must be prevented from developing adaptive immune responses to antigens on the grafted tissue—responses that can result in rejection of the transplant and in other complications.

During the past 50 years, solutions to these problems have been found, and organ transplantation has progressed from being an experimental procedure to the treatment of choice for a variety of conditions. In clinical practice, selective suppression of the response to the transplanted tissue has yet to be achieved and so nonspecific suppression is accomplished by using a variety of drugs and antibodies. In contrast to vaccination, which selectively stimulates immunity to a particular pathogen, transplantation involves manipulations that cause widespread inactivation of the immune response.

12-9 Transplant rejection and graft-versus-host reaction are immune responses caused by genetic differences between transplant donor and recipient

Immune responses against transplanted tissues or organs are caused by genetic differences between donor and recipient, the most important of which are antigenic differences in the highly polymorphic HLA molecules. This is why these antigens are generally known as the major histocompatibility antigens (**histocompatibility** means tissue compatibility), and the complex of genes that encodes them has the general name of the major histocompatibility complex (MHC), although it goes under different names in different species (for example, the HLA complex in humans). Antigens such as these, which vary between members of the same species, are known as **alloantigens**; the immune responses that they provoke are known as **alloreactions** (see Section 3-19, p. 93). A subfield of immunology, called **immunogenetics**, is devoted to the genetics of alloantigens. After transplantation of a solid organ, alloreactions developed by the recipient's immune system are directed at the cells of the graft and can kill them, a process called **transplant rejection**. In contrast, in bone marrow transplantation, the recipient's immune system has been destroyed, and the major type of alloreaction arises from mature T cells in the grafted bone marrow that attack the recipient's tissues. This type of alloreactive response by donor lymphocytes is called a **graft-versus-host reaction** (**GVHR**). It causes **graft-versus-host disease** (**GVHD**), which, with varying degrees of severity, affects almost all patients who have a bone marrow transplant (Figure 12.11).

Because people do not normally make an immune response to their own tissues, tissues transplanted from one site to another on the same person are not rejected. This type of transplant, called an **autograft**, is used to treat patients who have suffered burns. Skin from unaffected parts of the body is grafted into the burnt areas, where it facilitates wound healing. Immunogenetic differences are also avoided when tissue is transplanted between identical twins. The first successful kidney transplant, in 1954, involved the donation of a kidney by a healthy twin to his brother, who was suffering from kidney failure. A transplant between genetically identical individuals is called a **syngeneic** transplant or an **isograft**. A transplant made between two genetically different individuals is called an **allograft** or an **allogeneic** transplant.

12-11 Antibodies against A,B,O or HLA antigens cause hyperacute rejection of transplanted organs

A,B,O antigens are also expressed on the endothelial cells of blood vessels; this is an important factor in the transplantation of solid organs such as kidneys. For example, were a type O recipient to receive a kidney graft from a type A donor, then anti-A antibodies in the recipient's circulation would quickly bind to blood vessels throughout the graft. By fixing complement throughout the vasculature of the graft, the antibodies would produce a very rapid rejection of the graft (Figure 12.14). This type of rejection, called **hyperacute rejection**, can occur even before a transplanted patient has left the operating room. Hyperacute rejection is the most devastating form of rejection for organ grafts. It is directly comparable to type III hypersensitivity reactions (see Section 10-17, p. 333), in which immune-complex deposition causes complement activation within blood vessel walls. To avoid hyperacute rejection, transplant donors and recipients are typed and cross-matched for the A,B,O blood group antigens.

As HLA class I molecules are expressed constitutively on vascular endothelium, preexisting antibodies against HLA class I polymorphisms can also cause hyperacute rejection. It is therefore essential that transplant recipients do not have antibodies that bind to the HLA class I allotypes of the transplanted organ. To a lesser extent, antibodies against HLA class II can also contribute to hyperacute rejection. HLA class II molecules are not normally expressed on endothelium but can be induced by infection, inflammation, or trauma, all of which can occur during transplantation.

No reliable method of reversing hyperacute rejection has been found, so this type of rejection is avoided by choosing compatible transplant donors and recipients. Compatibility is assessed with a cross-match test in which blood serum from the prospective recipient is assessed for the presence of antibodies that bind to the white blood cells of the prospective donor. The traditional method for cross-match reveals antibodies in the patient's serum that can trigger complement-mediated lysis of the donor's lymphocytes. The assay is usually performed on separated B cells and T cells so that reactivities due to antibodies against HLA class I and II molecules can be distinguished: anti-HLA class I antibodies react with both B cells and T cells, whereas antibodies against HLA class II react only with B cells. A more sensitive cross-match assay uses flow cytometry to examine binding of the patient's antibodies to the prospective donor's lymphocytes (see Figure 2.13, p. 46).

12-12 Anti-HLA antibodies can arise from pregnancy, blood transfusion, or previous transplants

Circulating anti-HLA class I and II antibodies in prospective transplant patients can have arisen for several reasons. The most common is a previous pregnancy. The fetus is protected from the mother's immune system during gestation, but during the trauma associated with birth, cells of fetal origin can stimulate an immune response in the mother. Anti-HLA antibodies can be

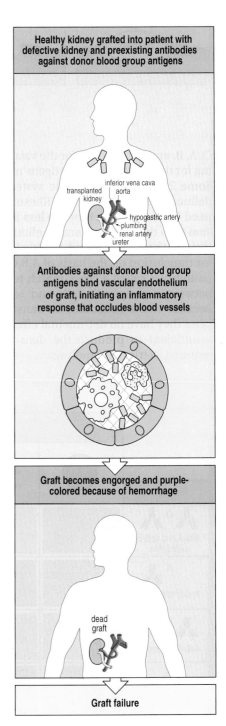

Figure 12.14 Hyperacute rejection is caused by preexisting antibodies binding to the graft. In some cases, recipients already have antibodies against donor antigens. When the donor organ is grafted into such recipients, these antibodies bind to vascular endothelium, initiating the complement and clotting cascades. Blood vessels in the graft become obstructed by clots and leak, causing hemorrhage of blood into the graft. This becomes engorged and turns purple from the presence of deoxygenated blood. The graft dies.

formed if the fetus expresses paternal HLA allotypes different from those of the mother. With successive pregnancies, increasing levels of anti-HLA antibodies can develop, and multiparous women are the main source of the anti-HLA sera used in HLA typing.

The fetus can be considered as an allograft within the mother's body (Figure 12.15) and any anti-HLA antibodies in a mother's circulation have the potential to bind to fetal cells and cause hyperacute rejection. That they never do so shows how well the fetus is protected from the maternal immune system.

Blood transfusions can also lead to the production of anti-HLA antibodies. In routine blood transfusion, no assessment of HLA type is performed and so although the donor and recipient are matched for A,B,O type they are not matched for HLA type. The infusion of HLA-incompatible leukocytes and platelets in a blood transfusion can therefore generate antibodies specific for the donor HLA allotypes. Patients who have had multiple blood transfusions have been stimulated by many HLA allotypes and can develop antibodies that react with the cells of most other people in the population. The degree to which a patient seeking a transplant has been sensitized to potential donors is assessed by testing their sera against a representative panel of individuals from the population and expressing the number of positive reactions as a percentage **panel reactive antibody** (**PRA**). The higher the value of a patient's PRA, the more difficult it is to find a suitable transplant donor.

A third way in which patients develop anti-HLA antibodies is from a previous transplant. Now that transplantation has been routine practice for more than 30 years, many patients have had more than one transplant. As with blood transfusions, the more transplants that a person has had, the greater their percentage PRA tends to be.

12-13 Organ transplantation involves procedures that inflame the donated organ and the transplant recipient

Patients receiving an organ transplant usually have a history of disease in which the organ to be replaced has gradually degenerated. Often there is an immune component to this decay, such as the deposition of immune complexes, which leads to kidney damage and failure. Before transplantation, patients with renal failure are maintained by dialysis, a procedure that induces inflammation through the interaction of dialysis membranes and serum proteins. Consequently, the transplant patient is already inflamed before transplantation, a state that is exacerbated by the damage and disruption caused by the surgery involved in transplantation. Thus, on receiving an organ transplant the recipient's body is both prepared and ready to direct innate and adaptive immunity towards the transplanted tissue.

Donated organs are also inflamed. In the case of cadaveric organs, that is those taken from a dead donor, the donor will usually have died in a violent or stressful manner, and the procedures used to collect organs and transport them to the transplant center add to the stress (Figure 12.16). During this time, the organs are deprived of blood, a state called **ischemia**. Ischemia causes damage to the blood vessels and tissue of the organs through activation of the endothelium and of the complement system, infiltration with inflammatory leukocytes, and cytokine production. The success of transplantation depends heavily on limiting the damage caused by ischemia. For kidney or liver transplantation it is possible to use a living healthy donor, and this confers a great advantage. Donation and transplantation can be performed at the same time and in the same place, with minimal time of ischemia. This approach has mainly been used for related donors and recipients, but has

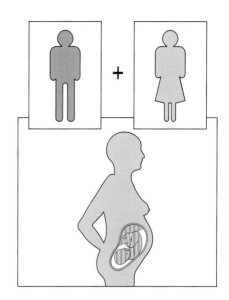

Figure 12.15 The fetus is an allograft that is protected from rejection. In human families the mother and father almost always have different HLA types. When the mother becomes pregnant she carries for nine months a fetus that expresses one HLA haplotype of maternal origin (pink) and one HLA haplotype of paternal origin (blue). Although the paternal HLA class I and II molecules expressed by the fetus are alloantigens against which the mother's immune system has the potential to respond, the fetus does not provoke such a response during pregnancy and is protected from preexisting alloreactive antibodies or T cells. The trauma associated with childbirth can lead to fetal stimulation of the mother's immune system.

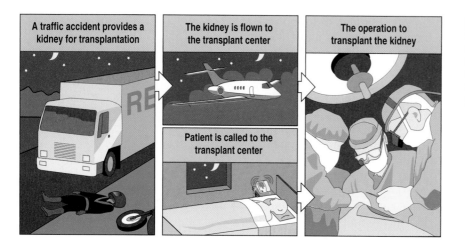

A traffic accident provides a kidney for transplantation	The kidney is flown to the transplant center	The operation to transplant the kidney
	Patient is called to the transplant center	

Figure 12.16 Clinical transplantation involves both a donated organ and a transplant recipient that are stressed and inflamed. This cartoon depicts a typical sequence of events that occurs before transplantation of a cadaveric organ transplant.

also been extended to unrelated family members, for example husbands and wives. Because of the reduced inflammation and tissue damage associated with organs from living healthy donors, the success of the transplantation is less sensitive to HLA mismatch than transplantation using cadaveric donors.

12-14 Acute rejection is caused by effector T cells responding to HLA differences between donor and recipient

Most organ transplants are performed across some HLA class I and/or II difference. In this situation the transplant recipient's T-cell population includes clones of alloreactive T cells that are specific for those HLA allotypes of the transplanted tissue that are not shared with the recipient. CD8 T cells respond to the HLA class I differences, and CD4 T cells respond to the HLA class II differences. The alloreactive T-cell response produces effector CD4 and CD8 T cells, both of which can attack the organ graft and destroy it (Figure 12.17). This is called **acute rejection**. Unlike hyperacute rejection, it takes days to develop and can be reduced or prevented by intervention. To prevent acute rejection all transplant patients are conditioned with immunosuppressive drugs before transplantation and maintained on them after transplantation. Patients are carefully monitored for early signs of acute rejection and treated with additional immunosuppressive drugs or anti-T-cell antibodies when it occurs. The effector mechanisms underlying acute rejection are just like those causing type IV hypersensitivity reactions (see Section 10-19, p. 336).

The inflamed state of the transplanted organ activates the organ's dendritic cells. These donor-derived dendritic cells migrate to the draining secondary lymphoid tissue, where they settle into the T-cell zone and present complexes of donor MHC and donor self-peptides to the recipient's circulating T cells. Because different HLA allotypes bind different sets of self-peptides, each allotype selects a different repertoire of T-cell receptors during thymic selection. Consequently, the T-cell repertoire selected by the recipient's HLA type contains numerous T-cell clones that can respond to the HLA:self-peptide complexes presented by donor cells of different HLA type. For this reason, alloreactive T-cell responses stimulated by HLA differences are stronger than T-cell responses to a vaccine or to a pathogen. Many clones of alloreactive T cells have a memory phenotype, revealing that they were originally stimulated and expanded in response to pathogens and are cross-reactive with allogeneic HLA. This type of alloreactive response, in which recipient T cells are stimulated by direct interaction of their receptors with the allogeneic HLA molecules expressed by donor dendritic cells, is called the **direct pathway of allorecognition** (Figure 12.18). It produces effector T cells that migrate to the

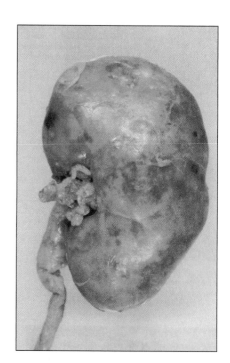

Figure 12.17 Gross appearance of an acutely rejected kidney. The rejected graft is swollen and has deep red areas of hemorrhage and gray areas of necrotic tissue. Courtesy of B.D. Kahan.

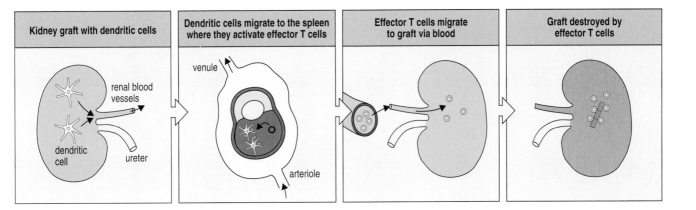

| Kidney graft with dendritic cells | Dendritic cells migrate to the spleen where they activate effector T cells | Effector T cells migrate to graft via blood | Graft destroyed by effector T cells |

transplanted tissue, where T_H1 cells activate the resident macrophages to inflame the tissue further and CD8 T cells systematically kill the cells of the transplanted tissue.

There is a second way in which HLA alloantigens on the transplant can be presented to the recipient's immune system and provoke a response. Some of the donor-derived dendritic cells that migrate to the draining lymphoid tissue die there by apoptosis. Membrane fragments containing HLA molecules from these apoptotic cells are taken up by dendritic cells of the transplant recipient and processed so that peptides derived from donor HLA allotypes are presented by the recipient's HLA allotypes. Because of the endocytic mode of uptake, most of these peptides, which can be derived from HLA class I or II, will be presented by the HLA class II allotypes of the recipient. These complexes will stimulate a CD4 T-cell alloreaction if the peptides are different in amino-acid sequence from those produced by degradation of the recipient's own HLA allotypes. These alloreactive CD4 T cells are specific for the complex of a peptide derived from a donor HLA allotype bound to a recipient HLA class II allotype. This way of stimulating alloreactive T cells is called **the indirect pathway of allorecognition** because the alloreactive T cells do not directly recognize the transplanted cells but recognize subcellular material that has been processed and presented by autologous cells (Figure 12.19).

The indirect pathway of allorecognition is a special case of the normal mechanism by which T cells recognize the protein antigens of pathogens; in transplantation, the foreign antigens come from another human body. T cells stimulated by the indirect pathway of allorecognition contribute to the acute rejection of transplanted organs, although they are usually less numerous than those stimulated by the direct pathway.

12-15 Chronic rejection of organ transplants is due to the indirect pathway of allorecognition

In addition to hyperacute and acute rejection, transplanted human organs can be rejected by a third mechanism called **chronic rejection**. This phenomenon, which occurs months or years after transplantation, is characterized by reactions in the vasculature of the graft that cause thickening of the vessel walls and a narrowing of their lumina (Figure 12.20). The blood supply to the graft gradually becomes inadequate, causing ischemia and loss of function of the graft, and the graft eventually dies. Chronic rejection causes the failure of more than half of all kidney and heart grafts within 10 years after transplantation. Chronic rejection is correlated with the presence of antibodies specific for the HLA class I molecules of the graft, and is rarely seen in their absence, thus implicating a B-cell response. Further evidence for the role of B

Figure 12.18 Acute rejection of a kidney graft through the direct pathway of allorecognition. Donor dendritic cells in the graft (in this case a kidney) are carried to a draining secondary lymphoid organ (the spleen is illustrated here), where they move to the T-cell area. Here, they activate recipient T lymphocytes whose receptors bind directly to an allogeneic HLA class I or class II molecule in combination with donor peptides. After activation, the effector T cells travel in the blood to the grafted organ, where they attack cells that express their specific HLA-molecule:peptide complexes.

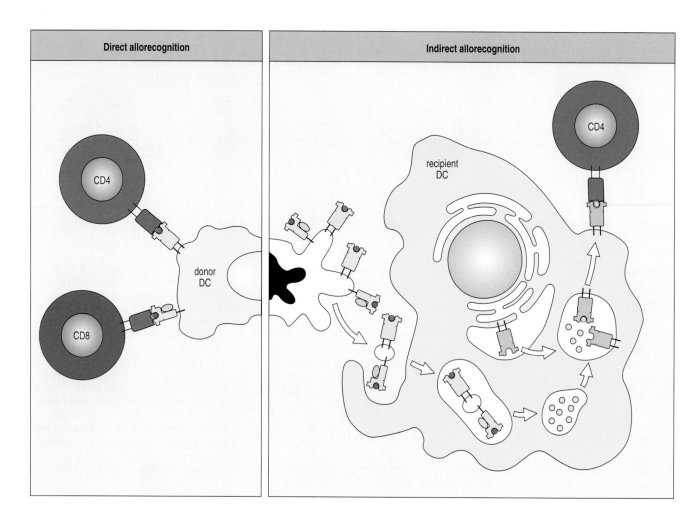

Direct allorecognition	Indirect allorecognition

cells is the infiltration of grafts undergoing chronic rejection with lymphocytes expressing CD40 (characteristic of B cells) and CD40 ligand (characteristic of helper T cells). Chronic rejection is likely to be caused by alloreactive T cells stimulated by the indirect pathway continuing to drive and expand an antibody response to HLA and other alloantigens of the transplanted organ.

The strength of the response from alloreactive T cells stimulated by the direct pathway wanes with time after transplantation. This is associated with the elimination of dendritic cells of donor origin and the repopulation of the transplanted organ with immature dendritic cells of recipient origin. These latter, however, can still increase the stimulation of alloreactive T cells through the indirect pathway of allorecognition. Transplant recipients are selected for not having serum antibody that reacts with the transplanted organ; this means they will not have memory B cells that can respond to the allogeneic HLA allotypes. They may, however, have naive B cells reactive for the allogeneic HLA antigens. After transplantation, the stimulation of helper CD4 T cells through the indirect pathway can thus initiate an antibody response against the allogeneic HLA allotypes (Figure 12.21). The stimulated helper CD4 T cells will help naive B cells that are specific for HLA allotypes of the graft and thus provoke a B-cell response in which alloantibodies specific for HLA class I are formed. The HLA-specific CD4 T cells will also provide help to B cells specific for other alloantigens that are incorporated into the HLA-containing subcellular fragments. This can lead to expansion and epitope spreading of the allogeneic B- and T-cell response and the gradual impairment of the function of the transplanted organ through the same mechanisms that operate in autoimmune disease (see Section 11-20, p. 370).

Figure 12.19 Direct and indirect pathways of allorecognition contribute to graft rejection. Shown here is how dendritic cells from an organ graft stimulate both the direct and indirect pathways of allorecognition when they travel from the graft to the draining lymphoid tissue. The left-hand panel shows how the expression of allogeneic HLA class I and II allotypes of donor type on a donor antigen-presenting cell (APC) will interact directly with the T-cell receptors of alloreactive CD4 and CD8 T cells of the recipient (direct allorecognition). The right-hand panel shows how the death of the same antigen-presenting cell produces membrane vesicles containing the allogeneic HLA class I and II allotypes, which are then endocytosed by the recipient's dendritic cells. Peptides derived from the donor's HLA molecules (yellow) can then be presented by the recipient's HLA class I and II molecules (orange) to peptide-specific CD4 and CD8 T cells (indirect allorecognition).

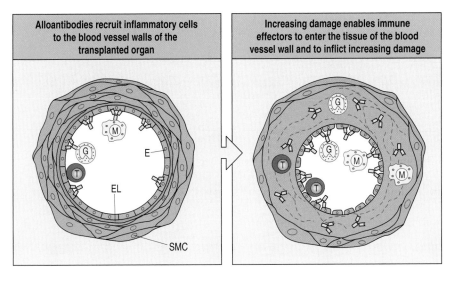

Figure 12.20 Schematic representation of chronic rejection in the blood vessels of a transplanted kidney. Left-hand panel: chronic rejection is initiated by the interaction of alloantibodies, principally anti-HLA class I, to the blood vessels of the transplanted organ. Antibodies bound to the endothelial cells (E) recruit Fc-receptor bearing monocytes and neutrophils. EL, internal elastic lamina; SMC, smooth muscle cells. Right-hand panel: in time the accumulating damage leads to thickening of the EL and to the underlying intimas becoming infiltrated by SMCs, macrophages (M), granulocytes (G), alloreactive T cells (T) and antibodies. The overall effect is to narrow the lumen of the blood vessel and to create a chronic state of inflammation that intensifies the remodeling of the tissue. Eventually the vessel becomes obstructed, ischemic and fibrotic.

The indirect pathway of allorecognition can also give rise to regulatory CD4 T cells that suppress alloreactive CD4 and CD8 effector T cells, and lead to an improved clinical outcome after transplantation. Such regulatory T cells seem active in patients who before the transplant had received blood transfusions from donors who, by chance, shared an HLA-DR allotype with the organ with which the patients were subsequently transplanted. This phenomenon of prior blood transfusions improving the outcome of organ transplantation is known as the **transfusion effect**.

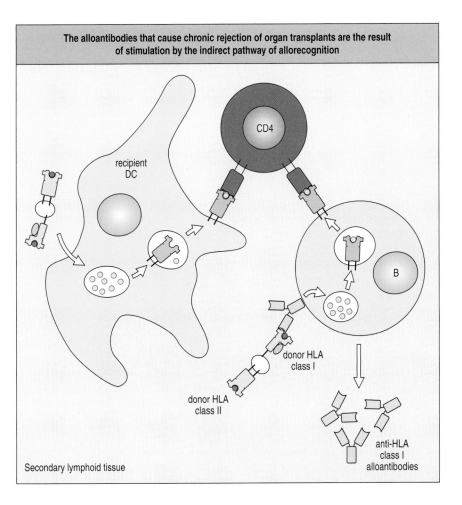

Figure 12.21 The indirect pathway of allorecognition is responsible for stimulating the production of the anti-HLA antibodies that cause chronic rejection. The processing and presentation of allogeneic HLA class I by a dendritic cell (DC) of the recipient is shown. The dendritic cell activates helper CD4 T cells, which in turn activate B cells that have bound and internalized allogeneic donor HLA molecules. Shown here is a cognate interaction that leads to the production of an anti-HLA class I antibody. Anti-HLA class II antibodies can be produced similarly. As activated endothelium expresses both HLA class I and II molecules, antibodies against both classes of HLA molecule can contribute to chronic rejection.

12-16 Matching donor and recipient for HLA class I and class II allotypes improves the outcome of transplantation

The first successful organ transplant was achieved with a kidney transplant in 1954. As the donor and recipient were identical twins there was no risk of alloreactivity leading to graft rejection. As very few patients have an identical twin, additional approaches were needed to make transplantation more generally available to patients with kidney disease. The combination of two complementary approaches proved successful. To reduce the genetic differences that stimulate rejection, HLA class I and II genes were identified as its major cause and methods for determining HLA type were used to assess the histocompatibility of prospective donors and recipients. To prevent the response to HLA and other alloantigenic differences, immunosuppressive drugs were sought, discovered, and used to interfere with the activation and proliferation of T cells.

Clinical transplantation was pioneered with the kidney for two principal reasons. First, patients whose kidneys had failed could be sustained by the well-established procedure of dialysis. This meant that graft failure or rejection did not inevitably lead to the patient's death. The second was the simple fact that everyone has two kidneys but can manage with one, so healthy relatives could donate a kidney to a needy patient. Immunogenetic differences within a family are much smaller than within the population at large, and so the probability of finding an HLA-matched person within the family is higher. Analysis of the fate of kidney transplants performed between HLA-identical and non-identical family members was instrumental in showing that the better the HLA class I and II match, the better the clinical outcome. Matching for HLA-A, HLA-B, and HLA-DR is the most important. Clinical HLA typing is currently performed by DNA analysis.

After the success of transplantation between living relatives, methods were developed to transplant kidneys from unrelated donors who had been killed in accidents (cadaveric donors). Over 100,000 kidney transplants have been performed worldwide and a statistical analysis of these data demonstrates that both graft performance and long-term health of the recipient increase with the degree of HLA match (Figure 12.22).

12-17 Allogeneic transplantation is made possible by the use of immunosuppressive drugs

The limited supply of donated organs and the diversity of HLA type in the population mean that a majority of patients receive organs that are mismatched at one or more HLA loci. Drugs are used to suppress the alloreactions that would

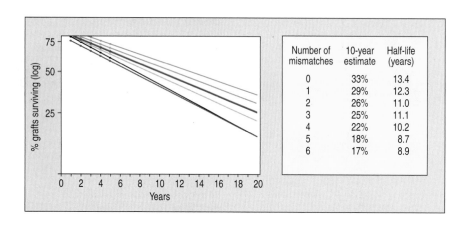

Figure 12.22 **HLA matching improves the survival of transplanted kidneys.** The colored lines in the left panel represent the actual (to 5 years) and projected survival rates of kidney grafts in patients with no (blue), 1 (orange), 2 (red), 3 (dark blue), 4 (green), 5 (black), and 6 (brown) HLA mismatches, plotted on a semi-logarithmic scale. Data courtesy of G. Opelz.

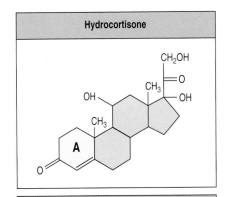

Figure 12.23 Chemical structures of hydrocortisone, prednisone, and prednisolone. Prednisone is a synthetic analog of the natural adrenocorticosteroid hydrocortisone, or cortisol. It is converted *in vivo* to the biologically active form, prednisolone. Introduction of the 1,2 double bond into the A ring increases anti-inflammatory potency approximately fourfold compared with hydrocortisone, without modifying the sodium-retaining activity of the compound.

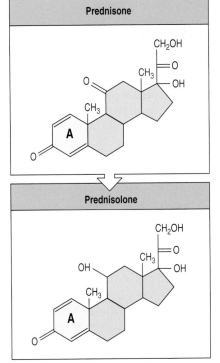

otherwise lead to transplant rejection. The **immunosuppressive** drugs used in clinical transplantation are of three kinds. The **corticosteroids** are steroids with anti-inflammatory properties like those of the natural glucocorticosteroid hormones made in the adrenal cortex. The second kind consists of cytotoxic drugs that, by interfering with DNA replication, kill proliferating lymphocytes activated by graft alloantigens. The third type of immunosuppressive drugs are microbial products that inhibit the signaling pathways of T-cell activation.

Any drug powerful enough to inhibit alloreactions also inhibits the normal immune responses to infecting pathogens. Consequently, administration of these drugs, which is greatest during the period immediately before and after transplantation, renders transplant patients highly susceptible to infection. Patients are initially cared for in conditions under which their exposure to pathogens is reduced. As their immune systems accommodate to the graft, the dose of immunosuppressive drugs is gradually reduced to 'maintenance levels' that prevent rejection while sustaining active defenses against infection. Aggressive early treatment has reduced the incidence of acute rejection in clinical transplantation. However, with the reduction of immunosuppression and the restoration of a patient's immunocompetence, the likelihood of chronic rejection increases.

All immunosuppressive drugs are also toxic to other tissues in varying degrees. Because these 'side-effects' vary for the different drugs, immunosuppressive drugs are generally used in combination so that their immunosuppressive effects are additive whereas their toxic effects are not. Certain side-effects emerge only after patients have taken immunosuppressive drugs for long periods. These include a higher incidence of certain types of malignant disease, particularly carcinomas of the skin and the genital tract, lymphoma, and Kaposi's sarcoma. The incidence of cancer in transplant recipients is on average three times that of similarly aged people who have not received a transplant.

12-18 Corticosteroids change patterns of gene expression

Hydrocortisone, also called cortisol, is the principal steroid made by the adrenal cortex; for more than 50 years it has been used clinically to reduce inflammation. The steroid commonly given to transplant patients is **prednisone**, a synthetic derivative of hydrocortisone that is about four times more potent in reducing inflammation. Prednisone has no biological activity until it is enzymatically converted *in vivo* to **prednisolone** (Figure 12.23). Prednisone is an example of a **pro-drug**, a name given to drugs that are given to patients in an inactive form and become chemically or enzymatically converted to the active form within the body. By itself, prednisone is insufficiently immunosuppressive to prevent graft rejection, but it works well in combination with a cytotoxic drug.

Corticosteroids have wide-ranging physiological effects and affect all white blood cells, not just lymphocytes, as well as other cells of the body. Unlike many other biologically active molecules, steroid hormones do not act on

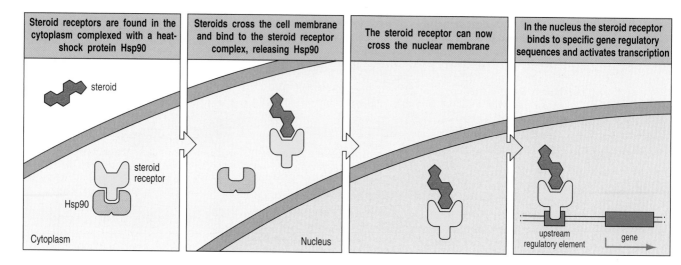

| Steroid receptors are found in the cytoplasm complexed with a heat-shock protein Hsp90 | Steroids cross the cell membrane and bind to the steroid receptor complex, releasing Hsp90 | The steroid receptor can now cross the nuclear membrane | In the nucleus the steroid receptor binds to specific gene regulatory sequences and activates transcription |

receptors at the cell surface but diffuse across the plasma membrane and bind to specific receptors in the cytoplasm. Before steroid binding, the receptors are associated with another cytoplasmic polypeptide called Hsp90 (*heat-shock protein* of 90 kDa molecular weight). Steroid binding induces a conformational change in the receptor, which then dissociates from Hsp90 and enters the nucleus. There, the complex of receptor and steroid binds selectively to certain genes, activating their transcription (Figure 12.24). The transcription of about 1% of a cell's genes can be influenced by corticosteroids.

In the context of their anti-inflammatory action, an important effect of corticosteroids is inhibition of the function of NFκB, a transcription factor important for cellular activation and cytokine production in the immune response. In quiescent cells, NFκB is held in the cytoplasm through its association with a protein called IκBα. On cellular activation, IκBα becomes phosphorylated, allowing NFκB to dissociate and enter the nucleus, where it initiates cytokine gene transcription. Corticosteroids increase the production of IκBα, thereby preventing NFκB from gaining access to the nucleus. By this mechanism they suppress the production of cytokines, such as IL-1 by monocytes, which stimulate inflammation and immune responses (Figure 12.25).

Because of their multifarious effects on gene expression and cellular metabolism, corticosteroid drugs have many adverse side-effects, including fluid retention, weight gain, diabetes, loss of bone mineral, and thinning of the skin. Because of their mechanism of action, corticosteroids are most effective as immunosuppressive drugs when they are first administered before transplantation. With this approach, the patterns of cytokine gene expression are already changed in the recipient's cells at the time of alloantigenic challenge. They are used as an acute immunosuppressive agent during episodes of rejection that are often caused by infection, but their continued use is to be avoided wherever possible.

Figure 12.24 Steroids act at intracellular receptors. Corticosteroids are lipid-soluble compounds that diffuse across the plasma membrane and bind to their receptors in the cytosol. The binding of corticosteroid to the receptor displaces a dimer of a heat-shock protein named Hsp90, exposing the DNA-binding region of the receptor, which then enters the nucleus and binds to specific DNA sequences in the promoter regions of steroid-responsive genes. Corticosteroids exert their effects by modulating the transcription of a wide variety of genes.

Figure 12.25 Effects of corticosteroids on the immune system. Corticosteroids regulate the expression of many genes, with a net anti-inflammatory effect. First, they reduce the production of inflammatory mediators, including some cytokines, prostaglandins, and nitric oxide (NO). As well as the cytokines listed here, corticosteroids also indirectly cause a decrease in IL-2 synthesis by activated lymphocytes, by their effects on other cytokines. Second, they inhibit inflammatory cell migration to sites of inflammation by inhibiting the expression of adhesion molecules. Third, corticosteroids promote the death by apoptosis of leukocytes and lymphocytes. NOS, nitric oxide synthase.

Corticosteroid therapy	
Activity	**Effect**
↓ IL-1, TNF-α, GM-CSF ↓ IL-3, IL-4, IL-5, CXCL8	↓ Inflammation ↓ caused by cytokines
↓ NOS	↓ NO
↓ Phospholipase A₂ ↓ Cyclo-oxygenase type 2 ↑ Lipocortin-1	↓ Prostaglandins ↓ Leukotrienes
↓ Adhesion molecules	Reduced emigration of leukocytes from vessels
Induction of endonucleases	Induction of apoptosis in lymphocytes and eosinophils

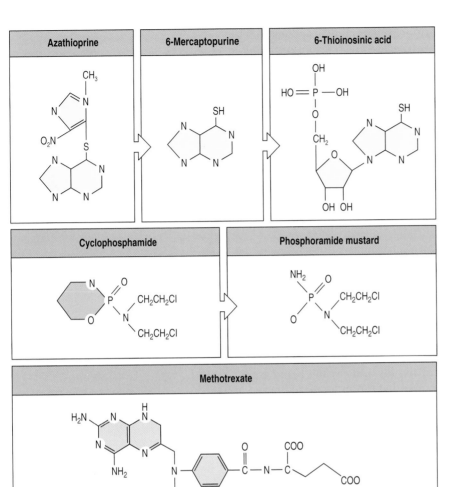

Figure 12.26 The chemical structures and metabolism of cytotoxic drugs. Azathioprine was developed as a modification of the anti-cancer drug 6-mercaptopurine; by blocking the reactive thiol group, the metabolism of this drug is slowed down. It is slowly converted *in vivo* to 6-mercaptopurine, which is then metabolized to 6-thioinosinic acid; this blocks the pathway of purine biosynthesis. Cyclophosphamide was similarly developed as a stable pro-drug, which is activated enzymatically in the body to phosphoramide mustard, a powerful and unstable DNA alkylating agent. Methotrexate blocks DNA synthesis by interfering with thymidine synthesis.

12-19 Cytotoxic drugs kill proliferating cells

A cytotoxic drug commonly used in solid organ transplantation is **azathioprine**, a pro-drug that is first converted *in vivo* to 6-mercaptopurine and then to 6-thioinosinic acid (Figure 12.26). The latter compound inhibits the production of inosinic acid, an intermediate in the biosynthesis of adenine and guanine nucleotides, which are essential components of DNA. The principal effect of azathioprine is therefore to inhibit DNA replication. Azathioprine has no effect on cells until they attempt to replicate their DNA, whereupon they die. While helpfully inhibiting the proliferation of alloantigen-activated lymphocytes, azathioprine, like other cytotoxic drugs, damages all tissues of the body that are normally active in cell division. Principally affected are the bone marrow, the intestinal epithelium, and hair follicles, leading to anemia, leukopenia, thrombocytopenia, intestinal damage, and loss of hair. When pregnant women have to take cytotoxic drugs, fetal development can be adversely affected. Because azathioprine cannot act until a patient's immune system has been stimulated by alloantigen, it need only be administered after transplantation.

Cyclophosphamide is one of the nitrogen mustard compounds that were developed as chemical weapons and saw heavy use during World War I. It is a pro-drug that is converted in the body to phosphoramide mustard; this alkylates and cross-links DNA molecules (see Figure 12.26). These covalent modifications render cells incapable of normal division and also affect transcription. Consequently, cyclophosphamide is equally immunosuppressive given before or after antigenic stimulation.

Cyclophosphamide has many toxic effects that limit its clinical application. In addition to side-effects shared with other cytotoxic drugs, cyclophosphamide specifically damages the bladder, sometimes causing cancer or a condition called hemorrhagic cystitis. Unlike azathioprine, cyclophosphamide is not particularly toxic to the liver, and for patients who have sustained liver damage or become otherwise sensitized to azathioprine, it is a useful alternative. Cyclophosphamide is most effective when used in short courses of treatment.

Methotrexate was one of the first cytotoxic drugs shown to be effective in treating cancer cells. It prevents DNA replication by inhibiting dihydrofolate reductase, an enzyme essential for the cellular synthesis of thymidine. Methotrexate is the drug of choice for inhibiting GVHR in bone marrow transplant recipients (see Figure 12.26).

12-20 Cyclosporin A, tacrolimus, and rapamycin selectively inhibit T-cell activation

During the 1960s and 1970s, transplant physicians depended on combinations of corticosteroids and cytotoxic drugs to prevent the rejection of transplanted organs. Toward the end of the 1970s, new kinds of immunosuppressive drug were introduced that selectively inhibited T-cell activation. These drugs had a marked impact on clinical transplantation in the 1980s and 1990s, leading to improved graft survival, a wider range of tissues and organs being transplanted, and transplantation being recommended for an increased range of diseases. This period became known as the cyclosporin era, because cyclosporin A was the first of these drugs to be introduced.

Cyclosporin A (also known as **cyclosporine**) is a cyclic decapeptide derived from a soil fungus, *Tolypocladium inflatum*, originally isolated in Norway. It inhibits the activation of T cells by antigen by disrupting the transduction of signals from the T-cell receptor. Signals from the T-cell receptor normally lead to hydrolysis of membrane lipids to give inositol trisphosphate and the consequent release of Ca^{2+} from intracellular stores. The elevation in cytosolic Ca^{2+} concentration activates the cytoplasmic serine/threonine phosphatase **calcineurin**, which in turn activates the transcription factor **NFAT** (see Section 6-6, p. 155). In resting T cells, NFAT is present in the cytoplasm in a phosphorylated form. Calcineurin removes the phosphate, permitting NFAT to enter the nucleus, where it binds to the transcription factor AP-1 to form a transcriptional regulatory complex that turns on transcription of the IL-2 gene.

Cyclosporin interferes with calcineurin activity. It diffuses across the plasma membrane into the cytosol, where it binds peptidyl-prolyl isomerase enzymes, which in this context are called **cyclophilins**. The complex of cyclosporin A and cyclophilin binds to calcineurin, inhibiting its phosphatase activity and preventing it from activating NFAT. In the presence of cyclosporin, therefore, IL-2 cannot be made and the program of T-cell activation, proliferation, and differentiation is shut down at a very early stage (Figure 12.27).

Tacrolimus, also called **FK506**, was isolated from the soil actinomycete *Streptomyces tsukabaensis*. It is a macrolide, a class of compound with structures based on a many-membered lactone ring that is attached to one or more deoxy sugars. Although structurally distinct from cyclosporin A, tacrolimus suppresses T-cell activation by the inhibition of calcineurin through a similar mechanism. The peptidyl-prolyl isomerases to which tacrolimus binds are distinct from the cyclophilins and are known as FK-binding proteins. The cyclophilins and **FK-binding proteins** are known collectively as **immunophilins**.

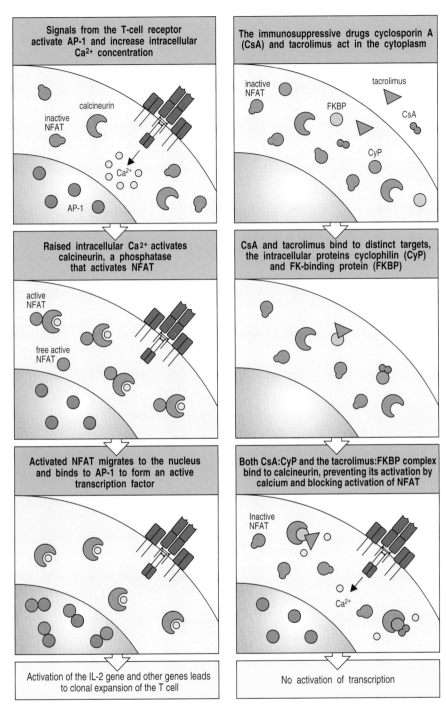

Figure 12.27 Cyclosporin A and tacrolimus inhibit T-cell activation by interfering with the serine/threonine phosphatase calcineurin. Signaling via T-cell receptor-associated tyrosine kinases leads to the activation and increased synthesis of the transcription factor AP-1, as well as increasing the concentration of Ca^{2+} in the cytoplasm (left panels). The Ca^{2+} binds to calcineurin and thereby activates it to dephosphorylate the cytoplasmic form of the nuclear factor of activated T cells (NFAT). Once dephosphorylated, the active NFAT migrates to the nucleus to form a complex with AP-1; the NFAT:AP-1 complex can then induce the transcription of genes required for T-cell activation, including the IL-2 gene. When cyclosporin A (CsA) or tacrolimus are present, they form complexes with their immunophilin targets, cyclophilin (CyP) and FK-binding protein (FKBP), respectively (right panels). The complex of cyclophilin with cyclosporin A can bind to calcineurin, blocking its ability to activate NFAT. The complex of tacrolimus with FKBP binds to calcineurin at the same site, also blocking its activity.

Although the principal effect of cyclosporin A and tacrolimus is to inhibit T-cell activation, the activation of B cells and granulocytes is also suppressed (Figure 12.28). A major advantage of these drugs is that they do not target proliferating cells, so the reduced hematopoiesis and intestinal damage seen with cytotoxic drugs does not occur. A side-effect associated with the continued administration of cyclosporin A or tacrolimus is nephrotoxicity, and some patients can no longer tolerate the drug—they are said to have become sensitized to it.

The success of cyclosporin A and tacrolimus in transplantation encouraged the search for similarly selective drugs. One find is an immunosuppressive macrolide called **rapamycin** (also called **sirolimus**), which was isolated from *Streptomyces hygroscopicus*, a soil bacterium found on Easter Island. The

Cell type	Effects
T lymphocyte	Reduced expression of IL-2, IL-3, IL-4, GM-CSF, TNF-α Reduced cell division because of decreased IL-2 Reduced Ca²⁺-dependent exocytosis of cytotoxic granules Inhibition of antigen-driven apoptosis
B lymphocyte	Inhibition of cell division because T-cell cytokines are absent Inhibition of antigen-driven cell division Induction of apoptosis following B-cell activation
Granulocyte	Reduced Ca²⁺-dependent exocytosis of granules

Figure 12.28 Immunological effects of cyclosporin A and tacrolimus.

island's Polynesian name, 'Rapa ui', was used to name the drug. Although rapamycin binds to FK-binding proteins, it does not interfere with calcineurin but blocks T-cell activation at a later stage by preventing signal transduction from the IL-2 receptor. Rapamycin is more toxic than either cyclosporin A or tacrolimus, but has become a useful component of combination therapy.

12-21 Antibodies specific for T cells are used to control acute rejection

After transplantation, patients are maintained on a combination of immunosuppressive drugs. Because of their toxicity and the immunodeficiency that these drugs cause, physicians seek to lower the dosage gradually to the minimum that will maintain tolerance of the transplant. Inevitably, there are times when the balance is upset and early symptoms of rejection appear (Figure 12.29). Such episodes can be treated with a 5–15-day course of daily injections of T-cell-specific antibodies, as well as an increased dose of immunosuppressive drugs.

Polyclonal anti-T-cell antibodies are made in sheep and goats that have been immunized with human thymocytes or lymphocytes. Antibody-containing fractions called **antithymocyte globulin** (**ATG**) or **antilymphocyte globulin** (**ALG**) are prepared from the animals' blood. Another source of anti-T-cell antibodies is hybridoma cell lines making mouse monoclonal antibodies specific for proteins present only on the T-cell surface, for example CD3.

Immunosuppressive antibodies work in one of two ways. ALG and ATG cause the destruction of the lymphocytes to which they bind, through complement fixation and phagocytosis. In contrast, monoclonal anti-CD3 interferes with the function of the T cells to which it binds, causing reduced expression of the CD3:T-cell receptor complex on the cell surface and depletion of these cells from the circulation. Because the immunosuppressive antibodies are **xenogeneic**, that is, they come from a nonhuman species, they tend to stimulate an antibody response against them. This reduces their immunosuppressive activity when used on subsequent occasions. In such situations the patient's antibodies form immune complexes with the immunosuppressive antibodies, clearing them from the circulation before they can bind to T cells. Such reactions can also lead to serum sickness (see Section 10-18, p. 334). For this reason, physicians generally use each xenogeneic immunosuppressive antibody to counter just one episode of acute rejection per patient.

In recent years, anti-CD3 and other immunosuppressive mouse monoclonal antibodies have been humanized by incorporating the minimal

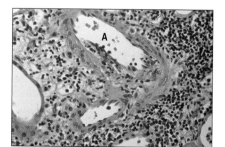

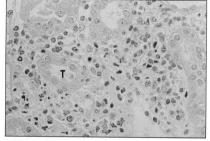

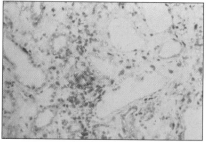

Figure 12.29 Acute rejection in a kidney graft. Top panel shows lymphocytes around an arteriole (A) in a kidney undergoing rejection. The middle panel shows lymphocytes surrounding the renal tubules (T) of the same kidney, and the bottom panel shows the staining of T lymphocytes with anti-CD3 (brown staining) in the same section. Photographs courtesy of F. Rosen.

Tissue transplanted	Number of transplants		Patients waiting for a transplant	
	Oct 1987 – Dec 2002	2002	August 1999	November 2003
Kidney	179,288	14,776	42,875	83,284
Kidney–pancreas	11,159	905	1929	2426
Pancreas	3479	554	502	1484
Liver	60,453	5329	13,698	17,237
Heart	33,911	2155	4287	3556
Heart–lung	821	33	218	184
Lung	10,823	1042	3343	3907

Figure 12.30 Organs commonly transplanted in medicine. The numbers of organs transplanted in the USA during the period 1987–2001 and in 2002 are shown. That the availability of organs and other factors are limiting transplantation is shown by the numbers of patients who could benefit from a transplant. Data courtesy of United Network for Organ Sharing.

mouse antigen-binding site into the framework of a human antibody (see Section 2-5, p. 45). This permits them to be given on multiple occasions to the same patient. This added flexibility means that immunosuppressive mono-clonal antibodies can be used prophylactically to prevent acute rejection as well as for treatment after transplantation. The use of a humanized mono-clonal antibody specific for the high-affinity IL-2 receptor reduces the inci-dence of acute rejection in kidney transplantation by 40%. An advantage in using an antibody specific for this IL-2 receptor is that it targets only activated T cells (see Section 6-7, p. 158), whereas anti-CD3 antibodies target all T cells.

12-22 Patients needing a transplant outnumber the available organs

Kidney transplantation procedures have developed to a point at which it is now possible to transplant cadaveric kidneys across considerable HLA mis-matches. This progress helped the development of heart transplantation, for which only cadaveric donors could be considered. Heart transplantation is inherently more difficult than kidney transplantation, principally because the failure of a grafted heart is fatal, whereas patients with failed kidney grafts can go back to dialysis. The use of cyclosporin A and tacrolimus has increased the success of heart transplantation, mainly by preventing death from acute rejection or infection during the first few months after transplantation. As a consequence, the number of heart transplants increased considerably after 1979. In the USA some 3600 patients each year receive a heart transplant (Fig-ure 12.30) and more than half of them are projected to be alive 10 years after the operation.

Liver transplantation has similarly progressed in the cyclosporin era, from being a relatively risky procedure to one offering considerable benefit. In 1979 only 30–40% of patients survived a liver transplant for more than a year; today 70–90% of patients are surviving after 1 year and 60% are alive after 5 years. A similar improvement has been seen for lung transplantation.

The very success of solid organ transplantation has created its own problem, namely that there are many more patients who could benefit from a kidney, heart, or liver transplant than there are organs available from live and cadav-eric donors (see Figure 12.30). Patients are therefore placed on waiting lists and chosen for transplantation on the basis of various criteria, including the severity of disease and the HLA match with available organs. To increase the organ supply, some countries have instituted a policy whereby cadaveric organs from accident victims become automatically available for clinical transplantation unless the person has deliberately opted out. Other countries retain the policy that organs can be used for transplantation only if an accident victim has deliberately opted in by previously signing a consent form. Even then, relatives can overrule the victim's wishes. The increasing demand for kidneys has inevitably led to an unregulated international trade in human kidneys, generally involving donors in poorer countries selling one of their kidneys for transplantation to rich patients.

The limiting supply of organs could be overcome by using organs from animals. This type of transplantation, in which donor and recipient are of different species, is called **xenotransplantation**, and the grafted tissue a **xenograft**. Pigs are considered the most suitable donor species for humans: first because pig and human organs are of similar sizes, and second because they are already farmed, slaughtered and consumed by humans in large numbers. The immunological barriers facing xenotransplantation are formidable. As a start, most humans have circulating antibodies that bind to pig endothelial cells and would cause hyperacute rejection. In this context these antibodies are called **xenoantibodies** and the carbohydrate antigens on pig endothelium to which they bind are called **xenoantigens**. As with the alloantibodies that humans make against A,B,O blood group alloantigens, the xenoantibodies are probably induced by infections with common bacteria whose surface carbohydrates resemble those of pig cells. Exacerbating the potential problems of hyperacute rejection is the fact that the complement regulatory proteins on the surface of pig cells (the pig versions of CD59, DAF, and MCP; see Section 7-26, p. 220) do not inhibit human complement.

To improve the compatibility of their organs with the human immune system, pigs would need to be genetically modified with several human genes. Over and above the immunological barrier is the concern that immunosuppressed patients receiving pig xenotransplants could provide a route for endogenous pig retroviruses to infect the human population and have an effect like HIV. As on many previous occasions, the development of procedures to save lives by organ transplantation raises complicated ethical questions.

12-23 Bone marrow transplantation is a treatment for genetic diseases of blood cells

Bone marrow transplantation, which permanently replaces an individual's entire hematopoietic system, including the immune system, is used as treatment for a variety of genetic diseases that impair the function of one or more types of hematopoietic cell (Figure 12.31). These include red-cell deficiencies, such as thalassemia major, sickle-cell anemia, and Fanconi's anemia, as well as a variety of conditions affecting leukocytes and causing immunodeficiency. In a bone marrow transplant, the important cells are the pluripotent stem cells that reconstitute the patient's immune system and also their red cells and platelets. In 2–3 weeks after a successful transplant, new circulating blood cells begin to be produced from the transplanted marrow. This is a sign that the pluripotent stem cells have colonized the bones, the process known as **engraftment**. With time the transplant replaces the defective hematopoietic system with one that is normal.

The logistics of bone marrow transplantation are different from those of organ transplantation, more resembling a blood transfusion. The donors are alive and healthy, and the transplanted tissue is given by an intravenous infusion that involves no surgery. The immunology of bone marrow transplantation is also different from that of other types of transplantation. This is because it involves transplantation of the immune system, cells of which are present in almost every organ and tissue of the body. Whereas kidney trans-

Genetic diseases treatable by bone marrow transplantation
SCID
Wiskott–Aldrich syndrome
Fanconi's anemia
Kostmann's syndrome
Chronic granulomatous disease
Osteopetrosis
Ataxia telangiectasia
Diamond–Blackfan syndrome
Mucocutaneous candidiasis
Chédiak–Higashi syndrome
Cartilage-hair hypoplasia
Mucopolysaccharidosis
Gaucher's disease
Thalassemia major
Sickle-cell anemia

Figure 12.31 Genetic diseases for which bone marrow transplantation is a therapy.

plantation involves suppression of the recipient's T-cell response to prevent rejection of the graft, in bone marrow transplantation the recipient's immune system is deliberately destroyed to the point where the patient would not survive without the transplant. This conditioning regimen is called **myeloablative therapy** and is accomplished by a combination of cytotoxic drugs and irradiation. Myeloablative therapy serves two purposes: the first is to prevent rejection of the grafted cells by the recipient's T cells; the second is to kill the hematopoietic cells within the recipient's bone marrow and provide room for the transplanted stem cells to interact with bone marrow stromal cells.

Within a few weeks after transplantation the hematopoietic system begins to be reconstituted. The cells of innate immunity, for example granulocytes and NK cells, recover more quickly than the B and T cells of adaptive immunity. When fully reconstituted the patient is a chimera in which the hematopoietic cells are of donor genotype and the rest of the cells are of recipient genotype. The patient's new immune system can then become tolerant of both donor and recipient HLA allotypes. A critical feature of the T cells of the patient's new immune system is that they are positively selected by thymic epithelial cells expressing the recipient's HLA allotypes. For those T cells to be activated by infections, however, they need to interact with the donor-derived dendritic cells, which present antigens on donor HLA allotypes (see Section 5-6, p. 132). An interaction will only occur when the recipient and donor have HLA allotypes in common, and the more HLA allotypes they have in common the better it works.

12-24 The alloreactions in bone marrow transplantation attack the patient, not the transplant

Because of the severe conditioning regimen, rejection mediated by recipient T cells responding to HLA allotypes of a bone marrow graft is not the problem that it is in kidney transplantation. Instead, the main cause of damaging alloreactions are mature T cells in the transplanted bone marrow that respond to the recipient's HLA allotypes in a graft-versus-host reaction (Figure 12.32). This GVHR is the major cause of morbidity and mortality after bone marrow transplantation. The condition it causes, acute graft-versus-host disease, can attack almost every tissue of the body, but principally involves the skin, the intestines, and the liver. In essence, GVHD is an acute autoimmune disease that can prove fatal. The severity of GVHD varies, four grades being defined in clinical diagnosis (Figure 12.33). The characteristic skin rash of GVHD tends to develop with the kinetics of a primary immune response during the 10–28 days after transplantation. The fine, diffuse erythematous rash begins on the palms of the hands, the soles of the feet, and on the head, and then spreads to the trunk. The intestinal reaction causes cramps and diarrhea, and inflammation of the bile ducts in the liver causes

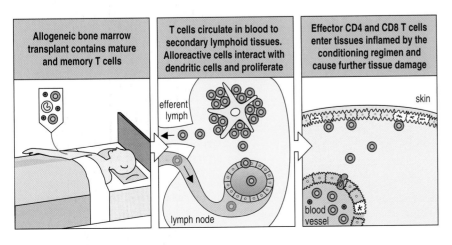

Figure 12.32 Graft-versus-host disease is due to donor T cells in the graft that attack the recipient's tissues. After bone marrow transplantation, any mature donor CD4 and CD8 T cells present in the graft that are specific for the recipient's HLA allotypes become activated in secondary lymphoid tissues. Effector CD4 and CD8 T cells move into the circulation and preferentially enter and attack tissues that have been most damaged by the conditioning regimen of chemotherapy and irradiation: skin, intestines and liver.

Tissue reactions in the four grades of graft-versus-host disease			
Grade	Skin	Liver	Gastrointestinal tract
I	Maculopapular rash on <25% of body surface	Serum bilirubin 2–3 mg dl^{-1}	>500 ml diarrhea day^{-1}
II	Maculopapular rash on <25–50% of body surface	Serum bilirubin 3–6 mg dl^{-1}	>1000 ml diarrhea day^{-1}
III	Generalized erythroderma	Serum bilirubin 6–15 mg dl^{-1}	>1500 ml diarrhea day^{-1}
IV	Generalized erythroderma with bullous formation and desquamation	Serum bilirubin 15 mg dl^{-1}	Severe abdominal pain with or without ileus

Figure 12.33 Characteristics of the four grades of graft-versus-host disease.

hyperbilirubinemia and a rise in the levels of liver enzymes in the blood. Methotrexate in combination with cyclosporin A is used to reduce the incidence and severity of GVHD.

Mature T cells from the transplant will circulate in the recipient's blood and enter secondary lymphoid tissues, where they interact with the recipient's dendritic cells. Alloreactive T cells will be stimulated to divide and differentiate into effector cells, which will travel via the lymph, and the blood to inflamed tissues. The conditioning regimen, which seeks to destroy the rapidly dividing bone marrow cells, also damages other tissues in which there is normally much cellular proliferation. Prominent among these are the skin, the intestinal epithelium, and the hepatocytes of the liver: the targets for GVHD. The conditioning regimen preferentially creates inflammation in these tissues, which has been described as a 'cytokine storm.' This environment causes dendritic cells to activate and migrate from the inflamed tissues to draining secondary lymphoid tissue, where they stimulate alloreactive T cells. More importantly, it makes these tissues more accessible to alloreactive effector T cells.

Almost all bone marrow transplant patients suffer to some extent from GVHD. The severity of GVHD correlates strongly with the degree of HLA mismatch, and, because of the potentially fatal consequences of GVHD, the clinical outcome of bone marrow transplantation is more sensitive to HLA mismatching than is solid organ transplantation (Figure 12.34). There are, however, large numbers of potential bone marrow donors, because bone marrow, like blood, can be donated by healthy individuals without compromising their immunological or hematological functions. Marrow is usually aspirated from the iliac crests of the pelvis under anesthesia. For patients

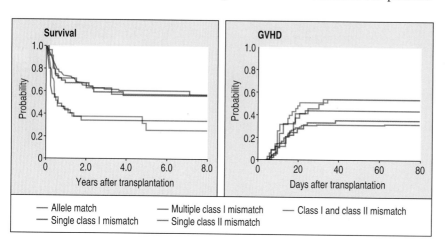

Figure 12.34 The clinical outcome of treatment with bone marrow transplantation correlates with the degree of HLA match. Two parameters of clinical outcome are shown: survival of the patient (top panel) and the probability of getting severe GVHD (lower panel). Severe GVHD is defined as grades III and IV (see Figure 12.33). Data courtesy of E. Petersdorf.

without HLA-identical siblings to donate bone marrow, there are national and international registries of HLA-typed people that are searched to identify histocompatible donors. Worldwide, some 5 million potential donors have been HLA typed for this purpose.

Because the number of mature T cells present in the transplant is limited, the duration of acute GVHD is usually restricted to the first few months after transplantation. The chronic rejection reactions seen in kidney transplant patients have their analog in the chronic GVHD experienced by 25–45% of bone marrow transplant patients who survive longer than 6 months after transplant. The course of this condition resembles that of an autoimmune disease, and the cumulative effect is to produce severe immunodeficiency, leading to recurrent life-threatening infections.

12-25 The impact of alloreactions on transplantation depends on the type of tissue or organ transplanted

The previous sections described how alloreactions directed at HLA alloantigens have very different effects in patients with bone marrow or kidney transplants. Tissue-specific differences in transplantation immunology are not unusual. Each tissue has a unique anatomy and vasculature, and these properties affect both the strength of antigenic stimulation that it gives upon transplantation and the nature of the alloreactive response. An extreme case is the cornea of the eye, which is not vascularized and can be transplanted successfully without immunosuppression.

The outcome of liver transplantation is not improved by matching the donor and recipient for HLA class I or II antigens. Indeed, it has been claimed that transplant outcome is inversely correlated with the degree of HLA match. As a consequence, HLA type or cross-match are not assessed before liver transplantation; A,B,O type is the only genetic factor affecting donor selection. Clinical experience indicates that the liver is relatively refractory to either acute or hyperacute rejection, yet the use of cyclosporin A and tacrolimus has markedly improved the success of liver transplants. The liver has a specialized architecture and vasculature, and hepatocytes express very low levels of HLA class I and no HLA class II. These properties, and the daily exposure of liver cells to the digestion products of a myriad of foreign proteins in the intestines, could all contribute to the distinct immunobiology of the transplanted allogeneic liver.

Summary

Many human diseases involve the malfunction of a single organ or tissue that can cause incapacitation or death. With transplantation, diseased tissues are replaced by healthy ones, leading to improved health and longer life. Of the many problems faced by clinical transplantation, the most challenging has been control of the alloreactive immune response, which rejects transplanted organs or causes graft-versus-host disease (GVHD) in bone marrow transplantation. These alloreactions are due to genetic differences between transplant donor and recipient, which determine the epitopes recognized by host or donor lymphocytes. Two complementary strategies are used to reduce the effects of alloreactions: the first is the matching of donor and recipient for the most critical genetic factors; the second is the administration of drugs that suppress the immune system.

A transplanted tissue can be rejected at various times after transplantation. Hyperacute rejection, which occurs immediately, is caused by preformed antibodies in the recipient that react with cells of the transplant, particularly the vascular endothelium. Common antibodies of this type are directed against A,B,O antigens or HLA class I molecules. Hyperacute rejection cannot be treated and it is therefore avoided by matching for A,B,O and performing a cross-match test. Acute rejection is caused by the recipient's mature CD4 and CD8 T cells, responding to HLA class I and II differences between transplant donor and recipient. Acute GVHD is similarly caused by mature immuno-competent T cells in the graft specific for antigens on the recipient's tissues. Acute rejection can be prevented by matching transplant donors and recipients for HLA class I and II allotypes and by treatment with immunosuppressive drugs and antibodies. After transplantation there is a gradual accommodation between the grafted tissue and the recipient that allows the doses of immunosuppressive drugs to be reduced. During this period, episodes of rejection can occur, which are treated with additional immuno-suppressive drugs and anti-T-cell antibodies. Chronic rejection and chronic GVHD are caused by antibody responses to the allogeneic HLA molecules that are initiated after transplantation and persist because of incomplete immunosuppression. With time, many patients can be maintained on a level of immunosuppression that allows the immune system to recover and provide defense against infection.

Cancer and its interactions with the immune system

Cancer is a diverse collection of life-threatening diseases that are caused by abnormal and invasive cell proliferation. It accounts for about 20% of deaths in the industrialized countries; worldwide there are some 6 million new cases of cancer each year, and half of these people will die from the disease. Cancer is largely a disease of older people, and in the industrialized societies where both life expectancy and the average age of the population are increasing, so is fear of cancer.

Cancer cells are very similar to normal cells, and the immune system seems unable to attack them effectively. In treating cancer, physicians resort to surgery, radiation, and cytotoxic drugs, sometimes referred to as strategies of 'slash, burn, and poison.' Although these treatments give remission or cure to some patients, more often they are limited by the incomplete elimination of cancer cells and the deleterious side-effects of the treatment. For more than a century, cancer immunologists have sought to harness a patient's immune system to augment conventional therapies. Although ideas for boosting cancer immunity have been regularly demonstrated to work to some extent in animal models, useful routine immunotherapies for human cancer have yet to emerge. In just the past few years, increased understanding of the mechanisms by which an immune response is started and sustained has led to several new initiatives that do hold promise for the treatment of cancer. In this part of the chapter we first consider the processes by which cancers emerge in the body. Then we examine the specific application of bone marrow transplantation as a treatment for cancer, and how the alloreactions it generates can work as immunotherapy. Finally, we turn to the antigens on tumor cells that can stimulate an autologous immune response and how that response might be quickened and strengthened.

12-26 Cancer results from mutations that cause uncontrolled cell growth

Maintaining the human body involves continuing cell division to replace worn-out cells, repair damaged tissue, and mount immune responses against invading pathogens. Of the order of 10^{16} cell divisions are estimated to occur in a human body during a lifetime. An essential preparation for cell division is DNA replication, which makes two identical copies of the diploid genome. Although the enzymes that replicate DNA are extremely accurate and use proof-reading mechanisms to correct mistakes, rare errors are still made. Additional changes in DNA arise from chemical damage that escapes the DNA repair machinery.

Changes in DNA are called **mutations**. They include the substitution, insertion, and deletion of nucleotides, recombination between different members of a gene family, and chromosomal rearrangements. Mutations in the germline—the eggs and sperm—provide the variation that allows the human species to diversify and evolve. In contrast, mutations in somatic cells affect only the individual in which they arise. Most somatic mutations are never noticed because their effects, if any, are manifest in a single cell. There are, however, certain mutations that abolish the normal controls on cell division and cell survival. When that happens, the mutant cell proliferates to form an expanding population of mutant cells that eventually disrupts the body's physiology and organ function, causing the diseases that are collectively called cancer.

Tumor, which means swelling, and **neoplasm**, which means new growth, are words used synonymously to describe tissue in which cells are multiplying abnormally. The branch of medicine that deals with tumors is called **oncology**, the prefix *onco* deriving from the Greek word for swelling, *ogkos*. Not all tumors are malignant: **benign tumors**, such as warts, are encapsulated, localized, and limited in size; the **malignant tumors**, in contrast, can continually increase their size by breaking through basal laminae and invading adjacent tissues (Figure 12.35).

The word **cancer**, meaning an evil spreading in the manner of a crab (Latin, *cancer*), is used to describe the diseases caused by malignant tumors. In addition to spreading out locally from their site of origin, cancer cells can be carried by lymph or blood to distant sites, where they initiate new foci of cancerous growth. This mode of spreading is called **metastasis**, the site of origin being called the primary tumor and the sites of spreading being called secondary tumors. Cancers arise most commonly in tissues that are actively undergoing cell division and are thus most likely to accumulate mutations due to errors in DNA replication. These tissues include the epithelial linings of the gastrointestinal tract, urogenital tract, and the mammary glands (Figure 12.36). Cancers of epithelial cells are known as **carcinomas**, cancers of other cell types as **sarcomas**. Cancers of immune system cells are known as

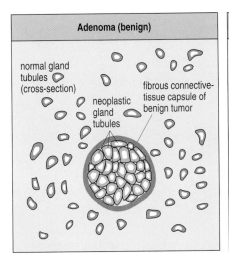
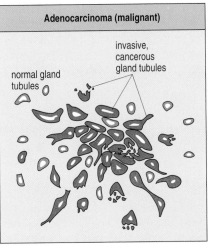

Figure 12.35 The contrast between benign and malignant tumors derived from the same tissue. The diagram illustrates tumors found in the breast. Adenoma is the general name given to a benign tumor of glandular tissue; malignant tumors of glandular tissues are called adenocarcinomas.

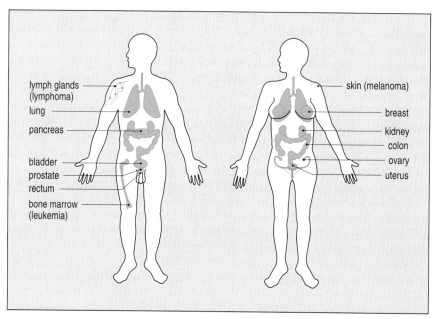

leukemias when they involve circulating cells, **lymphomas** when they involve solid lymphoid tumors, and **myelomas** when they involve bone marrow (see Sections 4-10, p. 115, and 5-12, p. 139).

12-27 A cancer arises from a single cell that has accumulated multiple mutations

The best defenses against cancer lie within each cell of the human body and not with the specialized cells of the immune system. The integrity of the body is so dependent on well-controlled cell division that many mechanisms have evolved to ensure this. These include the repair of certain types of DNA damage as well as mechanisms that prevent the survival and division of cells with badly damaged DNA. Consequently, the control of cell division never depends on the function of just one protein, and a cell cannot become cancerous by mutation in just one gene. For a cell to give rise to a cancer, it must first accumulate multiple mutations, and these must occur in genes concerned with the control of cell multiplication and cell survival. When a cell becomes able to form a cancer it is said to have undergone **malignant transformation**.

There are two main classes of gene that, if mutated or misexpressed, contribute to malignant transformation. **Proto-oncogenes** are genes that normally contribute positively to the initiation and execution of cell division. More than 100 human proto-oncogenes have been identified; they encode growth factors and their receptors, and also proteins involved in signal transduction and gene transcription. The mutant forms of proto-oncogenes that contribute to malignant transformation are called **oncogenes**.

The second class of genes involved in cellular transformation are called **tumor suppressor genes** because they encode proteins that prevent the unwanted proliferation of mutant cells. A typical tumor suppressor gene is that for the protein p53. This protein is expressed in response to DNA damage and causes the damaged cell to die by apoptosis. Loss of the p53 gene or mutations that interfere with its protective function are the most frequent mutations present in human cancers. Over 50% of cases of human cancer have a mutation in p53, revealing its importance in protecting the body from cancer. Indeed, p53 might well have evolved as an internal cellular defense against cancer.

Figure 12.37 A typical series of mutations acquired during the development of a cancer. This illustration shows the development of colorectal cancer by successive mutations in different genes. The morphological changes accompanying each change are indicated. RAS is an oncogene; APC, DCC, and p53 are tumor suppressor genes. Both copies of a tumor suppressor gene must be mutated to contribute to malignant transformation, so the number of mutations that has occurred here totals seven. The sequence of events shown here usually takes 10–20 years or more. After a cell has undergone the initial series of mutations that render it malignant, the tumor cells then rapidly accumulate more mutations. Some of these contribute to making the cancerous cells more invasive, but others are thought to be simply a consequence, rather than a cause, of the cell's becoming cancerous. Once a tumor has become genetically heterogeneous, selection acts to favor those cells that divide more rapidly and are more invasive.

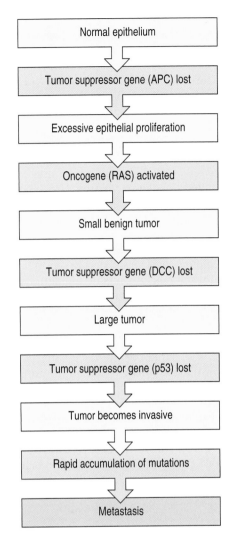

It has been estimated that a cell must accumulate at least five or six independent mutations before it can become cancerous. The actual number depends on the cell type and the particular genes in which the mutations occur (Figure 12.37). Because of the random manner in which mutations occur, the combination of genetic lesions in each cancer is unique.

The low frequency of mutation together with the requirement for multiple mutations means that each cancer inevitably arises from a single cell that has undergone malignant transformation. This explains why in paired organs such as the lungs, cancer initially affects only one of them. In the course of a lifetime, a person's cells accumulate mutations, and the probability that a cell somewhere in the body will have the right combination of mutations to cause a cancer increases in a nonlinear fashion. For this reason, the incidence of cancer increases with age, and cancers are largely diseases of older people.

12-28 Exposure to chemicals, radiation, and viruses can facilitate the progression to cancer

Cancer is not, however, an inevitable outcome of aging. Most people never suffer from cancer, even in old age. This reveals the existence of genetic and environmental factors that influence the risk of a cancer developing. Genetic factors include the possession of a germline mutation in one copy of a tumor suppressor gene. Such a mutation in the p53 gene is the underlying cause of Li–Fraumeni syndrome, a condition in which there is an unusually strong predisposition to multiple cancers that arise at a relatively early age. This is because only one new mutation is needed, in the good copy of p53, to abolish all p53 function. In other people at least two new mutations, in both copies of p53, would be needed.

Of greater importance to the population at large are environmental insults that increase the number of mutations suffered by the body. Chemical and physical agents that damage DNA in such a way as to cause an increased rate of mutation are called **mutagens**. Many known mutagens can increase the risk of cancer and are known as **carcinogens**. People who have had heavy or prolonged exposure to carcinogenic agents, which include certain chemicals, ultraviolet light, and other forms of radiation, are more at risk of developing cancer than those who have not been exposed. Most marked is the increased frequency of lung cancer among people who smoke.

Chemical carcinogens tend to cause mutations due to single nucleotide substitutions in DNA. Radiation, in contrast, tends to produce grosser forms of damage such as DNA breaks, cross-linked nucleotides, abnormal recombination, and chromosome translocations. The radiation released by the atomic bombs exploded at Hiroshima and Nagasaki in 1945 caused an increased incidence of leukemia in those who survived the acute effects of the blasts. Less marked, but more pervasive, is the increasing incidence of skin cancer caused by overexposure to ultraviolet radiation from the sun in the pursuit of work, fun, and fashion. However, despite the fear of cancer in developed societies and the well-known risk factors, people in those same societies still deliberately engage in behavior that they know increases their chances of developing cancer in later years.

Certain viruses also have the potential to transform cells, and viruses are associated with some 15% of human cancers (Figure 12.38). Such viruses are called **oncogenic viruses**. The known oncogenic viruses that affect humans are DNA viruses, except for the RNA retrovirus HTLV-1, which is associated with adult T-cell leukemia.

Human oncogenic viruses typically set up chronic infections in a cell, producing novel virally encoded proteins that override or interfere with the cell's normal mechanisms for regulating cell division. Infected cells therefore start to proliferate. For example, the Epstein–Barr virus induces infected B cells to divide repeatedly (see Section 9-4, p. 283), and if the infected cells are not cleared rapidly, a proportion of them go on to become transformed. Certain strains of papilloma virus predispose to cancer of the cervix. They encode proteins that prevent the normal tumor suppressor mechanisms from acting within the infected cells. Viral proteins bind to p53 protein and to another tumor suppressor protein called Rb, blocking their functions and enabling the virus-infected epithelial cells to proliferate. Other chronic infections can lead to cancer because the tissue damage that they cause necessitates a high rate of cell renewal and, as a consequence, mutations accumulate more rapidly. The liver cancers arising from hepatitis B and C virus infections might be of this type, as might the stomach cancers associated with ulcers caused by infection with the bacterium *Helicobacter pylori*.

Viruses associated with human cancers		
Virus	**Associated tumors**	**Areas of high incidence**
DNA viruses		
Papillomavirus (many distinct strains)	Warts (benign) Carcinoma of uterine cervix	Worldwide Worldwide
Hepatitis B virus	Liver cancer (hepatocellular carcinoma)	Southeast Asia Tropical Africa
Epstein–Barr virus	Burkitt's lymphoma (cancer of B lymphocytes). Nasopharyngeal carcinoma B-cell lymphoproliferative disease	West Africa Papua New Guinea Southern China Greenland (Inuit) Immunosuppressed or immunodeficient patients
RNA viruses		
Human T-cell leukemia virus type 1 (HTLV-1). Human immunodeficiency virus (HIV-1) and human herpes virus 8 (HHV8)	Adult T-cell leukemia/ lymphoma. Kaposi's sarcoma	Japan (Kyushu) West Indies Central Africa

Figure 12.38 Viruses associated with human cancers. For all the viruses listed here, the number of people infected is much larger than the number who develop cancer; the viruses must act in conjunction with other factors. Some of the viruses probably contribute to cancer only indirectly. For example, an increased incidence of Kaposi's sarcoma due to transformation of endothelial cells by HHV8 is seen not only in HIV-infected patients but also in other immunosuppressed patients. HIV-1, by obliterating cell-mediated immune defenses, is probably simply allowing the HHV8-transformed endothelial cells to thrive as a tumor instead of being destroyed by the immune system.

Figure 12.39 The growth and lifespan of a typical human tumor. The example shown is for a tumor of the breast. The diameter of the tumor is plotted on a logarithmic scale. Years can elapse before the tumor becomes noticeable to the patient or the patient's immune system.

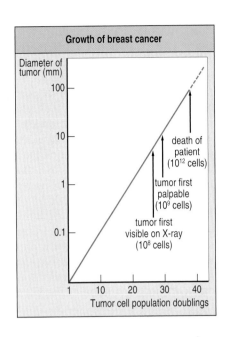

12-29 The immune system is insensitive to emerging cancer

The immune response to an infectious disease is started by inflammation produced at the initial site of infection. This makes innate immune mechanisms available within hours after the start of infection and adaptive immune responses are fully operational within 2 weeks. In comparison, the malignant transformation of a single cell that marks the start of a cancer goes unnoticed by the body. For years the growth of a cancer might cause no damage of the sort that would trigger inflammation and an immune response (Figure 12.39) By the time that the cancer's growth is damaging tissue and raising the alarm of inflammation, the tumor load is usually so great that the immune system is likely to be overwhelmed.

In the early twentieth century it was noted that some cancer patients contracting bacterial infections experienced regression of their tumors. A more recent example of the phenomenon is the established treatment of superficial bladder cancer in which chronic inflammation of the bladder is achieved by introduction of the BCG vaccine. The component of the vaccine that gives it this anti-tumor effect is the unmethylated CpG-containing DNA of the mycobacteria, the ligand for the Toll-like receptor TLR-9 (see Section 8-6, p. 238). It seems that when the immune system is triggered by a pathogen it can sometimes respond to an existing tumor.

Tumor cells are not inherently resistant to attack by the effector mechanisms of the immune system. Tumors of laboratory mice that grow when transplanted to mice of identical MHC type fail to take hold when transplanted into mice of different MHC type (Figure 12.40). In these animals the tumor cells are killed by alloreactive CD8 T cells that recognize the allogeneic MHC class I molecules. Human alloreactive CD8 T cells are equally good at killing human tumor cells in tissue culture. As we saw in Section 12-14, many alloreactive T cells are memory cells that were first stimulated by infection and are more numerous and more easily activated than naive T cells. As will be seen in the following three sections, alloreactive T cells are helpful participants in the treatment of human cancers by bone marrow transplantation.

12-30 Allogeneic bone marrow transplantation is the preferred treatment for many cancer patients

We have seen how allogeneic bone marrow transplantation is used to replace a defective hematopoietic system with a functional one (see Section 12-23).

Figure 12.40 When tumors are transplanted between MHC-incompatible mice they are rejected by the alloreactive response to MHC differences. The left panels show the transplantation of a tumor between two mice of the same MHC type. The tumor grows in the recipient. In the right panels, the tumor is transplanted to a mouse of a different MHC type and is rejected. The high polymorphism of HLA class genes in human populations is therefore a barrier that prevents any tumor cells that are passed from one person to another from growing. Situations in which such transfer could potentially occur are the same as those that serve to spread HIV: intimate contact, blood transfusion, and sharing of syringes and needles in the non-medical use of drugs. Thus cancer cells themselves are not infectious, although oncogenic viruses can be.

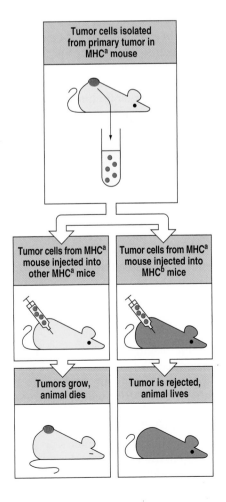

Bone marrow transplantation is also an important part of treatment for many cancer patients, particularly those with tumors of immune system cells. In treating cancer patients with chemotherapy and irradiation, the advantages gained by killing off the malignant cells have to be balanced against the damage caused to normally proliferating vital tissues. Of these the bone marrow is the most susceptible. Bone marrow transplantation offers the opportunity to increase treatment beyond the point where it is lethal, after which the patient is rescued with an allogeneic transplant from a healthy HLA-matched donor. Since the first bone marrow transplants were performed some 30 years ago the procedure has been used against an increasing range of cancers (Figure 12.41).

Despite millions of registered bone marrow donors, about 30% of patients who are clinically eligible for a transplant do not find a suitable HLA-matched donor. To help these patients a procedure called the **autologous bone marrow transplant** was devised. Here, samples of the patient's own bone marrow are taken before the remainder is destroyed by the treatments given for the cancer. After separation of the bone marrow stem cells from any remaining tumor cells, the stem cells are reinfused into the patient. Autologous transplants avoid the problems of histoincompatibility and GVHD and thus patients do not require immunosuppression. However, their use is limited by the rate of relapse of malignant disease, which is greater than with allogeneic transplants.

Hematopoietic stem cells for transplantation can now be obtained by less invasive procedures than bone marrow aspiration. Pluripotent hematopoietic stem cells are first mobilized from the donor's bone marrow into the peripheral blood by treatment with granulocyte colony-stimulating factor (G-CSF) and granulocyte–macrophage colony-stimulating factor (GM-CSF), which are sometimes combined with cyclophosphamide chemotherapy in autologous transplants. Leukocytes are then selectively removed from the blood by a process called leukapheresis, which involves a machine to which the donor's circulation is connected for several hours. The CD34 surface glycoprotein is used as a marker for hematopoietic stem cells. A cell fraction enriched for the CD34-expressing stem cells is isolated from the removed leukocytes and used as the transplant. Between a quarter and half a billion CD34-positive cells are needed to ensure prompt engraftment after transplantation.

Another source of hematopoietic stem cells is umbilical cord blood, obtained from placentas after birth. Umbilical cord blood is rich in stem cells and it also contains fewer of the cells that contribute to the GVHR. Because stem cells for transplantation are now being obtained from sources other than bone marrow, the terms **hematopoietic stem cell transplantation** and **hematopoietic cell transplantation** are beginning to be used instead of bone marrow transplantation

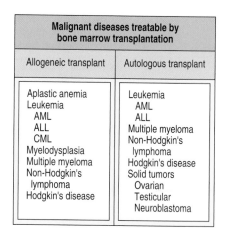

Malignant diseases treatable by bone marrow transplantation	
Allogeneic transplant	Autologous transplant

Aplastic anemia
Leukemia
  AML
  ALL
  CML
Myelodysplasia
Multiple myeloma
Non-Hodgkin's
  lymphoma
Hodgkin's disease
	Leukemia
	  AML
	  ALL
	Multiple myeloma
	Non-Hodgkin's
	  lymphoma
	Hodgkin's disease
	Solid tumors
	  Ovarian
	  Testicular
	  Neuroblastoma

Figure 12.41 Cancers that are treated with a hematopoietic stem cell transplant.

12-31 Patients receiving an HLA-identical bone marrow transplant can still get GVHD

When seeking an allogeneic bone marrow donor the gold standard is a healthy HLA-identical sibling. Thousands of such transplants have been performed. Despite the HLA compatibility, many of these patients are still affected by GVHD, which is particularly frequent in males who receive bone marrow transplants from their HLA-identical sisters. The cause of the GVHD is cytotoxic T cells that are restricted by an HLA class I allotype shared by the siblings but kill only male cells possessing that allotype. In the first example studied, the T cells were found to be specific for complexes of HLA-A2 and a peptide derived from a protein called SMCY that is encoded on the male-

Antigen name	HLA restriction	Gene name	Chromosome	Tissue distribution
HA-1	HLA-A2	KIAA0223	19p13.3 autosome	Hematopoiesis restricted
HA-1	HLA-B60	KIAA0223	19p13.3 autosome	Hematopoiesis restricted
HA-2	HLA-A2	Myosin related gene	6p21.3 autosome	Hematopoiesis restricted
HA-8	HLA-A2	KIAA0020	9 autosome	Ubiquitous
HB-1	HLA-B44	HB1	5q32 autosome	B-ALL
HY-A1	HLA-A1	DFFRY	Y	Ubiquitous
HY-A2	HLA-A2	SMCY	Y	Ubiquitous
HY-B7	HLA-B7	SMCY	Y	Ubiquitous
HY-B8	HLA-B8	UTY	Y	Ubiquitous
HY-B60	HLA-B60	UTY	Y	Ubiquitous
HY-DQ5	HLA-DQ5	DBY	Y	Ubiquitous

Figure 12.42 Human minor histocompatibility antigens. B-ALL, B-cell acute lymphoblastic leukemia. Data courtesy of J.H.F. Falkenburg.

specific Y chromosome. Because SMCY is never present in females, their T cells are not selected to be tolerant of it. Alloantigens of this type, in which the allogeneic difference is due to the bound peptide and not to the MHC molecule, are called **minor histocompatibility antigens**, and the genes encoding them **minor histocompatibility loci** (Figure 12.42). With one exception all the minor histocompatibility antigens are presented by HLA class I allotypes, consistent with their being the breakdown products of intracellular proteins. Minor histocompatibility antigens arise not only because of the differences between the sexes. All the many human proteins that exhibit inherited structural polymorphisms are a potential source of a peptide that can bind to an autologous HLA allotype and become a minor histocompatibility antigen (Figure 12.43).

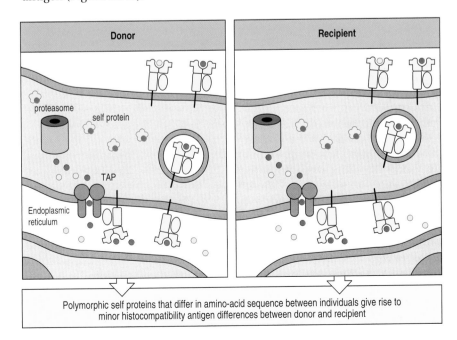

Polymorphic self proteins that differ in amino-acid sequence between individuals give rise to minor histocompatibility antigen differences between donor and recipient

Figure 12.43 Minor histocompatibility antigens are peptides derived from polymorphic proteins other than HLA class I and class II molecules. Self proteins are routinely digested by proteasomes within the cell's cytosol, and peptides derived from them are delivered to the endoplasmic reticulum, where they can bind to MHC class I molecules and be delivered to the cell surface. If a polymorphic protein differs between the graft donor (shown in red on the left) and the recipient (shown in blue on the right), it can give rise to an antigenic peptide (red on the donor cell) that can be recognized by the recipient's T cells as non-self and elicit an immune response. Such antigens are the minor histocompatibility antigens.

12-32 Some GVHD helps engraftment and prevents relapse of malignant disease

One way to reduce the severity of GVHD is to remove T cells from the bone marrow graft before it is given to the patient. This is done by using lectins or monoclonal antibodies that bind selectively to mature T cells. Although markedly reducing GVHD, this procedure leads to a higher incidence of graft failure and, with cancer, a higher incidence of recurrence of disease. This indicates that T-cell alloreactions help engraftment by subduing residual activity of the recipient's immune system, and help to eliminate cancer cells that have escaped the conditioning regimen. Consistent with this interpretation was the experience that graft failure and disease relapse were also higher for bone marrow transplants in which alloreactions cannot occur—autologous transplants and transplantation between identical twins. Also pointing to the same conclusion were observations that some acute GVHD, due to minor histocompatibility antigens, improves the long-term clinical outcome for transplants between HLA-identical siblings.

When alloreactive T cells in the graft help to rid the patient of residual leukemia cells it is called a **graft-versus-leukemia (GVL) effect** or a **graft-versus-tumor (GVT) effect**. A current movement in hematopoietic stem cell transplantation is to use new protocols that seek to promote the GVL reaction and to place less emphasis on using chemotherapy and irradiation to eliminate the tumor. This allows less severe conditioning regimens to be used that do not completely disable the patient's hematopoietic system; after treatment the patient is less severely immunocompromised and recovers more quickly. Patients treated with the older protocols would often need to be isolated for several weeks in intensive care, whereas patients receiving these so-called 'mini-transplants' need not be isolated, spend much less time in hospital, and some can even be treated as outpatients. The mini-transplant has the potential to provide treatment to many cancer patients who for reasons of age or previous treatment history are not likely to survive myeloablative therapy.

One approach to the manipulation of a GVL effect is to give patients transfusions of donor lymphocytes or T cells after they have received the hematopoietic stem cell graft. Such donor lymphocyte transfusions are given at a time when the inflammation caused by the conditioning regimen has subsided and the likelihood of severe GVHD is diminished.

12-33 NK cells can also mediate GVL effects

Although the trend in HLA-matched transplants is to use gentler protocols with nonmyeloablative conditioning regimens, a very different strategy is proving helpful to some of the 30% of patients who cannot find an HLA-matched donor. Most of these patients have a willing donor in their family who shares one HLA haplotype with the patient but differs for the second. Potential donors are mothers, fathers, and 50% of siblings. This type of transplant is called a **haploidentical transplant** (Figure 12.44). Incompatibility for a complete HLA haplotype has the potential to generate strong, fatal alloreactions. To prevent GVHD, the graft is stringently depleted of T cells and the recipient is infused with anti-T-cell antibodies. Graft rejection is prevented through a combination of an intensive conditioning regimen and a larger dose of hematopoietic stem cells than normal.

Patients given a haploidentical transplant get little or no GVHD and require no further immunosuppressive treatment after transplantation. As their immune systems reconstitute, alloreactive NK cells can emerge. These provide a GVL effect that reduces the incidence of leukemic relapse. The occurrence and

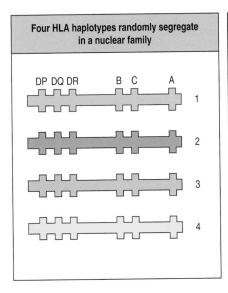

Four HLA haplotypes randomly segregate in a nuclear family

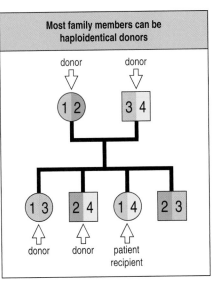

Most family members can be haploidentical donors

Figure 12.44 Almost all patients who need a hematopoietic cell transplant have an HLA-haploidentical family member who is willing to be the donor. In the vast majority of nuclear families there are four different HLA haplotypes (left panel). For any child in the family both parents and 50% of siblings can provide an HLA-haploidentical transplant (right panel). In comparison, neither parent and only 25% of siblings can provide an HLA-identical transplant.

specificity of the NK-cell-mediated alloreactions are determined by the interactions of inhibitory KIR receptors with HLA-B and HLA-C ligands and are predictable from the HLA types of the donor and recipient (see Sections 8-14 and 8-15, pp. 253 and 255) (Figure 12.45). As a rule, NK-cell alloreactions occur when the recipient's HLA class I allotypes provide ligands for fewer types of inhibitory KIR than the donor's HLA class I allotypes. Patients with acute myelogenous leukemia (AML) benefit from such NK-cell alloreactions, whereas patients with acute lymphocytic leukemia (ALL) do not. This difference correlates with the presence of LFA-1, an important adhesion molecule for NK cells, on AML but not ALL cells. The alloreactive NK-cell response wanes and is undetectable 4 months after transplantation. With full reconstitution of the immune system the NK-cell repertoire becomes tolerant of both recipient and donor cells.

12-34 Cancer cells continue to acquire mutations throughout the cancer's lifetime

Cells that become malignantly transformed have genomic differences that distinguish them from every other cell in the body. Cells transformed by oncogenic viruses are genetically the most distinctive, whereas at the other end of the spectrum are tumors arising from a transformed cell with point mutations in some half-dozen oncogenes or tumor suppressor genes. As a

Figure 12.45 Alloreactive NK cells can provide a graft-versus-leukemia effect in patients receiving a haploidentical hematopoietic cell transplant. The HLA class I genotypes of the donor and of the recipient, a patient with acute myelogenous leukemia, are shown in the left panel. The key difference is at the HLA-C locus. The donor has Cw1 with Asn80, and Cw2 with Lys80; the recipient has Cw2 and Cw4, both of which have Lys80. The NK cells of donor genotype that emerge soon after transplantation can be divided into two groups according to whether they are inhibited by Cw2 interacting with the NK-cell receptor KIR2DL1 or by Cw1 interacting with the receptor KIR2DL3 (center panel). Hematopoietic cells of the recipient, including residual leukemia cells, can inhibit only the first group of NK cells and not the second group, which kill the residual leukemia cells (right panel). The graft-versus-leukemia effect stems from the fact that the recipient's HLA type lacks an inhibitory HLA-C ligand that is part of the donor's HLA type.

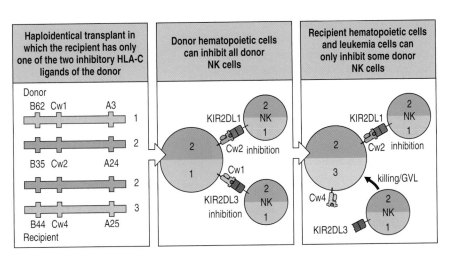

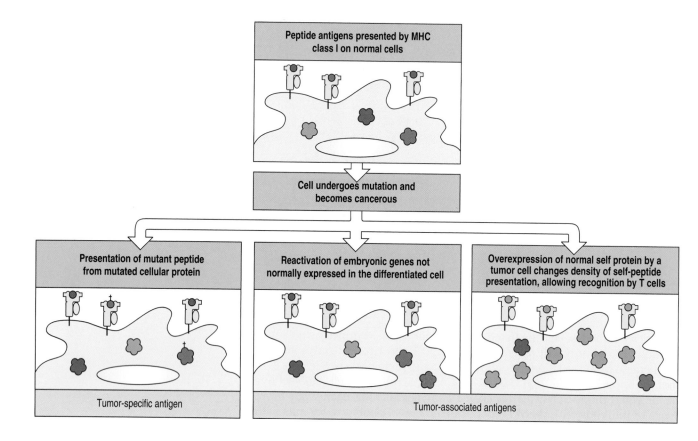

tumor grows, further mutations occur, introducing genetic heterogeneity into the tumor-cell population. Selection for variant cells with faster growth or increased metastatic potential causes the genomes of the tumor and host to diverge even further.

Some of the mutations in tumor cells produce antigenic changes on the tumor-cell surface that can be recognized by the immune system. The new antigens on tumor cells are called **tumor antigens**. Antigens present on tumor cells but not on normal cells are called **tumor-specific antigens**; antigens expressed on tumor cells but also found on certain normal cells, often in smaller amounts, are called **tumor-associated antigens** (Figure 12.46). Tumor-specific antigens have amino acid sequences not present in any normal cell; they can derive from viral proteins, the mutated parts of mutant cellular proteins, or amino-acid sequences spanning tumor-specific recombination sites between genes. Tumor-associated antigens derive from normal cellular proteins to which the immune system is not tolerant and which become immunogeneic when expressed by the tumor. These can be proteins that are normally made in immunologically privileged sites or proteins that are normally not produced in an amount sufficient to be seen by T cells (Figure 12.47).

When an immune response is made against a tumor it imposes selection upon the population of tumor cells. Variant cells that have low expression of tumor antigens or have mutant epitopes that are no longer recognized by effector T cells or antibodies may be able to evade the immune response. The longer a cancer grows, expands its population, and colonizes different sites and environments within the human body, the more genetic variation it acquires and the less likely it becomes that the immune response can delay or terminate the disease. The problem is compounded by the fact that most cancer patients have elderly immune systems that are long past their prime.

Figure 12.46 Sources of tumor-specific and tumor-associated antigens.

Class of antigen	Antigen	Nature of antigen	Tumor type
Embryonic	MAGE-1 MAGE-3	Normal testicular proteins	Melanoma Breast Glioma
Abnormal post-translational modification	MUC-1	Underglycosylated mucin	Breast Pancreas
Differentiation	Tyrosinase	Enzyme in pathway of melanin synthesis	Melanoma
	Surface immunoglobulin	Specific immunoglobulin after gene rearrangements in B-cell clone	Lymphoma
Mutated oncogene or tumor suppressor	Cyclin-dependent kinase 4	Cell-cycle regulator	Melanoma
	β-catenin	Relay in signal transduction pathway	Melanoma
	Caspase-8	Regulator of apoptosis	Squamous cell carcinoma
Oncoviral protein	HPV type 16, E6 and E7 proteins	Viral transforming gene products	Cervical carcinoma

Figure 12.47 Antigens recognized by tumor-specific cytotoxic CD8 T cells. The antigens listed here have all been found to be recognized by cytotoxic T cells raised from patients with the tumor type listed. MAGE, melanoma antigen encoding; MUC, mucin; HPV, human papilloma virus.

12-35 Vaccination with tumor antigens can produce regression of cancer

Immune responses to tumor antigens are made by many cancer patients, even though they have failed to eliminate the cancer or prevent its spread. Study of these tumor-specific CD8 T cells (isolated from tumor tissue, draining lymph node, or peripheral blood) and antibodies (present in serum) is identifying the antigens that stimulate anti-tumor responses and the genes that encode them. Many of the antigens so far identified correspond to human proteins that are normally expressed in testis, ovary, or trophoblast (the cells of the very early embryo that contribute to the placenta) and to which the immune system is not tolerant. The genes encoding these proteins are normally silent in other cell types, but become turned on in cancer cells and are recognized by T cells and B cells. This subset of **tumor-associated antigens** are called cancer/testis antigens or CT antigens. The genes encoding CT antigens are frequently members of gene families, and a disproportionate number are on the X chromosome (Figure 12.48).

Knowing that the immune system can make a response that will kill tumor cells but is poorly triggered to make such a response, tumor immunologists are exploring ways to stimulate and enhance human immune responses to cancer. They hope to devise cancer vaccines that can be used for both treatment and prophylaxis. Several studies and clinical trials have focused on patients with malignant melanoma, an aggressive disease for which there is no reliable treatment. Some CT antigens were discovered as targets for cytotoxic T cells obtained from melanoma patients and were called MAGE-1 and

MAGE-3 (*m*elanoma *a*ntigen *e*ncoding). Figure 12.49 shows the effects of vaccination with an epitope of the CT antigen MAGE-3.A1, which is presented by HLA-A1 to cytotoxic CD8 T cells. Before vaccination the patient had had surgery to remove a cutaneous melanoma and 2 months later developed metastases that expressed the MAGE-3.A1 antigen. Over a period of 2 years the patient was vaccinated 11 times with either a recombinant virus encoding the epitope or a synthetic peptide of the same sequence. As a result of the vaccination, the frequency of MAGE-3.A1-specific cytotoxic CD8 T cells was increased 30-fold, commensurate with a steady regression of the tumor to produce a complete remission that lasted for more than 2 years (see Figure 12.49). For reasons that are not understood, tumor regression was seen in only 20% of the vaccinated patients and remission was achieved only in 10%.

Cancer/testis (CT) antigens			
CT antigen family	Number of genes	Chromosome	Unmanipulated immunity in cancer patients
MAGE-A	15	Xq28	Cellular and humoral
MAGE-B	17	Xp21	Cellular
BAGE	2	4, 13	Cellular
GAGE-A	8	Xp11.4	Cellular
SSX-2	5	Xp11.2	Cellular and humoral
NY-ESO-1	3	Xq28	Cellular and humoral
SCP-1	3	1p12-p13	Humoral
CT7/MAGE-C1	7	Xq26-27	Humoral
HOM-TES-85	1	Xq24	Humoral
CT9/BRDT	1	lp22.1	Not known
CT10	1	Xq27	Humoral
CTp11/SPAN-X-C1	3	Xq27	Not known
SAGE	1	Xq28	Not known
OY-TES-1	1	12p13	Humoral
cTAGE-1	1	18p11.2	Humoral
CT15/Fertilin beta	30	8p11.2	Not known
CT16	2	Xp11.2	Not known
CT17	1	21q11	Not known
MMA-1	1	21q22.2	Not known
CAGE	1	Xp22	Humoral

Figure 12.48 The known cancer/testis antigens. Analyses using human genomics technologies are enabling tumor antigens to be identified. Listed here are the known cancer/testis (CT) antigens, the genes that encode them and the type of immunity they naturally provoke in patients with cancer. Data courtesy of M.J. Scanlan, A.O. Gure, A.A. Jungbluth, L.L. Old, and Y.T.Chen.

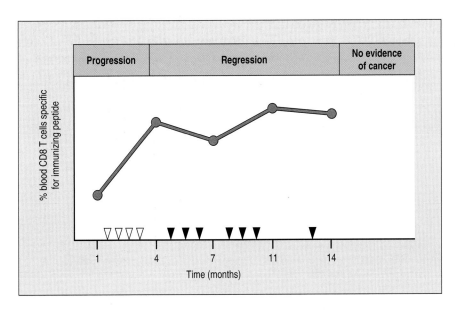

Figure 12.49 The results of vaccination of a cancer patient with a tumor antigen. A patient with metastatic melanoma was vaccinated with antigenic peptides from the MAGE-1 and MAGE-3 proteins that are presented by the patient's HLA-A1 allotype. The first four immunizations used a recombinant viral vaccine (open arrowheads); the latter seven immunizations were with synthetic peptides. The red line shows the percentage of total CD8 T cells in the peripheral blood that are specific for the peptide. The status of the cancer's growth is shown above the graph. Data courtesy of Pierre Coulie.

12-36 Tumors frequently evade immunity by downregulation of HLA class I

Cytotoxic CD8 T cells seem to be the best immune effector cells for ridding the body of tumor cells. One way in which tumor cells can evade recognition by a cytotoxic T cell is not to express the HLA class I molecule that presents tumor antigen to the T cell. Between one-third and one-half of human tumors have defective expression of one or more of their HLA class I allotypes (Figure 12.50). What this tells us is that many patients have made a CD8 T-cell response against their cancer but that variant cells lacking particular HLA class I allotypes escaped the immune response and expanded to keep the cancer going.

Loss of HLA class I expression could make a tumor susceptible to attack by NK cells. The killing of acute myelogenous leukemia cells by alloreactive NK cells in an HLA-haploidentical bone marrow transplant (see Section 12-33) is because the recipient's tumor cells lack an HLA class I allotype that the donor-derived NK cells see as self. This is an example of the NK cell attacking a cell that is lacking self (see Section 8-15, p. 255). Whether NK cells kill non-hematopoietic tumors that have lost HLA class I expression is a point of debate. In laboratory experiments, NK cells are much better killers of hematopoietic cells than of other cell types.

When infected, or stressed in other ways, human epithelial cells are induced to express the cell-surface MIC proteins, which are structurally related to MHC class I molecules. MIC proteins are ligands for the activating NKG2D receptor, which is expressed on all NK cells, γ:δ T cells, and cytotoxic CD8 T cells (see Section 8-14, p. 253). Expression of MIC by epithelial cells therefore facilitates their attack by these three types of lymphocyte. Many human cancers of epithelial cells exploit the expression of MIC to their own advantage.

Figure 12.50 Loss of HLA class I expression in a cancer of the prostate gland. The section is of a human prostate cancer that has been stained with a monoclonal antibody specific for HLA class I molecules. The antibody is conjugated to horseradish peroxidase, which produces a brown stain wherever the antibody binds. The stain and HLA class I molecules are not seen on the tumor mass, but are restricted to lymphocytes infiltrating the tumor and tissue stromal cells. Photograph courtesy of G. Stamp.

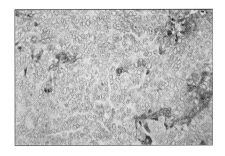

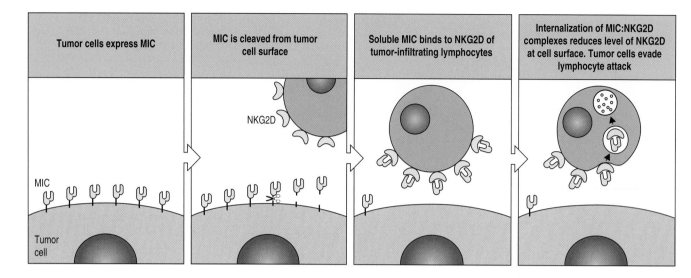

They use proteases to cleave MIC from the cell surface, producing a soluble form that binds to the NKG2D receptor of infiltrating lymphocytes. Binding of MIC to NKG2D induces receptor-mediated endodytosis that removes NKG2D from the surface and speeds its degradation. By removing MIC from their own surfaces and removing NKG2D from lymphocyte surfaces the tumor cells evade attack by NK cells, γ:δ T cells, and cytotoxic CD8 T cells (Figure 12.51). Attack by lymphocytes is likely to be the selective force for the emergence of tumor-cell variants that have these properties.

12-37 Heat-shock proteins can provide natural adjuvants of tumor immunity

The strategy for cancer vaccination described in previous sections is to target a few well-characterized antigens that are shared by many tumors, such as the CT antigens, and not antigens that are specific to a particular tumor. A drawback to this approach is that tumors can more easily escape the immune response produced by such vaccines, as has been observed in patients in whom vaccine-induced regression was temporary. A different approach to vaccine design seeks to embrace a wider range of tumor antigens, including those that have not been defined. It employs a group of ubiquitous cellular proteins called the heat-shock proteins.

Heat-shock proteins are soluble intracellular chaperone proteins that organize the folding, assembly, and degradation of cellular proteins. They were first discovered as proteins that were produced in cells in response to the stress of raised temperature—hence the name 'heat shock.' Heat-shock proteins bind peptides and disorganized parts of proteins in a manner akin to MHC molecules, but they bind with more promiscuous specificity. Some heat-shock proteins, for example calnexin and calreticulin, are familiar for their contribution to antigen processing and presentation (see Section 3-9, p. 78), but there are many others. Animal models of cancer vaccination have shown that immunization with heat-shock proteins extracted from cancer cells produces a powerful antitumor response. The heat-shock proteins bind and carry a potent cargo of protein and peptide tumor antigens. These they specifically deliver to dendritic cells for presentation to T cells (Figure 12.52). When taken up by dendritic cells, heat-shock proteins can deliver tumor antigens directly to the antigen-processing machinery of the dendritic cell. They may also activate dendritic cells through signaling receptors. Because of these functions, heat-shock proteins have been described as natural or internal adjuvants of the human immune system.

Figure 12.51 Human epithelial tumors can inhibit the response of lymphocytes expressing NKG2D. Many successful human tumors of epithelial tissue synthesize the MIC protein. Upon reaching the cell surface the MIC is cleaved to a soluble form, which is released into the extracellular milieu. Soluble MIC binds to the NKG2D receptor on lymphocytes that enter the tumor tissue and prevents them from attacking the tumor cells. NKG2D is expressed by all CD8 T cells, γ:δ T cells, and NK cells.

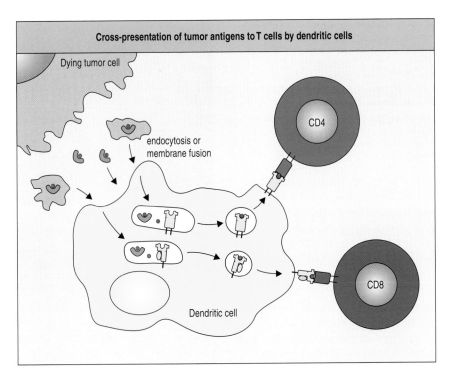

Cross-presentation of tumor antigens to T cells by dendritic cells

Dying tumor cell

endocytosis or membrane fusion

CD4

Dendritic cell

CD8

Figure 12.52 Cross-presentation of tumor antigens to T cells by dendritic cells. Complexes of heat-shock protein 70 (green) and peptides derived from tumor antigens (red and purple circles) are taken up taken up by dendritic cells, either as soluble complexes or associated with membrane vesicles. The bound peptides are delivered to the pathways of antigen processing where they are bound by MHC class I and II molecules, taken to the cell surface and presented to CD8 and CD4 T cells respectively.

That cancers are rarely terminated by the immune system is largely because of failures in starting the response early enough and making it strong enough. Manipulation of dendritic cells and heat-shock proteins might help to overcome these limitations. The basic idea is to isolate dendritic cells from a patient's blood and load them with heat-shock proteins isolated from the patient's tumor or with defined tumor antigens. Thus charged, the dendritic cells will be replaced in the patient's circulation, where they will home to the secondary lymphoid tissues and initiate a tumor-specific immune response.

12-38 Vaccination against oncogenic viruses

Several types of cancer are caused by chronic viral infection. Human papilloma virus strain 16 (HPV16) is a known cause of most types of cervical cancer, and the virus can be detected in 50% of cervical tumors. If HPV16 infection is necessary to produce the cancer, then vaccination against the virus should reduce the incidence of cervical cancer. Trials of a vaccine against HPV16 are now in progress. A group of over 2000 young women have been randomly given either the vaccine or a placebo. That none of the vaccinated women has become infected with HPV16, whereas some 4% of the unvaccinated women became infected in each of the first 2 years of the trial, shows that the vaccine is preventing infection. Its effect on the incidence of cancer will only be known with time and progression of the HPV16 infections.

Chronic hepatitis B virus (HBV) infection is associated with liver cancer. A vaccination program begun in Taiwan in 1986 is starting to show that the incidence of liver cancer in children decreases after vaccination. Vaccination of newborns in The Gambia, a small country in West Africa, has reduced acute childhood HBV infections by 83% and chronic infections by 95%. Again, the effect on the incidence of cancer will only be known once the cohort of vaccinated children reach that later stage of life when most liver cancer becomes manifest.

Figure 12.53 Detection of colon cancer using a radiolabeled antibody against carcinoembryonic antigen. Anterior projection of the pelvis showing uptake of anti-carcinoembryonic antigen (CEA) antibody in a pelvic tumor. A patient with a suspected recurrence of colon cancer was injected intravenously with ^{111}In (indium)-labeled monoclonal antibody against CEA. The tumor is seen as two dark red spots in the pelvic region. The blood vessels are faintly outlined by circulating antibody that has not bound to the tumor, but may have bound to soluble CEA released from the tumor and circulating in the blood. R, right side; L, left side. Photograph courtesy of A.M. Peters.

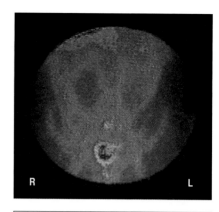

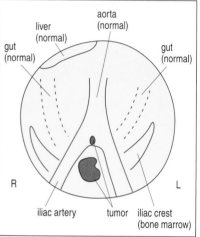

12-39 Monoclonal antibodies against cell-surface tumor antigens can be used for diagnosis and immunotherapy

Monoclonal antibodies specific for tumor antigens are used in both tumor analysis and therapy. The location of tumor cells within the body can be revealed by giving patients purified monoclonal antibody specific for a tumor antigen and covalently coupled to a radioactive isotope, for example ^{131}I. The binding of the antibodies to tumor cells concentrates the radioactivity at sites in the body containing tumors (Figure 12.53). In this manner, the size and location of a primary tumor can be determined, as can the extent of its metastasis. Such information can help to determine the type of therapy.

Monoclonal antibodies specific for tumor antigens can also be used to kill tumor cells. One approach, which has had some success in the treatment of B-cell lymphoma, relies on the natural effector functions of the antibody to direct the destruction of the tumor cells by phagocytes and complement. In the individual patient all the lymphoma cells carry the same immunoglobulin, whereas few healthy B cells express that same immunoglobulin. Thus the B-cell lymphoma immunoglobulin is practically a tumor-specific antigen. Mouse monoclonal antibodies specific for the unique antigen-binding site of the lymphoma immunoglobulin can be made and humanized. These are known as anti-idiotypic antibodies and are directed against epitopes determined by the CDR loops of the antigen-binding site. The advantage of targeting the surface immunoglobulin of B-cell lymphoma is that the destruction of all cells expressing a particular immunoglobulin poses no threat to the patient's immunocompetence. The disadvantage is that each patient's therapy requires a custom-made monoclonal antibody.

A second use of monoclonal antibodies in cancer therapy does not rely on the natural effector functions of the antibodies. Instead, the antibodies are used to deliver a toxic agent that kills the cells to which the antibody binds. Various types of toxic agent are being tried: radioactive isotopes such as iodine-131 or yttrium-99, chemical toxins such as those used in conventional chemotherapy, plant toxins such as ricin, and bacterial toxins such as *Pseudomonas* toxin. The advantage of immunotoxins over conventional irradiation or chemotherapy is that the destructive power of the toxin is more specifically targeted toward the tumor cells and away from proliferating normal tissues (Figure 12.54).

The conjugates of biological toxins and antibodies are called **immunotoxins**. In addition to the selectivity imposed by the binding of the monoclonal antibody to the tumor antigen, immunotoxins are activated only after their entry into the cell and therefore should not cause damage to healthy noncancerous cells. On binding to tumor cells, immunotoxins are endocytosed. Once within the endosomes, the toxin is cleaved from the antibody, allowing it to exert its toxic effect. A related approach uses a pro-drug that is administered

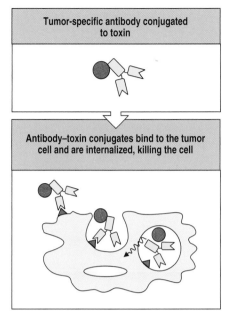

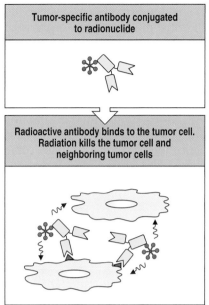

Figure 12.54 **Antibodies can target toxins or radioactive isotopes to the tumor-cell surface.** The tumor-specific antibody is coupled to either a toxin (red ball in left panels) or a radionuclide (red star in right panels). Toxins such as ricin exert a toxic effect on the cell only after internalization followed by proteolytic cleavage, which releases the toxic part of the molecule from the antibody.

in a conventional manner but can be activated only by an enzyme administered separately as a conjugate linked to a tumor-specific antibody.

Summary

Unlike infection, cancer is a disease that arises from within and is caused by cells almost identical to normal human cells. Each case of cancer derives from one cell that has undergone malignant transformation, a process requiring mutations in several genes important for cell survival and cell division. Cancer cells become divorced from the mechanisms that maintain tissue integrity, and they gradually compete with normal cells for food and space within the human body. In time, the effects of that competition become manifest as disease and ultimately death. Every cancer is unique and dies with the person in which it arose.

The principal defenses against cancer are present in every cell of the body. These are the normal cell processes that monitor DNA integrity, and control DNA replication and cell division; they ensure that multiple insults are required for malignant transformation. The immune system's response to cancer appears too little and too late. Although tumor-specific cytotoxic CD8 T cells and antibodies can be detected in patients with cancer, these immune responses do not enable them to control or eliminate the disease. Treatment of cancer with allogeneic hematopoietic cell transplantation provides a context in which powerful alloreactive responses can be aimed at eliminating tumor cells. Cancer vaccines seek to enhance the autologous response to tumor antigens.

Summary to Chapter 12

The prevention of infectious diseases by vaccination illustrates superbly how manipulation of the immune response can benefit public health. The pathogens for which effective vaccines have been found are those that cause acute infections and are not highly mutable. When these pathogens caused epidemic disease, many people survived and developed immunity, showing that the human immune system could respond to the infection in a productive way. Death, when it occurred, was because the immune system was just too slow. What vaccination achieved was to start the immune response ahead of the infection, giving the immune system an edge on the pathogen.

The pathogens that have foiled the vaccinologists are those that foil the immune system. Their infections are protracted and they cause chronic disease. For such diseases we know little about what constitutes a successful immune response. Against some of these pathogens the human immune response is uniformly unsuccessful; for others a minority of people seem to be able to clear the infection soon after infection; and for yet others, some immunity does develop but what it consists of, and when and how it is produced, are still unclear. A greater understanding of the successful immune responses against such pathogens should help in the development of vaccines against them.

Whereas vaccination provides a boost to immunity, in tissue transplantation the goal is immune suppression. Although the elimination of alloreactivity by exact MHC matching can be demonstrated in model systems, the complexity of HLA polymorphism has made this concept difficult to apply in the clinic. Instead, the practical approach has been to avoid hyperacute rejection by cross-matching, to reduce the T-cell alloreactions that cause acute rejection by matching for HLA where possible, and to suppress the immune response with the judicious use of nonspecific immunosuppressive drugs. The recent increased success of transplantation is due to the discovery of drugs that specifically suppress T-cell activation, and to the use of drug combinations in which immunosuppressive activities, but not toxicities, are additive.

In their interactions with the immune system, cancers are akin to the pathogens that cause chronic infections. Over long periods, cancer cells selfishly exploit the cellular society of the human body to expand their own populations. They largely avoid stimulating the immune system by being almost identical to normal cells. Even when the lymphocytes respond, the tumors can escape by changing their antigens. Although cancer immunotherapy is still an experimental field, a variety of approaches are showing some promise. In the past 50 years, clinical transplantation has benefited from the strategy of combining different approaches, each with its own advantages and limitations. Over the next 50 years it seems likely that a similar strategy will lead to advances in cancer immunotherapy.

Questions

Question 12–1
Differentiate between the following types of vaccines and give an example of each: (a) inactivated virus vaccines; (b) live-attenuated virus vaccines; (c) subunit vaccines; (d) toxoid vaccines; and (e) conjugate vaccines.

Question 12–2
A. "An antigen is not necessarily an immunogen." Explain this statement.
B. Explain why adjuvants are used in experimental immunology.
C. Which adjuvants are used for human vaccination? (Refer to Figure 12.4.)

Question 12–3
What risks are associated with live-attenuated virus vaccines?

Question 12–4
Explain why, in principle, an organ transplanted from any donor other than an identical twin is almost certain to be rejected in the absence of any other treatment. In your answer indicate the role of the HLA molecules in acute rejection.

Question 12–5
A. Explain why a transplant made between a donor whose blood group is AB and a recipient whose blood group is O will always be rejected, even if it is perfectly HLA matched and the recipient has been given immunosuppressant drugs. What is this type of rejection called?
B. Give another example of ABO incompatibility between donor and recipient that would lead to this type of rejection.

C. What other antigen incompatibilities, other than those of blood group, are most likely to provoke this type of rejection?

D. Which pre-surgical laboratory test should be performed to prevent this type of rejection? (Refer to Figures 12.13 and 12.14.)

Question 12–6
A. Identify three general classes of drugs used to suppress acute transplant rejection and provide examples of each class.

B. What side-effects and toxic effects are associated with each class of drug?

Question 12–7
Explain how cyclosporin A acts as an immunosuppressant drug. (Refer to Figure 12.27.)

Question 12–8
A. Explain how mouse monoclonal antibodies (MoAbs) can be used to suppress acute graft rejection.

B. What feature of these mouse antibodies compromises their effectiveness *in vivo* and limits their use?

Question 12–9
Many inherited immunodeficiency diseases affecting hematopoietic cells can be treated through bone marrow transplantation. Explain why it is important to maximize the degree of HLA relatedness between donor and recipient. (Refer to Figure 12.32.)

Question 12–10
A. Explain why a boy with leukemia who receives a bone marrow transplant from his sister that is perfectly matched for MHC class I and class II is still likely to get graft-versus-host disease.

B. Which effector T cells are usually involved in this reaction and why?

Question 12–11
A. What are tumor-specific antigens?

B. How do they originate?

C. Provide two examples and the relevant tumor type.

Question 12–12
A. What are tumor-associated antigens?

B. How do they originate?

C. Provide two examples and the relevant tumor type.

Question 12–13
Identify two *in vitro* strategies by which an anti-tumor immune response could be boosted in cancer patients without the use of modified tumor cells.

Question 12–14
How are mouse monoclonal antibodies (MoAbs) used by oncologists for the diagnosis and immunotherapy of cancer?

Answers

Chapter 1

Answer 1–1

A. The four classes of pathogen are: bacteria, viruses, fungi, and parasites (protozoa and worms).

B. Vaccines have been successful in controlling many bacterial and viral pathogens, including the variola virus, which causes smallpox, the bacterium *Corynebacterium diphtheriae*, which causes diphtheria, and the influenza virus, the orthomyxovirus that causes influenza. A vaccine provides protective immunity against a specific pathogen by stimulating an antigen-specific acquired immune response that results in immunological memory for that pathogen. When the pathogen is subsequently encountered in its natural, virulent form, immunological memory stimulates a swift, strong and effective immune response that destroys the pathogen before it can cause disease.

Answer 1–2

The skin and mucous membranes present physical, chemical and microbiological barriers to invasion. The outer layers of the epidermis of the skin consist of cells filled with keratin, which form a relatively impermeable physical barrier that microbes are unable to penetrate. The more vulnerable mucosal epithelia of the respiratory, gastrointestinal, and urogenital tracts are protected by a surface layer of mucus that contains enzymes, proteoglycans, glycoproteins, and antimicrobial peptides that help to minimize colonization by pathogens and prevent them from penetrating to the tissues underneath and establishing an infection. The resident harmless microbes that colonize all body surfaces also compete with pathogens and help to prevent them from becoming established.

Answer 1–3

The molecules released by activated macrophages have three principal effects. Some are direct antibacterial agents. Some cytokines act as chemoattractants and recruit other leukocytes into the infected tissue, for example, neutrophils, which efficiently phagocytose and kill bacteria, forming pus. Other cytokines act on the endothelial cells of local blood vessels to increase vascular permeability and vasodilation, thus initiating inflammation of the infected tissue.

Answer 1–4

Calor means heat, *dolor* means pain, *rubor* means redness, and *tumor* means swelling. During inflammation, blood vessels dilate (vasodilation) increasing the flow of blood to the infected site (heat and redness), vascular endothelium becomes permeable to the mobilization of neutrophils and monocytes into infected tissue (swelling), and fluid and cellular infiltrate exert pressure on local nerve endings (pain).

Answer 1–5

Innate immune responses are initiated almost immediately after infection, whereas adaptive immunity takes longer to develop. Innate immunity uses generalized and invariant mechanisms to recognize pathogens. Examples of these are the receptors on phagocytes that recognize surface molecules shared by many different pathogens and stimulate phagocytosis, and serum proteins such as mannose-binding lectin, which binds to bacterial carbohydrates. Innate immunity is often unable to eradicate the pathogen completely, and even when it does, it does not produce immunity to reinfection. An adaptive immune response, in contrast, involves specific recognition of the particular pathogen by specific antigen-recognition molecules selected from a pool of millions of different molecules. Adaptive immunity is often powerful enough to eradicate the infection and provides long-term protective immunity through immunological memory.

Answer 1–6

A. The two major progenitor subsets of leukocytes are the common lymphoid progenitor and the myeloid progenitor.

B. In adults all leukocytes originate in the bone marrow and are derived from pluripotent hematopoietic stem cells.

C. The common lymphoid progenitor differentiates into three cell types: B cells, T cells, and natural killer (NK) cells. The myeloid progenitor differentiates into six main cell types: basophils, eosinophils, neutrophils, mast cells, dendritic cells, and monocytes. Monocytes are circulating leukocytes that enter tissues, where they then differentiate into macrophages.

Answer 1–7

A. The primary (or central) lymphoid tissues are the bone marrow (and liver in the fetus) and the thymus. The main secondary (or peripheral) lymphoid tissues are the lymph nodes, spleen, and mucosa-associated lymphoid tissues (MALT). The latter include gut-associated lymphoid tissue (GALT), such as the tonsils, adenoids, appendix, and Peyer's patches, and bronchial-associated lymphoid tissues (BALT).

B. Primary (or central) lymphoid tissues are the anatomical locations where lymphocytes complete their development and reach the state of maturation required for recognition of and response to a potential pathogen. B cells mature in the bone marrow and fetal liver, and T cells mature in the thymus. Both lymphocyte lineages are derived from a common hematopoietic stem cell. Somatic recombination of antigen-receptor genes and negative selection of potentially autoreactive B and T cells (to produce self-tolerance) occurs in primary lymphoid tissues.

Secondary (or peripheral) lymphoid tissues provide the anatomical sites where lymphocytes encounter antigen and immune responses are induced. Antigen is delivered to the secondary lymphoid tissues through an afferent lymphatic vessel, where it is then filtered and retained for encounter with lymphocytes bearing antigen-specific receptors. Clonal selection and somatic hypermutation (in B cells) occur in secondary lymphoid tissues.

C. The spleen differs from the other secondary lymphoid tissues in that lymph does not filter through this organ. Instead, the spleen filters the blood and traps blood-borne pathogens.

Answer 1–8

A. (i) B cells recognize antigen through immunoglobulin on their surface. After activation, B cells become plasma cells, which secrete a soluble form of this immunoglobulin as antigen-specific antibodies. (ii) T cells recognize antigen through a different, although structurally related, type of molecule called the T-cell receptor on their cell surface. The T-cell receptor is not secreted.

B. Refer to Figure A1.8. Immunoglobulins can bind a wide range of antigens belonging to different chemical subclasses, such as proteins, glycoproteins, proteoglycans, lipids, or carbohydrates, and often recognize their specific antigenic determinant (epitope) in the intact foreign macromolecule or pathogen. In contrast, T-cell receptors can bind only to protein fragments (peptides) that are associated with an MHC molecule on the surface of another cell. Thus, T cells cannot

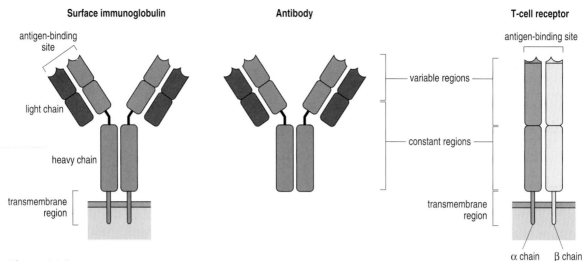

Figure A1.8

recognize intact antigens, require an ancillary cell to present the antigen to them, and can recognize only protein antigens, in the form of peptides.

C. Peptides are produced and presented by mechanisms known as antigen processing and antigen presentation, respectively. Proteins are denatured and degraded enzymatically, generating small peptide fragments that are able to bind to MHC molecules. Proteins degraded in the cytosol bind to MHC class I molecules, whereas proteins degraded in endocytic vesicles bind to MHC class II molecules. The ancillary cell that presents the MHC molecule:peptide complex to the T cell is referred to as the antigen-presenting cell.

Answer 1–9

(1) Neutralization. By binding to the surface of a pathogen antibodies interfere with the ability of the pathogen to grow and replicate. Antibody binding to a pathogen or a bacterial toxin can also inhibit its binding to receptors on host cells and therefore prevent its entry into cells. (2) Opsonization. Antibody coating the surface of a pathogen or toxin can promote phagocytosis of the antibody-covered particle. Antibodies acting in this way are known as opsonins. The antibody-bound material interacts with Fc receptors on the surface of phagocytic cells such as macrophages and neutrophils, which bind the constant region (the stem) of the antibody. Stimulation of Fc receptors in this way stimulates engulfment and degradation of antibody-coated material by the phagocyte. (3) Complement activation. IgG or IgM antibody bound to a pathogen stimulates activation of the complement system, leading to the deposition of complement proteins on the surface of the pathogen. Certain of these act as opsonins and bind to complement receptors on phagocytic cells to stimulate phagocytosis and destruction of the pathogen.

Answer 1–10

A. Immunodeficiency diseases are disorders in which some aspect of host immune defense is missing or defective.

B. Susceptibility to infection increases when either innate or acquired immune responses are not operating correctly. This may cause susceptibility to particular subsets of pathogens, or to all pathogens depending on the particular deficiency.

C. Immunodeficiency diseases may be caused by autosomal or sex-linked mutations affecting genes involved in innate or acquired immune responses. Alternatively, immunodeficiency diseases may arise through infection with pathogens that suppress or disrupt an otherwise healthy immune system.

Answer 1–11

(1) Allergy. IgE antibodies made against normally innocuous environmental antigens trigger widespread mast-cell activation. This can lead to allergic diseases such as asthma or to a potentially fatal anaphylactic reaction. (2) Autoimmune disease. Chronic immune responses by B cells or T cells to self antigens can cause tissue damage and chronic illnesses such as diabetes, multiple sclerosis, and myasthenia gravis. Autoimmunity is sometimes provoked as a consequence of an immune response to pathogen-derived antigen that cross-reacts on healthy host cells or tissue. (3) Transplant rejection. A person's immune system will make an immune response against the foreign MHC molecules on transplanted tissue that is MHC-incompatible.

Chapter 2

Answer 2–1

A. Immunoglobulins are the membrane-bound form of the antigen receptors of B cells. Antibodies are the secreted form of the same immunoglobulins.

B. Immature, mature, and memory B cells have membrane-bound immunoglobulin. Antibodies are produced by plasma cells.

Answer 2–2

An antibody molecule is made of four polypeptide chains—two identical heavy chains and two identical and smaller light chains, with a total molecular weight of approximately 150 kDa. Each chain is made up of a series of structurally similar domains known as immunoglobulin domains. The amino-terminal portion of each H chain combines with one L chain and the two carboxy-terminal portions of the H chains combine with each other, forming a Y-shaped quaternary structure. Disulfide bonds hold the H and L chains together, hold the two H chains together (interchain disulfide bonds) and stabilize the domain structure of the chains (intrachain disulfide bonds). The arms of the antibody molecule are called Fab (fragment antigen binding) and interact with antigen. The stalk is called Fc (fragment crystallizable) and is made up of H chains only.

The amino-terminal domains of an H and an L chain together make up a site that binds directly to antigen and which varies greatly between different antibodies. These domains are referred to as the variable region, and each antibody has two identical antigen-binding sites. The remaining domains of both H and L chains are the same in all antibodies of a given class (isotype). These domains are referred to as the constant region.

The variable region of each chain includes hypervariable regions of amino-acid sequences that differ the most between different antibodies. These are nested within less variable sequences known as the framework regions. The hypervariable regions make loops at one end of the domain structure and are also known as complementarity-determining regions because they confer specificity on the antigen-binding site.

Answer 2–3

A. An epitope is the specific part of the antigen that is recognized by an antibody and binds to the complementarity-determining regions in the antibody variable domains. Epitopes are sometimes referred to as antigenic determinants. Epitopes can be part of a protein or can be carbohydrate or lipid structures present in the glycoproteins, polysaccharides, glycolipids, and proteoglycans of pathogens.

B. Multivalent antigens are complex macromolecules that contain more than one epitope.

C. Linear epitopes are epitopes in proteins that comprise a contiguous amino-acid sequence. They are also called continuous epitopes. In contrast, a conformational epitope is formed by amino acids brought together as a result of protein folding and which are not adjacent to each other in the protein sequence. Conformational epitopes are also known as discontinuous epitopes.

D. Antibodies bind antigens via noncovalent bonding such as hydrogen bonds, hydrophobic interactions, van der Waals forces, and electrostatic attraction.

Answer 2–4

A. Polyclonal antibodies are a mixture of antibodies of different specificities and affinities for a particular antigen. They are the product of numerous different B cells. Monoclonal antibodies have a single specificity and affinity for a given antigen. They derive from a single B cell.

B. Polyclonal antibodies are produced *in vivo* by immunizing an animal with antigen, allowing sufficient time for an immune response to occur and then preparing antiserum containing the antibodies from the blood. Monoclonal antibodies are made *in vitro* from individual cell lines derived from single B cells. This is achieved by producing a hybrid immortalized cell line through fusion of an antibody-producing B cell with a myeloma tumor cell to produce an antibody-producing 'hybridoma'. A hybridoma producing the desired antibody can then be cloned and grown on to produce unlimited amounts of monoclonal antibody.

Answer 2–5

A. In developing B cells, gene rearrangements within the genetic loci for immunoglobulin light and heavy chains can produce an almost unlimited variety of different variable regions, and thus produce the huge repertoire of antibodies with different specificities for many types of antigens. This gene rearrangement mechanism is called somatic recombination. In the germline configuration, prior to gene rearrangement, the immunoglobulin loci in progenitor B cells are composed of sequences encoding the constant regions and families of gene segments encoding different portions of the variable region. Heavy-chain loci contain series of gene segments called variable (V), diversity (D), and joining (J). Light-chain loci contain only V and J gene segments. In somatic recombination in developing B cells, one of each family of gene segments is randomly selected and joined together to give a complete variable-region sequence, which is subsequently expressed as an immunoglobulin heavy or light chain. Immunoglobulin gene rearrangement is irreversible, leading to permanent alteration of the chromosome and occurs exclusively in B cells.

B. A D gene segment first joins to a J to form DJ, followed by a V becoming joined to DJ to form VDJ, which encodes a complete variable region.

C. The heavy-chain locus rearranges before the light-chain loci. For light chains in humans, the kappa locus rearranges first and is followed by the lambda locus only if both kappa loci fail to produce a successful rearrangement.

Answer 2–6

Gene rearrangement by somatic recombination involves recombination signal sequences (RSS) which flank V, D, and J segments and are recognized by the enzymes involved in cutting and rejoining the gene segments. An RSS is composed of a conserved nonamer sequence and heptamer sequence separated by a spacer region. There are two types of RSSs, one with a spacer of 12 bp and one with a spacer of 23 bp. To ensure that segments are brought together in the right order, an RSS with a 12-bp spacer is always brought together with one with a 23-bp spacer. This is called the 12/23 rule. This ensures that in the heavy-chain locus, V rearranges to DJ and not directly to J or another V, and in the light-chain locus, V rearranges to J and not to another V.

Answer 2–7

The rejoining and repair of DNA during the recombination process leads to additional variation in sequence at the junctions between the rearranged gene segments. This is called junctional diversity and contributes considerably to the final diversity of immunoglobulin specificities. Two sources of junctional diversity are introduced: P (palindromic) and N (nontemplated) nucleotides. P nucleotides are generated through endonuclease activity and repair around a hairpin loop at the ends of the gene segments to be joined. N nucleotides are nucleotides added at random at the junctions by terminal deoxynucleotidyl transferase (TdT) activity.

Answer 2–8

Naive B cells express IgM and IgD simultaneously through a mechanism involving alternative ways of processing the RNA transcript before translation. A primary transcript containing leader (L), V, D, J, C_μ and C_δ is produced first. This transcript contains two distinct polyadenylation signal sequences, one following the C_μ exons (pA1) and the other following the C_δ exons (pA2). Processing results in the removal of either C_μ or C_δ exons (plus introns) through alternative splicing. The resulting mRNAs, which encode either C_μ or C_δ, are polyadenylated at the pA1 or pA2 site, respectively.

Answer 2–9

Whether immunoglobulin is expressed as a transmembrane-anchored protein or a secreted protein is determined by alternative processing of the heavy-chain RNA transcript. All the heavy-chain C genes contain MC (membrane-coding) exons, which encode the transmembrane region and cytoplasmic tail and an SC (secretion-coding) exon, which encodes the carboxy terminus of the secreted antibody. The primary RNA transcript contains the MC and SC exons. In naive resting B cells or memory B cells, cleavage and polyadenylation of the transcript at a site (pAm) following the MC exons and deletion of the SC exon by RNA splicing produces the membrane-bound immunoglobulin. On B-cell activation and differentiation into plasma cells the SC exon is retained in the transcript, and a polyadenylation signal sequence, pAs, immediately following it is used to produce an mRNA encoding the secreted form of the heavy chain.

Answer 2–10

A. Affinity maturation is the phenomenon observed during a B-cell response in which antibodies with increasing affinity for the antigen are produced as the response proceeds. This occurs as a result of the process known as somatic hypermutation.

B. In somatic hypermutation, which occurs only in activated B cells, random point mutations are introduced into the rearranged V regions of H- and L-chain genes at a rate six orders of magnitude higher than spontaneous mutation. Some of these mutations give rise to immunoglobulin with higher affinity for the antigen than the original immunoglobulin. Those B cells producing higher-affinity surface immunoglobulin will be preferentially selected for activation by the antigen and will come to dominate the response, differentiating into plasma cells producing high-affinity antibodies.

Answer 2–11

A. Isotype switching is the process by which antibodies change their heavy-chain constant regions in order to acquire different effector functions, while preserving the variable region and antigen specificity. The light chain is unaffected.

B. The molecular mechanism involves a recombination between sequences, called switch regions, which lie upstream (5′ side) of heavy-chain C genes. All heavy-chain C genes except C_δ have a switch region. Recombination between two switch regions results in the excision of DNA (as a circular DNA molecule) between the two and the movement of the new heavy-chain C gene next to the preserved V region. Transcription will produce an mRNA encoding the same V-region sequence and the new C region. Switching can occur between the first switch region and any other switch region that lies downstream (3′ side). Isotype switching is not random but is influenced by T-cell cytokines.

C. Isotype switching is important because the different antibody isotypes have different effector functions and efficient immune responses rely upon the production of the most appropriate effector function to combat the particular pathogen.

Answer 2–12

A. Neutralization: IgM, IgG1, IgG2, IgG3, IgG4, IgA

B. Opsonization: IgG1, IgG2, IgG3, IgG4, IgA

C. Sensitization for killing by NK cells: IgG1, IgG3

D. Sensitization of mast cells: IgG1, IgG3, IgE

E. Activation of complement: IgM, IgG1, IgG2, IgG3, IgA

F. Transport across epithelium: IgM, IgA (dimer)

G. Transport across placenta: IgG1, IgG2, IgG3, IgG4

H. Diffusion into extravascular sites: IgM, IgG1, IgG2, IgG3, IgG4, IgA (monomer), IgE

Answer 2–13

(1) Gene rearrangements affect the variable region of immunoglobulins whereas isotype switching affects the constant region. (2) Different recombination-signal sequences and enzymes are utilized for the two processes. (3) Isotype switching occurs only after antigen stimulation, whereas gene rearrangement occurs only during B-cell maturation in the bone marrow. (4) All isotype switch recombinations are productive, but not all gene rearrangments are. (5) The direction of isotype switching is not random, but is directed by T-cell cytokines and interactions between cell-surface molecules (CD40/CD40 ligand). Somatic recombination is a random process involving the random joining of V, D, and J segments. (6) Only heavy chains are involved in isotype switching whereas both heavy- and light-chain genes are involved in somatic recombination.

Answer 2–14

Mice are used routinely to generate monoclonal antibodies. The constant regions of mouse antibodies are different enough from the constant regions of human antibodies in amino-acid composition that if mouse antibodies are infused into a patient, an immune response will be stimulated and directed against the mouse constant-region epitopes. This immune response neutralizes the monoclonal antibody and in practice limits its intended use to one effective dose. When monoclonal antibodies are used for serological or diagnostic purposes in the laboratory, the monoclonal antibodies do not need to be humanized because laboratory assays are carried out *in vitro*.

Answer 2–15

The B cells in a person carrying such a defect would be unable to switch antibody isotype and would be unable to produce any antibody other than IgM. As IgM antibodies can implement fewer effector functions than IgG antibodies, the main class of antibody produced in an adaptive immune response, one would expect that immunity would be impaired. In addition, no IgA antibodies could be produced, leaving the person highly vulnerable to infection through mucosal surfaces. There are, in fact, rare inherited genetic deficiencies that result in an inability to switch isotype. They are called hyper IgM immunodeficiencies because the patient is unable to produce any antibody other than IgM. The most frequent one affects the expression of a cell-surface molecule called CD40 ligand in T cells, which is required for the interaction between T cells and B cells that stimulates isotype switching, as we shall learn later in this book.

Chapter 3

Answer 3–1

A. Similar: (1) The T-cell receptor has a similar overall structure to the membrane-bound Fab fragment of immunoglobulin, containing an antigen-binding site, two variable domains and two constant domains. (2) T-cell receptors and immunoglobulins are both generated through somatic recombination of sets of gene segments. (3) The variable region of the T-cell receptor contains three complementarity-determining regions (CDRs) encoded by the V_α domain and three CDRs encoded by the V_β domain, analogous to the CDRs encoded by the V_H and V_L domains. (4) There is huge diversity in the T-cell receptor repertoire and it is generated in the same way as that in the B-cell repertoire (by combination of different gene segments, junctional diversity due to P- and N-nucleotides, combination of two different chains). (5) T-cell receptors are not expressed at the cell surface by themselves but require association with the CD3 γ, δ, ε, and ζ chains for stabilization and signal transduction, analogous to the Igα and Igβ chains required for immunoglobulin cell-surface expression and signal transduction.

B. Different: (1) A T-cell receptor has one antigen-binding site, an immunoglobulin has at least two. (2) T-cell receptors are never secreted. (3) T-cell receptors are generated in the thymus, not the bone marrow. (4) The constant region of the T-cell receptor has no effector function and it does not switch isotype. (5) T-cell receptors do not undergo somatic hypermutation.

Answer 3–2

First, T-cell receptors can only bind to one type of antigen, protein fragments called peptides. Immunoglobulins can bind to peptides, intact proteins, carbohydrates, and lipids. Second, unlike immunoglobulins, T-cell receptors cannot bind to a free antigen directly, but instead require accessory antigen-presenting cells that present the peptide antigens in association with cell-surface glycoproteins called MHC class I and class II molecules. Third, T-cell receptors possess a single antigen-binding site; immunoglobulins have at least two binding sites for antigen, more in the case of secreted dimeric IgA (four sites) and secreted pentameric IgM (10 sites).

Answer 3–3

The organization of the TCRα locus resembles that of an immunoglobulin light-chain locus, in that both contain V and J gene segments and no D gene segments. The TCRα locus on chromosome 14 contains ~80 V gene segments, 61 J gene segments, and one C gene. The immunoglobulin light-chain loci, lambda and kappa, are encoded on chromosomes 22 and 2, respectively. The lambda locus contains ~29 V gene segments, and four J gene segments, each paired with a C gene. The kappa locus contains ~40 V gene segments, five J segments, and one C gene segment. The arrangement of the kappa locus more closely resembles that of the TCRα locus except that there are more J segments in the T-cell receptor locus.

The organization of the TCRβ locus resembles that of the immunoglobulin heavy-chain locus; both contain V, D, and J gene segments. The TCRβ locus contains ~52 V gene segments, two D gene segments, 13 J gene segments, and two C genes, encoded on chromosome 7. Each C gene is associated with a set of D and J gene segments. The immunoglobulin heavy-chain locus on chromosome 14 contains ~62 V segments, 27 D segments, and six J segments, followed by nine C genes, each specifying a different immunoglobulin isotype. The heavy-chain C genes determine the effector function of the antibody.

Answer 3–4

Figure A3.4 below shows the number of J gene segments in each locus. This difference may have arisen because T-cell receptors, unlike immunoglobulins, do not continue to diversify after antigen recognition; affinity maturation is restricted to B cells. T-cell receptors may have evolved larger numbers of J segments to compensate for the lack of somatic hypermutation and to provide sufficient diversity in the T-cell repertoire.

Immunoglobulin			T-cell receptor	
Heavy chain	Light chain lambda (λ)	Light chain kappa (κ)	Alpha (α)	Beta (β)
6	4	5	61	13

Figure A3.4

Answer 3–5

Secreted antibodies are the effector molecules of the B-cell response. When bound to antigen, their constant regions activate phagocytosis and complement activation. Different constant regions have different effector functions so isotype switching by antibodies ensures that the antibody acquires the suitable effector function required for the particular immune response. T-cell receptors, on the other hand, bind to antigen but their constant regions do not contribute to T-cell effector function. Other molecules secreted by T cells are used for effector functions. There is therefore no need for isotype switching in T cells and the T-cell receptor loci do not contain numerous alternative C genes.

Answer 3–6

A. Antigen processing is the intracellular breakdown of pathogen-derived proteins into peptide fragments which are of the appropriate size and specificity required to bind to MHC molecules.

B. Antigen presentation is the assembly of peptides with MHC molecules and the display of these complexes on the surface of antigen-presenting cells.

C. Antigen processing and presentation must occur for T cells to be activated because: (1) T-cell receptors cannot bind to intact protein, only to peptides; and (2) T-cell receptors do not bind antigen directly, but rather must recognize antigen bound to MHC molecules on the surface of antigen-presenting cells.

Answer 3–7

A. (i) Pathogens that are propagating freely within cells (for example, viruses), are eradicated by the actions of cytotoxic T cells. (ii) Cytotoxic T cells express a glycoprotein called CD8, a T-cell co-receptor that interacts with (iii) MHC class I on antigen-presenting cells. (iv) Once activated, cytotoxic T cells kill cells infected with the pathogen, which are displaying pathogen peptides on MHC class I molecules, and thereby inhibit further replication of the pathogen and infection of neighboring cells.

B. (i) Pathogens that reproduce in extracellular spaces, for example, encapsulated bacteria like *Streptococcus pneumoniae*, are eradicated following the activation of other cell types by helper T cells, namely the classes T_H1 and T_H2. (ii) T_H1 and T_H2 cells express a glycoprotein called CD4, a T-cell co-receptor that interacts with (iii) MHC class II molecules on antigen-presenting cells. (iv) T_H1 cells activate macrophages that are displaying pathogen peptides (derived from phagocytosed pathogen) on MHC class II molecules on their surface. This stimulates increased phagocytosis by the macrophage and destruction of pathogens inside phagolysosomes. Activated macrophages also secrete inflammatory mediators that play

an important part in eradicating the infection by helping to induce inflammation which recruits phagocytic cells and effector lymphocytes to the site of infection. T_H1 cells also induce switching of B cells to certain antibody isotypes. T_H2 cells activate B cells displaying antigen-derived peptides on MHC class II molecules, resulting in the differentiation of the B cells into plasma cells and the production of antibodies which remove the extracellular pathogen or its toxins as a result of neutralization, opsonization, and complement activation.

Answer 3–8

A. (i) The complete MHC class I molecule is a heterodimer made up of one α chain and a smaller chain called β_2-microglobulin. The α chain consists of three extracellular domains—α_1, α_2, and α_3—a transmembrane region and a cytoplasmic tail. β_2-microglobulin is a single-domain protein noncovalently associated with the extracellular portion of the α chain, providing support and stability. (ii) The polymorphic class I molecules in humans are called HLA-A, HLA-B, and HLA-C. The α chain is encoded in the MHC region by an MHC class I gene. The gene for β_2-microglobulin is elsewhere in the genome. (iii) The antigen-binding site is formed by the α_1 and α_2 domains, the ones farthest from the membrane, which create a peptide-binding groove. The region of the MHC molecule that binds to the T-cell receptor encompasses the α helices of the α_1 and α_2 domains that make up the outer surfaces of the peptide-binding groove. The α_3 domain binds to the T-cell co-receptor CD8. (iv) The most polymorphic parts of the α chain are the regions of the α_1 and α_2 domains that bind antigen and the T-cell receptor. β_2-microglobulin is invariant; that is, it is the same in all individuals.

B. (i) MHC class II molecules are heterodimers made up of an α chain and a β chain. The α chain consists of α_1 and α_2 extracellular domains, a transmembrane region, and a cytoplasmic tail. The β chain contains β_1 and β_2 extracellular domains, a transmembrane region, and a cytoplasmic tail. (ii) In humans there are three polymorphic MHC class II molecules called HLA-DP, HLA-DQ, and HLA-DR. Both chains of an MHC class II molecule are encoded by genes in the MHC region. (iii) Antigen binds in the peptide-binding groove formed by the α_1 and β_1 domains. The α helices of the α_1 and β_1 domains interact with the T-cell receptor. The β_2 domain binds to the T-cell co-receptor CD4. (iv) With the exception of HLA-DRα, which is monomorphic, both the α and β chains of MHC class II molecules are polymorphic. Polymorphism is concentrated around the regions that bind antigen and the T-cell receptor in the α_1 and β_1 domains.

Answer 3–9

A. There are three MHC class I genes in humans (HLA-A, HLA-B, and HLA-C) and they are expressed from both chromosomes. Assuming that each gene is heterozygous, the maximum number of different MHC class I α chains that could be expressed is 6. As β_2-microglobulin is invariant, this means that six different MHC class I molecules could be produced. For MHC class II molecules, assuming complete heterozygosity and the presence of two functional DRB genes (DRB1 and DRB3, 4, or 5) on both chromosomes, the maximum number of MHC class II molecules that could be expressed is 12 (Figure A3.9). The total number of different HLA molecules that can be expressed is 18.

B. MHC molecules have degenerate binding specificity, which means that one MHC molecule is able to bind a wide range of peptides of different sequence. For all MHC molecules, only a few of the amino acids in the antigen peptide are critical for binding to amino acids in the peptide-binding groove. The critical amino acids in the peptide are called anchor residues; they will be the same or similar in all peptides that bind to a given MHC molecule. The other amino-acid residues in the peptides can be different. The pattern of anchor residues that binds to a given MHC molecule is called the peptide-binding motif. Hence, a very large number of discrete peptides can bind to each MHC isoform, the only constraint being the possession of the correct anchor residues at the appropriate positions in the peptide. MHC class I molecules also bind peptides that are typically nine amino acids long, whereas MHC class II molecules bind longer peptides with a range of lengths.

MHC class I		MHC class II	
α chain		**α chain**	**β chain**
HLA-A HLA-A m* HLA-A p*	β_2-microglobulin β_2-microglobulin	HLA-DR DRA-m or p DRA-m or p DRA-m or p DRA-m or p	DRB1-m DRB1-p DRB$_{3,4, or 5}$-m DRB$_{3,4, or 5}$-p
HLA-B HLA-B m HLA-B p	β_2-microglobulin β_2-microglobulin	HLA-DQ DQA-m DQA-p DQA-m DQA-p	DQB-m DQB-p DQB-p DQB-m
HLA-C HLA-C m HLA-C p	β_2-microglobulin β_2-microglobulin	HLA-DP DPA-m DPA-p DPA-m DPA-p	DPB-m DPB-p DPB-p DPB-m
Total	6	12	

Figure A3.9 The number of HLA molecules that can be expressed in a single individual. *m = maternal chromosome; p = paternal chromosome.

Answer 3–10

A. Polygeny refers to the presence of multiple genes for MHC class I and MHC class II molecules in the genome, encoding a set of structurally similar proteins with similar functions. MHC polymorphism is the presence of multiple alleles (in some cases several hundreds) of most of the MHC class I and class II genes in the human population.

B. T cells recognize peptide antigens in the form of MHC:peptide complexes, which they bind using their T-cell receptors. To bind specifically, the T-cell receptor must fit both the peptide and the part of the MHC molecule surrounding it in the peptide-binding groove. (i) Because each individual produces a number of different MHC molecules from their polygenic MHC class I and class II genes, the T-cell receptor repertoire is not restricted to recognizing peptides that bind to just one MHC molecule (and thus must all have the same peptide-binding motif). Instead, the T-cell receptor repertoire can recognize peptides with different peptide-binding motifs during an immune response, increasing the likelihood of antigen recognition and, hence, T-cell activation. (ii) The polymorphism in MHC molecules is localized in the regions affecting T-cell receptor and peptide binding. Thus, a T-cell receptor that recognizes a given peptide bound to variant 'a' of a particular MHC molecule is likely not to recognize the same peptide bound to variant 'b' of the same MHC molecule. Polymorphism also means that the MHC molecules of one person will bind a different set of peptides from those in another person. Taken together, this means that because of MHC polymorphism, each individual recognizes a somewhat different range of peptide antigens using a different repertoire of T-cell receptors.

Answer 3–11

A. Proteins from pathogens growing in the cytosol are broken down into small peptide fragments in proteasomes. The peptides are transported into the lumen of the endoplasmic reticulum (ER) using TAP (transporter associated with antigen processing) which is a heterodimer of TAP-1 and TAP-2 proteins located in the ER membrane. Peptides bearing the appropriate peptide-binding motif bind to MHC class I molecules already delivered into the ER. MHC class I α chains are bound to the chaperone calnexin until β_2-microglobulin binds, and then are bound by the chaperones calreticulin and tapasin until peptide binds. Tapasin binds to TAP-1, positioning the MHC class I molecule near the peptide source. MHC class I molecules bound to peptide dissociate from the chaperone molecules and progress to the Golgi apparatus for completion of glycosylation and transport to the cell surface in membrane-bound vesicles.

B. (i) If an MHC class I α chain is unable to bind β_2-microglobulin, it will be retained in the ER and will not be transported to the cell surface. It will remain bound to calnexin and will not fold into the conformation needed to bind to peptide. Thus, antigens will not be presented using that particular MHC class I molecule. (ii) If TAP-1 or TAP-2 proteins are mutated and not expressed, then peptides will not be transported into the lumen of the ER. Without peptide, an MHC class I molecule cannot complete its assembly and will not leave the ER. A rare immunodeficiency disease called bare lymphocyte syndrome (MHC class I immunodeficiency) is characterized by a defective TAP protein, causing less than 1% of MHC class I molecules to be expressed on the cell surface compared to normal. Thus, T-cell responses to all pathogen antigens that would normally be recognized on MHC class I molecules will be impaired.

Answer 3–12

A. Extracellular pathogens are taken up by endocytosis or phagocytosis and degraded by enzymes into smaller peptide fragments inside acidified intracellular vesicles called phagolysomes. MHC class II molecules delivered into the ER and being transported to the cell surface intersect with the phagolysosomes, where these peptides are encountered and loaded into the antigen-binding groove. To prevent MHC class II molecules from binding to peptides prematurely, invariant chain (Ii) binds to the MHC class II antigen-binding site in the ER. Ii is also involved in transporting MHC class II molecules to the phagolysosomes via the Golgi as part of the interconnected vesicle system. Ii chain is removed from MHC class II molecules once the phagolysosome is reached. Removal is achieved in two steps: (1) proteolysis cleaves Ii into smaller fragments, leaving a small peptide called CLIP (class II-associated invariant chain peptide) in the antigen-binding groove of the MHC class II molecule; and (2) CLIP is then released by HLA-DM catalysis. Once CLIP is removed, peptides derived from endocytosed material will bind (if the correct peptide-binding motif is possessed) and the peptide:MHC class II molecule complex will progress to the cell surface.

B. (i) Defects in the invariant chain would impair normal MHC class II function because invariant chain not only protects the peptide-binding groove from binding prematurely to peptides present in the ER but is also required for transport of MHC class II molecules to the phagolysosome. (ii) If HLA-DM were not expressed, most MHC class II molecules on the cell surface would be occupied by CLIP rather than endocytosed material. This would compromise the presentation of extracellular antigens at the threshold levels required for T-cell activation.

Chapter 4

Answer 4–1

A. B-1 cells are activated early during an immune response, do not produce immunological memory, and do not carry out affinity maturation. Individual B-1 cells are polyspecific for antigen; that is, their immunoglobulins will bind to a number of different antigens. They produce primarily antibodies of the IgM isotype. In contrast, B-2 cells are activated later in immune responses (owing to the requirement for T-cell help and antigen processing), give rise to immunological memory, and carry out somatic hypermutation. Individual B-2 cells are monospecific for antigen. In addition, B-2 cells can express different isotypes of antibodies through the process of isotype switching, which involves T-cell help. B-1 cells are unable to switch isotype.

B. The antigens recognized by most B-1 cells are bacterial polysaccharides and other carbohydrate antigens. Helper T cells recognize protein antigens, not carbohydrates and therefore are unable to provide help for B-1 cells.

C. B-1 cells are probably best associated with innate immune responses because of their rapid response to antigen, limited diversity, lack of memory, and polyspecificity.

Answer 4–2

A. Histogram. Your histogram should look like the one in the left panel of Figure A4.2, with axes labeled "cell number" (*y*) versus "relative fluorescence intensity" (*x*) depicting total bone marrow cells stained with anti-CD19. Two peaks will be observed. One is the CD19-negative population and the other is the CD19-positive population. Gate the CD19-positive population (developing and mature B cells) for two-dimensional dot-plot analysis.

B. Two-dimensional dot plot. Your dot plot should look like the one in the right panel of Figure A4.2, with relative fluorescence intensity of anti-IgD versus anti-IgM. Three populations of B cells will be distinguished: (1) IgM-positive, IgD-negative, which represent immature B cells; (2) IgM-positive, IgD-positive, which represent mature B cells; and (3) IgM-weakly positive, IgD-positive, which represent anergic B cells.

Answer 4–3

In order to survive, circulating B cells are required to enter primary follicles where survival signals are delivered by cells in the follicles including follicular dendritic cells (which are the stromal cells of primary lymphoid follicles). Circulating B cells that fail to enter follicles in secondary lymphoid tissues will die in the peripheral circulation with a half-life of about 3 days. Anergic B cells that enter secondary lymphoid organs are withheld in the T-cell areas adjacent to primary follicles and are thus not permitted to penetrate the primary follicle. This is probably due to the lack of T cells in the T-cell zone that are specific for the same soluble self-antigen (the T cells would have been deleted in the thymus). As a result, anergic B cells fail to receive the necessary stimulatory signal for survival and "permission" to proceed to the primary follicle is denied. Instead, anergic B cells will undergo apoptosis in the T-cell zone. This is an efficient cleansing mechanism and serves to delete potentially autoreactive B cells from the circulation.

Answer 4–4

A. They both have limited life-spans, express decreased levels of IgM on the cell surface and are nonresponsive to antigen and T-cell help.

B. Anergic B cells do not secrete antibody. Plasma cells, in contrast, secrete very large amounts of antibody.

Answer 4–5

A. Memory enables faster, more efficient recall responses when antigen is encountered subsequently. This enables the body to get rid of a pathogen before it has time to cause disease.

B. Immunoglobulin produced during a primary immune response is primarily IgM, in low concentration (titer) and of low affinity for the antigen. Immunoglobulin expressed during a secondary immune response has undergone isotype switching and is often of the IgG isotype. It also has a higher titer and, through the process of somatic hypermutation, will have a higher affinity for its corresponding antigen.

Answer 4–6

Checkpoint 1 is the formation of a complex of a μ heavy-chain complexed with the surrogate light-chain VpreBλ5, Igα and Igβ. This delivers an important signal to the cell verifying that a functional heavy-chain has been made. It triggers cessation of heavy-chain gene rearrangement followed by inactivation of surrogate light-chain synthesis. Thus, only one heavy-chain locus ends up producing a product. As surrogate light-chain becomes unavailable, μ accumulates and is retained in the endoplasmic reticulum, ready to bind to functional light chain when that is synthesized following successful light-chain gene rearrangement.

Checkpoint 2 is when a complete B-cell receptor, comprising μ heavy chains, κ or λ light chains, and Igα and Igβ chains is expressed on the B-cell surface. This signals cessation of light-chain rearrangement. Thus only one light-chain locus out of the possible four produces a functional product.

Figure A4.2

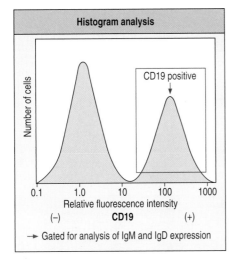

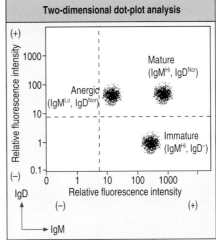

Answer 4–7

N nucleotides would be added at the VJ joints of all rearranged light-chain genes during gene rearrangement (instead of about half), resulting in an increase in immunoglobulin diversity. It is interesting to note that because TdT is not expressed until after birth, B-1 cells that are generated prenatally lack N nucleotides in the VD and DJ junctions of their rearranged heavy-chain genes as well as in the VJ junctions of all light-chain genes.

Answer 4–8

A. Bone marrow stromal cells provide the necessary environment for B-cell development by expressing secreted products and membrane-bound adhesion molecules. For example VCAM-1 adhesion molecule binds to the integrin VLA-4 on early B-cell progenitors. Cytokines such as IL-7 play an important role in later stages of B-cell development, serving to stimulate growth and cell division of late pro-B and pre-B cells.

B. If anti-IL-7 antibody were introduced into this environment, developing B cells would be arrested at the late pro-B or pre-B cell stage and would not be able to progress normally to the immature B-cell stage. Interestingly, in transgenic mice over-expressing IL-7, significant increases in pre-B cells are observed in the bone marrow and secondary lymphoid organs, while in IL-7 knockout mice (where the IL-7 gene locus is interrupted and no IL-7 is produced) early B-cell expansion is significantly impaired. These experiments in mice demonstrate clearly the importance of IL-7 in B-cell maturation.

Answer 4–9

A. A B-cell tumor comprises cells that derive from a single cell that has undergone transformation that resulted in uncontrolled growth. No further maturation of the B cell occurs following transformation. If the B cell has rearranged the heavy- and light-chain genes prior to transformation, then immunoglobulin will be expressed at the cell surface. Because all of the cells in the tumor belong to the same clone, the immunoglobulin on all of them will be made of the same heavy and light chains.

B. Pre-B cell leukemia is characterized by transformation prior to the rearrangement of light-chain genes. If transformation occurs at the large pre-B cell stage, then the immunoglobulin on the cell surface will be composed of μ:VpreBλ5. If transformation occurs at the small pre-B cell stage, then there would be little or no immunoglobulin on the cell surface because surrogate light chain expression is turned off at this stage and the μ heavy chains are retained in the endoplasmic reticulum. Normal immature B cells that have not undergone transformation express IgM containing μ plus a κ or λ light chain.

Answer 4–10

A. Normally light-chain specificity is polyclonal because a diverse array of plasma cells produce different light chains. In multiple myeloma, the tumor originates from a single plasma cell expressing heavy and light chains with specificity for a single antigen (clonotypic immunoglobulin). Because B cells express only kappa or lambda light chains, the tumor will also express only kappa or lambda, but not both. Therefore, Bence-Jones protein for a given patient will be of one type or another, not both.

B. Although patients will have elevated immunoglobulin levels (usually IgG or IgA), most will be produced by the myeloma cells and will be monospecific. Hence, normal concentrations of polyclonal immunoglobulin will be severely compromised. Pyogenic infections caused by encapsulated bacteria require humoral immune responses involving mechanisms that utilize antibody-mediated complement activation and phagocytosis. Insufficient polyclonal immunoglobulins put these patients at risk.

Chapter 5

Answer 5–1

The analog of VpreB:λ5 in developing T cells is preTα (pTα) which combines with the T-cell receptor β chain, which is the first of the two T-cell receptor chains to be expressed, to form the pre-T cell receptor. The β chain, like the immunoglobulin heavy chain, contains V, D, and J segments. pTα also binds CD3 and ζ components to this complex and the assembly of the complete complex induces T-cell proliferation and the cessation of rearrangement at the TCRβ loci (leading to allelic exclusion). Formation of the analogous pre-B cell receptor complex of VpreB:λ5 and heavy chain with Igα and Igβ in B cells similarly prevents further rearrangement of the heavy-chain loci.

Answer 5–2

A. Successive gene rearrangement is possible at a TCRα locus because there are numerous V and J gene segments available. As long as there are V segments upstream and J segments downstream of the initial unproductive rearrangement, a second rearrangement can occur. This will involve deletion of the unproductive rearrangement. If the second rearrangement is productive, then rearrangement at this locus terminates. If it is unproductive, then another rearrangement is permitted, providing there are V and J segments still available. If the V and/or J segment pool is depleted at one locus, then the other TCRα locus can begin to rearrange.

B. Successive gene rearrangement is possible at a TCRβ locus because there are two sets of D, J, and C gene segments downstream of the cluster of V gene segments: $(V_\beta)_n...D_\beta 1...(J_\beta 1)_n...C_\beta 1...D_\beta 2...(J_\beta 2)_n...C_\beta 2$. If a first rearrangement involving $D_\beta 1$ and a $J_\beta 1$ segment is unproductive, an upstream V gene segment can rearrange to the second D gene segment and an associated J segment. If this is unproductive, no more rearrangements can be made.

C. Successive gene rearrangement is possible at the κ light-chain locus because there are numerous V and J gene segments available. As with the TCRα locus, as long as there are available V and J segments, further rearrangements can take place, resulting in the excision of the unproductive rearrangement product. Each κ locus has the potential of rearranging a maximum of five times because there are five J segments.

D. Successive gene rearrangement is possible at a λ light-chain locus because there are four sets of one J and one C gene segment downstream of the V gene segment cluster: $(V_\lambda)_{-29}...J_\lambda 1...C_\lambda 1...J_\lambda 2...C_\lambda 2...J_\lambda 3...C_\lambda 3...J_\lambda 4...C_\lambda 4$. Each λ locus thus has the potential to make a maximum of four rearrangements.

E. The heavy-chain locus has the following configuration: $(V)_n$-heptamer...23 spacer...nonamer...nonamer...12 spacer...heptamer-$(D)_n$-heptamer...12 spacer...nonamer...nonamer...23 spacer...heptamer $(J)_n...C_\mu$. After the first DJ rearrangement, the intervening D segments between the chosen D and J will be deleted. Following VDJ rearrangement, the D segments that lie between the chosen V and DJ will be deleted. Therefore, no unrearranged D segments remain following these two rearrangement events. V and J cannot rearrange directly because the recombination signal sequences are not paired appropriately and do not follow the 23/12 rule, but rather they both contain 23-spacers with their recombination signal sequences. Successive gene rearrangement is thus not possible at a heavy-chain locus.

Answer 5–3

A. Cells that express MHC class II are the professional antigen-presenting cells (B cells, macrophages, and dendritic cells), thymic epithelial cells, neural microglia, and activated T cells (in humans).

B. Macrophages, dendritic cells, and thymic epithelial cells either populate the thymus or circulate through it. Cortical thymic epithelial cells participate in positive selection by presenting MHC class I and class II molecules with self-peptides to double-positive (CD4$^+$/CD8$^+$) thymocytes. These are developing T cells that have successfully rearranged the TCRα and TCRβ genes. Only T cells that have T-cell receptors that can interact with self-MHC are positively selected, thus shaping a T-cell repertoire that is specific for self-MHC molecules. If the affinity for self-MHC:self-peptide is too weak, the T cells will die by neglect through apoptosis. Cortical and medullary thymic epithelium may also participate in negative selection by inducing apoptosis of thymocytes that bear a T-cell receptor with high affinity for self-MHC, self-peptide or a combination of the two. Thymic epithelium, circulating macrophages, and dendritic cells found primarily at the cortico-medullary junction participate in negative selection and aid in the elimination of potentially self-reactive T cells bearing high-affinity T-cell receptors for complexes of self-MHC:self-peptides. Because these cells are circulating between tissues, organs and the thymus, a heterogeneous array of self-peptides will be transported to the thymus and displayed. This is important because there are some self-peptides that are expressed in locations other than the thymus and which may be encountered in secondary lymphoid organs.

C. Although circulating dendritic cells and macrophages are effective at endocytosing cellular debris in extra-thymic locations and then presenting self-peptides derived from this debris to thymocytes, there are some self-proteins that are excluded from this housekeeping function. Some self-proteins are located in immunologically privileged sites where leukocytes do not usually circulate. These sites are normally MHC class II negative and this is important because this avoids the presentation of self-peptides with MHC class II that were not presented in the thymus during negative selection. It should be pointed out, however, that some of these tissues can be induced to express MHC class II molecules under the influence of certain cytokines, for example, interferon-γ, during an inflammatory response. This is believed to be one mechanism by which tolerance can be broken, resulting in autoimmunity.

Answer 5–4

A. Only one of the receptors will actually have to be positively selected for the cell to get the survival signals necessary for it to pass on to the next stage. Even if the other receptor does not react with self-MHC this will have no effect.

B. In contrast, both receptors will have to pass the negative selection test for the T cell to survive, as if only one of them fails it, the cell will die.

C. Yes. Imagine this situation. The T cell with dual specificity could be activated appropriately during a *bona fide* infection by a professional antigen-presenting cell plus foreign antigen 1 using T-cell receptor 1. But that same T cell, because it is now an activated effector T cell, would also be able to respond to a second peptide, which might be a self-peptide, using T-cell receptor 2, without requiring the co-stimulatory signals only professional antigen-presenting cells deliver. Thus it could cause a reaction against a self-tissue, either directly, if it is a CD8 cytotoxic T cell, or indirectly, if it is a CD4 T cell, by activating potentially autoreactive B cells.

   Furthermore, interferon-γ produced in the response against foreign antigen 1 could activate nonprofessional antigen-presenting cells nearby, inducing expression of MHC class II with presentation of the self-peptide above. Effector T cells with T-cell receptor 2 could make an autoimmune response against it.

Answer 5–5

Because T cells drive almost all immune responses, once activated, their receptors must continue to recognize the exact complex of foreign antigen and MHC molecule (which does not itself change) that activated them. Because of this requirement for dual recognition (MHC restriction), somatic hypermutation would more likely than not change the T-cell receptor to make it unable to recognize either the peptide or the MHC molecule, or the combination of both, thus rendering it unable to give help to B cells or to attack infected cells, depending on the type of T cell. This would destroy both the primary immune response and the development of immunity. Even changes that simply increased the affinity of the T cell for its antigen would have no real advantage, as it would not make the immune response any stronger or improve immunological memory in the same way that affinity maturation of B cells does. Also, if somatic hypermutation changed the specificity of the T-cell receptor so that it now recognized a self-peptide, this could result in an autoimmune reaction. These considerations do not apply to B cells, as they require T-cell help to produce antibody and will only receive it if their B-cell receptor still recognizes the original antigen.

Answer 5–6

A. MHC class II deficiency affects the development of CD4 T cells in the thymus. If the thymic epithelium lacks MHC class II, then positive selection of CD4 T cells will not take place. CD8 T cells are not affected because MHC class I expression is unaffected by this defect.

B. In order to produce antibody B cells require T-cell help in the form of cytokines produced by CD4 T$_H$2 cells. Low immunoglobulin levels (hypogammaglobulinemia) are attributed to the inability of B lymphocytes to proliferate and differentiate into plasma cells in the absence of T$_H$2 cytokines.

Answer 5–7

A. Following irradiation and chemotherapy, the patient would be infused with bone marrow obtained from a donor with the best HLA match possible.

B. Peripheral blood mononuclear cells (PBMCs) from the recipient of the transplant will be collected before transplantation for comparison at a later date. At a sufficient time after transplantation, the patient's reconstituted PBMCs will be harvested and, together with the pre-transplant sample, stained with two antibodies conjugated to different fluorescent dyes. One antibody will be specific for CD3 and the other will be specific for the α chain of the mismatched HLA class I antigen. Dot plots showing that donor-derived T cells (positive for both CD3 and donor HLA class I) in the post-transplant sample only (Figure A5.7) will demonstrate that donor-derived hematopoietic stem cells did in fact reconstitute the T-cell repertoire in this patient.

Answer 5–8

Beyond a certain number of isotypes, the T-cell repertoire will be decreased, owing to the disproportionate increase in negative selection events as the number of different isotypes increases. Each additional isotype will decrease the number of T cells exported to the periphery, compromising the diversity of the T-cell population.

Answer 5–9

Thymic maturation of T cells occurs within a short time period; thymocytes not rescued promptly during positive selection in the cortex are destined to die by apoptosis. For this reason thymocytes do not undergo significant proliferation between the early and late progenitor stages of development and there is little opportunity for mutations to accumulate at intermediate stages. Therefore, transformation of T cells occurs either at an early lymphoid progenitor stage of thymocyte differentiation in the thymus or at the post-maturational stage in the periphery. Examples include common acute lymphoblastic leukemia (C-ALL), and chronic lymphocytic leukemia (CLL), respectively.

Answer 5–10

A. Southern blotting is a molecular technique based on DNA hybridization, which can be used to detect and characterize a

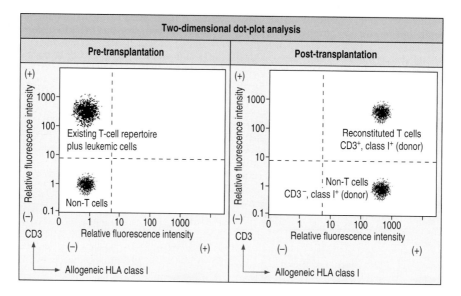

Figure A5.7

T-cell tumor by demonstrating that the transformed cells all contain the same rearranged T-cell receptor gene. DNA is extracted from the desired cells, digested with a restriction endonuclease to generate DNA fragments of varying lengths, electrophoresed in an agarose gel, denatured and then transferred to a nitrocellulose filter or nylon membrane. Because the tumor cells all contain the same rearranged β gene, a distinct band corresponding to the rearranged gene can be visualized after hybridizing the nitrocellulose filter with a radioactive, single-stranded probe complementary to the T-cell receptor β-locus C-region followed by autoradiography. This is in addition to bands corresponding to the unrearranged β locus, which will also be visible.

B. (i) The spermatocytes will all contain T-cell receptor β loci in the germline configuration, and so the sample will only give the two bands corresponding to the $C_\beta 1$ and $C_\beta 2$ genes in the unrearranged locus when probed with a C-region probe. (ii) Normal PBMCs (which will contain T cells in addition to other mononuclear cells) will give the same two bands and a barely detectable smear corresponding to the numerous different rearranged β genes in the heterogeneous polyclonal population of T cells. (iii) The PMBCs from the patient with T-cell leukemia in contrast, will contain a much larger proportion of T cells than normal, and the majority will harbor the same rearranged β gene, as a result of outgrowth of a single transformed T cell. On the Southern blot, this patient's sample will have a discrete band migrating at a location distinct from the bands representing the germline configuration. This band represents the numerous identical rearranged β genes in the sample. The intensity of the band will correlate with the proportion of transformed cells in the sample.

Chapter 6

Answer 6–1

A. Naive T cells encounter antigen, and start the primary immune response, in a secondary lymphoid tissue (for example, lymph nodes, spleen, Peyer's patches, tonsils).

B. (i) Lymph nodes. The pathogen, and dendritic cells that have ingested the pathogen, are carried to the nearest lymph node in the afferent lymph. (ii) Gut-associated lymphoid tissues (GALT) such as Peyer's patches. Pathogens enter GALT via specialized cells (M cells) in the gut epithelium. (iii) Spleen. Pathogens circulating in the blood enter the spleen directly from the blood vessels that feed it.

C. T cells are delivered to all secondary lymphoid organs from the blood.

D. After activation by antigen only CD8+ T cells and T_H1 cells exit the lymphoid tissue (via efferent lymph which delivers them eventually into the blood) in search of infected tissues. Antigen-activated T_H2 cells remain in the lymphoid tissue where they provide help to antigen-specific B cells.

Answer 6–2

First, antigen needs to be transported to a nearby secondary lymphoid tissue, processed, and presented by antigen-presenting cells to naive cytotoxic T cells or helper T cells in order to activate them. Second, the number of T cells specific for a given pathogen will be only around 1 in 10,000 to 1 in 100,000 (10^{-4} to 10^{-6}) of the circulating T-cell repertoire; thus it can take some time before the relevant T cells circulating through the secondary lymphoid tissues get to the tissue that contains the antigen that will activate them. Finally, it takes several days for an activated T cell to proliferate and differentiate into a large clone of fully functional effector T cells.

Answer 6–3

A. T cells (and B cells) express L-selectin which binds to sulfated carbohydrates of mucin-like vascular addressins in HEVs. Three types of mucin-like vascular addressins are involved: GlyCAM-1 and CD34 expressed on lymph node HEVs, and MAdCAM-1, expressed on mucosal endothelium.

B. Chemokines made by endothelium and bound to its extracellular matrix induce T cells to express LFA-1, an integrin. LFA-1 binds with high affinity to an intercellular adhesion molecule, ICAM-1, expressed on the endothelium. Finally, the T cell squeezes between the endothelium junctions, a process called diapedesis, and hence gains entry into the lymph node.

Answer 6–4

A. The three types of professional antigen-presenting cells are dendritic cells, macrophages and B cells.

B. Dendritic cells are found in T-cell rich areas of secondary lymphoid tissue; macrophages are distributed throughout the tissue; B cells are localized in lymphoid follicles.

C. Dendritic cells present all types of antigen, but present viral antigens particularly efficiently. Macrophages present bacterial antigens well because they bear generalized receptors that can bind and internalize many different bacteria. B cells present peptides of soluble protein antigens, such as protein toxins, that they have internalized via their antigen receptors.

Answer 6–5

A. Expression of B7, a co-stimulator molecule, distinguishes professional-antigen presenting cells from other cells.

B. When B7 binds to CD28, the B7 receptor expressed earliest on T cells, an activation signal is delivered and T cells undergo clonal expansion and differentiation. This interaction requires, of course, that the T-cell receptor and CD4 co-receptor are engaged specifically with a peptide:MHC class II complex. The second B7 receptor, CTLA-4, binds B7 with ~20-fold higher affinity than does CD28. An inhibitory signal is delivered to the T cell when B7 on the professional antigen-presenting cell binds to CTLA-4. This mechanism serves to regulate T-cell proliferation and to suppress T-cell activation after an immune response.

C. If they engage antigen in the absence of B7 expression and, hence, co-stimulation, T cells will become irreversibly nonresponsive (anergic) instead of activated. This is one mechanism by which T-cell tolerance may be achieved.

Answer 6–6

A. Immature dendritic cells are very efficient at phagocytosis owing to expression of the receptor DEC 205. They also take up extracellular material indiscriminately by macropinocytosis. They have specialized pathways of antigen processing for extracellular antigens that can present these antigens on both MHC class I and class II molecules. They do not express co-stimulatory molecules. Immature dendritic cells migrate to nearby lymphoid tissue following antigen ingestion. Upon arrival in the lymphoid tissue, they differentiate into mature dendritic cells. These are non-phagocytic and express the co-stimulatory molecules B7.1 and B7.2.

B. Immature dendritic cells need to be phagocytic because they are located in sites susceptible to infection. Expression of B7 in non-lymphoid tissue is not required because this is not where T cells circulate and sample MHC:peptide complexes. Once outside the infected tissue, mature dendritic cells no longer need to phagocytose material. They do, however, need to express B7 molecules, because without co-stimulation, T cells do not receive the necessary activation signal for differentiation into effector T cells.

C. An example of an immature dendritic cell is the Langerhans' cell of the skin which matures into an interdigitating reticular cell in the lymph node.

Answer 6–7

A. Cytotoxic T cells recognize antigen on the surface of a cell presenting antigen bound to MHC class I molecules. The cytotoxic T cell responds by killing these target cells by inducing an apoptotic pathway. T_H1 cells recognize antigen bound to MHC class II molecules on the surface of an antigen-presenting cell such as a macrophage. The T_H1 cell responds by activating the macrophage to destroy intravesicular bacteria and increase phagocytosis of extracellular bacteria. T_H2 cells recognize antigen bound to MHC class II molecules on the surface of B cells. The T_H2 cell responds by secreting cytokines that will activate B cells to differentiate into antibody-producing plasma cells.

B. Examples: An example of a cytotoxic T-cell antigen is a virus replicating in the cytosol of the target cell. An example of a T_H1-cell antigen is a protein encoded by *Mycobacterium tuberculosis*. An example of T_H2-cell antigen is diphtheria toxin produced by *Corynebacterium diphtheriae*.

Answer 6–8

A. Cytotoxic T cells focus their killing machinery on target cells through a process called polarization. The cytoskeleton and the cytoplasmic vesicles containing lytic granules are oriented toward the area on the target cell where MHC class I:peptide complexes are engaging T-cell receptors. In the T cell, the microtubule-organizing center, Golgi apparatus and lytic granules, which contain cytotoxins, align towards the target cells. The lytic granules then fuse with the cell membrane, releasing their contents into the small gap between the T cell and the target cell, resulting in deposition of cytotoxins on the surface of the target cell. The cytotoxic T cell is not killed in this process, and will continue to make cytotoxins for release onto other target cells, thereby killing numerous target cells in a localized area in succession.

B. The cytotoxins include perforin, granzymes and granulysin, molecules which induce apoptosis (programmed cell death) of the target cell.

Answer 6–9

(i) The CD3 subunits γ, δ, and ε associated with the antigen-binding T-cell receptor help transmit the signal from the T-cell receptor:MHC:peptide interaction at the cell surface into the interior of the cell through immunoreceptor tyrosine-based activation motifs (ITAMs) present on their cytoplasmic tails. These are phosphorylated by associated protein tyrosine kinases, such as Fyn, when the antigen receptor is activated, and in turn activate further molecules of the signaling pathway. (ii) Lck associates with the tails of the CD4 and CD8 co-receptors. When these participate in binding to MHC:peptide complexes, Lck is activated and phosphorylates ZAP-70, a cytoplasmic protein tyrosine kinase. (iii) CD45 is a cell-surface protein phosphatase that helps activate Lck and other kinases by removing inhibitory phosphate groups from their tails. (iv) When ZAP-70 is phosphorylated it binds to the phosphorylated ITAMs of (v) the ζ chain, which initiates the signal transduction cascade by activating phospholipase C-γ (PLC-γ) and guanine-exchange factors. (vi) IP_3, which is produced by the action of PLC-γ on membrane inositol phospholipids, causes an increase in intracellular Ca^{2+} levels, which leads to the activation of the protein calcineurin. (vii) Calcineurin activates the transcription factor NFAT by removing an inhibitory phosphate group. Activated NFAT enters the nucleus, and together with the transcription factors NFκB and AP-1 will initiate the transcription of genes that lead to T-cell proliferation and differentiation.

Answer 6–10

A. A granuloma contains at its center macrophages and giant multinucleated cells created by macrophage fusion, infected with bacteria which are replicating intracellularly. Epithelioid cells surround the core, which consists of non-fused macrophages. Epithelioid cells are surrounded by activated CD4 T cells.

B. Chronic infections resistant to macrophage killing mechanisms can lead to the formation of granulomas, for example in tuberculosis, caused by *Mycobacterium tuberculosis* growing inside intracellular vesicles.

C. Granulomas surrounded by CD4 T cells cut off the infection from the blood supply and the cells in the core of the granuloma will die from oxygen deprivation and toxic products of the macrophages. In tuberculosis, the dead tissue is referred to as caseation necrosis because it has a cheesy consistency. If the infection were not localized in this fashion, it could disseminate systemically to other anatomical locations.

Answer 6–11

In an immune response, the binding of IL-2 to a high-affinity IL-2 receptor composed of α, β, and γ chains drives the proliferation and differentiation of T cells that occurs after they have encountered their specific antigen. The activated T cell will divide two to three times daily for about a week, producing a clone of thousands of identical antigen-specific, effector T cells. The α chain is required to form a complex with pre-existing γ and β chains to make the high-affinity receptor. The receptor formed by the γ and β chains on their own is of low affinity for IL-2. In the absence of IL-2 and its high-affinity receptor, T cells will not be fully activated, will not differentiate and will not undergo clonal expansion. Thus, by preventing the production of IL-2 and its receptor, cyclosporin prevents the clonal expansion of T cells specific for the foreign antigens on the graft and their differentiation into effector T cells, and thus suppresses the immune response directed against the graft.

Answer 6–12

Effective immune responses against intravesicular pathogens living in macrophages are mediated by T_H1 cells rather than T_H2 cells. In tuberculoid leprosy, the predominant effector T cells produced after infection are T_H1 cells. These are effective in containing the infection, although they do not clear it completely. The disease is chronic, progresses slowly, and the damage to skin and peripheral nerves is caused mainly by the inflammatory responses initiated by activated macrophages. In lepromatous leprosy, on the other hand, the predominant T cells produced are T_H2 cells. Humoral immunity is induced, which results in the production of antibodies which are ineffective against intracellular bacteria. As a result, *M. leprae* replicates unchecked, causing severe tissue destruction and eventually the death of the patient.

Many factors influence the differentiation of CD4 T cells into T_H1 or T_H2 cells, including the cytokines produced by the antigen-presenting cells and leukocytes involved in the innate immune responses, the antigen concentration and MHC:peptide density, T-cell receptor affinity for MHC:peptide, and the cytokines produced by T_H1 and T_H2 cells themselves. If T_H1 cells dominate an immune response, then a cell-mediated immune response is favored. If T_H2 cells dominate, then a humoral immune response is favored.

Answer 6–13

A. Many bacteria are surrounded by a polysaccharide capsule, and in some cases, antibodies against the capsular polysaccharides give protective immunity against the pathogen. Antibodies produced against polysaccharide antigens are generally restricted to the IgM isotype, as the help needed to switch isotypes to IgG is provided by T cells, which only recognize peptide antigens. Adult humans make effective immune responses to polysaccharides alone and thus can be protected by subunit vaccines made from the capsular polysaccharides of encapsulated bacteria. Their antibody responses are polysaccharide-specific, T-cell independent and IgM. In contrast, children do not make effective immune responses to polysaccharides alone and thus cannot be immunized with such vaccines.

However, if the polysaccharide is conjugated to a protein, peptides from the protein part of the molecule can activate specific T_H2 cells. B cells specific for polysaccharide will bind and internalize the whole antigen via their antigen receptors, process it, and then present peptides from the protein part on their surface. T cells specific for these peptides will interact with the B cell, delivering the necessary cytokines (such as IL-4) and the CD40–CD40-ligand signal required for isotype switching. The B cell will then produce IgG anti-polysaccharide antibodies. This type of vaccine can be used to immunize children in order to induce protective anti-polysaccharide antibodies.

B. A vaccine of this type has been produced against *Haemophilus influenzae* B (HiBC), which can cause pneumonia and meningitis. The conjugate vaccine is composed of a capsular polysaccharide of *H. influenzae* conjugated to tetanus or diphtheria toxoid (a protein). The antibody response is polysaccharide-specific, T-cell dependent, and IgG, and children are protected from meningitis caused by this microorganism.

Chapter 7

Answer 7–1

The B-cell co-receptor, made up of CD21 (complement receptor 2), CD19 and CD81 (TAPA-1), cooperates with the B-cell receptor in B-cell activation and increases the sensitivity of the B cell to antigen by

1000-10,000 fold. This becomes important when antigen concentrations are low. CD21 binds to complement components that have been deposited on the surface of pathogens, for example, C3d. CD19 provides the long cytoplasmic tail involved in signaling. When the co-receptor and B-cell receptor are both ligated on C3d and antigen, respectively, Lyn and the cytoplasmic tail of CD19 become close together. Lyn is a protein tyrosine kinase bound to the immunoreceptor tyrosine-based activation motifs (ITAMs) of Igα, which phosphorylates CD19 when it is nearby. The phosphorylated CD19 tail will initiate activation signals that complement those generated by the B-cell receptor complex.

Depending on the characteristics of the antigen, another signal provided by CD4⁺ T_H2 cells may also be needed. T_H2 cells express CD40 ligand on their surface, which binds to CD40 on the B cell and delivers a stimulatory signal. They also secrete the cytokines IL-4, -5, and -6. By binding to specific receptors these help stimulate B cells to proliferate and differentiate to plasma cells.

Answer 7–2

A. Thymus-dependent antigens can only induce the generation of antibodies by B cells when T-cell help is available. They induce immunological memory and affinity mutation after somatic hypermutation. Thymus-independent (TI) antigens are able to induce antibody generation in the absence of T-cell help even in DiGeorge syndrome patients who lack a thymus. Unlike thymus-dependent antigens, thymus-independent antigens do not induce immunological memory, do not activate T cells and affinity maturation is not observed owing to the absence of somatic hypermutation.

B. TI-1 antigens bind to B-cell receptors and other cell-surface receptors and intrinsically activate B cells to proliferate and differentiate into plasma cells. At high concentration, TI-1 antigens stimulate antigen-specific and antigen nonspecific antibody production. These polyclonal activators of B-cell mitosis are referred to as B-cell mitogens. An example is lipopolysaccharide (LPS), which is found in the cell wall of Gram-negative bacteria like *Neisseria meningitidis*. In the absence of T-cell cytokines needed for isotype switching, only IgM antibodies are made.

C. TI-2 antigens are highly repetitive protein or carbohydrate epitopes on the surface of pathogens. An example is the polysaccharide found in capsules of *Streptococcus pneumoniae*. The B cells responding to TI-2 antigen are often B-1 cells, and both IgM and IgG antibodies can be produced; however, predominantly IgM is made.

D. TI-1 and TI-2 by-pass the requirement for T-cell help because the number of B-cell receptors occupied by these antigens is very high and IgM cross-linking is efficient, thereby obviating the need for cognate T-cell involvement.

Answer 7–3

A. False. A plasma cell is a terminally differentiated B cell that is not dividing and cannot change its antibody specificity. Somatic hypermutation and selection of B cells with higher-affinity immunoglobulin receptors occurs in activated B cells before they differentiate into plasma cells,

B. True.

C. True.

D. False. TI-2 polysaccharide antigens activate only mature B cells. Often these B cells are B-1 cells which do not acquire full function until the child is about 5 years old. Therefore, TI-2 antigens do not stimulate efficient antibody responses in infants. Recall from Chapter 6 that vaccines administered to children to prevent meningitis caused by *Haemophilus influenzae* type b are conjugate vaccines composed of polysaccharide coupled to diphtheria or tetanus toxoid. The toxoid component stimulates T cells which then provide the necessary help to activate T-dependent antigen responses to polysaccharide in B cells of infants and young children.

Answer 7–4

CD40 ligand on T cells binds to CD40 on B cells, signaling B cells to activate NFκB. NFκB is a transcription factor that upregulates expression of ICAM-1, an adhesion molecule, on the surface of B cells. ICAM-1 prolongs interaction between the B and T cell, trapping the B cell in the T-cell zone, causing a primary focus of proliferating B cells (B lymphoblasts) to form.

Answer 7–5

A. IgM antibody, which is always the first antibody to be made in an adaptive immune response, will fix complement, which can bind to CR2 of FDCs.

B. It is the role of FDCs to supply a source of intact antigen to B cells by holding immune complexes on the cell surface and shedding immune complexes in the form of iccosomes (immune-complex coated bodies) from their cell surfaces. B cells survey immune complexes and iccosomes as they circulate through secondary lymphoid tissue. If the FcγR and CR2 on FDCs were internalized through receptor-mediated endocytosis, then cell-surface concentrations of immune complexes would diminish and iccosomes would not form. As a result, antigen recognition and uptake by antigen-specific B cells would not occur and the B cell would not be able to interact with helper T (T_H2) cells in the outer region of the light zone of peripheral lymphoid tissue. Without this interaction, expression of Bcl-x_L protein will not be induced, thus the B cell would not be rescued from apoptosis. IgG and C3b are important opsonins for macrophages, targeting immune complexes for uptake by macrophage receptors; after uptake, degradation and presentation of antigens to T_H1 cells using MHC class II molecules occurs. In this way, the receptors of FDCs and macrophages function differently to achieve different outcomes and stimulate different branches of acquired immunity (humoral versus cell-mediated).

Answer 7–6

A. The main effector functions of IgM include neutralization and complement activation, and to a lesser extent transport across epithelium and diffusion into extravascular sites.

B. (i) IgM is the first antibody to be secreted by mature B cells during a primary antibody response. IgM production accompanies T-dependent as well as T-independent antigen exposure, thereby generating antibodies specific for foreign proteins, carbohydrates and lipids. Neutralization of blood-borne pathogens and their toxins can be mediated by IgM. Because of the large size of pentameric IgM, it is most effective in the blood and ineffective in infected tissues where it cannot penetrate well. (ii) IgM is secreted by plasma cells as a pentamer, and because complement activation requires two Fc regions in order for C1qrs to bind and become activated in the complement cascade, a single pentamer of IgM can achieve complement activation. In comparison, IgG is secreted as a monomer, and would require two IgG antibodies in close proximity to each other to bind to C1qrs. (iii) There are Fc receptors for IgG (FcγR) and IgA (FcαR), but not for IgM. Depending on the Fc receptor type and which cell type expresses the receptor, a variety of effects can be induced when the receptor is bound to antibody. These include antigen uptake, activation of respiratory burst, granule release and killing. In some cases, Fc receptors can inhibit stimulation and serve to dampen down immune responses as antigen load is cleared. Because there is not an FcμR, IgM:antigen immune complexes cannot be opsonized by macrophages through receptor-mediated endocytosis, as is the case with IgG and IgA antibodies. If complement has been activated, and the alternative opsonin C3b has been deposited on the pathogen surface, the IgM:antigen:C3b complex can be taken up by the macrophage after binding to complement receptors.

Answer 7–7

A. Dimeric IgA is made in mucosal lymphoid tissues (MALT) and is transported across the barrier of the mucosal epithelium. First, dimeric IgA binds to the poly-Ig receptor on the basolateral surface of an epithelial cell followed by uptake through receptor-mediated endocytosis into an endocytic vesicle. Upon reaching the other face of the cell, the apical surface, the vesicle fuses with the membrane. Here the poly-Ig receptor is cleaved proteolytically between the membrane-anchoring and the IgA-binding regions, thus releasing IgA into mucus secretions. The dimeric IgA remains attached to a small piece of the poly-Ig receptor, called the secretory component, which holds the IgA at the epithelial surface through interactions with molecules in the mucus. The rest of the poly-Ig is degraded and serves no purpose.

B. Antibody is transported into the gastrointestinal, urogenital, and respiratory tracts, the eyes, nose and throat, and mammary glands (where newborn babies can receive IgA through breast milk).

Answer 7–8

A. Receptor-mediated endocytosis transports IgG antibody across endothelial barriers (capillary walls) from the blood into extracellular spaces surrounding tissues. IgG binds to two Brambell receptors (FcRB) on the apical side of the endothelium in the lumen of the capillary followed by receptor-mediated endocytosis into an endocytic vesicle. Upon reaching the other side of the endothelium, the basal surface, the vesicle fuses with the membrane and IgG is released into the extracellular space.

B. Antibody is transported into infected tissues and across the placenta into the fetal circulation during pregnancy.

Answer 7–9

A. Similarities: (1) Activation of both mast cells and NK cells requires that their Fc receptors be bound to antigen:antibody complexes. (2) When cross-linking occurs, both mast cells and NK cells release the contents of granules through exocytosis, which involves fusion of vesicles containing preformed proteins with the cell membrane.

B. Differences: (1) Exocytosis of granules from mast cells occurs at random around the cell membrane. Exocytosis of granules from NK cells is highly polarized, focusing only on the target cell to minimize damage to neighboring cells. (2) Mast cells bind IgE, whereas NK cells bind IgG. (3) IgE binds to FcεRI with high affinity in the absence of antigen; mast cells become activated when antigen becomes available and binds to the receptor-bound IgE. NK cells bind IgG with low affinity, and only bind IgG effectively when it is already bound to multivalent antigen. (4) Mast cells release inflammatory mediators (histamine, serotonin) that affect other cells, for example endothelium, causing increased vascular permeability and vasodilation. NK cells release apoptosis-inducing compounds (perforin and granzyme/fragmentin) that kill target cells directly. (5) Antibody-dependent cell-mediated cytotoxicity (ADCC) carried out by NK cells could be achieved in newborns by maternal IgG acquired transplacentally. Newborns do not activate mast cells due to passive acquisition of maternal IgE because IgE cannot be transferred across the placenta.

Answer 7–10

A. Microorganisms are recognized in three different ways, each initiating a different pathway of complement activation. (1) The classical pathway is dependent on the presence of antibody bound to the surface of the microorganism, for example, IgM bound to lipopolysaccharide (LPS) of Gram-negative bacteria. (2) The lectin pathway requires the presence of mannose-binding protein, an acute-phase protein made by the liver in response to IL-6 (secreted by activated macrophages) and which accumulates in plasma during infection. (3) The alternative pathway requires an activating surface of a pathogen which stabilizes complement components.

B. Only the classical pathway is considered part of the acquired immune response because of the requirement for antibody. The other two pathways are initiated independently of antibody and require something other than antibody to be bound to, or part of, the pathogen surface (for example, LPS on bacterial surfaces, and mannose-binding protein bound to mannose-containing glycans on the surfaces of bacteria and yeast). For this reason, the lectin and alternative pathways are classified as innate immune responses.

Answer 7–11

A. The cleavage of C3 into C3a and C3b is shared by the classical, lectin and alternative pathways of complement activation.

B. The enzyme responsible for cleaving C3 into C3a and C3b is called C3 convertase and it differs in composition depending on the particular complement pathway. The classical and lectin pathways utilize the classical C3 convertase (C4b2a) while the alternative pathway uses the alternative convertase (C3bBb).

C. C3 is the most abundant complement component in the plasma and circulates as a zymogen, an inactive enzyme. When cleaved into C3a and C3b, three different effector mechanisms are armed: (1) C3b binds to and tags pathogens for destruction by phagocytes through binding to a C3b receptor, CR1; (2) C3b contributes to a multicomponent enzyme, C5 convertase, which catalyzes the assembly of the terminal complement components and the formation of the membrane-attack complex; and (3) C3a is an inflammatory mediator which serves as a chemoattractant and recruits inflammatory cells to the infection site.

Answer 7–12

B lymphoblasts that have bound specific antigen and encountered their cognate T cells in T-cell areas move from a primary focus, where they have started to proliferate, into primary follicles, which are primarily B-cell areas, where they become centroblasts—large, metabolically active, dividing cells. As centroblasts accumulate and proliferate, the primary follicle enlarges and changes morphologically into a germinal center. Centroblasts undergo somatic hypermutation while dividing in the germinal center, producing centrocytes with mutated surface immunoglobulin. Only cells with mutated surface immunoglobulin that can take up antigen efficiently through receptor-mediated endocytosis and present it to helper T cells (T_H2) will be selected to differentiate into plasma cells or memory cells. Antigen will be encountered at the surface of follicular dendritic cells as an immune complex. If B cells do not encounter their specific antigen, they will undergo apoptosis and then be ingested and cleared by tingible body macrophages. This process takes ~7 days after an infection begins, and the increase in cell numbers due to lymphocyte proliferation accounts for the swollen lymph nodes that drain an infected area.

Answer 7–13

A. Passive transfer of immunity refers to the process of transferring preformed immunity from an immune subject to a non-immune subject. This can be achieved by transferring whole serum (antiserum), purified antibody, monoclonal antibody, or intact effector or memory lymphocytes (adoptive transfer).

B. (i) IgG antibodies transported transplacentally provide passive protection in the bloodstream and extracellular spaces of tissues until the newborn can begin making its own antibodies, after which time maternal IgG levels decrease transiently. IgG1 is transported most effectively. (ii) IgA is transferred passively in breast milk and protects the newborn's gastrointestinal epithelia from colonization and invasion by microorganisms which may be ingested.

C. It is possible for autoreactive antibodies to be transferred passively to a fetus via the placenta if the isotype is IgG (see Figure 2.29). Any reaction will only persist as long as the antibodies are present. Maternal antibodies can be removed by plasmapheresis, a procedure which involves replacing blood plasma, and consequently the removal of maternal immunoglobulins, or will eventually be degraded by serum proteases.

Answer 7–14

(1) Deficiency of C3. These patients are more susceptible to bacterial infections. (2) C4 deficiencies. More than 30% of the human population is deficient for either the C4A or C4B gene, which encode two different types of C4, making this the most common human immunodeficiency. Like C3, deficiency of C4B is associated with increased susceptibility to bacterial infections. C4A deficiency is also linked to increased susceptibility to systemic lupus erythematosus (SLE). SLE is an autoimmune disease characterized by anti-DNA, anti-RNA and anti-nucleoprotein antibodies which form immune complexes that are not cleared efficiently and can cause immune complex-related complications (vasculitis, glomerulonephritis and arthritis).

Chapter 8

Answer 8–1

A. Direct pathogenic mechanisms of tissue damage include exotoxin release, endotoxin release and direct cytopathic effects.

B. Indirect pathogenic mechanisms of tissue damage include the formation of immune complexes, the production of cross-reactive antibodies that bind to pathogen and host tissue, and cell-mediated immune responses.

C. **Exotoxin release:** *Vibrio cholerae*, cholera. (Others are: *Streptococcus pyogenes* (strep throat (tonsillitis) and scarlet fever); *Staphylococcus aureus* (boils, toxic shock syndrome); *Corynebacterium diphtheriae* (diphtheria); *Clostridium tetani* (tetanus)). **Endotoxin release:** *Yersinia pestis*, plague (Others are: *Salmonella typhi* (typhoid fever); *Escherichia coli* (Gram-negative sepsis); *Haemophilus influenzae* (meningitis, pneumonia); *Salmonella typhi* (typhoid fever); *Shigella flexneri* (dysentery); *Pseudomonas aeruginosa* (wound infection)). **Direct cytopathic effects:** variola, smallpox. (Others are: varicella-zoster (chickenpox); hepatitis B virus (hepatitis); polio virus (polio); measles virus (measles); influenza virus (influenza); herpes simplex virus (cold sores)). **Immune complexes:** *Treponema pallidum*, kidney damage in secondary syphilis. (Others are: hepatitis B virus (kidney disease); *Streptococcus pyogenes* (vascular deposits, glomerulonephritis)). **Cross-reactive antibodies:** *Streptococcus pyogenes*, rheumatic fever. (Others are: *Mycoplasma pneumoniae* (hemolytic anemia)). **Cell-mediated immune responses:** *Mycobacterium tuberculosis*, tuberculosis. (Others are: *Mycobacterium leprae* (tuberculoid leprosy); *Borrelia burgdorferi* (Lyme arthritis); lymphocytic choriomeningitis virus (aseptic meningitis); *Schistosoma mansoni* (schistosomiasis); herpes simplex virus (herpes stromal keratitis)).

Answer 8–2

A. Extracellular: interstitial spaces, blood, lymph, epithelial surfaces. Intracellular: cytoplasm, membrane-bounded vesicles.

B. Virus particles outside cells are bound by antibodies, which neutralize them so that they cannot infect cells. The virus–antibody complex is ingested and destroyed by phagocytic cells. It also binds complement, which facilitates uptake by phagocytosis. Viruses inside cells are destroyed by the killing of the virus-infected cells by cytotoxic T cells and NK cells.

C. Membrane-bounded vesicles. Trypanosomes inside macrophage vesicles can be killed by activation of the macrophage through interaction with T cells or NK cells.

D. IgA antibodies. *Candida albicans*.

Answer 8–3

A. **Mechanical (physical) barriers:** Tight junctions between the epithelial cells prevent the penetration of pathogens between the cells to underlying tissues. In addition, there is a flow of air and fluid over epithelial surfaces, serving to oxygenate and flush the surface, which prevents anaerobic bacterial growth and transient adhesion. On ciliated epithelial surfaces, such as those of the respiratory tract, the formation of a layer of mucus that is kept in continual movement by the beating cilia inhibits colonization and invasion by microorganisms. **Chemical barriers:** The epithelium produces a variety of chemical substances that interfere with the adherence of microorganisms to epithelium and with their replication. The skin produces fatty acids in sebaceous glands, which helps to create an acid environment inhibitory to the growth of many bacteria. Lysozyme, an enzyme that inhibits cell-wall formation in bacteria, is secreted in tears, saliva, and sweat. The stomach produces industrial-strength hydrochloric acid, creating a highly acidic and formidable environment, which when combined with the stomach enzyme pepsin (an acid protease) poses one of the most inhospitable environments for microbial growth in our bodies. Bacteria escaping the stomach are confronted with antibacterial peptides called cryptidins, which are secreted by the Paneth cells of the epithelium of the small intestine. **Microbiological barriers:** A flora of non-pathogenic microorganisms colonizes many epithelial surfaces and provides an additional barrier to infection. They compete with pathogenic microbes for space and nutrients, and sometimes produce antibacterial proteins that further inhibit attachment to epithelium. For example, *Escherichia coli* in the large intestine produce colicins, which prevent colonization by other bacteria.

B. Antibiotics attack the microbiological barriers of intestinal epithelia. The normal flora sensitive to the antibiotics are killed off and the intestine can then be re-colonized and overgrown by microorganisms that in normal circumstances are present in very small numbers and thus do not cause a problem. An example is a condition called pseudomembranous colitis caused by the overgrowth of *Clostridium difficile*. A membrane-like substance is produced in the large intestine, causing an obstruction that can block intestinal flow and usually requires surgical removal.

Answer 8–4

A. (i) Both are phagocytic white blood cells (leukocytes) produced by the bone marrow. Both ingest extracellular pathogens and destroy them intracellularly. They both carry receptors on their surface that recognize pathogens and their components and facilitate pathogen uptake. (ii) Macrophages are resident in the tissues; neutrophils circulate in the blood and enter tissues only after an infection has become established. Macrophages are long-lived and have important functions other than the uptake and killing of pathogens: once activated by the presence of pathogens, they produce cytokines that induce an inflammatory response, help attract neutrophils to the site of infection, and help initiate an adaptive immune response. Also, once activated, they produce additional cell-surface molecules that enable them to act as professional antigen-presenting cells. In contrast, once in the tissues, neutrophils are short-lived cells. Their only function is to take up and kill extracellular microorganisms.

B. Neutrophils and macrophages ingest microorganisms by phagocytosis, taking them up into phagolysosomes, where they are destroyed by bactericidal substances produced by the phagocyte. These include toxic oxygen derivatives such as superoxide, hydrogen peroxide, singlet oxygen, hydroxyl radical, hypohalite and nitric oxide generated by NADPH-dependent oxidases, nitric oxide synthase and myeloperoxidase. Peptides called defensins and cationic proteins also inhibit microbial growth, while lysozyme inhibits cell wall formation.

Phagolysosomes have a pH of ~3.5–4.0, a hostile environment for bacteria, fungi and some enveloped viruses, and one that favors the activity of acid proteases and hydrolases, which degrade the phagocytosed material. Extracellular secretion of lactoferrin serves to compete for essential iron otherwise bound by bacterial siderophores.

Answer 8–5

Step 1: Dilation of blood capillaries, combined with the binding of neutrophil **sialyl-Lewis^x** carbohydrate to selectins, a type of **adhesion molecule**, expressed on activated endothelium, slows down neutrophils as they roll along the endothelium, binding reversibly to the endothelial selectins, a process called **rolling adhesion**. Two types of selectin are expressed on activated endothelium in response to inflammatory mediators. The first, P-selectin, is stored in preformed granules called **Weibel–Palade bodies** and is expressed soon after endothelial activation. Later, E-selectin will appear on the surface of activated endothelium in response to TNF-α or LPS. Step 2: In response to the **chemokine** CXCL8, neutrophil integrins LFA-1 and CR3 increase their affinity for the endothelial adhesion molecule ICAM-1 through conformational changes. Leukocytes are now engaged in **tight binding** to the endothelium and the rolling stops. Step 3: The neutrophil squeezes between the endothelial cells. When the basement membrane is encountered, proteases are secreted by the neutrophil, the basement membrane is degraded, and the process of **diapedesis** is complete. Step 4: Migration to the focus of infection is mediated through a chemokine (CXCL8) gradient. Neutrophils bearing IL-8 receptors will travel toward the highest concentration of IL-8 until they arrive at the focus of infection.

Answer 8–6

A.

1, 5	a. Activation of blood vessel endothelium
1, 2	b. Lymphocyte activation
1, 2, 5	c. Fever
1	d. Induction of IL-6 synthesis
5	e. Increase in vascular permeability
1	f. Localized tissue destruction
2	g. Production of acute-phase proteins by hepatocytes
6	h. Induction of resistance to viral replication
6	i. Increase in levels of MHC class I molecules on cell surfaces
4, 6	j. Activation of NK cells
4	k. Biased differentiation of naive T cells into T_H1 cells
2	l. Enhanced antibody production
3	m. Leukocyte chemotaxis
3	n. Activation of binding by β_2 integrins (LFA-1, CR3)
5	o. Septic shock
5	p. Mobilization of metabolites

B. The cytokines IL-1, IL-6, CXCL8, IL-12, and TNF-α are produced by macrophages. Type I interferons can be produced by many different types of cell when infected with a virus. Specialized cells called interferon-producing cells (IPCs) or natural interferon-producing cells (NIPCs) produce large amounts of interferon.

Answer 8–7

A. Type I interferon genes (for interferons-α and -β) are transcribed as a result of the presence of double-stranded RNA.

B. Normal cells not infected with virus do not contain double-stranded RNA; however, cells infected with virus often do. Some viruses either have double-stranded RNA genomes or use double-stranded RNA as an intermediate in the replication cycle.

C. Type I interferons (IFN-α and -β) have three major effects. First, they block virus replication in infected cells and protect uninfected cells nearby from becoming infected. This is accomplished by: (1) inducing cellular genes that destroy viral

RNA through endonuclease attack; and (2) inhibiting protein synthesis of viral mRNA by modifying initiation factors required for protein synthesis.

Second, IFN-α and -β secreted by infected cells cause an increase in the numbers of MHC class I molecules on the surface of any cell bearing receptors for IFN-α and -β. This ensures that if these cells are infected with virus, effective presentation of viral peptides to MHC class I-restricted cytotoxic T cells will occur and virus-infected cells will be recognized and killed. IFN-α and -β also induce increased transcription of the TAP transporter and LMP2/LMP7 proteasome subunit genes, helping to ensure that the peptides generated are tailored for MHC class I binding.

Finally, IFN-α and -β activate natural killer (NK) cells. NK cells will kill virus-infected cells that have fewer or no MHC class I molecules on their surface. A lack of MHC class I is often a general consequence of viral infection, in which the synthesis of host proteins is blocked, or can be caused by selective block of MHC class I export to the cell surface by some viruses.

Answer 8–8

A. The killing activity of the NK cell. Like cytotoxic T cells, NK cells can kill other cells by releasing molecules that induce apoptosis. When an activating receptor on an NK cell recognizes its ligand on the surface of a target cell, this tends to activate the killing function of the NK cell. But when an inhibitory receptor also recognizes its ligand on the target cell, this tends to inhibit the killing activity of the NK cell, even if activating receptors are also engaged. Whether the NK cell kills the target cell depends on the balance between the activating and inhibiting signals. The known ligands for the inhibitory receptors are MHC class I molecules, and in the presence of normal levels of these molecules on the target-cell surface, the cell is not killed.

B. Virus-infected cells often have reduced levels of MHC class I molecules on their surface. This feature is thought to be exploited by NK cells, which are continually monitoring levels of MHC class I molecules on host cells. When an NK cell finds a cell that lacks or has decreased MHC class I on its surface, the signals from the activating receptors predominate over those from the inhibitory receptors and the target cell is killed. Some viruses also encode proteins that mimic MHC class I molecules and interfere with NK-cell attack by ligation of the NK cells' inhibitory receptors.

C. The actions of NK cells are considered part of innate immunity because NK cells can, in principle, act against any virus-infected cell, because their killing activity is not dependent on the recognition of viral protein epitopes. Also, NK cells are already present and ready to act immediately after they encounter an infected cell. Although there are large numbers of different inhibitory and activating receptors in the NK-cell repertoires, each of these receptors is invariant, and in most cases they specifically recognize HLA allotypes and not the peptides bound. Of the three classes of inhibitory receptor found on human NK cells, two recognize HLA-A and HLA-B allotypes, irrespective of the peptide bound. The other class recognizes a complex of the invariant class I molecule HLA-E with bound peptides derived from the relatively invariant leader regions of the other class I molecules.

D. Individual inhibitory NK-cell receptors are specific for particular allotypes of a given MHC class I molecule. The NK-cell repertoire of a person seems to be tailored to their own MHC tissue type, so that all NK cells in that person will carry at least one inhibitory receptor that recognizes one of the person's own HLA class I molecules. This ensures that NK cells do not attack the healthy tissues of their own body. However, because MHC class I molecules are highly polymorphic, one person may have HLA class I allotypes that are not recognized by all the NK cells from another individual. If a tissue transplant is not matched exactly for HLA class I, therefore, some of the recipient's NK cells may not recognize the HLA class I molecules on the transplanted tissue and will attack it.

Answer 8–9

A. (i) Short-term immunological memory operates shortly after an adaptive immune response has cleared the infection in an individual and while the pathogen is still present in the community. If the individual is reexposed and reinfected, antibodies generated in the first round of infection can bind immediately to the pathogen, blocking its action by neutralization and mediating its removal and destruction by complement fixation and phagocytosis. In addition, any remaining activated T and B cells can respond straight away to the presence of antigen. Together, these activities ensure that the infection does not reestablish itself and generate a fresh supply of antibodies and effector cells. (ii) Long-term immunological memory is mediated through long-lived memory lymphocytes that are generated in the primary immune response. These are cells that can be rapidly stimulated by reexposure to the same antigen to produce a strong and effective immune response that rapidly clears the pathogen.

B. Antigen-specific memory B and T cells are present in greater numbers and can be activated more quickly than antigen-specific naive lymphocytes are in the primary response. As a result of undergoing somatic hypermutation and affinity maturation in the primary immune response, memory B cells have B-cell receptors of higher affinity for the antigen. They can thus start to make a response to lower concentrations of pathogen antigens than in the primary response. They have greater numbers of MHC class II molecules on their surface than naive B cells and are thus more efficient at presenting antigen to their cognate helper T cells. Many memory B cells will also have switched isotype. Thus, the antibody response produced in a secondary immune response starts sooner after infection, and rapidly produces large amounts of high-affinity antibody of various isotypes, which are more effective in clearing the pathogen than the low-affinity IgM produced at the start of a primary immune response.

Answer 8–10

Memory T cells express raised levels of certain cell-surface molecules compared with naive T cells, including LFA-1 and LFA-3, VLA-4 (α4 integrins), CD2, and CD44. Elevated levels of these molecules provide more efficient adhesion, homing to infected sites, activation, and signaling. Perhaps the most marked difference is seen with CD45. Memory and naive T cells express different isoforms of CD45, produced through alternative mRNA splicing. Memory T cells express CD45RO, a low molecular weight isoform that more effectively transduces signals than CD45RA, the high molecular weight isoform expressed on naive T cells.

Answer 8–11

Toxic oxygen species including superoxide, hydrogen peroxide, singlet oxygen, hydroxyl radical, hypohalite, and nitric oxide are produced during the respiratory burst in macrophages and neutrophils. Simultaneous extraphagosomal production of enzymes that neutralize these compounds occurs. Specifically, superoxide dismutase metabolizes superoxide to hydrogen peroxide, which is further metabolized by catalase to innocuous water and molecular oxygen.

Answer 8–12

A. TNF-α induces the activation of endothelial cells lining blood vessels. This in turn results in the dilation of blood vessels, increased vascular permeability, and the subsequent leakage of fluid from the blood across the endothelium and into the interstitial spaces of tissue. The fluid consists of complement components and antibodies, which are released in response to TNF-α to help to combat infection in tissues. But when large amounts of TNF-α are produced throughout the body as the result of a systemic infection, blood pressure is compromised; blood volume decreases, leading to shock associated with low blood pressure (hypovolemic shock).

B. TNF-α also triggers the release of platelet-activating factor from activated endothelium. Platelet-activating factor initiates the blood-clotting cascade, consuming blood-clotting factors and causing localized blood vessel blockage at sites of TNF-α production by resident macrophages in various organs, for example the Kupffer cells of the liver and splenic macrophages. When this occurs on a large scale, it is known as disseminated intravascular coagulation.

C. The blood supply to organs is choked off as a result of disseminated intravascular coagulation, leading to organ failure.

D. As blood-clotting factors are consumed by disseminated intravascular coagulation, they are not available elsewhere to help repair broken blood vessels. As a result, hemorrhaging occurs, causing petechiae to develop, which present as small purple spots under the skin.

Answer 8–13

A. The phenomenon of original antigenic sin is related to the observation that B-cell secondary adaptive immune responses involve only those B cells that participated in the primary immune response, that is, the memory B lymphocytes. Naive lymphocytes are suppressed by antibody made in the primary immune response. This latter observation has been exploited to prevent hemolytic anemia of the newborn.

Hemolytic anemia of the newborn results when a Rh-negative mother carrying a Rh-positive fetus makes an immune response against the Rh antigens on the fetal red blood cells and produces anti-Rh IgG antibodies. These maternal IgG antibodies can cross the placenta during pregnancy, bind to fetal red blood cells, and target them for uptake and destruction via macrophages; this causes a hemolytic anemia, which in severe cases leads to the death of the fetus *in utero* or soon after birth. Maternal anti-Rh IgG antibodies are made if the mother has been exposed to Rh-positive red blood cells that entered her circulation, for example during the delivery of a previous Rh-positive baby or an incorrect blood transfusion. The condition can be prevented by preventing the mother from making any immune response to her first and subsequent exposures to Rh antigen. This is now done routinely by infusing Rh-negative mothers with anti-Rh antibodies (RhoGAM) during late pregnancy and shortly after delivery of the baby. RhoGAM will bind to any fetal red blood cells that enter the maternal circulation, forming an immune complex. Maternal naive B cells bearing Fc receptors for IgG will bind to the immune complex. If the B-cell receptor also binds to Rh antigen, a negative signal will be delivered and the naive B-cell response to this and any subsequent encounters with the antigen will be suppressed. The anti-Rh antibody prophylaxis is effective only if administered before the mother has made an anti-Rh immune response and made memory B cells, as Rh-reactive memory cells are not suppressed by exposure to anti-Rh antibody.

B. In this situation, the two receptors that need to be cross-linked on naive B cells to suppress their differentiation into antibody-producing cells are the FcγRIIB-1 and the B-cell receptor.

Chapter 9

Answer 9–1

A. Some pathogens exist in a number of antigenically different strains known as serotypes. Antibodies and memory B cells generated during an infection with one serotype will protect the host from reinfection with the same serotype, but will not protect against a first infection with a different serotype of the same pathogen in which different epitopes are expressed. Thus, the immunity generated is serotype-specific.

B. *Streptococcus pneumoniae* has evolved at least 90 different serotypes that differ in their capsular polysaccharide antigens. A new infection with a different serotype provokes a new primary response rather than the more effective secondary immune response. This is advantageous to the pathogen because it prolongs the period of survival in the host, and hence increases the likelihood of transmission to a new host.

Answer 9–2

A. Influenza hemagglutinin (HA) and neuraminidase (NA) bear the main epitopes against which protective antibodies are made.

B. Antigenic drift is due to the frequent point mutations in the RNA genome of influenza virus. Mutants that acquire changes in the HA and NA epitopes are selected because they are not subject to the serotype-specific immunity generated against the original strain. Because point mutation causes only relatively small changes at a time, this process is known as antigenic drift.

Antigenic shift is due to recombination between a human influenza virus and an avian influenza virus, leading to replacement of the human NA and/or HA with the avian type. The genome of the influenza virus is composed of eight separate pieces of RNA. When a human influenza virus and an avian influenza virus infect the same cell (usually in domestic livestock such as pigs, ducks or chickens), reassortment of the RNAs can lead to the generation of a new human virus encoding the avian HA and/or NA which has never before been encountered by the human population.

C. Worldwide pandemics are caused by influenza viruses that have undergone antigenic shift, because no human population has any immunity to them. In antigenic drift, where the changes are more gradual, a human population will contain some people who are immune to some epitopes and some to others. Epidemics due to antigenic drift are usually relatively mild and limited.

D. The phenomenon of 'original antigenic sin' means that when a person is reinfected with a new strain of influenza virus that shares some epitopes with the virus they were first infected by, they will make an immune response (which will be a secondary immune response) only to the shared epitopes and not to the new epitopes. This is because when the virus binds via a new epitope (say HA*) to the B-cell receptor of a HA*-specific naive B cell, preexisting IgG antibodies against other epitopes on the virus can bind via their Fc regions to FcγRII-B1 receptors on the surface of the same cell. Cross-linking of the B-cell receptor and Fc receptor delivers an inhibitory signal to the B cell, thus blocking the production of a HA*-specific antibody response. As an influenza virus undergoes antigenic drift, some HA and NA epitopes change while others remain the same, so the person retains some immunity. However, when a completely new HA or NA is produced as a result of antigenic shift, and it shares no epitopes with the original antigens, the body sees the virus as a completely new infection, and disease results while the body is making a new primary response against the virus.

Answer 9–3

The periodic rise and fall in the number of parasites is caused by a phenomenon known as antigenic variation in the trypanosomes. Trypanosomes express surface antigens called variable surface glycoproteins (VSGs) of which there are over a 1000 different types encoded in the trypanosome genome. Only one VSG gene is expressed at any given time, that gene being placed in the single 'expression site.' Through a process known as gene conversion, a different VSG gene can relocate (rearrange) to the expression site, replacing the original gene, and will be expressed instead. This process occurs at random in the life cycle of trypanosomes. Antibodies made by the host against the VSG expressed on the infecting trypanosomes will start to suppress the infection and parasite numbers will fall. But if some parasites switch VSG, these antibodies will be ineffective, and parasites bearing the new VSG will multiply. Parasite numbers will rise for a time until the new immune response begins to suppress them in turn. By then a third VSG gene is likely to have been expressed, and parasites bearing that one will begin to predominate in the host, and so the cycle continues.

Answer 9–4

7	a.	Variant pilin protein expression
10	b.	Induction of quiescent (latent) state in neurons
6	c.	Reactivation of infected ganglia after stress or immunosuppression
3	d.	Alternate expression of two antigenic forms of flagellin
4	e.	Recombination of RNA genomes of avian and human origins
9	f.	Escape from phagosome and growth and replication in cytosol
2	g.	Survival in a membrane-bound vesicle resistant to fusion with other cellular vesicles
8	h.	Coating its surface with human proteins
5	i.	Inhibiting fusion of phagosome with lysosome and survival in the host cell's vesicular system
1	j.	Immunosuppression caused by nonspecific proliferation and apoptosis of T cells

Answer 9–5

A. Human immunodeficiency virus (HIV) infects cells bearing CD4 molecules on their surface. These include T_H1 and T_H2 cells, macrophages and dendritic cells. CD4 acts as a receptor involved in binding the gp120 envelope glycoprotein of HIV. A co-receptor is also required for virus entry. Two are used by HIV—the chemokine receptors CCR5 and CXCR4. CCR5 is expressed on all CD4$^+$ cells, and CXCR4 is restricted to T cells. After binding to the co-receptor, another viral envelope glycoprotein, gp41, facilitates fusion between the host-cell plasma membrane and viral envelope with the release of viral components into the cytoplasm.

B. The cell tropism of HIV depends on which co-receptor is used for entry. Macrophage-tropic HIV, associated with early infection, uses the CCR5 co-receptor. Lymphocyte-tropic HIV, associated with a phenotypic change in late infection in about 50% of cases, uses CXCR4.

C. About 1% of the caucasoid population is homozogous for a mutant form of CCR5 that cannot be used by HIV as a co-receptor. This renders such individuals resistant to primary infection by macrophage-tropic variants of HIV.

Answer 9–6

A. Seroconversion is the point during an infection when antibodies specific for the pathogen can first be detected in the blood serum. Antibody is produced when HIV virions are being released from infected cells and are thus accessible to B-cell antigen receptors, a period referred to as acute viremia. During clinical latency the level of infectious virus in plasma decreases markedly, but does not disappear.

B. Immediately after seroconversion the levels of free virus fall and T-cell numbers rise and for a time the infection is held in check. Eventually, however, virus levels start to rise again and T-cell numbers decrease, leading eventually to AIDS. The level of virus in the blood after seroconversion is positively correlated with the severity and progression of the disease: the more virus that remains at this time, the more rapid is the progress to AIDS.

Answer 9–7

A. 1. X-linked agammaglobulinemia. (i) No antibody at all. (ii) It is caused by a defect in the tyrosine kinase Btk, which is necessary for B-cell development and is encoded on the X chromosome. No mature B cells develop. 2. X-linked hyper-IgM syndrome. (i) Large amounts of IgM antibody, but no antibodies of isotypes other than IgM are produced. Virtually no antibodies are made against T-cell dependent antigens. (ii) It is caused by a defect in the molecule CD40 ligand, which is encoded on the X chromosome and is expressed on T cells. T cells lacking CD40 ligand cannot give help to B cells, which thus cannot respond to most protein antigens and cannot switch isotype.

Other diseases in which defects in antibodies seem to be the main deficiency are: common variable immunodeficiency (defective antibody production; cause unknown); selective IgA and/or IgG deficiency (no IgA or IgG synthesis; cause unknown).

B. Defective antibody responses lead to increased susceptibility to extracellular bacteria and to some viruses.

Answer 9–8

A. A lack of C3 or C4 function means that immune complexes of antibody and antigen do not bind C3 or C4 and thus cannot bind to the complement receptors on phagocytes that facilitate their clearance from the circulation.

B. Immune complexes accumulate in the circulation and are deposited in tissues. They damage tissues directly and also activate phagocytes (via the binding of antibody Fc regions to Fc receptors), causing inflammation and further tissue damage.

C. Increased susceptibility to bacteria of the genus *Neisseria*. C5–C9 are required to form the membrane-attack complex on the bacterial membrane that leads to bacterial lysis. An absence of any of these components means that the complex cannot be formed. Complement-mediated lysis seems to be an important means of protection against *Neisseria*.

Answer 9–9

A. Chronic granulomatous disease. Chédiak–Higashi syndrome. Leukocyte adhesion deficiency. Others are: G6PD deficiency and myeloperoxidase deficiency.

B. Chronic granulomatous disease is caused by defects in phagocyte NADPH oxidase. Phagocytes with this defect cannot produce superoxide radicals and are less effective at intracellular killing of ingested bacteria. Macrophages infected with bacteria that they cannot kill form the granulomas characteristic of this condition.

C. Persistent infections with bacteria, especially capsulated bacteria, and fungi.

Answer 9–10

A. Severe combined immune deficiency (SCID), characterized by an almost complete absence of T-cell and B-cell function and the inability to make any effective immune responses. Because the child cannot make functional receptors for these important cytokines, their lymphocytes cannot respond to them. The inability of naive T cells to respond to IL-2 in particular blocks the proliferation and differentiation of T cells and thus all cell-mediated immune responses and T-dependent antibody responses. Without treatment, children with SCID die early in infancy from common bacterial or viral infections.

B. Jak3 kinase is part of the intracellular signaling pathway that leads from activated cytokine receptors. Thus, a lack of Jak3 kinase also leads to SCID.

C. Reconstitution of a functioning immune system by bone marrow transplantation from a healthy donor can treat those types of SCID in which the genetic defect is intrinsic to lymphocytes or other bone marrow-derived cells.

Answer 9–11

HIV is a retrovirus that uses the enzyme reverse transcriptase to complete its life cycle. This enzyme uses the single-stranded RNA genome of HIV as a template to make a double-stranded DNA (complementary DNA or cDNA) that integrates into the host genome as a provirus. Reverse transcriptase is an error-prone DNA polymerase that lacks proof-reading capabilities. Mutations are therefore introduced into the HIV genome each time it is replicated. Such rapid mutation leads to variant viruses with modified antigens, which thus escape detection by antibodies or cytotoxic T cells specific for the original epitopes.

The high rate of mutation in HIV also causes the emergence of resistance to the antiviral drugs used to treat HIV infection, which are mainly drugs that inhibit viral reverse transcriptase and protease. Mutation produces viruses with mutant enzymes that are not blocked by the drugs. Multi-drug regimes are used to try to eliminate the virus before the multiple mutations needed to resist all of the drugs have accumulated.

Answer 9–12

Some staphylococci produce toxins that act as superantigens, for example, the staphylococcal enterotoxins and toxic-shock syndrome toxin-1. Superantigens can bind simultaneously to MHC class II molecules and to T-cell receptors that possess certain types of V_β region, resulting in activation of the T cell carrying such a receptor. Superantigens thus nonspecifically activate a large proportion of the $CD4^+$ T-cell repertoire, resulting in the production of large quantities of cytokines from the activated T cells and from macrophages activated by these cells. The systemic production of cytokines (for example, TNF-α) causes increased vascular permeability and vasodilation throughout the body, leading to loss of fluid into the tissues, rapid collapse of the circulatory system (systemic shock), and organ failure. Toxic shock therefore requires immediate medical attention, primarily to replace the intravascular fluid.

Answer 9–13

Toxic-shock syndrome is caused when superantigens (such as toxic-shock syndrome toxin-1 (TSST-1) produced by some strains of *Staphylococcus*) nonspecifically activate large numbers of $CD4^+$ T cells. When a T cell is activated, the transcription factor NFκB is activated and switches on genes involved in the normal T-cell response. However, NFκB can also bind to the promoter of the HIV provirus to initiate its transcription and subsequent viral replication. Thus, if a T cell activated during toxic shock coincidentally contains an HIV provirus, NFκB will initiate the replication of HIV and the cell will die. If a significant proportion of the person's T cells harbor the provirus, one would predict widespread reactivation of the virus and a sharp rise in the amount of HIV in the blood (viremia) after the onset of toxic shock.

Chapter 10

Answer 10–1

(1) Inhalation: house dust mite feces, animal dander. (2) Injection: drugs administered intravenously, wasp venom. (3) Ingestion: peanuts, drugs administered orally. (4) Contact with skin: poison ivy, nickel in jewelry.

Answer 10–2

A. (i) Type I hypersensitivity reactions involve IgE, whereas type II and type III use IgG. (ii) Type I, type II and type III hypersensitivity reactions involve humoral immune responses (antibodies and complement), whereas type IV involves cell-mediated immune responses.

B. (i) Soluble antigen is involved in type I and type III hypersensitivity reactions, whereas cell-surface or matrix-associated antigens or receptors are involved in type II hypersensitivity reactions. (ii) T_H1- and T_H2-dependent type IV hypersensitivity reactions involve soluble antigen, whereas type IV reactions depending on cytotoxic lymphocytes involve cell-associated antigen.

Answer 10–3

A. If an individual becomes sensitized to antigen by making antibodies of the IgE isotype during first exposure, then a type I

hypersensitivity reaction may result if antigen is encountered again. IgE made initially binds to and is stabilized on very high-affinity FcεRI receptors on mast-cell surfaces, a tripartite Fc receptor made up of α, β and γ chains ($\alpha\beta\gamma_2$). The α chain of FcεRI binds to IgE antibody, and when antigen binds to IgE, cross-linking of FcεRI occurs, delivering an intracellular signal in the mast cell via the β- and γ- chain components.

B. Mast cells contain preformed granules containing a wide range of inflammatory mediators that are triggered to be released extracellularly though an exocytic mechanism called degranulation. The inflammatory mediators contained in the granules and released immediately include histamine, heparin, TNF-α, and proteases involved in the remodeling of connective tissue matrix. The proteases include tryptase and chymotryptase (expressed by mucosal and connective mast cells, respectively), cathepsin G, and carboxypeptidase. Additional inflammatory mediators are generated after mast-cell activation, including IL-3, IL-4, IL-5 and IL-13, GM-CSF, CCL3, leukotrienes C4 and D4, and platelet-activating factor.

Answer 10–4

A. In allergic individuals penicillin typically causes a type II hypersensitivity reaction. Such reactions are mediated by IgG antibodies directed towards cell- or matrix-associated antigens. Production of IgG antibody requires isotype switching, which is dependent on help from T_H2 cells. T_H2 cells cannot be activated by non-protein drug molecules such as penicillin. They can, however, become activated by cell-surface proteins that have been modified by penicillin and thus generate foreign epitopes to which the host is not tolerant.

B. Penicillin binds to surface proteins of red blood cells by forming a covalent bond through the highly reactive bond of the β-lactam ring of penicillin. Meanwhile, the bacterial infection for which the drug had been administered has activated the alternative pathway of complement. As a side-effect, red blood cells become coated with C3b through the action of alternative C3b convertase (C3bBb). Red blood cells coated with penicillin and the opsonin C3b are ingested by macrophages bearing receptors for C3b (CR1). Macrophages process the penicillin-bound red blood cell proteins and present penicillin-modified protein epitopes to T_H2 cells. T_H2 effector cells are now available to provide help to penicillin-specific B cells. Differentiation of penicillin-specific B cells to plasma cells producing penicillin-specific antibodies results in the uptake of penicillin-coated red blood cells by B cells through receptor-mediated endocytosis; antigen processing of penicillin-modified red blood cell surface proteins; and presentation of antigen to effector T_H2 cells. T_H2 cytokines and CD40 ligand will direct isotype switching and the subsequent production of anti-penicillin IgG .

   IgG antibodies bind to penicillin-coated red blood cells and activate the classical pathway of complement, leading to destruction of penicillin-coated red blood cells through opsonization (FcγR and CR1 on macrophages) and lysis mediated through formation of the membrane-attack complex.

C. If administered in large doses, penicillin can cause a form of 'serum sickness,' a type III hypersensitivity reaction in which large amounts of immune complexes of penicillin-modified proteins and antibodies are deposited in tissues, causing a widespread inflammatory response that leads, for example, to rashes, fever, chills, vasculitis and sometimes glomerulonephritis. Penicillin can cause serum sickness even in a non-allergic individual.

Answer 10–5

Farmer's lung is a type III hypersensitivity reaction caused by the repeated inhalation of large quantities of hay dust or mold spores to which an IgG response is mounted. Small immune complexes form in the presence of excess antigen and limiting antibody. Instead of being cleared from the circulation, as are large immune complexes, the small immune complexes become deposited at the alveolar capillary interface. These deposits result in complement activation and inflammation leading to the accumulation of exudates, which compromises gas exchange and causes difficulty in breathing and long-term irreversible lung damage.

Answer 10–6

Type III hypersensitivity reactions differ from type I in several important ways. First, IgG, not IgE, antibody is secreted by activated B cells after isotype switching, a process regulated by T-cell cytokines; for example, IFN-γ favors IgG whereas IL-4 favors IgE. Second, large amounts of antigen are required to form the small immune complexes that become deposited at various anatomical sites such as blood vessel walls, renal glomeruli, joint spaces, perivascular sites, and the walls of the alveoli. Type I reactions, in contrast, can be provoked with small amounts of allergen. Third, unlike type I reactions, type III reactions do not cause systemic anaphylaxis. Finally, type III reactions are often self-limiting and dissipate as antibody titer increases, leading to the formation of large immune complexes which are cleared efficiently from the circulation. Type I reactions, however, are exacerbated by elevated IgE levels and sensitize the individual for subsequent exposure to allergen.

Answer 10–7

A. The tuberculin test is used to determine whether someone has been infected with *Mycobacterium tuberculosis*. Protein derived from *M. tuberculosis* is injected subcutaneously. Dendritic cells and macrophages nearby process and present the mycobacterial protein. If memory or effector T cells against *M. tuberculosis* are present in the circulation from a previous or existing infection, these will be activated if they enter the site of injection. Activated T cells produce cytokines that activate vascular endothelium and initiate a local inflammatory reaction that eventually produces a raised red lesion at the site of antigen injection. The lesion may take 1–3 days to develop because the number of preexisting antigen-specific memory or effector T cells will be low, and because they will not be specifically attracted to the site of injection by a preexisting inflammatory reaction. This is why the reaction is referred to as delayed-type hypersensitivity. Once effector T cells are generated, these produce cytokines, which induce the activation of vascular endothelium at the site of injection. This results in increased vascular permeability and blood cell chemotaxis, culminating in edema and infiltration of cells into tissues, and, finally, the formation of a visible skin lesion.

B. The tuberculin test is a type IV delayed-type hypersensitivity reaction induced in sensitized individuals.

C. If an individual has been immunized against *M. tuberculosis*, with the BCG vaccine used commonly in the United Kingdom, for example, then a positive skin lesion will form after the injection of mycobacterial protein. The memory T cells cannot discriminate between vaccine proteins and proteins encountered during a genuine infection, and in these circumstances the tuberculin test is of no value in determining whether the individual has been naturally infected or not. In the United States the diagnostic value of the tuberculin skin test for existing infection currently outweighs the benefits of vaccination against tuberculosis and hence vaccination is not routine.

Answer 10–8

A. Pentadecacatechol is a low-molecular-weight lipid-like molecule found in the leaves and roots of poison ivy. Its lipid-like nature enables it to penetrate the epidermal layer of the skin readily and to cross the lipid bilayer of the cell plasma membrane where it forms covalent bonds with intracellular proteins in the cytosol. Chemically modified intracellular proteins are processed via the cytosolic pathway involving proteasomes.

B. These antigenic peptides are presented to cytotoxic T cells via MHC class I molecules. Activated cytotoxic T cells kill all cells

presenting these peptides on their cell surface, causing the characteristic skin lesions of poison ivy rash.

Answer 10–9

A. When smooth muscle cells bind histamine using H1 receptors, they contract. In combination with increased mucus production by mucosal epithelium this produces a variety of effects, for example, wheezing due to bronchial constriction, coughing, sneezing, watery eyes, nasal discharge, itchiness, and, if the reaction occurs in the gut, vomiting and diarrhea. When endothelial cells of blood vessels bind histamine, an increase in vascular permeability enables the entry of fluid (edema) and leukocytes into affected tissues.

B. Antihistamines are used as pharmacological inhibitors of histamine binding to H1 receptors on vascular endothelium. This prevents elevated vascular permeability and thus decreases rhinitis and fluid accumulation in the respiratory tract. If the allergen is delivered to the skin by the bloodstream, antihistamines will also prevent urticaria (hives).

Answer 10–10

A. Desensitization involves the deliberate subcutaneous injection of allergen into sensitized individuals with the aim of skewing the immune response from an IgE to an IgG isotype. This is achieved by gradually increasing the subcutaneous allergen concentration over time, which favors IgG over IgE production. When antigen is encountered subsequently, IgG will compete with IgE for binding and inhibit IgE cross-linking on mast-cell surfaces.

B. A potential risk of administering allergy shots is the possibility of activating a systemic anaphylactic response after mast-cell activation. Because the patient was previously sensitized, IgE antibodies against allergen are present and, if bound to mast cells, will induce mast-cell degranulation. In the event of an anaphylactic response to an allergy shot, the patient will have epinephrine administered immediately by the attending physician or nurse practitioner.

Chapter 11

Answer 11–1

A. Autoimmune diseases are distinguished on the basis of the type of immune reaction that is responsible for the disease. Three types of autoimmune response parallel the effector mechanisms described in Chapter 10 for hypersensitivity reactions type II, III, and IV. The same system used for differentiating hypersensitivity reactions is also used for autoimmune diseases.

Specifically, type II autoimmunity is caused by antibodies directed against cell-surface or extracellular matrix self antigens. Type III autoimmunity is the result of deposition of small, soluble immune complexes in tissues. The third type of autoimmune disease that corresponds to a hypersensitivity reaction is type IV, mediated by effector T cells.

B. There is no type I autoimmune disease in this categorization because IgE is not involved in autoimmunity and hence no autoimmune disease corresponds to a type I hypersensitivity reaction.

Answer 11–2

a. Rheumatoid arthritis	7, 11	IV
b. Subacute bacterial endocarditis	4, 13	III
c. Autoimmune hemolytic anemia	8, 10	II
d. Mixed essential cryoglobulinemia	5, 17	III
e. Multiple sclerosis	1, 18	IV
f. Systemic lupus erythematosus	2, 16	III
g. Insulin-dependent diabetes mellitus	9, 12	IV
h. Graves' disease	3, 14	II
i. Pemphigus foliaceus	6, 15	II

Answer 11–3

(1) Red blood cells coated with anti-red blood cell antibodies bind to splenic macrophages via Fc receptors, inducing phagocytosis of the red blood cell by receptor-mediated endocytosis. (2) Red blood cells bound by anti-red blood cell antibodies fix complement, which causes deposition of C3b on the red blood cell surface. C3b then binds to the receptor CR1 on splenic macrophages, inducing red blood cell phagocytosis. (3) Anti-red blood cell antibody triggers the complement cascade and the formation of membrane-attack complexes on the red blood cell, leading to cell lysis.

Answer 11–4

Antibodies and complement bound to neutrophil surfaces do not impair the normal function of these cells. In addition, white blood cells are far less susceptible to the formation of membrane-attack complexes, owing to continual production of complement regulatory proteins in these nucleated cells. Hence, the mechanism that destroys the antibody-targeted neutrophils is phagocytosis by splenic macrophages via Fc receptors and the complement receptor CR1. Splenectomy would reduce the rate at which neutrophils are destroyed because splenic macrophages are the major site of antibody and complement-coated cell clearance.

Answer 11–5

First, endocrine glands synthesize tissue-specific proteins unique to that gland. These proteins are not normally found in primary lymphoid organs where lymphocyte maturation occurs. Hence, the population of T and B lymphocytes is not tolerant to some endocrine gland-specific proteins, and these self proteins are thus recognized as foreign antigens. Second, endocrine glands are highly vascularized because their products need to gain access to the circulation. This feature gives leukocytes relatively easy access to endocrine tissue.

Answer 11–6

Both Hashimoto's and Graves' diseases disrupt normal production of the thyroid hormones tri-iodothyronine (T3) and thyroxine (T4), derived from thyroglobulin in thyroid follicles. Formation of T3 and T4 requires engagement of the thyroid-stimulating hormone receptor (TSHR) with thyroid-stimulating hormone (TSH) secreted from the pituitary gland, a key step in the regulation of thyroid hormone production. This step malfunctions for these two diseases.

In Hashimoto's disease, anti-thyroid antigen antibodies and T_H1 effector cells are involved. Large numbers of lymphocytes take up residence in the gland tissue, establishing germinal centers that resemble those in lymph nodes. Eventually the thyroid tissue is destroyed and thyroid follicles are no longer able to respond to TSH and make T3 or T4, a condition called hypothyroidism.

Graves' disease, in contrast, results in hyperthyroidism. Anti-TSHR antibodies act agonistically mimicking TSH, even in the absence of TSH. The thyroid follicle is chronically overstimulated by these antibodies and overproduces T3 and T4. The effector T cells are of the T_H2 type and the absence of lymphocyte infiltration retains the thyroid gland in operable condition. Therefore T3 and T4, no longer regulated by TSH, are secreted continuously in excess of concentrations required by the body.

Answer 11–7

Hashimoto's and Graves' diseases are treated differently because, although they both affect the thyroid gland, they exert opposing effects on the production of thyroid hormones—inhibition and overproduction, respectively. Hashimoto's disease is treated by administering synthetic thyroid hormones orally to replace the deficiencies in T3 and T4. Graves' disease is treated either by suppressing thyroid function with inhibitory drugs, or by thyroidectomy combined with thyroid hormone replacement therapy.

Answer 11–8

A. The negative selection of developing autoreactive T cells in the thymus.

B. The underlying genetic defect is in a gene encoding a protein called AIRE (autoimmune regulator). This is a transcription factor that, when working normally, causes several hundreds of proteins otherwise expressed only in particular peripheral tissues to be expressed by medullary epithelial cells in the thymus. This induces the negative selection of T cells specific for these proteins and their deletion from the T-cell repertoire. The T cells emerging from the thymus are therefore tolerant to a large number of antigens found primarily on organs and tissues elsewhere in the body. When AIRE is defective and these proteins are not expressed in the thymus, the population of naive T cells that develops will contain T cells reactive against these antigens of peripheral tissues.

Answer 11–9

These cells are called regulatory CD4 T cells (T_R). When activated by encounter with their corresponding self antigen they become able to suppress the activation of naive autoreactive T cells. This active suppression of autoreactive T cells in the periphery is now thought to be an important method of preventing autoimmune reactions.

Answer 11–10

A. Genes of the HLA complex, specifically the HLA class I and class II genes. Associations with HLA class II genes are most common. Various alleles of these genes are associated with a higher or lower susceptibility to particular autoimmune diseases compared with the incidence of the diseases in the population as a whole.

B. The polymorphic HLA genes encode the proteins that present peptide antigens to T cells. It has been proposed that particular alleles of these genes are associated with particular autoimmune diseases because of their ability to present the required peptide epitope(s) to autoreactive T cells. Associations with HLA class II genes are more common than those with HLA class I because CD4 T cells rather than CD8 T cells are most commonly involved in autoimmunity.

Answer 11–11

The condition is precipitated by food antigens—the gluten proteins in wheat, barley, and rye—which makes it similar to a hypersensitivity reaction. But, unlike other common food allergies, the condition does not involve sensitization to the antigen and production of IgE antibodies. Instead, the disease is considered to result from a loss of the normal oral tolerance to gluten. The inflammatory reaction in the gut could be caused both by the indirect effects of the reaction to gluten (as in a type IV hypersensitivity reaction) and by autoimmune reactions mounted against gut tissue antigens. As in a hypersensitivity reaction, T_H1 cells specific for peptides from the gluten proteins are found in the inflammatory lesions. However, autoantibodies against a tissue enzyme—tissue transglutaminase— are also produced.

Answer 11–12

A. Molecular mimicry refers to the phenomenon in which a pathogen expresses an antigen that bears a chemical similarity to a host-cell antigen. Once pathogen-specific antibodies or effector T cells are generated, they have the potential to cross-react with self antigen.

B. An autoimmune disease involving molecular mimicry is rheumatic fever. Infection with *Streptococcus pyogenes* (for example 'strep throat') results in the production of antibodies specific for the bacterial cell-wall proteins. These antibodies cross-react with chemically similar (but not identical) self antigen expressed on heart tissue. This is followed by complement activation and production of inflammatory mediators, which cause damage to heart tissue and valves and the formation of scar tissue, which can lead to cardiovascular complications later in life. This type of immunological aftermath can be avoided if antibiotics are administered early during infection.

Answer 11–13

A. Epitope spreading is the phenomenon in which the immune response, having initially targeted a particular epitope on an antigen molecule, progressively involves other cross-reactive epitopes on the same molecule.

B. Pemphigus foliaceus. Autoantibody production against the protein desmoglein, a component of desmosomes, initially involves epitopes that do not result in any tissue damage or symptoms. Only when additional desmoglein epitopes become involved, as a result of epitope spreading, does the autoimmune response cause damage to desmosomes and the resulting lesions in the skin (the disease pemphigus).

Chapter 12

Answer 12–1

A. Inactivated virus vaccines are made of virus particles that are not able to replicate because they have been chemically or physically treated (for example, heat treatment) in a way that inactivates the nucleic acid. Examples: Salk polio vaccine, rabies vaccine, influenza vaccine.

B. Live-attenuated virus vaccines are made of viruses that have lost their pathogenicity and ability to reproduce efficiently in human cells through mutations accumulated as a result of growing the virus in non-human cells. Examples: Sabin polio vaccine (oral), measles vaccine, mumps vaccine, yellow fever vaccine, rubella vaccine, varicella vaccine.

C. Subunit vaccines are composed only of particular antigenic pathogen components known to induce protective immune responses. Recombinant DNA technology enables the production of antigenic proteins in the absence of other pathogen gene products. Examples: hepatitis A vaccine, hepatitis B vaccine, pertussis vaccine.

D. Toxoid vaccines are made from chemically inactivated toxins purified from pathogenic bacteria. Toxin activity is eliminated but not antigenic activity, so an immune response is generated in the absence of pathological damage. Examples: diphtheria vaccine, tetanus vaccine.

E. Conjugate vaccines are made by covalently coupling antigenic polysaccharide found in bacterial capsules to a carrier protein (often a toxoid). This converts the otherwise T-independent bacterial polysaccharide antigen into a T-dependent antigen. T cells respond to an epitope on the protein carrier, whereas B cells respond to epitopes on the polysaccharide portion of the conjugate. This ensures that T-cell help is provided to B cells making anti-capsule antibodies. Example: *Haemophilus influenzae* type B vaccine (HIB)

Answer 12–2

A. Some potential antigens will evoke a very weak or no immune response when injected into an animal or human in their pure form. Immunogens, in contrast, are antigens that are able to evoke a strong immune response when introduced into an animal or human.

B. Adjuvants are substances that when mixed with any antigen increase its immunogenicity. They are commonly used in experimental immunology and in vaccines. Adjuvants work by inducing a nonspecific antigen-independent inflammation, which helps to drive the immune response forward. Adjuvants delay the release of antigen at the injection site, and prevent its rapid clearance from the body, by converting soluble antigen into particulate material. The most effective adjuvants also activate Toll-like receptors on macrophages and tissue dendritic cells, which then produce inflammatory cytokines, chemokines, and co-stimulatory molecules (B7) that help drive the immune response to the antigen itself.

C. The following adjuvants are used in humans: alum, a form of aluminum hydroxide; MF59, a squalene–oil–water emulsion; and bacterial components included as part of some vaccines, for example, whole *Bordetella pertussis* as part of the DTP (diphtheria, tetanus, and pertussis) vaccine.

Immune stimulatory complexes (ISCOMS) are non-toxic adjuvants that might one day be used in humans. They are composed of lipid micelles that fuse with plasma membranes and deliver proteins enclosed within the micelles to the cytosol of antigen-presenting cells. Other powerful adjuvants, such as Freund's complete adjuvant, are not used in humans because of their adverse side-effects.

Answer 12–3

Live-attenuated virus vaccines are mutant viruses that can replicate, albeit inefficiently, in human cells, thus simulating conditions of a normal viral infection. The attenuated vaccine strains of virus have been obtained by growing the virus over many generations in non-human cells (for example, monkey cells), so that it acquires multiple mutations that allow it to replicate but prevent it from spreading in the human body and causing disease. When introduced into humans as a vaccine, there is small chance that some or all of the mutations may revert to the original nucleotide sequence, restoring the properties of the virulent strain of the virus. Very rarely, this occurs with the poliovirus used in the trivalent oral polio vaccine (TVOP) and, now that polio is very rare in the United States, this vaccine is no longer recommended and an inactivated polio virus vaccine is used instead.

The more rounds of replication the vaccine virus undergoes in the human host before being contained by the immune response, the greater is the potential for genetic reversion. This is why individuals who suffer from inherited or acquired immunodeficiencies should never receive live-attenuated virus vaccines.

Answer 12–4

Acute rejection is due chiefly to immune responses made by the recipient's T cells against HLA class I and II molecules of the graft that are different from those of the recipient and which the recipient's immune system perceives as 'foreign.' The differences can be due to the HLA molecules, the self-peptides they bind or both. Transplantation between identical twins and transplantation of autografts are the only situations where the graft and the recipient are genetically identical and there are no differencs in either the HLA molecules or the bound peptides. In these situations graft rejection does not occur. Transplantation between donors and recipients who have identical HLA class I and II molecules, usually HLA-identical siblings, almost always involves differences in the peptides that are bound by the HLA molecules. These differences trigger peptide-specific alloreactive T cells to cause graft rejection through the direct pathway of allorecognition. Although it is possible to match donor and recipient for many HLA class I and II allotypes, in practice most clinical transplants involve one or more mismatched HLA loci. For these differences in HLA type, alloreactive T-cell clones activated by either the direct or indirect pathway of allorecognition cause graft rejection. Destruction of the grafted organ is effected through a type IV delayed-type hypersensitivity response.

Answer 12–5

A. The organ would be rejected immediately by the process of hyperacute rejection as a result of the presence in the recipient's blood of preformed antibodies against the A and B blood group antigens present on the tissues of the graft. Such antibodies are made early in life as a result of exposure to common bacteria that carry surface carbohydrates similar to those on human cells. A person of O blood group would have made antibodies against bacterial 'A' and 'B' antigens, because the person does not have these antigens on their own cells and is thus not tolerant to them. These preexisting anti-A and anti-B antibodies in the recipient's blood will immediately attack the endothelium of blood vessels throughout the transplant, which expresses the A and B blood group antigens. Blood vessels become occluded through thrombus formation. The graft is deprived of oxygen and becomes engorged with blood hemorrhaging from leaky blood vessels. Hyperacute rejection occurs almost immediately after transplantation and cannot be treated once it has started.

B. Other examples of combinations that will induce hyperacute rejection: O recipient, A donor; O recipient, B donor; A recipient, B donor; B recipient, A donor; A recipient, AB donor; and B recipient, AB donor.

C. A recipient's preformed antibodies against an HLA class I antigen expressed on the endothelial cells of the transplant can also cause hyperacute rejection. Such antibodies can be generated in pregnancies in which the fetus expresses a paternal HLA allotype different from the maternal HLA allotype. These antibodies can also arise from HLA-incompatible blood transfusions or previous transplants.

D. Hyperacute rejection can be prevented by typing and cross-matching donor and recipient for the A,B,O blood groups and HLA antigens. The recipient's serum antibodies are assayed *in vitro* for their ability to bind to donor white blood cells.

Answer 12–6

A. Class 1: Corticosteroids. Examples: hydrocortisone, prednisone.
Class 2: Cytotoxic drugs. Examples: azathioprine, cyclophosphamide, methotrexate.
Class 3: T-cell activation inhibitors. Examples: cyclosporin A, tacrolimus (FK506), rapamycin.

B. Class 1: Side/toxic effects: fluid retention, weight gain, diabetes, bone demineralization, thinning of the skin.
Class 2: Side/toxic effects: nonspecifically prevent DNA replication in all mitotic cells causing, for example, diarrhea, hair loss. More specific effects are liver damage caused by azathioprine, and bladder damage caused by cyclophosphamide.
Class 3: Side/toxic effects: nephrotoxicity, suppression of B-cell and granulocyte activation.

Answer 12–7

Cyclosporin A prevents the production of IL-2 and its high-affinity receptor, and thus prevents the activation of T cells and their proliferation and differentiation. It acts by inactivating the protein calcineurin. Calcineurin is a serine/threonine protein phosphatase that is activated by the first part of the T-cell receptor pathway and that dephosphorylates the transcription factor NFAT. This modification is necessary for NFAT, which normally resides in the cytoplasm, to enter the nucleus and stimulate transcription of the genes for IL-2 and the IL-2 receptor α chain. Inactivation of calcineurin by cyclosporin thus prevents the production of IL-2 and its high-affinity receptor.

Answer 12–8

A. Anti-CD3 MoAbs are often administered to patients to suppress T-cell activity when signs of graft rejection are observed. Because CD3 is expressed only on T lymphocytes, this therapy is extremely specific. Anti-CD3 antibodies cross-link CD3:T-cell receptor complexes, leading to a reduction in the number of these complexes on the cell surface and a reduction in the number of T cells in the circulation. Suppression of effector T-cell activity protects the graft.

B. Mouse MoAbs are antigenic in species other than mice and stimulate anti-MoAb responses. Repeated doses will exacerbate this situation and lead to the formation and clearance of MoAb:anti-MoAb immune complexes before the antibody can bind to the T cells, thus rendering the mouse antibody ineffective. A type III hypersensitivity reaction resembling serum sickness can also result when small immune complexes are formed, that is, when MoAb levels exceed anti-MoAb levels. Hence, repeated doses are discouraged, and physicians must restrict this form of immunosuppressive therapy to one episode of rejection.

Answer 12–9

Matching HLA between donor and recipient will reduce undesirable and harmful alloreactions that may cause graft-versus-host disease (GVHD) when mature donor-derived T cells in transplanted bone marrow react against the recipient's alloantigens in tissues and organs. In addition, the thymic epithelium of the recipient, and thus recipient MHC molecules, will positively select a new T-cell repertoire derived from the donor bone marrow. Therefore, matching is essential to ensure that the donor T cells are selected on the same MHC molecules that will be used to present foreign antigen during infection in peripheral lymphoid organs. Foreign antigen will be presented by donor-derived professional antigen-presenting cells. Failure to match MHC molecules between donor and recipient will result in the development of a useless T-cell repertoire in the host after transplantation.

Answer 12–10

A. Graft-versus-host disease is caused by T cells in the transplanted bone marrow making an immune response against antigens on the recipient's tissues. This can happen even though donor and recipient are HLA matched, because there are proteins other than HLA antigens that can differ between people and provoke an immune response. Such antigens are known as minor histocompatibility antigens. In a bone marrow transplant from a female to a male, the minor histocompatibility antigens most likely to cause a problem are male-specific proteins (which are encoded on the Y chromosome) which a female's T cells will not be tolerant to and will see as 'foreign' or non-self.

B. CD8 cytotoxic T cells. The proteins that act as minor histocompatibility antigens are mainly intracellular proteins. Intracellular proteins of the recipient's cells are processed into peptides by proteasomes as part of normal protein degradation and turnover. These peptides are transported into the endoplasmic reticulum and thus are eventually presented on the surface of the recipient's cells by HLA class I molecules. Any peptides that are different from those in the donor may be recognized as non-self by the donor's cytotoxic T cells, which recognize peptides bound to HLA class I molecules. The naive CD8 T cells in the bone marrow can be activated to effector status by the presentation of minor histocompatibility peptides by dendritic cells in secondary lymphoid organs.

As the brother and sister share HLA class I type, the sister's T cells will be able to recognize non-self peptides presented by her brother's HLA molecules.

Answer 12–11

A. Tumor-specific antigens are antigens found exclusively on tumor cells and are not expressed by normal cells.

B. They can originate as the result of: (1) mutations in normal genes in the tumor cell that cause changes in amino acid sequence that generate new epitopes; (2) the generation of hybrid genes through genetic recombination, with the consequent production of a new protein unique to the tumor cell; or (3) viral proteins expressed as a result of viral infection or integration in the host cell genome.

C. Examples: Mucin-1 (MUC-1)—breast and pancreatic cancer; human papilloma virus—cervical carcinoma.

Answer 12–12

A. Tumor-associated antigens are antigens expressed in tumor cells as well as some normal cells, but often at higher levels in tumor cells.

B. They can originate from: (1) proteins involved in mitosis that are produced at higher concentrations because the cells are continually dividing; (2) proteins expressed during embryogenesis that have become deregulated and reactivated transcriptionally; or (3) proteins expressed continuously at high levels in tumor cells that are normally expressed at low levels or transiently.

C. Examples: Melanoma Antigen Encoding -1 and -3 (MAGE-1 and MAGE-3)—melanoma, glioma, and breast cancer; cyclin-dependent kinase 4—melanoma.

Answer 12–13

First, the patient's tumor-specific T cells could be expanded *in vitro* to generate a large number of tumor-specific effector T cells. To achieve this, the patient's tumor-specific T cells would be co-cultured with tumor antigen, antigen-presenting cells, and cytokines. Second, the patient's purified dendritic cells could be loaded with tumor antigen *in vitro* and then be placed back in the patient, where they would stimulate tumor-specific T cells *in vivo*. Both of these techniques require the tumor antigen to have been characterized and purified so that it can be used in the co-culture *in vitro*.

Answer 12–14

Tumor-specific MoAbs are useful tools for the oncologist because of their ability to home in on tumor cells *in vivo*. The size, location, and extent of metastasis can be evaluated. The MoAb can be coupled covalently to a radioactive tag (for example, iodine-131) that is detectable by diagnostic imaging equipment.

MoAbs can be used not only to identify and characterize tumors but also to target the tumor for destruction. This can occur through several different mechanisms. First, if MoAbs are coupled to radioisotopes (iodine-131 or yttrium-99), the tumor cell's DNA is damaged and the cell dies. Coupling MoAbs to cytotoxic drugs is another way of destroying tumor cells. Drugs coupled to antibodies are cleaved from their MoAb transporter only after they have bound to and been internalized by the tumor cell. Finally, alerting the host immune system to tumors coated with MoAb can lead to the destruction of tumors. MoAb bound to tumor cell surfaces will activate the classical complement pathway, which will induce opsonization by phagocytes, lead to the production of inflammatory mediators which will recruit additional phagocytes, and form membrane-attack complexes on tumor cell surfaces. NK cells can also participate in tumor eradication through antibody-dependent cell-mediated cytotoxicity (ADCC).

Glossary

α:β T cells T cells that express an antigen receptor made up of α and β chains. This majority population of T cells includes all T cells that recognize peptide antigen presented by MHC class I and II molecules.

α:β T-cell receptor *see* **T-cell receptor**

Acquired immune deficiency syndrome (AIDS) the disease caused by infection with the human immunodeficiency virus (HIV). It involves a gradual destruction of the CD4 T-cell population and increasing susceptibility to infection.

Acquired immunity *see* **adaptive immunity**

Activated dendritic cells *see* **dendritic cells**

Acute rejection rejection of transplanted cells, tissues, or organs that is due to a T-cell response stimulated by the transplant.

Acute-phase proteins plasma proteins made by the liver whose synthesis is rapidly increased in response to infection. They include mannose-binding lectin (MBL), C-reactive protein (CRP), and fibrinogen.

Acute-phase response a response of innate immunity that occurs soon after the start of an infection and involves the synthesis of acute-phase proteins by the liver and their secretion into the blood.

ADA *see* **adenosine deaminase**

Adaptive immune response the response of antigen-specific B and T lymphocytes to antigen, including the development of immunological memory.

Adaptive immunity the state of resistance to infection that is produced by the adaptive immune response.

ADCC *see* **antibody-dependent cell-mediated cytotoxicity**

Adenoids mucosa-associated secondary lymphoid tissues located in the nasal cavity.

Adenosine deaminase (ADA) enzyme involved in purine breakdown. Its absence leads to the accumulation of toxic purine nucleosides and nucleotides, resulting in the death of most developing lymphocytes within the thymus. The disease that results from a genetically determined lack of adenosine deaminase is called **adenosine deaminase deficiency**. Its symptoms are severe combined immunodeficiency.

Adjuvant substance used to enhance the adaptive immune response to any antigen. To exert its effect an adjuvant must be mixed with the antigen before injection or vaccination.

Afferent lymphatic vessels the several vessels that bring lymph draining from connective tissue into a lymph node en route to the blood.

Affinity a measure of the strength with which one molecule binds to another via a single binding site.

Affinity maturation the increase in affinity of the antigen-binding sites of antibodies for the antigen that occurs during the course of an adaptive immune response.

Agglutination the clumping together of particles, usually caused by antibody or some other multivalent binding molecule interacting with antigens on the surfaces of adjacent particles. Such particles are said to be **agglutinated**. When the particles are red blood cells, the phenomenon is called hemagglutination. When they are white blood cells it is called leukoagglutination.

Agonist any molecule that binds to a receptor and causes it to function.

ALG *see* **antilymphocyte globulin**

Alleles the natural variants of a single gene.

Allelic exclusion in reference to antibody production, the case that, in a heterozygous individual, only one of the two C-region alleles at the immunoglobulin heavy-chain or light-chain loci is expressed in each B cell. In the B-cell population, for each locus roughly half of the cells express one allele and the remaining cells express the other.

Allergen an antigen that elicits hypersensitivity or allergic reactions. Allergens are usually innocuous proteins that do not inherently threaten the integrity of the body.

Allergic asthma disease caused by an allergic reaction to inhaled antigen, in which the bronchi constrict and the patient has difficulty breathing.

Allergic conjunctivitis allergic reactions in the conjunctiva of the eye, usually caused by airborne antigens.

Allergic reaction the result of a secondary immune response to an otherwise innocuous environmental antigen, or allergen. Allergic reactions can involve either antibodies or effector T cells.

Allergic rhinitis an allergic reaction in the nasal mucosa that causes runny nose, sneezing, and tears. It is also known as hay fever.

Allergy a state of hypersensitivity to a normally innocuous environmental antigen. It results from the interaction between the antigen and antibodies or T cells produced by earlier exposure to the same antigen.

Alloantibody antibody that is made by immunization of one member of a species with antigen derived from another member of the same species. Alloantibodies recognize antigens that are the result of allelic variation at polymorphic genes. Common types of alloantibody are those recognizing blood group antigens and HLA class I and class II molecules.

Alloantigen an antigen that differs between members of the same species, such as HLA molecules and blood group antigens. Alloantigens are determined by the different alleles of polymorphic genes.

Allogeneic describes two members of the same species who are genetically different.

Allograft a tissue graft made between genetically non-identical members of the same species.

Alloreaction an adaptive immune response made by one member of a species against an allogeneic antigen from another member of the same species.

Alloreactive T cell T cell in one member of a species that responds to an allogeneic antigen from another member of the same species.

Allotypes natural protein variants that are encoded by the alleles of a gene.

Alternative C3 convertase the **C3 convertase** of the alternative pathway of complement activation. It is composed of C3 bound to proteolytically active Bb (C3Bb) and cleaves C3 into C3a and C3b.

Alternative C5 convertase the **C5 convertase** of the alternative pathway of complement activation. It is composed of two molecules of C3b bound to Bb (C3b$_2$Bb) and cleaves C5 into C5a and C5b.

Alternative pathway of complement activation one of three pathways of complement activation. It is triggered by the presence of infection but does not involve antibody. *See also* **classical pathway of complement activation**; **lectin pathway of complement activation**.

Anaphylactic shock IgE-mediated allergic reaction to systemically administered antigen that causes circulatory collapse and suffocation due to tracheal swelling. Also called systemic anaphylaxis.

Anaphylactoid reactions reactions that are clinically similar to anaphylactic shock but do not involve IgE.

Anaphylatoxins complement fragments C5a, C4a, and C3a, which are produced during complement activation. They recruit fluid and inflammatory cells to sites of antigen deposition.

Anchor residues residues in MHC-binding peptides that interact with pockets in the peptide-binding groove on the MHC molecule. Peptides that bind to a given MHC allotype have the same or very similar anchor residues.

Anergy a state of non-responsiveness to antigen. People are said to be anergic when they cannot mount delayed-type hypersensitivity reactions on challenge with an antigen. T and B cells are said to be anergic when they cannot respond to their specific antigen.

Angioedema swelling of the skin as a result of IgE-mediated allergic reactions, which cause increased permeability of subcutaneous blood vessels with consequent leakage of fluid into the skin.

Antagonist any molecule that binds to a receptor and prevents its function.

Antibody the secreted form of the immunoglobulin made by a B cell.

Antibody repertoire the total variety of antibodies that a person can make.

Antibody-dependent cell-mediated cytotoxicity (ADCC) the killing of antibody-coated target cells by NK cells having the FcγRIII receptor (CD16) that recognizes the Fc region of the bound antibody.

Antigen originally defined as any molecule that binds specifically to an antibody, the term now also refers to any molecule that can produce peptides that bind specifically to a T-cell receptor. *See also* **epitope**.

Antigen-binding site the site on an immunoglobulin or T-cell receptor molecule that binds specific antigen.

Antigen presentation the display of antigen as peptide fragments bound to MHC molecules on the surface of cells. This is the form in which antigen is recognized by most T cells.

Antigen-presenting cells cells that express either MHC class I and/or MHC class II molecules and thus display complexes of MHC molecule and peptide antigen on their surfaces.

Antigen processing the intracellular degradation of proteins into peptides that bind to MHC molecules for presentation to T cells.

Antigen receptor for a B cell, the antigen receptor is its cell-surface immunoglobulin; for a T cell the antigen receptor is a rather similar molecule called the T-cell receptor. Each individual lymphocyte bears receptors of a single antigen specificity.

Antigenic determinant *see* **epitope**

Antigenic drift a process by which point mutations in influenza virus genes cause differences in the structure of viral surface antigens. This causes year-to-year antigenic differences in strains of influenza virus.

Antigenic shift a process by which influenza viruses reassort their segmented genomes and change their surface antigens radically. New viruses arising by antigenic shift are the usual cause of influenza pandemics.

Antilymphocyte globulin (ALG) a preparation of antibodies made by immunizing animals with human lymphocytes. It is used clinically to prevent rejection in organ transplantation.

Antiserum (plural **antisera**) the fluid component of clotted blood from an immune individual that contains antibodies against an antigen. An antiserum contains a heterogeneous collection of antibodies that bind the antigen.

Antithymocyte globulin (ATG) a preparation of antibodies made by immunizing animals with human thymocytes. It is used clinically to prevent rejection in organ transplantation.

AP-1 a family of transcription factors, some of which participate in lymphocyte activation.

APD *see* **autoimmune polyendocrinopathy–candidiasis–ectodermal dystrophy**

APECED *see* **autoimmune polyendocrinopathy–candidiasis–ectodermal dystrophy**

Apoptosis a mechanism of cell death in which the cells to be killed are induced to degrade themselves from within, in a tidy manner. Also called programmed cell death.

Appendix a gut-associated secondary lymphoid tissue located at the beginning of the colon.

Arthus reaction an immune reaction in the skin caused by the injection of antigen into the dermis. The antigen reacts with specific IgG antibodies in the extracellular spaces, activating complement and phagocytic cells to produce a local inflammatory response.

ATG *see* **antithymocyte globulin**

Atopic allergy (atopy) the genetically determined tendency of some people to produce IgE-mediated hypersensitivity reactions against innocuous substances.

Atopic dermatitis *see* **eczema**

Autoantibody an antibody produced by an individual against an antigen produced by their own body.

Autoantigen a self antigen, an antigenic component of the body that provokes an immune response by the individual's own immune system.

Autocrine describes a cytokine or other secreted molecule that acts on the same type of cell as the one that secreted it.

Autograft a graft of tissue made from one anatomical site to another on the same individual.

Autoimmune disease disease in which the pathology is caused by an immune response to normal components of healthy tissue.

Autoimmune hemolytic anemia a disease characterized by reduced numbers of red blood cells (anemia). The reduction is caused by autoantibodies that bind to surface antigens of red blood cells and destroy them.

Autoimmune polyendocrinopathy–candidiasis–ectodermal dystrophy (APECED) an autoimmune disease caused by a lack of a protein called **autoimmune regulator (AIRE)**, which results in the production of T cells reactive against a number of tissues in the body. Also known as **autoimmune polyglandular disease (APD)**.

Autoimmune response an adaptive immune response directed at an antigenic component of the individual's own body.

Autoimmunity adaptive immunity specific for an antigenic component of the individual's own body.

Autologous describes cells, HLA molecules, and so on that derive from the individual in question.

Autologous bone marrow transplant a bone marrow transplant in which the donor and recipient are the same person.

Azathioprine an immunosuppressive drug that kills dividing cells. It is used in transplantation.

β_2-microglobulin (β_2m) the invariant polypeptide that is common to all MHC class I molecules. Also called the light chain of MHC class I molecules.

B cells lymphocytes that are dedicated to making immunoglobulins and antibodies.

B-cell receptor the antigen receptor on B cells. Each B cell is programmed to make a single type of immunoglobulin. The cell-surface form of this immunoglobulin serves as the B-cell receptor for specific antigen. Associated in the membrane with the immunoglobulin are the signal transduction molecules Igα and Igβ.

B-cell co-receptor a complex of the CD19, TAPA-1, and CR2 polypeptides that associates with the B-cell receptor and augments its response to specific antigen.

B-1 cells the minority population of B cells. They express the CD5 glycoprotein and make antibodies of broad specificities. They are also known as CD5 B cells.

B-2 cells the majority population of B cells. They do not express the CD5 glycoprotein and make antibodies of narrow specificities.

B7 molecules the B7.1 and B7.2 proteins, which are co-stimulatory molecules present on the surface of professional antigen-presenting cells.

B lymphocytes *see* **B cells**

Bacteria diverse prokaryotic microorganisms that are responsible for many infectious diseases of humans and other animals.

Balancing selection type of evolutionary selection that acts to maintain a variety of phenotypes (for example MHC isoforms) in a population.

BALT *see* **bronchial-associated lymphoid tissue**

Bare lymphocyte syndromes genetically determined diseases in which MHC class I or II molecules are not expressed on cells. They can be caused by various different regulatory gene defects, and their effect is severe immunodeficiency.

Basophil white blood cell present in small numbers in the blood; it is one of the three types of granulocyte. Basophils contain granules that stain with basic dyes.

Benign tumor a growth due to abnormal proliferation of cells that is localized and contained by epithelial barriers.

Bone marrow the tissue in the center of certain bones that is the major site of generation of all the cellular elements of blood (hematopoiesis).

Brambell receptor (FcRB) an Fc receptor that transports IgG across epithelia and has a structure resembling an MHC class I molecule.

Bronchial-associated lymphoid tissues (BALT) the lymphoid cells and organized lymphoid tissues in the respiratory tract.

Bronchiectasis chronic inflammation of the bronchioles of the lung.

C domain *see* **constant domain**

C region *see* **constant region**

C-reactive protein (CRP) an acute-phase protein that binds to phosphocholine, a surface constituent of various bacteria. CRP binds bacteria, opsonizing them for uptake by phagocytes.

C1 inhibitor (C1INH) a regulatory protein in plasma that inhibits the activity of activated complement component C1. C1INH deficiency causes the disease hereditary angioneurotic edema, in which spontaneous complement activation causes episodes of epiglottal swelling and suffocation.

C3 convertases proteolytic enzymes that are formed during complement activation and cleave complement component C3 to C3b and C3a, thereby enabling C3b to bond covalently to antigens. *See* **alternative C3 convertase**, **classical C3 convertase**.

C4-binding protein (C4BP) a regulatory protein in plasma that inactivates the classical C3 convertase by binding to C4b and displacing C2b.

C5 convertases proteolytic enzymes that are formed during complement activation and cleave complement component C5 to form C5a and C5b.

Calcineurin a cytosolic serine/threonine phosphatase that contributes to T-cell activation. The immunosuppressive drugs cyclosporin A and tacrolimus act by inhibiting calcineurin.

Calnexin a membrane protein in the endoplasmic reticulum that facilitates the folding of MHC molecules and other glycoproteins.

Calreticulin a soluble protein in the endoplasmic reticulum that is structurally related to calnexin and helps MHC molecules and other glycoproteins to fold properly.

CAM *see* **cell adhesion molecule**

Cancer diseases caused by abnormal and invasive cell proliferation.

Carcinogen chemical or physical agent that increases the risk of cancer.

Carcinoma cancer of epithelial cells.

Carrier foreign protein to which small non-immunogenic antigens, or haptens, can be coupled to render the hapten immunogenic. *In vivo*, self proteins can also serve as carriers if they are suitably modified by the hapten; this is important in allergy to drugs.

Caseation necrosis a form of necrosis seen in the center of some large granulomas. The term comes from the white cheesy appearance of the central necrotic area.

Catalytic antibody an antibody that binds an antigen, chemically changes it, and then releases it.

CCL2 a chemokine produced by activated T cells that attracts macrophages into a site of infection. Formerly known as macrophage chemotactic protein (MCP).

CCL21 a chemokine made by vascular endothelial cells and which is involved in extravasation of leukocytes. Formerly known as SLC.

CD2 an adhesion molecule of T cells that binds to the LFA-3 adhesion molecule of antigen-presenting cells. Also called LFA-2.

CD3 complex a complex of signaling molecules that associates with T-cell receptors. It consists of CD3γ, δ, and ϵ chains, and ζ chains.

CD4 a cell-surface glycoprotein on some T cells that recognize antigens presented by MHC class II molecules. CD4 binds to MHC class II molecules on the antigen-presenting cell and acts as a co-receptor to augment the T cell's response to antigen.

CD4 T cells the subset of T cells that express the CD4 co-receptor and recognize peptide antigens presented by MHC class II molecules.

CD5 B cells *see* **B-1 cells**

CD8 a cell-surface glycoprotein on some T cells that recognize antigens presented by MHC class I molecules. CD8 binds to MHC class I molecules on the antigen-presenting cell and acts as a co-receptor to augment the T-cell's response to antigen.

CD8 T cells the subset of T cells that express the CD8 co-receptor and recognize peptide antigens presented by MHC class I molecules.

CD19 a component of the B-cell co-receptor.

CD21 another name for complement receptor 2 (CR2). It is a component of the B-cell receptor.

CD28 the low-affinity receptor on T cells that interacts with B7 co-stimulatory molecules to promote T-cell activation.

CD34 a vascular addressin expressed on high endothelial venules in lymph nodes, and which is involved in extravasation of white blood cells.

CD40 cell-surface glycoprotein on B cells whose interaction with **CD40 ligand** on T cells triggers B-cell proliferation.

CD81 a component of the B-cell receptor that is also a cellular receptor for hepatitis C virus. Also called TAPA-1.

CDR *see* **complementarity-determining regions**

Celiac disease inflammatory autoimmune disease of the gut mucosa caused by an immune response to the gluten proteins present in some cereals such as wheat.

Cell adhesion molecules (**CAMs**) cell-surface proteins that enable cells to bind to each other.

Cell-mediated immunity (**cellular immunity**) any adaptive immune response in which antigen-specific effector T cells dominate. It is defined operationally as all adaptive immunity that cannot be transferred to a naive recipient with serum antibody.

Central lymphoid tissues the anatomical sites of lymphocyte development. In humans, B lymphocytes develop in bone marrow, whereas T lymphocytes develop in the thymus. *See also* **primary lymphoid tissues**.

Central MHC the class III region of the MHC, located between the class I and II regions.

Centroblast large dividing B cell present in germinal centers. Somatic hypermutation occurs in centroblasts, and antibody-secreting and memory B cells derive from them.

Centrocyte nondividing B cell in germinal centers. Centrocytes have undergone isotype switching and somatic hypermutation.

Chediak–Higashi syndrome a genetic disease in which phagocytes malfunction. Their lysosomes fail to fuse properly with phagosomes, and killing of ingested bacteria is impaired.

Chemokines large group of small proteins involved in guiding white blood cells to sites where their functions are needed. They have a central role in inflammatory responses.

Chronic asthma a disease characterized by chronic inflammation of the airways and difficulty in breathing. Although probably initiated by exposure to an allergen, chronic asthma can be perpetuated in its absence.

Chronic granulomatous disease an immunodeficiency disease in which multiple granulomas form as a result of defective elimination of bacteria by phagocytic cells. It is caused by a defect in the NADPH oxidase system of enzymes, which generates the superoxide radical involved in bacterial killing.

Chronic rejection rejection of organ grafts that occurs years after transplantation and is characterized by degeneration and occlusion of the blood vessels in the graft. It is caused by an antibody response to the HLA class I alloantigens of the graft.

Chronic thyroiditis an autoimmune disease that causes progressive destruction of the thyroid gland. Also called Hashimoto's thyroiditis or Hashimoto's disease.

Class I region the part of the major histocompatibility complex that contains the MHC class I heavy-chain genes.

Class II region the part of the major histocompatibility complex that contains the MHC class II α- and β-chain genes.

Class II-associated invariant chain peptide (**CLIP**) a peptide of variable length cleaved from the class II invariant chain by proteases in the endosomal pathway. It remains associated with the MHC class II molecule in an unstable form until it is removed by the HLA-DM protein.

Class III region a region of the major histocompatibility complex between the class I and II regions. Also known as the **central MHC**, it contains no genes for class I or II MHC molecules.

Class switching *see* **isotype switching**

Class *see* **isotype**

Classical C3 convertase a surface-associated serine protease of the classical pathway of complement activation. It is composed of the complement components C4b2a and cleaves C3 into C3a and C3b.

Classical C5 convertase C5-cleaving enzyme consisting of the complement fragments C4b, C2b, and C3b.

Classical pathway of complement activation one of three pathways of complement activation. It is activated by antibody bound to antigen, and involves complement components C1, C4, and C2 in the generation of the C3 and C5 convertases. *See also* **alternative pathway of complement activation**; **lectin pathway of complement activation**.

Clonal deletion the elimination of immature lymphocytes that bind to self antigens. Clonal deletion is the main mechanism that produces self-tolerance.

Clonal selection the central principle of adaptive immunity. It is the mechanism by which adaptive immune responses derive only from individual antigen-specific lymphocytes, which are stimulated by the antigen to proliferate and differentiate into antigen-specific effector cells.

Co-receptor a cell-surface protein that increases the sensitivity of an antigen receptor to its antigen. A co-receptor can accomplish this by increasing adhesive interactions between the interacting cells, and/or by enhancing signal transduction from the main receptor.

Co-stimulator molecule a molecule on an antigen-presenting cell that delivers signals to an interacting naive lymphocyte that are required in addition to the antigen-binding signal for the lymphocyte to respond. Co-stimulator molecules include the B7.1 and B7.2 proteins on professional antigen-presenting cells, which engage molecules CD28 and CTLA-4 on T cells. CD40 ligand serves a co-stimulatory role when it binds to CD40 on B cells.

Cognate interactions cell–cell interactions between B and T lymphocytes specific for the same antigen.

Collectins a family of calcium-dependent sugar-binding proteins or lectins containing collagen-like sequences. An example is mannose-binding lectin.

Combination therapy antiviral therapy (for HIV, for example) in which several antiviral drugs are used together to try and avoid the rapid generation of mutant viruses resistant to one of the drugs alone.

Combination vaccine vaccine that contains antigens derived from more than one pathogen and is designed to provide protection against more than one disease.

Common lymphoid progenitor stem cell that gives rise to all lymphocytes and is derived from a pluripotent hematopoietic stem cell.

Complement a set of plasma proteins that act in a cascade of reactions to attack extracellular forms of pathogens. As a result of complement activation, pathogens become coated with complement components, which can either kill the pathogen directly or cause its engulfment and destruction by phagocytes.

Complement activation the initiation by pathogens of a series of reactions involving the complement components of plasma, leading to the death and elimination of the pathogen. *See also* **alternative pathway of complement activation; classical pathway of complement activation; lectin pathway of complement activation**.

Complement control protein (CCP) modules family of structurally similar protein modules found in many of the proteins that regulate complement activity.

Complement receptors (**CR**) cell-surface proteins on various cell types that recognize and bind complement proteins bound to antigens. Complement receptors on phagocytes facilitate the phagocytic engulfment of pathogens coated with complement. Complement receptors include **CR1, CR2, CR3, CR4**, and the receptor for C1q.

Complement system *see* **complement**

Complementarity-determining regions (**CDRs**) the localized regions of immunoglobulin and T-cell receptor chains that determine the antigenic specificity and bind to the antigen. The CDRs are the most variable parts of the variable domains and are also called hypervariable regions.

Conformational epitopes epitopes on a protein antigen that are formed from several separate regions in the primary sequence of a protein brought together by protein folding. Antibodies that bind conformational epitopes bind only to native folded proteins. Also called **discontinuous epitopes**.

Conjugate vaccine a vaccine made from capsular polysaccharides bound to an immunogenic protein such as tetanus toxoid. The protein provides peptide epitopes that stimulate CD4 T cells to help B cells specific for the polysaccharide.

Constant domains (C domains) the constituent domains of the constant regions of immunoglobulin and T-cell receptor polypeptides.

Constant region (C region) the part of an immunoglobulin or T-cell receptor (or of its constituent polypeptide chains) that is of identical amino-acid sequence in molecules of the same isotype but different antigen-binding specificities.

Contact sensitivity a form of delayed-type hypersensitivity in which T cells respond to antigens that are introduced into the body by contact with the skin.

Corticosteroids compounds related to the steroid hormones produced in the adrenal cortex, such as cortisone. They suppress immune responses and are widely used as anti-inflammatory and immunosuppressive agents in medicine.

CR, CR1, CR2, CR3, CR4 *see* **complement receptors**

Cross-match testing a test used in blood typing and histocompatibility testing to determine whether donor and recipient have antibodies against each other's cells that might interfere with successful transfusion or transplantation.

CRP *see* **C-reactive protein**

Cryptic epitope antigenic determinant on a molecule that is normally hidden from the immune system but becomes revealed under conditions of infection or inflammation.

CTLA-4 a high-affinity inhibitory receptor on T cells that interacts with B7 co-stimulatory molecules.

Cutaneous T-cell lymphoma a malignant growth of T cells that home to the skin. Also known as mycosis fungoides.

CXCL8 a chemokine involved in the extravasation of neutrophils. Formerly known as IL-8.

Cyclophilins family of cytoplasmic proteins that bind to the immunosuppressive drugs cyclosporin A and tacrolimus. The complex of drug and cyclophilin binds calcineurin and this prevents T-cell activation.

Cyclophosphamide an alkylating agent used as an immunosuppressive drug. It acts by killing rapidly dividing cells, including lymphocytes proliferating in response to antigen.

Cyclosporin A an immunosuppressive drug that specifically prevents T-cell activation and effector function. Also called cyclosporine.

Cytokines proteins made by cells that affect the behavior of other cells. Cytokines made by lymphocytes are often called lymphokines or interleukins (abbreviated IL). Cytokines bind to specific receptors on their target cells.

Cytotoxic CD8 T cells the subset of T cells that express the CD8 co-receptor and recognize peptide antigen presented by MHC class I molecules.

Cytotoxic T cells T cells that can kill other cells. Almost all cytotoxic T cells are CD8 T cells. Cytotoxic T cells are important in host defense against viruses and other cytosolic pathogens, because they recognize and kill the infected cells.

Cytotoxins proteins made by cytotoxic T cells that participate in the destruction of target cells. Perforins, granzymes or fragmentins, and granulysin are examples of cytotoxins.

Dark zone the part of a germinal center in secondary lymphoid tissue that contains dividing centroblasts.

Decay-accelerating factor (DAF) a cell-surface protein that prevents complement activation on human cells. DAF binds to C3 convertases of both the alternative and classical pathways of complement activation and, by displacing Bb and C2b respectively, prevents their action.

Degenerate binding specificity the type of antigen-binding specificity exhibited by MHC class I and II molecules, in which each MHC allotype can bind numerous peptides of different amino acid sequences.

Delayed-type hypersensitivity (**DTH**) a form of cell-mediated immunity elicited by antigen in the skin and mediated by CD4 T_H1 cells. It is called delayed-type hypersensitivity because the reaction appears hours to days after antigen is injected.

Dendritic cells professional antigen-presenting cells with a branched, dendritic morphology. They are the most potent stimulators of T-cell responses. Also known as interdigitating reticular cells, they are derived from the bone marrow and are distinct from the follicular dendritic cell that presents antigen to B cells. **Immature dendritic cells** take up and process antigens but cannot yet stimulate T cells. **Mature** or **activated dendritic cells** are present in secondary lymphoid tissues and are able to stimulate T cells.

Dendritic epidermal T cells (**dETC**) a specialized class of γ:δ T cells found in the skin of mice and some other species, but not humans. All dETCs have the same γ:δ T-cell receptor; their function is unknown.

Desensitization a therapeutic procedure in which an allergic individual is exposed to increasing doses of allergen with the goal of inhibiting their allergic reactions. It probably works by shifting the balance between CD4 T_H1 and T_H2 cells and thus changing the antibody produced from IgE to IgG.

Diapedesis the movement of white blood cells from the blood across blood vessel walls into tissues.

DiGeorge syndrome a recessive genetic immunodeficiency disease in which thymic epithelium fails to develop.

Diphtheria toxin the toxin secreted by the bacterium *Corynebacterium diphtheriae*, the cause of diphtheria, which causes the disease symptoms. The diphtheria vaccine consists of an inactive form of the toxin called diphtheria toxoid.

Direct pathway of allorecognition type of alloreactive response in which T cells of the recipient of a transplant are stimulated by direct interaction of their receptors with the allogeneic HLA molecules expressed by dendritic cells from the donor, present in the transplant.

Directional selection type of natural selection that replaces older alleles with newer variants (for example in the MHC). Its characteristic outcome is change.

Discontinuous epitopes *see* **conformational epitopes**

Diversity (D) gene segments short DNA sequences present in immunoglobulin heavy-chain loci and in T-cell receptor β- and δ-chain loci. In the rearranged functional genes at these loci, a D region connects the V and J regions.

Double-negative thymocyte immature T cell within the thymus that expresses neither CD4 nor CD8.

Double-positive thymocyte T cell at an intermediate stage of development in the thymus. It expresses both CD4 and CD8.

Draining lymph node the lymph node to which extracellular fluid collected at a site of infection first travels.

DTH *see* **delayed-type hypensensitivity**

E-selectin *see* **selectins**

Early pro-B cell *see* **pro-B cell**

Eczema a common skin disease of children, appearing as scaly, reddened and itchy patches on the skin. Its etiology is poorly understood. Also called atopic dermatitis.

Edema abnormal accumulation of fluid in connective tissue, leading to swelling.

Effector cells lymphocytes that can mediate the removal of pathogens from the body without the need for further differentiation.

Effector mechanisms the physiological and cellular processes used by the immune system to destroy pathogens and remove them from the body.

Efferent lymphatic vessel the single vessel in which lymph and lymphocytes leave a lymph node en route to the blood.

Encapsulated bacteria bacteria that possess thick carbohydrate coats that protect them from phagocytosis. Encapsulated bacteria cause extracellular infections and can be dealt with by phagocytes only if the bacteria are first coated with antibody and complement.

Endocytic vesicle membrane vesicle that is pinched off from the plasma membrane and takes extracellular material into cells.

Endocytosis the uptake of extracellular material into cells by endocytic vesicles that form by pinching off pieces of plasma membrane.

Engraftment the time at which a bone marrow transplant is making new blood cells.

Eosinophil white blood cell that is one of the three types of granulocyte. It contains granules that stain with eosin and whose contents are secreted when the cell is stimulated. Eosinophils contribute chiefly to defense against parasitic infections.

Eotaxin-1 and **eotaxin-2** chemokines that act specifically on eosinophils. Now called CXCL11 and CXCL24, respectively.

Epidemic an outbreak of infectious disease that affects many individuals within a population.

Epitope the portion of an antigenic molecule that is bound by an antibody or gives rise to the MHC-binding peptide that is recognized by a T-cell receptor. Also called an antigenic determinant.

Erythrocyte red blood cell.

Extravasation the movement of cells or fluid from within blood vessels to the surrounding tissues.

F(ab')₂ a proteolytic fragment of IgG that consists of the two Fab arms held together by a disulfide bond. It is produced by digesting IgG with pepsin.

Fab fragment a proteolytic fragment of IgG that consists of the light chain and the amino-terminal half of the heavy chain held together by an interchain disulfide bond. It is called Fab because it is the **F**ragment with **a**ntigen **b**inding specificity. In the intact IgG molecule the parts corresponding to the Fab fragment are often called Fab or Fab arms.

Factor B plasma protein that binds to C3(OH) or C3b and is cleaved to form part of the C3 convertases (C3(OH),Bb and C3b,Bb) in the alternative pathway of complement activation.

Factor D plasma protease that cleaves factor B to Bb and Ba in the alternative pathway of complement activation.

Factor H a complement regulatory protein of plasma that inactivates the C3 convertase of the alternative pathway and C5 convertases by binding to C3b and rendering it susceptible to cleavage by factor I to produce inactive iC3b.

Factor I a regulatory protease of complement that can cleave C3b and C4b into inactive forms.

Factor P *see* **properdin**

Farmer's lung a hypersensitivity disease caused by the interaction of IgG antibodies with large particles of inhaled allergen in the alveolar wall of the lung. The resulting inflammation compromises gas exchange by the lungs. The disease is caused by chronic exposure to large quantities of the allergen.

Fas a member of the TNF receptor family that is expressed on certain cells and makes them susceptible to killing by cells expressing **Fas ligand**, a cell-surface member of the TNF family of proteins. Binding of Fas ligand to Fas triggers apoptosis in the Fas-bearing cell.

Fc fragment a fragment of an antibody, resulting from proteolytic cleavage, that consists of the carboxy-terminal halves of the two heavy chains disulfide-bonded to each other by the residual hinge region. It is called Fc because it was the fragment that was most readily crystallized in early studies of IgG antibody structure. In an intact antibody the part corresponding to the Fc fragment is called **Fc**, **Fc region**, or **Fc piece**.

Fc receptors cell-surface receptors for the Fc portion of some immunoglobulin isotypes. They include the Fcγ and Fcε receptors.

Fc region *see* **Fc fragment**

Fcε receptor (FcεRI) a receptor present on the surface of mast cells, basophils and activated eosinophils that binds free IgE with very high affinity. When antigen binds to IgE and cross-links FcεRI it causes cellular activation and degranulation.

Fcγ receptors receptors present on various cell types that are specific for the Fc regions of IgG antibodies. There are different receptors for different subclasses of IgG.

FDC *see* **follicular dendritic cells**

Fever a rise of body temperature above the normal. It is caused by cytokines produced in response to infection.

FK-binding proteins another name for the cyclophilins that bind tacrolimus (FK506). *See also* **cyclophilins**.

FK506 *see* **tacrolimus**

Follicular center cell lymphoma a malignancy of mature B cells.

Follicular dendritic cells (FDCs) characteristic nonlymphoid cells of follicles in secondary lymphoid tissues. They have long branching processes that make intimate contact with B cells and have Fc and complement receptors that hold antigen:antibody:complement complexes on their surfaces for long periods. These cells are crucial in selecting antigen-binding B cells during antibody responses.

Framework regions relatively invariant regions within the variable domains of immunoglobulins and T-cell receptors that provide a protein scaffold for the hypervariable regions.

Freund's complete adjuvant an adjuvant consisting of an emulsion of mineral oil and water containing killed mycobacteria.

Fungi single-celled and multicellular eukaryotic organisms, including the yeasts and molds, that can cause a variety of diseases. Immunity to fungi involves both humoral and cell-mediated responses.

γ:δ T cells a minority population of T cells that express receptors made up of γ and δ chains. The antigen specificities and functions of these cells are uncertain.

GALT *see* **gut-associated lymphoid tissues**

Gene family a set of genes encoding proteins of similar structure, and often of similar function, such as the MHC class I genes.

Gene segments multiple short DNA sequences in the immunoglobulin and T-cell receptor genes. These can be rearranged in many different combinations to produce the vast diversity of immunoglobulin or T-cell receptor polypeptide chains. *See also* **D gene segments; J gene segments; V gene segments**.

Genetic polymorphism variation in a population due to the existence of two or more alleles of a gene.

Germinal center area in secondary lymphoid tissue that is a site of intense B-cell proliferation, selection, maturation, and death. Germinal centers form around follicular dendritic cell networks when activated B cells migrate into lymphoid follicles.

Germline configuration the organization of the immunoglobulin and T-cell receptor genes in the DNA of germ cells and in the vast majority of somatic cells that do not undergo somatic recombination.

GlyCAM-1 a vascular addressin found on the high endothelial venules of lymphoid tissues. It is an important ligand for the L-selectin molecule on naive lymphocytes and directs these cells to leave the blood and enter the lymphoid tissues.

GM-CSF *see* **granulocyte–macrophage colony-stimulating factor**

Goodpasture's syndrome an autoimmune disease in which autoantibodies against type IV collagen of the basement membrane of blood vessel endothelium cause extensive vasculitis.

Graft-versus-host disease (GVHD) pathological condition caused by the **graft-versus-host reaction (GVHR)**, which is the response of mature donor-derived T cells in transplanted bone marrow against the alloantigens of the recipient's tissues.

Graft-versus-leukemia (GVL) effect an effect in bone marrow transplantation as therapy for leukemia where some degree of genetic incompatibility between donor and recipient is thought to help T cells or NK cells from the transplant to eliminate residual leukemia cells in the recipient. Also known as a **graft-versus-tumor effect**.

Granulocyte–macrophage colony-stimulating factor (**GM-CSF**) a cytokine involved in the growth and differentiation of cells in the granulocyte and monocyte lineages.

Granulocytes white blood cells with irregularly shaped, multilobed nuclei and cytoplasmic granules. There are three types of granulocyte: neutrophils, eosinophils, and basophils. They are also known as polymorphonuclear leukocytes.

Granuloma a site of chronic inflammation usually triggered by persistent infectious agents such as mycobacteria, or by a non-degradable foreign body. Granulomas have a central area of macrophages, often fused into multinucleate giant cells, surrounded by T lymphocytes.

Granulysin a membrane-perturbing protein present in the granules of cytotoxic T cells. With perforin it is thought to make pores in the target cell's membrane.

Granzymes serine esterases present in the granules of cytotoxic T cells and NK cells. On entering the cytosol of a target cell, granzymes induce apoptosis of the target. Also called fragmentins.

Graves' disease an autoimmune disease in which antibodies against the thyroid-stimulating hormone receptor cause the overproduction of thyroid hormone and the symptoms of hyperthyroidism.

Gut-associated lymphoid tissues (**GALT**) lymphoid tissues closely associated with the gastrointestinal tract, including the palatine tonsils, Peyer's patches in the intestine, and layers of intraepithelial lymphocytes.

H chain *see* **heavy chain**

HAART *see* **highly active anti-retroviral therapy**

HANE *see* **hereditary angioneurotic edema**

Haploidentical transplant a transplant from a donor who shares one HLA haplotype with the patient but differs in the second.

Haplotype in respect of a linked cluster of polymorphic genes, the set of alleles carried on a single chromosome is called a haplotype. Every person inherits two haplotypes, one from each parent. The term was first used in connection with the genes of the major histocompatibility complex.

Hashimoto's disease *see* **Hashimoto's thyroiditis**

Hashimoto's thyroiditis an autoimmune disease characterized by persistent high levels of antibodies against thyroid-specific antigens. These antibodies recruit NK cells to the thyroid, leading to damage and inflammation.

Heavy chain (H chain) the larger of the two component polypeptides of an immunoglobulin molecule. Heavy chains come in a variety of heavy-chain classes or isotypes, each of which confers a distinctive effector function on the antibody molecule.

Helper CD4 T cells CD4 T cells are sometimes generally known as helper cells because their function is to help other cell types to perform their functions. The term helper T cell sometimes refers to T_H2 cells only, the cells that help B cells to produce antibody.

Hematopoiesis the generation of the cellular elements of blood, including the red blood cells, white blood cells, and platelets. These cells all originate from pluripotent **hematopoietic stem cells** whose differentiated progeny divide under the influence of various **hematopoietic growth factors**.

Hematopoietic stem cell transplantation any transplantation in which the role of the graft is to replace the hematopoietic system. Sources of hematopoietic stem cells include bone marrow and umbilical cord blood.

Hemolytic anemia of the newborn a potentially fatal disease caused by maternal IgG antibodies directed towards paternal antigens expressed on fetal red blood cells. The usual target of this response is the Rh blood group antigen. Maternal anti-Rh IgG antibodies cross the placenta to attack the fetal red blood cells. Also called erythroblastosis fetalis.

Herd immunity the phenomenon whereby those people in a population who have no protective immunity against a pathogen are largely protected from infection when the majority of the population is resistant to the pathogen.

Hereditary angioneurotic edema (HANE) a genetic disease due to deficiency of the C1 inhibitor of the complement system. In the absence of C1 inhibitor, spontaneous activation of the complement system causes diffuse fluid leakage from blood vessels, the most serious consequence of which is epiglottal swelling leading to suffocation.

Heterozygous describes an individual who has inherited different forms (alleles) of a given gene from their two parents.

Highly active anti-retroviral therapy (HAART) combination therapy for HIV infection, in which several antiviral drugs are used together to try and avoid the rapid generation of drug-resistant mutant viruses that occurs when one of the drugs is used alone.

Highly polymorphic genes that have many alleles and for which most individuals in a population are heterozygotes.

Histamine a vasoactive amine stored in mast cell granules. Histamine is released when antigen binds to IgE molecules on mast cells and causes dilation of local blood vessels and contraction of smooth muscle, producing some of the symptoms of immediate hypersensitivity reactions. Anti-histamines are drugs that counter histamine action.

Histocompatibility literally, the ability of tissues (Greek *histo*) to get along with each other. The term is used in immunology to describe the genetic systems that determine the rejection of tissue and organ grafts as a result of the immunological recognition of alloantigens (known in this context as **histocompatibility antigens**).

HIV *see* **human immunodeficiency virus**

Hives itchy red swellings in the skin caused by IgE-mediated reactions. Also called urticaria or nettle rash.

HLA the acronym for Human Leukocyte Antigen. It is the genetic designation for the human MHC. Individual loci are designated by capital letters, as in HLA-A, and alleles are designated by numbers, as in HLA-A*0201.

HLA class I molecules the name for the human version of the **MHC class I molecules**.

HLA class II molecules the name for the human version of the **MHC class II molecules**.

HLA-A, **HLA-B**, and **HLA-C** the highly polymorphic human MHC class I genes.

HLA-DM an invariant MHC class II molecule in humans that is involved in the intracellular loading of MHC class II molecules with peptides.

HLA-DO a relatively invariant human MHC class II molecule.

Although not precisely defined, the function of HLA-DO is thought to modify that of HLA-DM.

HLA-DP, **HLA-DQ**, and **HLA-DR** the highly polymorphic human MHC class II molecules. Each class II molecule is made from α and β chains encoded by A and B genes respectively. For example, the HLA-DPα and HLA-DPβ chains are encoded by the HLA-DPA and HLA-DPB genes respectively. All the genes are in the MHC.

HLA-E, HLA-G relatively invariant human MHC class I molecules that form ligands for NK-cell receptors.

HLA-F a monomorphic human MHC class I molecule of unknown function.

HLA type the combination of HLA class I and class II allotypes that a person expresses.

Hodgkin's disease a malignant disease caused by transformed germinal center B cells.

Homing the movement of naive T cells into secondary lymphoid tissues, or of effector T cells to an effector site.

Homozygous describes an individual who has inherited the same form (allele) of a given gene from both parents.

Human immunodeficiency virus (HIV) the causative agent of the acquired immune deficiency syndrome (AIDS). HIV is a retrovirus of the lentivirus family that infects CD4 T cells, leading to their slow depletion, which eventually results in immunodeficiency.

Human leukocyte antigen (HLA) complex the human major histocompatibility complex (MHC).

Humanize to replace, by means of genetic engineering, the CDR loops in a human antibody with the corresponding CDR sequences of a desired specificity from a mouse antibody.

Humoral immunity immunity that is mediated by antibodies and can therefore be transferred to a non-immune recipient by serum.

HV regions *see* **hypervariable regions**

Hybridomas hybrid cell lines that make monoclonal antibodies of defined specificity. They are formed by fusing a specific antibody-producing B lymphocyte with a myeloma cell that grows in tissue culture and does not make any immunoglobulin chains of its own.

Hygiene hypothesis hypothesis advanced to explain the increasing incidence of hypersensitivity and autoimmune diseases in developed countries, in which the increase is proposed to be due to widespread hygiene, vaccination, and antibiotic therapy, preventing children's immune systems from becoming used to dealing correctly with infections.

Hyperacute rejection rejection of an allogeneic tissue graft as a result of preformed antibodies that react against A,B,O blood group antigens or HLA class I antigens on the graft. The antibodies bind to endothelium and trigger the blood clotting cascade, leading to ischemia and death of the transplanted organ or tissue.

Hyper-IgM syndrome a genetically determined X-linked immunodeficiency disease in which B cells cannot switch their immunoglobulin heavy-chain isotype. It arises from the absence of CD40 ligand.

Hypereosinophila abnormally high numbers of eosinophils in the blood.

Hypersensitivity reactions immune responses to innocuous antigens that lead to symptomatic reactions on reexposure. These can cause **hypersensitivity diseases** if they occur repetitively. This state of heightened reactivity to an antigen is called **hypersensitivity**. Hypersensitivity reactions are classified by mechanism: type I hypersensitivity reactions involve the triggering of mast cells by IgE antibodies; type II hypersensitivity reactions involve IgG antibodies against cell-surface or matrix antigens; type III hypersensitivity reactions involve antigen:antibody complexes; and type IV hypersensitivity reactions are mediated by effector T cells.

Hyperthyroid refers to an abnormally high production of thyroid hormones by the thyroid gland.

Hypervariable regions (HV regions) small regions of high amino-acid sequence diversity within the variable regions of immunoglobulin and T-cell receptors. They correspond to the complementarity-determining regions.

Hypothyroid refers to an abnormally low production of thyroid hormone by the thyroid gland.

ICAM-1, ICAM-2, ICAM-3 the three different types of ICAM. *See also* **intercellular adhesion molecules**.

Iccosomes immune-complex coated bodies. Small fragments of membrane coated with immune complexes that bud off from the processes of follicular dendritic cells in lymphoid follicles.

IDDM *see* **insulin-dependent diabetes mellitus**

IFN-α, IFN-β, IFN-γ *see* **interferons**

Ig domain *see* **immunoglobulin domains**

Igα, Igβ *see* **B-cell antigen receptor**

IgA the class of immunoglobulin having α heavy chains. IgA antibodies in dimeric form are the antibodies present in mucosal secretions. IgA in monomeric form is present in the blood.

IgD the class of immunoglobulin having δ heavy chains. It appears as surface immunoglobulin on mature naive B cells but its function is unknown.

IgE the class of immunoglobulin having ε heavy chains. It is involved in allergic reactions.

IgG the class of immunoglobulin having γ heavy chains. It is the most abundant class of immunoglobulin in plasma.

IgM the class of immunoglobulin having μ heavy chains. It is the first immunoglobulin to appear on the surface of B cells and the first antibody secreted during an immune response. It is secreted in pentameric form.

Ii *see* **invariant chain**

IL-1 (interleukin-1) a cytokine released by macrophages, which with IL-6 and TNF-α induces a wide range of inflammatory responses at early times in infection.

IL-2 (interleukin-2) a cytokine produced by activated T cells, which is essential for the development of adaptive immune responses.

IL-3 (interleukin-3) a growth factor for hematopoietic progenitor cells.

IL-4 (interleukin-4) a cytokine secreted by CD4 T_H2 cells, which helps to initiate the proliferation and clonal expansion of B cells.

IL-6 (interleukin-6) a cytokine released by macrophages, which with IL-1 and TNF-α induces a wide range of inflammatory responses at early times in infection.

IL-10 (interleukin-10) a cytokine released by CD4 T_H2 cells.

IL-12 (interleukin-12) a cytokine released by macrophages, which activates NK cells.

IL-13 (interleukin-13) a cytokine secreted by CD4 T_H2 cells.

Immature B cells B cells that have rearranged a heavy- and a light-chain variable-region gene and express surface IgM but not IgD.

Immature dendritic cells dendritic cells present in tissues, which take up antigen but do not express co-stimulatory molecules and cannot yet act as professional antigen-presenting cells to naive T cells.

Immediate reactions reactions occurring within minutes of exposure to antigen. They are mediated by preexisting antibodies in the circulation. Also called immediate hypersensitivity reactions.

Immune resistant to infection.

Immune complex the protein complex formed from the binding of antibodies to soluble antigens. The size of the immune complexes formed depends on the relative concentrations of antigen and antibody. Large immune complexes are cleared by phagocytes bearing Fc and complement receptors. Small soluble immune complexes tend to be deposited on the walls of small blood vessels, where they can activate complement and cause damage.

Immune-complex coated bodies *see* **iccosomes**

Immune stimulatory complexes (ISCOMs) lipid carriers that act as adjuvants.

Immune system the tissues, cells, and molecules involved in host defense mechanisms, primarily against infectious agents.

Immunity the ability to resist infection.

Immunization the deliberate provocation of an adaptive immune response by introducing antigen into the body.

Immunodeficiency diseases a group of inherited or acquired disorders in which some part or parts of host defense are either absent or defective.

Immunogenetics a subfield of immunology that was originally concerned with the analysis of genetic traits by means of antibodies against genetically polymorphic molecules such as blood group antigens and MHC proteins. Immunogenetics now encompasses the genetic analysis, by any technique, of molecules that are of specific importance to the immune system.

Immunoglobulin A *see* **IgA**

Immunoglobulin D *see* **IgD**

Immunoglobulin domains (Ig domains) components of protein structure consisting of about 100 amino acids that fold into a sandwich of two β sheets held together by a disulfide bond. Immunoglobulin heavy and light chains are made up of a series of immunoglobulin domains. Similar domains are present in many other proteins.

Immunoglobulin E *see* **IgE**

Immunoglobulin G *see* **IgG**

Immunoglobulin M *see* **IgM**

Immunoglobulin superfamily (Ig superfamily) the name given to all the proteins that contain one or more immunoglobulin or immunoglobulin-like domains.

Immunoglobulin-like domains protein domains that resemble the immunoglobulin domain in structure but are present in a variety of other proteins.

Immunoglobulins (Ig) the antigen-binding molecules of B cells.

Immunological memory the capacity of the immune system to make quicker and stronger adaptive immune responses to successive encounters with an antigen. Immunological memory is specific for a particular antigen and is long-lived.

Immunophilins intracellular proteins with peptidyl–prolyl *cis–trans* isomerase activity that bind the immunosuppressive drugs cyclosporin A, tacrolimus, and sirolimus (rapamycin).

Immunoreceptor tyrosine-based activation motifs (ITAMs) sequences in the cytoplasmic domains of membrane receptors that are sites of tyrosine phosphorylation and of association with tyrosine kinases and phosphotyrosine-binding proteins involved in signal transduction. Related motifs with opposing effects are **immunoreceptor tyrosine-based inhibitory motifs (ITIMs)**, which recruit phosphatases that remove the phosphate groups added by tyrosine kinases.

Immunosuppressive drugs compounds that inhibit adaptive immune responses.

Immunotoxins conjugates made of a specific antibody that is chemically coupled to a toxic protein usually derived from a plant or microbe. The antibody delivers the toxin moiety to the surface of the target cells.

Inactivated virus vaccines *see* **killed virus vaccines**

Indirect pathway of allorecognition one means by which alloreactive T cells in a transplant recipient can be stimulated to react against the transplant. The alloreactive T cells do not directly recognize the transplanted cells but recognize subcellular material that has been processed and presented by autologous antigen-presenting cells.

Inflammation a general term for the local accumulation of fluid, plasma proteins, and white blood cells that is initiated by physical injury, infection, or a local immune response. This is also known as an **inflammatory response**. The cells that invade tissues undergoing inflammatory responses are often called **inflammatory cells** or an **inflammatory infiltrate**.

Inflammatory mediators a variety of substances released by various cell types that contribute to the production of inflammation at a site of infection or trauma.

Influenza hemagglutinin a glycoprotein of the influenza virus coat that binds to certain carbohydrates on human cells, the first step in viral infection. Changes in the hemagglutinin are the major source of antigenic shift.

Innate immunity the host defense mechanisms that act from the start of an infection and do not adapt to a particular pathogen. Also called the **innate immune response**.

Insulin-dependent diabetes mellitus (IDDM) an autoimmune disease in which insulin-secreting β cells of the pancreatic islets of Langerhans are gradually destroyed.

Insulitis an infiltration of the pancreatic islets of Langerhans with lymphocytes and other leukocytes. This is a symptom of incipient diabetes.

Integrins a class of cell-surface glycoproteins that mediate adhesive interactions between cells and the extracellular matrix.

Interallelic conversion a mechanism of genetic recombination between two alleles of a locus in which a segment of one allele is replaced with the homologous segment from the other. This mechanism is used to generate new HLA class I and II alleles.

Intercellular adhesion molecules (ICAMs) ligands for the leukocyte integrins that enable lymphocytes and other leukocytes to bind to other cells, including antigen-presenting cells and endothelial cells.

Interdigitating reticular cells *see* **dendritic cells**

Interferons cytokines that help cells to resist viral infection. **Interferon-α (IFN-α)** and **interferon-β (IFN-β)** are produced by leukocytes and fibroblasts respectively, as well as by other cells, whereas **interferon-γ (IFN-γ)** is a product of CD4 T_H1 cells, CD8 T cells, and NK cells. IFN-γ acts principally to activate macrophages.

Interferon response changes in the expression of a variety of human genes in cells exposed to interferon.

Interferon-producing cells (IPCs) specialized lymphocyte-like cells in the blood that secrete up to 1000-fold more type I interferon than other cells. *See also* **plasmacytoid dendritic cells**.

Interleukin (IL) a generic term used for many of the cytokines produced by leukocytes. *See also* **IL-1, IL-2, IL-3**, etc.

Intermolecular epitope spreading the process by which the immune response initially reacts against an epitope of one antigenic molecule and then progresses to epitopes on different proteins.

Intramolecular epitope spreading the process by which the immune response initially reacts against epitopes in one part of an antigenic molecule and then progresses to other, non-cross-reactive, epitopes of the same molecule.

Invariant chain (Ii) polypeptide that associates with major histocompatibility complex (MHC) class II proteins in the endoplasmic reticulum and prevents them from binding peptides there. It guides the MHC class II molecules to endosomes where Ii is degraded, enabling MHC class II molecules to bind peptides present in the endosomes.

Ischemia deficient blood supply due to obstructed blood vessels.

ISCOMs *see* **immune stimulatory complexes**

Islets of Langerhans the endocrine hormone-producing tissue of the pancreas, which includes the β cells that produce insulin.

Isoforms the different forms of a protein that are encoded by the alleles of a gene or by different but closely related genes.

Isograft a tissue or organ graft from one genetically identical individual to another.

Isotype switching the process by which a B cell changes the class of immunoglobulin made while preserving the antigenic specificity of the immunoglobulin. Isotype switching involves somatic recombination that attaches a different heavy-chain constant-region gene to the variable-region exon.

Isotypes classes of immunoglobulin—IgM, IgG, IgD, IgA, and IgE—each of which has a distinct heavy-chain constant region encoded by a different constant-region gene. The heavy-chain constant region determines the effector properties of each antibody class.

ITAMs *see* **immunoreceptor tyrosine-based activation motifs**

ITIMs immunoreceptor tyrosine-based inhibitory motifs. *See also* **immunoreceptor tyrosine-based activation motifs**.

J gene segments *see* **joining (J) gene segments**

Janus kinases (JAKs) a family of tyrosine kinases that transduce activating signals from cytokine receptors.

Joining (J) gene segments one of the types of gene segment in immunoglobulin and T-cell receptor genes that is rearranged to make functional variable-region exons.

Junctional diversity diversity present in immunoglobulin and T-cell receptor polypeptides that is created during the process of gene rearrangement by the addition of nucleotides into the junctions between gene segments.

kappa (κ) one of the two types of immunoglobulin light chain.

Killed virus vaccines vaccines that contain viral particles that have been deliberately killed by heat, chemicals, or radiation.

Killer cell immunoglobulin-like receptors (KIRs) receptors on NK cells that bind to MHC class I molecules and can send either activating or inhibitory signals to the NK cell.

Kupffer cells phagocytic cells of the liver that line the hepatic sinusoids.

λ5 *see* **pre-B-cell receptor**

L chain *see* **light chain**

L-selectin an adhesion molecule of the selectin family found on lymphocytes. It binds to CD34 and GlyCAM-1 on high endothelial venules to initiate the migration of naive lymphocytes into secondary lymphoid tissue.

lambda (λ) one of the two types of immunoglobulin light chain.

Langerhans' cells phagocytic dendritic cells found in the epidermis. They can migrate in lymph from the epidermis to lymph nodes, where they differentiate into dendritic cells.

Large granular lymphocytes *see* **natural killer cells**

Large pre-B cells immature B cells that have a cell-surface pre-B-cell receptor.

Late pro-B cell *see* **pro-B cell**

Late-phase reaction that part of a type 1 immediate hypersensitivity reaction that occurs 7–12 hours after contact with antigen and is resistant to treatment with anti-histamine.

Latency a state adopted by some viruses in which they have entered cells but do not replicate.

Lck a protein tyrosine kinase associated with the CD4 and CD8 co-receptors of T cells.

Lectin pathway of complement activation one of the three pathways of complement activation. It is activated by the binding of a mannose-binding lectin present in blood plasma to mannose-containing peptidoglycans on bacterial surfaces. *See also* **alternative pathway of complement activation; classical pathway of complement activation**.

Lentiviruses group of slow retroviruses that includes the human immunodeficiency virus (HIV). They have a long incubation period and disease can take years to become apparent.

Leukemia the unrestrained proliferation of a malignant white blood cell characterized by abnormally high numbers of white cells in the blood. A leukemia can be lymphocytic, myelocytic, or monocytic depending on the type of white blood cell that became malignant.

Leukocyte a general term for a white blood cell. Lymphocytes, granulocytes, and monocytes are all leukocytes.

Leukocyte adhesion deficiency an immunodeficiency disease in which the common β chain of the leukocyte integrins is absent. This mainly affects the ability of leukocytes to enter sites infected with extracellular pathogens, so such infections cannot be effectively eradicated.

Leukocytosis increased numbers of leukocytes in the blood. It is commonly seen in acute infection.

LFA-1 *see* **lymphocyte function-associated antigen-1**

LFA-3 *see* **lymphocyte function-associated antigen-3**

Light chain (L chain) the smaller of the two types of polypeptide chain that make up immunoglobulins. It consists of one variable and one constant domain, and in the immunoglobulin molecule it is disulfide-bonded to a heavy chain. There are two classes of light chain, known as κ and λ.

Light zone the part of a germinal center in secondary lymphoid tissue that contains non-dividing centrocytes interacting with follicular dendritic cells.

Linear epitope epitope of a protein recognized by antibody that consists of a linear sequence of amino acids within the protein's primary structure.

Linkage disequilibrium (LD) the situation when particular alleles of two or more polymorphic genes (for example those comprising an HLA haplotype) are inherited together at frequencies higher than expected by chance.

Live-attenuated viral vaccines vaccines composed of live viruses that have an accumulation of mutations that impedes their growth in human cells and their capacity to cause disease.

LPS-binding protein plasma protein that binds bacterial lipopolysaccharide (LPS) and delivers it to LPS receptors on neutrophils and macrophages.

LT *see* **lymphotoxin**

Lymph mixture of extracellular fluid and cells that is carried by the lymphatic system.

Lymph nodes a type of secondary lymphoid tissue found at many sites in the body where lymphatic vessels converge. Antigens are delivered by the lymph and presented to lymphocytes within the lymph node where adaptive immune responses are initiated.

Lymphatic vessels (lymphatics) thin-walled vessels that carry lymph from tissues to secondary lymphoid tissues (with the exception of the spleen) and from secondary lymphoid tissues to the thoracic duct.

Lymphocyte function-associated antigen-1 (LFA-1) one of the leukocyte integrins that mediate the adhesion of lymphocytes, especially T cells, to endothelial cells and antigen-presenting cells.

Lymphocyte function-associated antigen-3 (LFA-3) cell adhesion molecule of the immunoglobulin superfamily that is expressed on antigen-presenting cells (and other types of cell) and mediates their adhesion to T cells.

Lymphocytes a class of white blood cells that consist of small and large lymphocytes. The small lymphocytes bear variable cell-surface receptors for antigen and are responsible for adaptive immune responses. There are two main classes of small lymphocyte—B lymphocytes (B cells) and T lymphocytes (T cells). Large granular lymphocytes are natural killer (NK) cells, lymphocytes of innate immunity.

Lymphoid containing lymphocytes, or pertaining to lymphocytes.

Lymphoid lineage all types of lymphocyte, and the bone marrow cells that give rise to them.

Lymphoid organs (lymphoid tissues) organized tissues that contain very large numbers of lymphocytes held in a non-lymphoid stroma. The primary lymphoid organs, where lymphocytes are generated, are the thymus and bone marrow. The main secondary lymphoid tissues, in which adaptive immune responses are initiated, are the lymph nodes, spleen, and mucosa-associated lymphoid tissues such as tonsils, Peyer's patches, and the appendix.

Lymphoid progenitor stem cell in bone marrow cell that gives rise to all lymphocytes.

Lymphokines cytokines produced by lymphocytes.

Lymphomas tumors of lymphocytes that grow in lymphoid and other tissues but do not enter the blood in large numbers.

Lymphotoxin (LT) a cytotoxic cytokine secreted by some CD4 T cells. Also known as tumor necrosis factor-β (TNF-β).

Lytic granules intracellular storage granules of cytotoxic T cells and NK cells that contain perforin and granzymes.

M cells specialized cells in intestinal epithelium through which antigens and pathogens enter gut-associated lymphoid tissue from the intestines. Short for microfold cells.

Macrophage activation stimulation of macrophages, which increases their phagocytic, antigen-presenting, and bacterial killing functions. It occurs in the course of infection.

Macrophages large mononuclear phagocytic cells resident in most tissues. They are derived from blood monocytes and contribute to innate immunity and early nonadaptive phases of host defense. They function as professional antigen-presenting cells and as effector cells in humoral and cell-mediated immunity.

MAdCAM-1 mucosal cell adhesion molecule-1, a mucosal addressin that is recognized by the lymphocyte surface proteins L-selectin and VLA-4. This interaction mediates the specific homing of lymphocytes to mucosal tissues.

Major basic protein a constituent of eosinophil granules that is released on eosinophil activation. It acts on mast cells to cause their degranulation.

Major histocompatibility complex (MHC) a cluster of genes on the short arm of human chromosome 6 that encodes a set of polymorphic membrane glycoproteins called the MHC molecules, which are involved in presenting peptide antigens to T cells.

Major histocompatibility complex (MHC) molecules *see* **MHC molecules**

Malignant transformation the changes that occur in a cell to make it cancerous.

Malignant tumors tumors that are capable of uncontrolled and invasive growth.

MALT *see* **mucosa-associated lymphoid tissues**

Mannose-binding lectin (MBL) an acute-phase protein in the blood that binds to mannose residues on pathogen surfaces and when bound activates the complement system. It is also known as **mannose-binding protein (MBP)** and **mannan-binding lectin**.

Mannose-binding protein pathway *see* **lectin-mediated pathway**

Mantle zone in lymphoid follicles, a rim of B lymphocytes that surrounds a follicle.

Mast cells large bone marrow derived cells found resident in connective tissues throughout the body. They contain large granules that store a variety of chemical mediators including histamine. Mast cells have high-affinity Fcε receptors (FcεRI) that bind free IgE. Antigen binding to mast cell associated IgE triggers mast cell activation and degranulation, producing a local or systemic immediate hypersensitivity reaction. Mast cells have a crucial role in allergic reactions.

Mature B cells B cells that have IgM and IgD on their surface and are able to respond to antigen.

Mature dendritic cells dendritic cells in secondary lymphoid tissues that express co-stimulatory molecules and other cell-surface molecules that enable them to present antigen to naive T cells and activate them.

Megakaryocytes large cells of the erythroid lineage, produced in the bone marrow and resident there. They produce platelets.

Membrane co-factor protein (MCP) complement regulatory protein on human cells that promotes the inactivation of C3b and C4b by factor I.

Membrane-attack complex the complex of terminal complement components that forms a pore in the membrane of the target cell, damaging the membrane and leading to cell lysis.

Memory B cells long-lived antigen-specific B cells that are produced from activated naive B cells during the primary immune response to an antigen. On subsequent exposure to their specific antigen they are reactivated to differentiate into plasma cells as part of the secondary and subsequent immune responses.

Memory cells general term for lymphocytes that are responsible for the phenomenon of immunological memory and protective immunity.

Memory T cells long-lived antigen-specific T cells that are activated in secondary and subsequent immune responses to an antigen.

Metastasis the spread of a tumor from its site of origin to other tissues. Some cells from the primary tumor invade other tissues and grow and divide to become secondary tumors.

Methotrexate cytotoxic drug used to inhibit graft-versus-host reactions in bone marrow transplant recipients.

MHC *see* **major histocompatibility complex**

MHC class I molecules the class of MHC molecules that present peptides generated in the cytosol to CD8 T cells. They consist of a heterodimer of a class I heavy chain associated with β_2-microglobulin.

MHC class II molecules the class of MHC molecules that present peptides generated in intracellular vesicles to CD4 T cells. They consist of a heterodimer of class II α and β chains.

MHC class II transactivator (CIITA) a transcriptional activator of MHC class II genes. When defective it causes a type of bare lymphocyte syndrome.

MHC molecules major histocompatibility complex molecules. Highly polymorphic glycoproteins encoded by the major histocompatibility complex (MHC). They form complexes with peptides and present peptide antigens to T cells. There are two classes—MHC class I and MHC class II molecules—with different roles in the immune response. They are also known as major histocompatibility antigens because they are the main alloantigens involved in the rejection of transplanted tissues.

MHC restriction the fact that a given T-cell receptor will recognize its peptide antigen only when the peptide is bound to a particular form of MHC molecule.

Microfold cells *see* **M cells**

Minor histocompatibility antigens (minor H antigens) peptides of polymorphic cellular proteins that can lead to graft rejection when they are bound by MHC molecules and recognized by T cells.

Minor histocompatibility loci the genes encoding proteins that can act as minor histocompatibility antigens.

Mixed lymphocyte reaction cellular assay for detecting MHC differences between two individuals. The T cells from one individual proliferate in response to allogeneic MHC molecules on the cells of the other individual.

Molecular mimicry antigenic similarity between a pathogen antigen and a cellular antigen, which results in the induction of antibodies or T cells that act against the pathogen but also cross-react with the self antigen.

Monoclonal antibodies antibodies produced by a single clone of B lymphocytes and that are therefore identical in structure and antigen specificity.

Monocytes white blood cells with a bean-shaped nucleus. They are the precursors of tissue macrophages.

Monocyte chemoattractant protein (MCP-1) *see* **CCL2**

Monomorphic having only one form. The term is used, for example, for genes that have only one allele.

Mucosa the mucus-secreting epithelia that line the respiratory, intestinal, and urogenital tracts. The conjunctiva of the eye and the mammary glands are also in this category.

Mucosa-associated lymphoid tissue (MALT) aggregations of lymphoid cells in mucosal epithelia and in the lamina propria beneath.

The main mucosa-associated lymphoid tissues are the gut-associated lymphoid tissues (GALT) and the bronchial-associated lymphoid tissues (BALT).

Mucosal surfaces *see* **mucosa**

Mucus slimy protective secretion produced by many internal epithelia.

Multiple sclerosis a chronic progressive neurological disease characterized by patches of demyelination in the central nervous system and lymphocyte infiltration into the brain. It is believed to be an autoimmune disease.

Multivalent having more than one binding site for the same or different ligands.

Mutagen any agent, such as a chemical or radiation, that can cause a mutation.

Mutation an alteration in the DNA sequence of a gene.

Myasthenia gravis an autoimmune disease in which autoantibodies against the acetylcholine receptor on skeletal muscle cells cause a block of signal transmission from nerve to muscle at neuromuscular junctions, leading to progressive muscle weakness and eventually to death.

Myeloablative therapy a conditioning regime used before a bone marrow transplant. The recipient's immune system is destroyed with a combination of cytotoxic drugs and irradiation.

Myeloid lineage a subset of bone-marrow derived cells comprising granulocytes, monocytes, and macrophages.

Myeloid progenitors stem cells in the bone marrow that give rise to granulocytes, monocytes, and macrophages.

Myelomas tumors of plasma cells resident in bone marrow.

Myelopoiesis the production of monocytes and granulocytes in the bone marrow.

N-nucleotides nucleotides added at the junctions between gene segments of T-cell receptor and immunoglobulin heavy-chain variable-region sequences during somatic recombination, which contribute to the diversity of these molecules. They are not encoded in the gene segments but are inserted by the enzyme terminal deoxynucleotidyltransferase (TdT).

Naive B cell a mature B cell that has left the bone marrow but has not yet encountered its specific antigen.

Naive T cell a mature T cell that has left the thymus but not yet encountered its specific antigen.

Natural cytotoxicity receptors (NCRs) a group of activating NK-cell receptors of the immunoglobulin superfamily.

Natural interferon-producing cells (NIPCs) *see* **interferon-producing cells**

Natural killer cells (NK cells) large, granular, cytotoxic lymphocytes that circulate in the blood. NK cells are important in innate immunity to viruses and other intracellular pathogens and also kill certain tumor cells. They are the cytotoxic cells in antibody-dependent cell-mediated cytotoxicity (ADCC).

NCR *see* **natural cytotoxicity receptor**

Necrosis the death of cells by lysis that results from chemical or physical injury. It leaves extensive cellular debris that must be removed by phagocytes. Neighboring tissue is also damaged by the molecules released from necrotic cells.

Negative selection process in the thymus whereby developing T cells that recognize self antigens are induced to die by apoptosis.

Neoplasm a tumor, which may be benign or malignant.

Neutralization the mechanism by which antibodies binding to sites on pathogens prevent growth of the pathogen and/or its entry into cells. The toxicity of bacterial toxins can similarly be **neutralized** by bound antibody.

Neutropenia a deficiency of neutrophils in the blood.

Neutrophils phagocytic white blood cells that enter infected tissues in large numbers. They contain granules that stain with neutral dyes. After their entry into infected tissues, neutrophils engulf and kill extracellular pathogens in large numbers. They are also known as neutrophilic polymorphonuclear leukocytes and are a type of granulocyte. The most abundant white blood cell.

NFAT (nuclear factor of activated T cells) a transcription factor that is activated as a result of signaling from the T-cell receptor. It functions as a complex of the NFAT protein with the dimer of Fos and Jun proteins known as AP-1.

NFκB nuclear factor κB, a transcription factor that helps turn on the expression of many immune system genes.

NIPC *see* **interferon-producing cell**

NK cells *see* **natural killer cells**

NK-cell lectin-like receptors one of the two main structural classes of NK-cell receptors. They include both activating and inhibitory receptors.

Non-obese diabetic (NOD) mice a strain of mice in which diabetes arises spontaneously and has clinical and immunological characteristics similar to those in human insulin-dependent diabetes mellitus.

Oligomorphic describes genes that have only a small number of different alleles in the population.

Oncogenes genes involved in the control of cell growth. When these genes are defective in either structure or expression, they can cause cells to proliferate abnormally and form a tumor.

Oncogenic viruses viruses that are involved in causing cancer.

Oncology the clinical discipline that deals with the study, diagnosis, and treatment of cancer.

Opportunistic pathogen a microorganism that causes disease only in individuals whose immune systems are in some way compromised.

Opsonins antibodies and complement components that bind to pathogens and facilitate their phagocytosis by neutrophils or macrophages.

Opsonization the coating of the surface of a pathogen or other particle with any molecule that makes it more readily ingested by phagocytes. Antibody and complement opsonize extracellular bacteria for phagocytosis by neutrophils and macrophages because the phagocytic cells carry receptors for these molecules.

Oral tolerance the fact that harmless antigens, such as those in food, taken into the body by mouth do not usually cause an immune response.

Organ-specific autoimmune diseases autoimmune diseases targeted at a particular organ, such as the thyroid in Graves' disease.

Original antigenic sin a bias seen in successive immune responses to structurally related antigens such as those on different strains of influenza virus. On infection for the second time with influenza, the antibody response is restricted to epitopes that the second strain shares with the first strain to which the person was exposed. Other highly immunogenic epitopes on the second and subsequent viruses are ignored.

P-nucleotides nucleotides added into the junctions between gene segments during the somatic recombination that generates a rearranged variable-region sequence. They are an inverse repeat (a palindrome) of the nucleotide sequence at the end of the adjacent gene segment, which gives them their name of palindromic or P-nucleotides.

P-selectin *see* **selectins**

Pandemic outbreak of an infectious disease that spreads worldwide.

Panel reactive antibody (PRA) a measure of the likelihood that a patient seeking a transplant is sensitized to potential donors. The patient's serum is tested against tissue from a representative panel of individuals for antibodies that would cause immediate hyperacute rejection of a graft. The PRA is the percentage of individuals in the panel whose cells react with the patient's antibodies.

Paracrine term applied to a cytokine that is released from one type of cell and acts on other cells nearby.

Parasites the unicellular protozoa and multicellular worms that infect animals and humans and live within them.

Paroxysmal nocturnal hemoglobinuria (PNH) a disease in which the complement regulatory proteins CD59 and DAF are defective, so that complement activation leads to episodes of spontaneous hemolysis. The defect is in the attachment of CD59 and DAF to cell membranes via a glycolipid anchor.

Passive immunization the injection of specific antibodies to provide protection against a pathogen or toxin. The administered antibodies may derive from human blood donors, immunized animals or hybridoma cell lines.

Passive transfer of immunity transfer of immunity to a non-immune individual by the injection of specific antibody, immune serum, or T cells.

Pathogen an organism, most commonly a microorganism, that can cause disease.

Pathology the scientific study of disease or detectable damage to tissue caused by disease.

Pemphigus foliaceus and **pemphigus vulgaris** autoimmune diseases involving blistering of the skin.

Pentraxins a family of acute-phase proteins to which C-reactive protein belongs. They are formed of five identical subunits.

Peptide-binding motif of an MHC isoform, the combination of anchor residues that are common to the amino-acid sequences of peptides that bind to the isoform.

Perforin one of the proteins released by cytotoxic T cells on contact with their target cells. It forms pores in the target cell membrane that contribute to cell killing.

Peripheral lymphoid tissues *see* **secondary lymphoid tissues**

Peyer's patches gut-associated lymphoid tissue present in the wall of the small intestine, especially the ileum.

Phagocyte a cell specialized to perform phagocytosis. The principal phagocytic cells in mammals are neutrophils and macrophages.

Phagocytosis cellular internalization of particulate matter, such as bacteria, by means of endocytosis. Cells specialized in phagocytosis are known as phagocytes.

Phagolysosome intracellular vesicle formed by fusion of a phagosome with a lysosome, in which the phagocytosed material is broken down by degradative lysosomal enzymes.

Phagosome intracellular vesicle containing material taken up by phagocytosis.

Plasma cells terminally differentiated B lymphocytes that secrete antibody.

Plasmablast an activated B cell differentiating within a lymphoid follicle at the stage at which it shows some features of a plasma cell.

Plasmacytoid dendritic cells (PDCs) dendritic-like cells derived from interferon-producing cells cultured with microbial products and inflammatory cytokines.

Pluripotent hematopoietic stem cell stem cell in bone marrow that gives rise to all the cellular elements of the blood.

PNH *see* **paroxysmal nocturnal hemoglobinuria**

Poly-Ig receptor a receptor present on the basolateral membrane of epithelial cells that binds polymeric immunoglobulins, especially dimeric IgA, and transports them across the epithelium by transcytosis.

Polymorphism the existence of different variants of a gene or trait in the population. Genetic polymorphism is defined as the existence of two or more forms (alleles) of a given gene within the population, with the variant alleles each occurring at a frequency greater than 1%.

Polymorphonuclear leukocytes *see* **granulocytes**

Polyspecificity the ability to bind to many different antigens, a property shown by some antibodies. It is also known as polyreactivity.

Positive selection a process in the thymus that selects immature T cells with receptors that recognize peptide antigens presented by self-MHC molecules. Only cells that are positively selected are allowed to continue their maturation.

PRA *see* **panel reactive antibody**

Pre-B cells a stage in B-cell development at which the B cells have rearranged their heavy-chain genes but not their light-chain genes.

Pre-B-cell receptor immunoglobulin-like receptor that is expressed on the surface of pre-B cells. It consists of μ heavy chains in association with surrogate light chains. Its appearance signals the cessation of heavy-chain gene rearrangement.

Pre-T-cell receptor receptor that is present on the surface of some immature thymocytes. It consists of a T-cell receptor β chain associated with a surrogate α chain called pTα.

Prednisolone the biologically active compound to which the immunosuppressive drug prednisone is converted *in vivo*.

Prednisone a synthetic steroid drug with potent anti-inflammatory and immunosuppressive activity. Prescribed to patients who have had organ transplants.

Present cells carrying cell-surface complexes of peptide antigens and MHC molecules are said to present these antigens to T lymphocytes.

Primary immune response or **primary response** the adaptive immune response that follows a person's first exposure to an antigen.

Primary lymphoid follicles the name given to the B-cell areas of secondary lymphoid tissues in the absence of an immune response. They contain resting B lymphocytes.

Primary lymphoid tissues anatomical sites of lymphocyte development. The bone marrow and the thymus gland.

Pro-B cells an early stage in B-cell development at which B-cell precursors express B-cell marker proteins and rearrange their heavy-chain genes. D_H to J_H joining occurs at the **early pro-B cell** stage, followed by V_H to DJ_H joining at the **late pro-B cell** stage.

Pro-drug biologically inactive compound that is metabolized in the body to produce an active drug.

Productive rearrangements DNA rearrangements within immunoglobulin and T-cell receptor genes that lead to a gene that can direct the synthesis of a functional polypeptide chain.

Professional antigen-presenting cells cells that can present antigen to naive T cells and activate them. A professional antigen-presenting cell not only displays peptide antigens bound to appropriate MHC molecules but also has co-stimulatory molecules on its surface that are needed to activate the T cell. Only dendritic cells, macrophages, and B cells can be professional antigen-presenting cells.

Programmed cell death *see* **apoptosis**

Properdin a blood protein that helps to activate the alternative pathway of complement activation. It binds to and stabilizes the alternative pathway C3 and C5 convertases on the surfaces of bacterial cells. It is also known as factor P.

Proteasome large multisubunit protease present in the cytosol of all cells that degrades cytoplasmic proteins. It generates the peptides presented by MHC class I molecules.

Protective immunity the specific immunological resistance to a pathogen that follows either from specific vaccination or recovery from an infection with the pathogen.

Protein tyrosine kinases enzymes that phosphorylate tyrosine residues on proteins. They are involved in signal transduction from many types of cell-surface receptor, including the antigen receptors on B cells and T cells.

Proto-oncogenes cellular genes that regulate the control of cell growth and division. When mutated or aberrantly expressed, they contribute to the malignant transformation of cells, which leads to cancer.

Provirus the DNA form of a retrovirus when it is integrated into the host cell genome. In this state it can remain transcriptionally inactive for a long time.

pTα *see* **pre-T-cell receptor**

Purine nucleotide phophorylase (PNP) enzyme involved in purine metabolism. Its deficiency results in the accumulation of purine nucleosides, which are toxic for developing T cells. This leads to severe combined immunodeficiency.

Pus thick yellowish-white fluid that is formed in infected wounds. It is composed of dead and dying white blood cells (principally neutrophils), tissue debris, and dead microorganisms.

Pyogenic bacteria extracellular encapsulated bacteria that cause the formation of pus at sites of infection.

RAG-1, RAG-2 *see* **recombination activation genes**

Rapamycin *see* **sirolimus**

Receptor editing a process in developing B cells that have generated a self-reactive receptor by which further rounds of light-chain gene rearrangement can produce a new light chain to replace the self-reactive one.

Recirculation of lymphocytes, their continual movement from blood to secondary lymphoid tissues to lymph and back to the blood. An exception to this pattern is traffic to the spleen; lymphocytes both enter and leave the spleen in the blood.

Recombination activation genes (**RAG-1**, **RAG-2**) two genes whose expression is required for immunoglobulin and T-cell receptor gene rearrangement in B cells and T cells.

Recombination signal sequences (RSSs) short stretches of DNA that flank the gene segments that are rearranged to generate V-region exons. They are the sites at which somatic recombination occurs.

Regulators of complement activation (RCA) proteins that regulate the activity of complement and contain a particular structural motif called the CCP motif.

Regulatory CD4 T cells (T$_R$) antigen-specific CD4 T cells whose actions can suppress immune responses.

Respiratory burst metabolic change accompanied by a transient increase in oxygen consumption that occurs in neutrophils and macrophages when they have taken up opsonized particles. It leads to the generation of toxic oxygen metabolites and other anti-bacterial substances that attack the phagocytosed material.

Respiratory syncytial virus (RSV) a paramyxovirus that is a common cause of severe chest infection in young children. Often associated with wheezing.

Rev protein product of the rev gene of the human immunodeficiency virus (HIV). The protein promotes the passage of viral RNA from nucleus to cytoplasm during HIV replication.

Rheumatic fever autoimmune disease involving inflammation of the heart, joints, and kidneys, which can follow 2–3 weeks after a throat infection with certain strains of *Streptococcus pyogenes*. It is due to antibodies made against bacterial antigens cross-reacting with components of heart tissue, and the continued deposition of immune complexes.

Rheumatoid arthritis a common inflammatory disease of joints that is due to an autoimmune response.

Rheumatoid factor an IgM antibody with specificity for human IgG that is produced in some people with rheumatoid arthritis.

RSS *see* **recombination signal sequences**

RSV *see* **respiratory syncytial virus**

Sarcoma a tumor that arises from a cell of connective tissue.

SCID, scid *see* **severe combined immunodeficiency**

SE *see* **staphylococcal enterotoxins**

Secondary immune response or **secondary response** the adaptive immune response provoked by a second exposure to an antigen. It differs from the primary response by starting sooner and building more quickly.

Secondary lymphoid follicle the name given to the B-cell areas of secondary lymphoid tissues that are responding to antigen. They contain proliferating B cells.

Secondary lymphoid tissues the lymph nodes, spleen and mucosa-associated lymphoid tissues. These are the tissues in which immune responses are initiated. The more highly organized tissues such as lymph nodes and spleen are also often known as **secondary lymphoid organs**.

Secretory component (secretory piece) fragment of the poly-Ig receptor left attached to dimeric IgA after its transport across epithelial cells.

Segmental exchange *see* **gene conversion**

Selectins family of adhesion molecules present on the surfaces of leukocytes and endothelial cells. They are lectins and bind to sugar moieties on certain mucin-like glycoproteins.

Self antigens a term used to describe all the normal constituents of the body to which the immune system would respond were it not for the mechanisms of tolerance that destroy or inactivate self-reactive B and T cells.

Self-MHC a person's own MHC molecules.

Self peptides peptides produced from the body's own proteins. In the absence of infection these peptides occupy the peptide-binding sites of MHC molecules on cell surfaces.

Self-renewal the ability of a population of cells to renew itself.

Self-tolerance the normal situation whereby a person's immune system does not respond to constituents of the person's body.

Sensitization in connection with allergy, the first exposure to an allergen that elicits an IgE response. Allergic reactions occur only in individuals who have already been **sensitized**.

Sepsis the toxic effects of infection of the bloodstream. Usually caused by Gram-negative bacteria. *See also* **septic shock**.

Septic shock shock syndrome that is frequently fatal, caused by the systemic release of the cytokine TNF-α after bacterial infection of the bloodstream, usually with Gram-negative bacteria.

Septicemia bacterial invasion of the blood.

Seroconversion the phase of an infection when antibodies against the infecting agent are first detectable in the blood.

Serotypes antigenically different strains of a bacterium or other pathogen that can be distinguished by immunological means, for example by antibody-based detection tests. Also used to describe human alloantigens such as HLA and blood group antigens.

Serum sickness the collection of symptoms that follow the injection of large amounts of a foreign molecule (such as a serum protein) into a person. It is caused by the immune complexes formed between the injected protein and the antibodies that are made against it. Characterized by fever, arthralgias, and nephritis.

Severe combined immune deficiency (SCID) immune deficiency disease in which neither antibody nor T-cell responses are made. It is usually the result of genetic defects that lead to T-cell deficiencies, and is fatal in childhood if not treated.

Single-positive thymocytes a late stage of T-cell development in the thymus characterized by the expression of either the CD4 or the CD8 co-receptor on the cell surface.

Sirolimus the proprietary name for the immunosuppressive drug rapamycin. Both names are in common use.

SLE *see* **systemic lupus erythematosus**

Small lymphocytes the general name for resting T and B lymphocytes as they recirculate.

Small pre-B cells pre-B cells in which light-chain gene rearrangements are being made.

Somatic gene therapy a possible therapy for curing inherited immunodeficiency diseases. Stem cells isolated from the patient are transfected with a normal copy of the gene that is defective in the patient. The stem cells, which now express a good copy of the formerly defective gene, are reinfused into the patient's circulation.

Somatic hypermutation mutation that occurs at high frequency in the rearranged variable-region DNA of immunoglobulin genes in activated B cells, resulting in the production of variant antibodies, some of which have a higher affinity for the antigen.

Somatic recombination DNA recombination that occurs between gene segments in the immunoglobulin genes and T-cell receptor genes in developing B cells and T cells respectively. It generates a complete exon that encodes the variable region of an immunoglobulin or T-cell receptor polypeptide chain.

Specificity the property of antibodies and other antigen-binding molecules for selective interaction with only one or a few types of molecule or cell.

Spleen organ situated adjacent to the cardiac end of the stomach. One function of the spleen is to remove old or damaged red blood cells from the circulation; the other is as a secondary lymphoid organ that responds to blood-borne pathogens and antigens.

Staphylococcal enterotoxins (SEs) toxins secreted by strains of staphylococcus that act on the gut and cause the symptoms of food poisoning. They are also superantigens.

STATs signal transducers and activators of transcription. Proteins that are activated to become transcription factors through the Janus kinase signaling pathway from cytokine receptors.

Stromal cells cells that provide the supporting framework of a tissue or organ.

Subunit vaccines vaccines composed only of isolated antigenic components of a pathogen and not the pathogen itself, either alive or dead.

Superantigens molecules that, by binding non-specifically to MHC class II molecules and T-cell receptors, stimulate the polyclonal activation of T cells.

Suppressor CD4 T cells *see* **regulatory CD4 T cells**.

Surrogate light chain a protein that mimics an immunoglobulin light chain and made up of two subunits, VpreB and λ5. It is produced by pro-B cells and together with μ heavy chain forms the pre-B-cell receptor.

Switch regions (switch sequences) DNA sequences preceding heavy-chain constant-region genes at which somatic recombination occurs when B cells switch from producing one immunoglobulin isotype to another.

Sympathetic ophthalmia an autoimmune response that sometimes follows damage to an eye and affects both the damaged eye and the healthy eye.

Syngeneic genetically identical. A syngeneic tissue graft is one made between two genetically identical individuals. It does not provoke an immune response or a rejection reaction.

Systemic anaphylaxis a rapid-onset and potentially fatal form of IgE-mediated allergic reaction, in which antigen in the bloodstream triggers the activation of mast cells throughout the body, causing circulatory collapse and suffocation due to tracheal swelling.

Systemic autoimmune disease (systemic autoimmunity) autoimmunity that involves reactions to common components of the body and whose effects are not confined to one particular organ.

Systemic lupus erythematosus a systemic autoimmune disease in which autoantibodies made against DNA, RNA, and nucleoprotein particles form immune complexes that damage small blood vessels.

T cells (T lymphocytes) lymphocytes that develop in the thymus and are responsible for cell-mediated immunity. Their cell-surface antigen receptor is called the T-cell receptor.

T-cell activation the stimulation of mature naive T cells by antigen presented to them by professional antigen-presenting cells. It leads to their proliferation and differentiation into effector T cells.

T-cell areas parts of secondary lymphoid tissues where the lymphocytes are predominantly T cells.

T-cell priming the activation of mature naive T cells by antigen presented to them by professional antigen-presenting cells.

T-cell receptor α chain (TCRα) one of the two polypeptide chains that make up the most common form of T-cell receptor.

T-cell receptor β chain (TCRβ) one of the two polypeptide chains that make up the most common form of T-cell receptor.

T-cell receptor complex the complex of T-cell receptor α and β chains and the invariant CD3 and ζ chains that makes up a functional antigen receptor on the T-cell surface.

T-cell receptor the highly variable antigen receptor of T lymphocytes. On most T cells it is composed of a variable α chain and β chain and is known as the α:β T-cell receptor. On a minority of T cells, the variable chains are γ and δ chains, and this receptor is known as the γ:δ T-cell receptor. Both types of receptor are present at the cell surface in association with the complex of invariant CD3 chains and ζ chains, which have a signaling function.

T lymphocyte *see* **T cells**

Tacrolimus immunosuppressive polypeptide drug that inactivates T cells by inhibiting signal transduction from the T-cell receptor. It is widely used to suppress transplant rejection. It is also known as FK506.

TAP *see* **transporter of antigen processing**

Tapasin TAP-associated protein, a chaperone protein involved in the assembly of peptide–MHC class I molecule complexes in the endoplasmic reticulum.

Target cell any cell that is acted on directly by effector T cells, effector cells or molecules. For example, virus-infected cells are the targets of cytotoxic T cells, which kill them, and naive B cells are the targets of effector CD4 T cells, which help to stimulate them to produce antibodies.

Tat protein product of the tat gene of the human immunodeficiency virus (HIV), which enhances the rate of transcription of viral RNA.

TD antigens *see* **thymus-dependent antigens**

Terminal deoxynucleotidyltransferase (**TdT**) enzyme that inserts nontemplated nucleotides (N-nucleotides) into the junctions between gene segments during the rearrangement of T-cell receptor and immunoglobulin heavy-chain genes.

Tetanus toxin protein neurotoxin produced by the bacterium *Clostridium tetani*. The cause of the disease tetanus.

TGF-β *see* **transforming growth factor-β**

T$_H$1 cells a subset of CD4 T cells that are characterized by the cytokines they produce. They are involved mainly in activating macrophages. Also called inflammatory T cells.

T$_H$2 cells a subset of CD4 T cells that are characterized by the cytokines they produce. They are involved mainly in stimulating B cells to produce antibody. Also called helper T cells.

Thymectomy the surgical removal of the thymus gland.

Thymic anlage the tissue from which the thymic stroma develops during embryogenesis.

Thymic stroma the reticular epithelial cells and connective tissue of the thymus, which form the essential microenvironment for T-cell development.

Thymocytes developing T cells in the thymus.

Thymus a lymphoepithelial organ in the upper part of the middle of the chest, just behind the breastbone. It is the site of T-cell development.

Thymus-dependent antigens (**TD antigens**) antigens for which the elicitation of an antibody response on immunization or vaccination is dependent upon help provided by antigen-specific T cells. Most protein antigens are of this type.

Thymus-dependent lymphocyte *see* **T cells**

Thymus-independent antigens (**TI antigens**) antigens that can elicit antibody production in the absence of T cells. There are two types of TI antigen: TI-1 antigens, which have intrinsic B-cell activating activity, and TI-2 antigens, which have multiple identical epitopes that cross-link B-cell receptors.

TI-1 antigens, TI-2 antigens *see* **thymus-independent antigens**

Tingible body macrophages phagocytic cells that are seen in histological sections engulfing apoptotic B cells in germinal centers.

TNF-α *see* **tumor necrosis factor-α**

TNF-β *see* **lymphotoxin**

Tolerance when the immune system of a person does not or cannot respond to an antigen. In such cases the individual is said to be **tolerant** of the antigen.

Tonsils large aggregates of lymphoid cells lying on each side of the pharynx.

Toxic shock syndrome a systemic toxic reaction caused by the overproduction of cytokines by CD4 T cells activated by the bacterial superantigen **toxic shock syndrome toxin-1** (**TSST-1**), which is secreted by *Staphylococcus aureus*.

Toxoids toxins that have been deliberately inactivated by heat or chemical treatment so that they are no longer toxic but can still provoke a protective immune response on vaccination.

Toxoplasmosis the disease caused by the protozoan parasite *Toxoplasma gondii*.

Transcytosis the transport of molecules from one side of an epithelium to the other by endocytosis into vesicles within the epithelial cells at one face of the epithelium and release of the vesicles at the other.

Transforming growth factor-β (**TGF-β**) a multifunctional cytokine produced by T$_H$2 cells and other cell types.

Transfusion effect improved outcome of transplantation if the recipient has been given prior blood transfusions from people who share an HLA-DR allotype with the organ with which the patient is subsequently transplanted.

Translocation a chromosomal abnormality in which a piece of one chromosome has become joined to another chromosome. The rearranging genes of B and T cells are often the sites of translocation in B-cell and T-cell tumors.

Transplant rejection immune reaction that is directed towards transplanted tissue and leads to death of the graft.

Transplantation the grafting of organs or tissues from one individual to another.

Transporter associated with antigen processing (**TAP**) an ATP-binding protein in the endoplasmic reticulum membrane that transports peptides from the cytosol to the endoplasmic reticulum lumen. It is composed of two subunits, TAP-1 and TAP-2. It supplies MHC class I molecules with peptides.

Tuberculin test clinical test to detect exposure to *Mycobacterium tuberculosis*, the causal agent of tuberculosis. Subcutaneous injection of purified protein derivative (PPD) of *M. tuberculosis* elicits a delayed-type hypersensitivity reaction in individuals who have had tuberculosis or have been immunized against it.

Tumor a growth arising from uncontrolled cell proliferation. It may be benign and self-limiting, or malignant and invasive.

Tumor antigens cell-surface components of tumor cells that can elicit an immune response in the affected individuals. *See also* **tumor-associated antigens; tumor-specific antigens**.

Tumor-associated antigens antigens characteristic of certain tumor cells that are also present on some types of normal cells.

Tumor necrosis factor-α (**TNF-α**) a cytokine produced by macrophages and T cells that has several functions in the immune response and is the prototype of the TNF family of cytokines. These cytokines function as cell-associated or secreted proteins that interact with receptors of the tumor necrosis factor receptor (TNFR) family.

Tumor-specific antigens antigens that are characteristic of tumor cells and are not expressed by normal cells.

Tumor suppressor genes genes for cellular proteins that function to prevent cells from becoming cancerous.

Type I, type II, type III, type IV hypersensitivity reactions *see* **hypersensitivity reactions**

Type I interferon cytokines (interferons-α and -β) produced by virus-infected cells that interfere with viral replication by the infected cell, and also signal neighboring uninfected cells to prepare for infection.

Unproductive rearrangements DNA rearrangements of immunoglobulin and T-cell receptor genes that produce a gene that cannot make a functional polypeptide chain.

Urticaria the technical term for hives, which are red, itchy skin welts usually brought on by an allergic reaction.

V domain *see* **variable domain**

V gene segment *see* **variable (V) gene segment**

V(D)J recombinase the set of enzymes needed to recombine V, D, and J segments.

V region *see* **variable region**.

Vaccination the deliberate induction of protective immunity to a pathogen by the administration of killed or non-pathogenic forms of the pathogen, or its antigens, to induce an immune response.

Vaccine any preparation made from a pathogen that is used for vaccination and provides protective immunity against infection with the pathogen.

Vaccinia the cowpox virus. It causes a limited infection in humans that leads to immunity to the human smallpox virus. Was used as a vaccine against smallpox.

Variable domain (V domain) the amino-terminal domain of immunoglobulin and T-cell receptor polypeptide chains. Paired variable domains make up the antigen-binding site.

Variable (V) gene segment DNA sequence in the immunoglobulin or T-cell receptor genes that encodes the first 95 or so amino acids of the V domain. There are multiple different V gene segments in the germline genome. To produce a complete exon encoding a V domain, one V gene segment must be rearranged to join up with a J or a rearranged DJ gene segment.

Variable region (V region) the part of an immunoglobulin or T-cell receptor, or of its constituent polypeptide chains, that varies in amino acid sequence between isoforms having different antigenic specificities. It is responsible for determining antigen specificity.

Variable surface glycoproteins (VSGs) glycoproteins that form the surface coat of African trypanosomes. The trypanosome can repeatedly change its glycoprotein coat by expressing different glycoprotein genes using a process akin to gene conversion.

Variola name given to both the smallpox virus and the disease it causes: smallpox.

Variolation historical procedure for immunization against smallpox in which a small amount of live smallpox virus was introduced through scarification of the skin.

Vascular addressins mucin-like cell adhesion molecules on endothelial cells to which some leukocyte adhesion molecules bind. They determine the selective homing of leukocytes to particular sites in the body.

Viruses submicroscopic pathogens composed of a nucleic acid genome enclosed in a protein coat. They replicate only in a living cell because they do not possess all the metabolic machinery required for independent life. A viral particle is called a virion.

VLA-4 integrin present on the surface of T lymphocytes that binds to mucosal cell adhesion molecules.

VpreB *see* **pre-B-cell receptor**

Weibel–Palade bodies granules containing P-selectin, which are present inside endothelial cells.

Wheal-and-flare reaction the reaction observed when small amounts of allergen are injected into the dermis of an individual who is allergic to the antigen. It consists of a raised area of skin containing fluid (the wheal) with a spreading, red, itchy reaction around it (the flare).

Wiskott–Aldrich syndrome a genetic disease in which interactions of T cells and B cells are defective. Antibody responses against polysaccharide antigens are deficient, and patients are susceptible to infection with pyogenic bacteria.

X-linked agammaglobulinemia (XLA) a genetic disorder in which B-cell development is arrested at the pre-B-cell stage and neither mature B cells or antibodies are formed. The disease is due to a defect in the gene encoding the protein tyrosine kinase Btk.

X-linked hyper-IgM syndrome *see* **hyper IgM syndrome**

X-linked lymphoproliferative syndrome immunodeficiency due to a defect in the SH2D1A gene on the X chromosome, resulting in an inability to control EBV infections.

Xenoantibodies antibodies produced in one species against the antigens of another species.

Xenoantigens antigens on the tissues of one species that provoke an immune response in members of another species.

Xenogeneic coming from a different species.

Xenograft an organ from one species transplanted into another species.

Xenotransplantation the transplantation of organs from one species into another. It is being explored as a possible solution to the shortage of human organs for clinical transplantation.

ZAP-70 a cytoplasmic tyrosine kinase in T cells that forms part of the signal transduction pathway from T-cell receptors.

Zymogen the inactive form in which some enzymes are synthesized. Active enzyme is usually produced by proteolytic cleavage of the zymogen.

Figure Acknowledgments

Chapter 1

Opener image: Colored SEM of macrophage engulfing protozoan. © Science Photo Library.

Figure 1.3 panel a © Sinclair Stammers/Science Photo Library; panel b, d, e, g h, k, l © Eye of Science/Science Photo Library; panel c © A.B. Dowsett/Science Photo Library; panel f © Omikron/Science Photo Library; panel i © Eddy Gray/Science Photo Library; panel j © Philipe Gontier/Science Photo Library.

Figure 1.19 reprinted from P. R. Wheater et al., *Functional Histology* 2nd Edition. © (1987) with permission from Elsevier.

Chapter 2

Opener image: Crystal from which the first antibody structure was detected. Reprinted by permission from L. Harris et al., *Nature,* **360**:369–372. © (1992) Macmillan Magazines Ltd.

Figure 2.5 panel a reprinted by permission from L. Harris et al., *Nature,* **360**:369–372. © (1992) Macmillan Magazines Ltd.

Figure 2.10 panel a from R. L. Stanfield et al., *Science,* **248**:712–719. © (1990) American Association for the Advancement of Science.

Figure 2.26 from A. C. Davis et al., *The European Journal of Immunology,* **18**:1001–1008. © (1988) Wiley-VCH.

Chapter 3

Opener image: Activated T cell looking for invading cancer cells. Courtesy of James Allison.

Figure 3.2 from K. C. Garcia et al., *Science,* **274**:209–219. © (1996) American Association for the Advancement of Science.

Figure 3.4 reprinted from I. Roitt et al., *Immunology* 5th Edition © (1998) with permission from Elsevier.

Figure 3.21 from K. C. Garcia et al., *Science,* **274**:209–219. © (1996) American Association for the Advancement of Science.

Chapter 4

Opener image: Plasma Cell. © Donald Fawcett Visuals Unlimited.

Figure 4.5 panel b from P. L. Witte et al., *The European Journal of Immunology,* **17**:1473–1484. © (1987) Wiley-VCH.

Chapter 5

Opener image: Three-dimensional rotation of a live T cell dendritic cell conjugate. Courtesy of Tomasz Zal and Nicholas Gascoigne.

Figure 5.5 reprinted by permission from C. D. Surh and J. Sprent, *Nature,* **372**:100–103. © (1994) Macmillan Magazines Ltd.

Chapter 6

Opener image: TCRs, visualized by direct fusion of CD3ζ with the green fluorescent protein, coalesce into the mature immunological synapse. Courtesy of Matthew Krummel at UCSF and Mark Davis at Stanford University.

Figure 6.2 reprinted by permission from P. Pierre and S.J. Turley et al., *Nature,* **388**:787–792. © (1997) Macmillan Magazines Ltd.

Figure 6.21 left panel from G. Kaplan and Z. A. Cohn, *International Review of Experimental Pathology,* **28**:45–78. © (1986) with permission from Elsevier.

Figure 6.28 panel c from P.A. Henkart and E. Martx (eds), *Second International Workshop on Cell Mediated Cytotoxicity* © (1985) Kluwer/Plenum Publishers.

Figure 6.32 from H. D. Oche et al., *Primary Immunodeficiency Diseases: A Molecular and Genetic Approach.* © (1998) by permission Oxford University Press, Inc. Also Chapter 28, J. Puck et al., *Inherited Disorders with Autoimmunity and Defective Lymphocyte Regulation.* (1999) Oxford University Press.

Chapter 7

Opener image: Scanning electron micrograph of *Cryptococcus neoformans* interacting with macrophages. Courtesy of Drs C. P. Taborda and A. Casadevall. Also C. P. Taborda et al., *Immunity,* **16**(6). © (2002) with permission from Elsevier.

Figure 7.4 from D. Zucker-Franklin et al., *Atlas of Blood Cells: Function and Pathology* 2nd Edition. Milan, Italy (1988) Lippincott, Williams and Wilkins.

Figure 7.11 left panel reprinted with permission from A. K. Szakal et al., *The Journal of Immunology,* **134**:1349–1359. © (1985) The American Association of Immunologists, Inc.

Figure 7.11 middle and right panel reprinted with permission from A. K. Szakal et al., *Annual Review of Immunology,* **7**:91–109. © (1989) by Annual Reviews (www.annualreviews.org).

Figure 7.31 from S. Thiel and K.M. Reid, *FEBS Letters,* **250**:78-84. © (1989) with permission from Elsevier.

Figure 7.45 from S. Bhakdi et al., *Blut,* **60**(6):309–318. © (1990) Springer-Verlag.

Chapter 8

Opener image: Herpes Simplex Virus. © Dr Linda Stannard, UCT/Science Photo Library.

Figure 8.27 reproduced with permission from F. P. Siegal et al., *American Scientist,* **284**(5421):1855–1837. © (1999) American Association for the Advancement of Science.

Chapter 9

Opener image: Influenza A virus (RNA virus, Orthomyxoviridae family). © Dennis Kunkel Microscopy, Inc.

Chapter 10

Opener image: Back-scattered electron microscope image of pollen grains from *Lilium auratum* (oriental lily). Courtesy of Science Centre Immaginario Scientifico. Image by Rippel Electron Microscope Facility, Dartmouth College, Hanover, New Hampshire/Irina Sancin, Laboratorio dell'Immaginario Scientifico, Trieste, Italy.

Figure 10.1 pollen © E. Gueho/CNRI/Science Photo Library; house dust mite © K.H. Kjeldsen/Science Photo Library; wasp © Claude Nuridsany and Marie Perennon/Science Photo Library; vaccine © Chris Priest and Mark Clarke/Science Photo Library; peanuts © Adrienne Hart-Davis/Science Photo Library; shellfish © Sheila Terry/Science Photo Library; poison ivy © Ken Samuelson/Tony Stone Images, Images ® Copyright 1999 Photodisc, Inc.

Chapter 11

Opener image: X-Ray of hand with rheumatoid arthritis. Courtesy of Medical Illustration, Barts and the London NHS Trust.

Figure 11.36 reprinted from C. M. Weyand et al., *Experimental Gerontology*, **38**:833–841. © (2003) with permission from Elsevier.

Chapter 12

Opener image: Chronic transplant rejection with fibrinoid necrosis. Courtesy of UK Pathology Videodisc Project. © BioMed Image Archive, University of Bristol.

Figure 12.6 reprinted with permission from V. A. A. Jansen et al., *Science*, **301**(804). © (2003) American Association for the Advancement of Science.

Figure 12.17 from B. D. Kahn and C. Ponticelli, *Principles and Practice of Renal Transplantation* 1st edition (2000). Reproduced with permission Thomson Publishing.

Figure 12.19 from B. D. Kahn and C. Ponticelli, *Principles and Practice of Renal Transplantation* 1st edition (2000). Reproduced with permission Thomson Publishing.

Figure 12.22 from G. Opelz et al., *Review of Immunogenetics*, **1**(3):334–42 (1999). By permission of Blackwell Publishing Ltd.

Figure 12.34 from E. W. Petersdorf et al., *Blood*, **92**:3515–3520. © (1998) American Society of Hematology, used with permission, and P. Parham and K. McQueen, *Nature Reviews Immunology* **3**,(2):108–122 (2003). By permission of Macmillan Magazines Ltd.

Figure 12.42 Reproduced from J. H. F. Falkenburg et al, *Experimental Hematology*, **31**:743–751. © (2003) International Society for Experimental Hematology.

Figure 12.48 from M. J. Scanlan et al., *Immunological Reviews*, **188**:22–32 (2002). By permission of Blackwell Publishing Ltd.

Figure 12.49 from P. Coulie et al., *Immunological Reviews*, **188**:22–32 (2002). By permission of Blackwell Publishing Ltd.

Index